CLIMATE IN
A SMALL AREA

CLIMATE IN A SMALL AREA

An Introduction to Local Meteorology

MASATOSHI M. YOSHINO

UNIVERSITY OF TOKYO PRESS

© UNIVERSITY OF TOKYO PRESS, 1975
UTP 3044-66375-5149
ISBN 0-86008-144-3
Printed in Japan

Dedicated to Midori

PREFACE

From experience we know that weather and climate change from place to place even in a small area. The local phenomena of weather and climate are apt to be thought of as "special phenomena" in a restricted area, because of their local nature. When the synoptic meteorological situation and the topographical conditions are similar, however, similar phenomena occur in any place and in any season. This book is a completely revised and enlarged edition of my earlier work entitled *Shôkiko* (Microclimate), which was written in Japanese in 1961. Here the emphasis is largely on the following points: (1) analysis of the horizontal structures of the phenomena; (2) comparison of as many examples as possible; (3) introduction to the synoptic climatological method; and finally, (4) arriving at general rules. In short, it may be said that the methodology here is not very physical in the description.

After a definition of local and microclimate and the histories of the studies, the phenomena on various ground surfaces such as grassland, cultivated fields, urban areas, industrial regions, forests, seashores, and lakeshores are dealt with. Then phenomena occurring under the influence of topographical conditions are discussed. These include the vertical and horizontal distributions of temperature, winds, precipitation, sunshine, and other climatic elements on mountain tops and slopes, in hilly regions, and in basins and valleys.

The third part concerns local meteorological evidence obtained through synoptic climatological methods, that is, (a) wind, precipitation, and air temperature distributions in relation to the local airstream conditions, (b) the micro-scale and meso-scale discontinuous lines and the local weather associated with them, (c) the föhn, bora, chinook, and other famous local winds in the world, (d) the cold air drainage and the cold air lakes appearing at night, and (e) climatological features of inversion layers as one of the controlling factors of local weather conditions.

The last part describes, within the framework of applied local climatology, some of the plant ecological and geomorphological facts under the influence of local climatological conditions. For example, wind-shaped trees are a type of plant ecological evidence whose distribution is greatly controlled by the local climatological conditions. This relation can be used in a survey of local wind conditions. Thus, using the wind-shaped trees as an indicator of wind conditions, the local wind conditions can be inferred from an extensive observation of the wind-shaped trees in a small area. Some local aspects of periglacial evidence can also be used in the same way.

The earlier edition of this book (in Japanese) contained a chapter on local climate and agriculture, forestry, rural houses, disasters, or nature modification, but space does not allow discussion of these topics in this book.

I would like to express my sincere thanks to Dr. Eiichiro Fukui, emeritus professor at Tokyo University of Education, who first guided me to the study of climatology when I was a student; to Dr. Carl Troll, emeritus professor at the University of Bonn, who recommended that I translate the Japanese edition into English; to Dr. Hermann Flohn, professor of meteorology at the University of Bonn, who has always encouraged me to continue climatological research; to Dr. Rudolf Geiger, emeritus professor at the University of München, from whom I got most valuable suggestions through personal conversations as well as from his book.

I am also indebted to Professor Kenzo Kihara and Assistant Professor Yoshihisa Miyakawa, both of Ochanomizu Women's University, and to Miss Kumiko Shinozuka, lecturer at the Japan College of Health and Physical Education, who all helped me in translating this book into English. I am grateful to Dr. Shûji Yamashita, Aichi University, who read and checked the manuscript thoroughly from the climatological viewpoint; to Miss Keiko Kai, Miss Mitsuko Hoshino and Mr. Yasuo Noguchi for their expert typewriting; and to Miss Keiko Kai for her technical assistance in completing the manuscript. Without help by these people, this book would not have been accomplished. Finally, partial financial support was given by the Ministry of Education for the publication of this volume.

June   1974

MASATOSHI M. YOSHINO

Professor of Climatology and
Meteorology
University of Tsukuba

CONTENTS

Preface

CHAPTER 1. LOCAL CLIMATE AND MICROCLIMATE

1.1. Definition of Local Climate and Microclimate 3

1.2. Concept of Local Climate 4
  1.2.1. Scale of Climatic Phenomena 4
  1.2.2. Microclimate and Local Climate 5
  1.2.3. Topoclimate and Other Climatic Phenomena on an
       Intermediate Scale 6
  1.2.4. Relation to the Meteorological Scale 7
  1.2.5. Relation to the Concept of Geographical Region 9
  1.2.6. Scale and Concept of Local Climate 11

1.3. Local Climatology and Microclimatology 13

CHAPTER 2. HISTORY OF RESEARCH

2.1. Before Instrumental Observations 15
  2.1.1. Cities 15
  2.1.2. Arable Land 16
  2.1.3. Seashores and Lakeshores 18
  2.1.4. Mountains 20
  2.1.5. Forests 21

2.2. Instrumental Observation (I) 22
  2.2.1. Cities 23
    (A) City Temperature or Heat Island 23
    (B) City Fog 24
    (C) City Wind 24
    (D) Insolation, Sunshine, and Air Pollution 25
    (E) Study on the City Climate in the 19th Century: A Summary ... 25
  2.2.2. Flat, Open Land 26
    (A) Air and Ground Temperatures 26
    (B) Heat Balance at the Ground Surface 26
    (C) Vertical Distribution of Wind Velocity 26
    (D) Snow 26

2.2.3. Seashores and Lakeshores 27
(A) Air Temperature 27
(B) Land and Sea Breezes: Facts 27
(C) Land and Sea Breezes: Theory 28
2.2.4. Mountains 28
(A) Atmospheric Pressure and Air Temperature 28
(B) Insolation 30
(C) Rain and Clouds 30
(D) Snow 31
(E) Wind 31
(F) Outline of Studies on Mountain Climate up to the 19th Century. 32
2.2.5. Forests 32
(A) Effect of Forests on Climate 32
(B) Air Temperature, Humidity, and Ground Temperature 33
(C) Wind and Precipitation 33

2.3. Instrumental Observation (II) 34

2.4. Development of Research Methods 39

CHAPTER 3. GROUND SURFACE AND CLIMATE IN A SMALL AREA

3.1. Flat, Open Land 41
3.1.1. Radiation Balance 41
(A) Heat Balance Equation 41
(B) Albedo 43
(C) Effective Radiation 45
(D) Distribution of Heat Balance and Radiation Balance in a
Small Area 46
3.1.2. Ground Temperature 51
3.1.3. Air Temperature 55
3.1.4. Wind Velocity 59
(A) Power Law 60
(B) Power Law and Logarithmic Law 60
(C) Logarithmic Law Taking Stability into Consideration 62
3.1.5. Exchange Coefficient and Some Other Characteristic Numbers
or Factors 66
3.1.6. Some Other Phenomena 73
(A) Thermic 73
(B) Diffusion 75
(C) The Temperature on Snow Cover 76
(D) CO_2 Concentration 78

3.2. City 80
3.2.1. Air Temperature 80

(A) Distribution of Air Temperature 80
(B) Three Dimensional Structure of Heat Island 85
(C) The Cause of the City Temperature 88
(D) Meteorological Elements Related with City Temperature,
    Heat Islands 91
(E) Relation to the Population 97
3.2.2. Humidity 98
3.2.3. Precipitation 100
3.2.4. Fog 104
3.2.5. Frost 105
3.2.6. Wind 106
3.2.7. Air Pollution 110
3.2.8. Insolation 116

3.3. Forest 119
3.3.1. Radiation, Heat, and Energy Balance 120
(A) Solar Radiation 120
(B) Light 122
(C) Radiation Balance 124
(D) Energy Budget 124
3.3.2. Temperature 126
(A) Air Temperature Inside and Outside a Forest 126
(B) Vertical Distribution of Air Temperature within a Forest 128
(C) Ground Temperature 130
3.3.3. Humidity 130
3.3.4. Evaporation 131
3.3.5. Precipitation 133
(A) Interception of Precipitation by a Forest 133
(B) Forest and Snow Cover 135
(C) Fog Precipitation or Kisame 138
3.3.6. Fog Prevention Forest 139
3.3.7. Wind 140
(A) Wind Velocity in a Forest 140
(B) Shelterbelts and Wind Breaks 142
3.3.8. The Forest and Run-off 147
3.3.9. Influence of Forests on Climate 149

3.4. Seashores, Lakeshores, and Riverbanks 151
3.4.1. Air Temperature 151
(A) Air Temperature on the Seashore 152
(B) Air Temperature on Lakeshores and Riverbanks 154
3.4.2. Winds 158
(A) Land and Sea Breezes: Facts 158
(B) Theories of Land and Sea Breezes 161
(C) Wind on the Coast 163
(D) Lake Breezes: Facts 165

(E) The Sea Breezes and the Air Temperature on the Coastal Plains. . 170
(F) The Sea Breeze Front 173
3.4.3. Fog 176
(A) Sea Fog 176
(B) River and Lake Fogs 176
3.4.4. Amount of Precipitation 179
3.4.5. Salty Wind 181

CHAPTER 4.   TOPOGRAPHY AND CLIMATE IN A SMALL AREA

4.1. Mountains 183
4.1.1. Air Pressure 183
4.1.2. Insolation and Sunshine 185
4.1.3. Air Temperature 189
(A) Temperature Lapse Rate 189
(B) Diurnal Change 192
(C) Annual Range and Monthly Range 193
(D) Daily Range 196
4.1.4. Ground Temperature 196
4.1.5. Humidity 202
4.1.6. Fog and Clouds 204
4.1.7. Precipitation 208
(A) Rainfall Zone 208
(B) Precipitation Distribution 211
(C) Empirical and Theoretical Representation 214
(D) Rainfall Intensity 216
4.1.8. Snow 218
(A) Snowfall and Snow Accumulation 218
(B) Distribution of Snow Accumulation 220
(C) Snow Line 224
(D) Avalanche and Remaining Snow 226
4.1.9. Thunderstorms 231
4.1.10. Wind 232
4.1.11. Icing and Hoarfrost 237

4.2. Hills, Basins, Valleys, and Bases of Mountain 239
4.2.1. Insolation and Sunshine 239
(A) Insolation 239
(B) Sunshine 242
4.2.2. Air Temperature 245
(A) Minimum Temperature or Nighttime Temperature 245
(B) Maximum Temperature or Daytime Temperature 249
(C) Diurnal Variation of Air Temperature Distribution 252
(D) Distribution of Frost 256
4.2.3. Ground Temperature 261

4.2.4. Wind 262
  (A) Wind Distribution in Mountains and Hilly Regions 263
  (B) Influence of Topography on Wind Structure near the Ground ... 268
  (C) Mountain and Valley Breezes 276
4.2.5. Precipitation 285
  (A) Distribution of Rainfall in Mountainous Regions 285
  (B) Area Rainfall 291
  (C) Snow Accumulation 293
4.2.6. Evaporation 296
4.2.7. Fog 299
4.2.8. Phenology 302

Chapter 5. LOCAL AIRSTREAMS AND WEATHER

5.1. Local Airstreams and Distribution of Climatological Elements 307
  5.1.1. Local Airstreams and Distribution of Wind Direction 307
  5.1.2. Local Airstreams and Distribution of Precipitation 312
    (A) Synoptic Climatological Treatment 312
    (B) Relation to Cyclones 315
    (C) Band Structure 319
    (D) Relation to Local Convergences, Fronts, or Instability Lines 323
    (E) Topographic Effects 325
    (F) Weather Divides 328
  5.1.3. Local Air Currents and Temperature Distribution 330

5.2. Local Anticyclones, Cyclones, and Discontinuous Lines............ 333
  5.2.1. Local Anticyclones 333
  5.2.2. Local Cyclones 336
  5.2.3. Local Fronts on a Meso-scale........................... 338
    (A) Hokuriku Front 338
    (B) Wakasa-wan Front 344
    (C) San'in Front 344
    (D) Bôsô Front 345
    (E) Utsunomiya Front 347
    (F) Ishikari Front 349
    (G) Other Local Fronts 351
  5.2.4. Local Fronts on a Micro-scale 354

5.3. Local Winds 355
  5.3.1. Föhn, Bora, and Similar Winds 355
    (A) Föhn 355
    (B) Chinook 358
    (C) Bora 361
    (D) Fall Wind; Oroshi 368
    (E) Katabatic Wind 372

5.3.2.   Strong or Characteristic Local Winds 374
(A) Catalogue and Distribution of Names 374
(B) Examples in the World 379
(C) Examples in Japan 383
5.3.3.   Classification and Schema 391
(A) Föhn and Bora 391
(B) Scheme of Fall Wind 394
(C) Classification of Local Winds 395
5.3.4.   Air Flow over Mountains and Lee Waves 396
(A) Observations 396
(B) Theory 400
(C) Model Experiments 403

5.4. Formation and Run-off of Cold Air at Night.................... 407
5.4.1.   Cold Air Drainage at Night 407
5.4.2.   Velocity of Cold Airstream 410
5.4.3.   Circulation System of Cold Air and Cold Air Lakes 412

5.5. Inversion Layer 416
5.5.1.   Ground Inversion 417
5.5.2.   High-level Inversion 418
5.5.3.   Formation and Dissipation of Inversions 419
5.5.4.   Seasonal Change of Occurrence Frequency and Intensity of
         Inversions 422
5.5.5.   Development of the Inversion Layers and Topography 425
5.5.6.   Thermal Belts on Slopes 429
5.5.7.   Inversion Layers and Local Weather 433

CHAPTER 6.   LOCAL AND MICROCLIMATE AND NATURE

6.1. Plant Ecology 437
6.1.1.   Forest Structure and Microclimate 437
6.1.2.   Relicts and Local and Microclimate 444
6.1.3.   Wind-shaped Trees 445
(A) Classification of Wind-shaped Trees 446
(B) Scale of Wind-shaped Trees 449
(C) Wind and Tree Deformation 451
(D) Local Distribution of Wind-shaped Trees: Examples 453
6.1.4.   Local Plant Ecology and Local Climate 459
(A) Wind-shaped Trees and Lichen in Mountainous Regions 459
(B) Wind and Plant Distribution 461
(C) Light and Radiation 462
(D) Temperature and Humidity 463
(E) Snow and Plant Distribution 465

6.2. Effects of Climate on Micro-topography 469
   6.2.1. The Asymmetry of Valley Topography and Microclimate 469
     (A) On a Steep North-facing Slope 469
     (B) On a Steep South-facing Slope 470
   6.2.2. Landslides, Landslips, and Erosion by Rainfall 472
   6.2.3. Snow and Frost in Relation to Micro-topography 474
   6.2.4. Wind and Micro-topography 477
     (A) Sand Dunes and Related Phenomena 477
     (B) Periglacial Wind Conditions and Micro-topography 479
     (C) Oriented Lakes 480

References 481
Abbreviation of Names of Periodicals 527
Author Index 531
Subject Index 537
Geographic Index 543

Notes

a) Figures and tables with caption (Yoshino, without published year) were prepared for this book.

b) Time is expressed by local standard time in terms of h and m. For instance; 2 h or 2 h 00 m means 2 a.m. and 13 h 05 m means 5 minutes past 1 p.m.

c) hr means hour(s) and min minute(s).

d) Spelling of the place and region names in the non-English-spoken countries is in English, if they are used commonly. For instance; Vienna (Wien), Munich (München) or Belgrade (Beograd).

e) N is north, northern or northerly; E east, estern or esterly; S south, southern or southerly; W west, western or westerly and so on.

f) a.s.l. means above sea level.

CLIMATE IN
A SMALL AREA

Chapter 1

LOCAL CLIMATE AND MICROCLIMATE

1.1. DEFINITION OF LOCAL CLIMATE AND MICROCLIMATE

A climate can be defined as a state of the atmosphere, repeating an annual cycle, at various points on the earth's surface. Although there are some inter-annual fluctuations in actual figures, statistically speaking, an atmospheric state shows approximately constant figures. Climatology, the science of climate and a branch of meteorology, is characterized by its special interest in meteorological differences caused by regional variations.

Although the divisions of the earth's surface can be either small or large according to the point of view, the expanse of any area can be determined according to the climatological phenomenon under consideration. For instance, in the case of observation of a climatic zone, the entire global distribution must be considered, but in the case of observation of the climate in the mountains, beaches, and cities, the climatic distribution in a small area of a given region is studied.

It has been customary to call the climate of a broad region "macroclimate," the climate of a medium size region "mesoclimate," and the climate of a small area "local climate" or "microclimate", the last term being also used to denote the climate meaning the smallest unit of space, such as the climate in a room or a greenhouse. Terminology is discussed below. There are different opinions among research workers, but the definitions of these terms shown in Table 1.1 are used here.

Of course, certain regions have their own peculiar climatic conditions; this fact causes slight differences from the climatic and/or meteorological phenomena shown in Table 1.1. For instance, the local climatic region in the Great Plain

Table 1.1 Scale of climate and its corresponding meteorological phenomena (Yoshino, 1961).

Climate	Horizontal distribution	Vertical distribution	Example of climatic phenomena	Life time of corresponding meteorological phenomena
Microclimate	10^{-2}–10^2 m	10^{-2}–10^1 m	Climate of greenhouse	10^{-1}–10^1 sec
Local climate	10^2–10^4 m	10^{-1}–10^3 m	Thermal belt of slope	10^1–10^4 sec
Mesoclimate	10^3–$2\cdot10^5$ m	10^0–$6\cdot10^3$ m	Climate of basin	10^4–10^5 sec
Macroclimate	$2\cdot10^5$–$5\cdot10^7$ m	10^0–10^5 m	Climatic zone, Monsoon region	10^5–10^6 sec

in North America covers a large area, even though a local climatic region ordi-
narily covers a smaller area in places like Japanese mountain regions where the
topography is complicated. It can be said, however, that on the whole a microclimate involves regions of 1 cm to several hundred meters, and a local climate
10 m to an order of 10 km. Since the vertical distribution will have to be taken
into special consideration in dealing with the microclimate of the atmospheric
layer near the ground, the limit of the height of a microclimatic region must be
defined. Generally, for local climate the limit may be set as one-tenth of the
horizontal distance.

1.2. CONCEPT OF LOCAL CLIMATE

1.2.1. Scale of Climatic Phenomena

Local climate has been the subject of study for a number of years (for a
history, see Chapter 2). It was only in the late 1920s that a rather clear-cut concept of a local climate was formulated. About fifty years have passed since the
publication of Geiger's "Das Klima der bodennahen Luftschicht" (1927).

First of all, there are no significant differences of opinion among scholars
concerning the concepts of macroclimate and microclimate. Microclimate involves the climate of the air layer near the earth's surface, where the condition
of the ground plays an important role. It may be said that the concept of a
microclimate is comparatively consistent among various researchers, since they
must base their judgement on physical examination of the vertical gradient near
the ground surface. In the case of microclimate, the vertical distribution, rather
than the horizontal expanse, becomes of primary interest, whereas the horizontal expanse is a primary interest in local climate. Hence, differences are considerable about the difinitions of microclimate and local climate in the treatment
and solution of problems.

The most complicated questions concern local climate and mesoclimate.
Geiger and Schmidt (1934) used the term *Kleinklima* to refer to the climatic
phenomena in areas spanning between macroclimate and microclimate sub-
classes, and they touched upon the climatic distribution in small regions. However, Geiger (1950) did not seem to like the term *Kleinklima*, and instead liked
the term *mesoclimat* used by Scaëtta (1935). However, at that time the latter
term had not been extensively used. Even currently there are some scholars who
insist on adopting *Mesoklima* for *Kleinklima* which corresponds to local climate
(Keil, 1950).

In the English-speaking countries, on the other hand, the term "local
climate" had been used to refer to local climatic differences that cannot possibly
be explained in macroclimatic terms. The best example of this argument is the
subclass called "local climate" (Landsberg, 1941, 1960). Landsberg also used
the term "spot climate" to refer to microclimate, but many scholars in England

and America in the 1930s used "microclimate" for the subclass corresponding to *Kleinklima* in Germany.

Baum and Court (1949) have defined microclimatology as a science that deals with the geographical distribution of the horizontal and vertical structures of atmospheric layers near the ground together with the physics of the layers. That is, microclimatology in a broader sense includes local climatology. This is one of the leading opinions in English-speaking countries. In a meteorological glossary published by the American Society of Meteorology (Huschke, 1959), mesoclimate and *Kleinklima* are regarded as synonyms.

In Japan, Fukui (1938), Suzuki (1944, 1951), Sasakura (1950), and Sekiguti (1950, 1952), roughly following Geiger and Schmidt (1934) in their classification of climatic phenomena, identified *Kleinklima* as the local climate. Yoshino (1961) summarized the divisions of the scale of climatic phenomena and came to the conclusion that the scale could be divided into four parts. That is: (1) microclimate (*Mikroklima*), (2) localclimate (*Lokalklima* or *Kleinklima*), (3) mesoclimate (*Mesoklima*) and (4) macroclimate (*Makroklima*). The horizontal and vertical ranges of each scale were roughly the same as shown in Table 1.1. Fukui (1962) made the category "local climate" between microclimate and mesoclimate. According to his criteria, the vertical range of local climate is limited to the lower friction layer (50–100 m) where surface friction outweighs the Coriolis effect, whereas mesoclimate includes the upper friction layer (up to about 1 km). Mäde (1964), in an opening address at a symposium on local climatology in Leipzig in 1963, suggested that *Geländeklima* (*Kleinklima*) must be placed in a position between microclimate and mesoclimate. Prošek (1970) also classified the climatic categories into micro-, local, meso-, and macroclimate, after reviewing the studies in Czechoslovakia, the USSR, and other countries.

1.2.2. Microclimate and Local Climate

In the Soviet Union, it has been pointed out that the range of local climate is very important in understanding climatic phenomena (Sapozhnikova, 1950; Chromow, 1952). Sapozhnikova has stated that the difference in scales for phenomena is determined by the scales of climatic factors, as well as by the proximity of the air layers to the active surface, the source of heat and moisture, over which surface the basic transport method of air (turbulent exchange) is on a scale hundreds or thousands of times smaller than in the free atmosphere. The determining climatic factors in local climate are meso-relief, vegetation masses, etc. The characteristics of local climate are manifested in a layer of air observed at places tens and even hundreds of meters high, but are diminished by height. Okołowicz (1960), after reviewing thoroughly macro-, meso-, and microclimates, defined mesoclimatic phenomena as those "limited to the size of objects representing single independent units in geographical taxonomy", that is, different, independent types of local climate. In the Soviet Union there are two different

views on scale: The first view is that local climate is a subclass between micro-
and macroclimate, as mentioned above. The second view is that local climate is
nothing but subclass of a microclimate in broader sense in which local climate
and microclimate are combined. An example of the second view can be seen in a
monograph entitled *Microclimate of the USSR* by Gol'tsberg (1969). Gol'tsberg
asserts that local climate should be included in microclimate in a broad sense
and that the term "microclimate" should denote the climate of a small area. He
points out further that microclimatic features are prominent in the upper layers
of the soil and in the lower air layer up to a height of several meters or tens of
meters, frequently to an altitude of 100–150 m, whereas local climate develops
under the effects of hills and valleys, land and water surfaces, which are seen up
to 800–1,000 m.

Berényi (1967), with Sapozhnikova, has laid down a scale for mesoclimate
between micro- and macroclimate. Mesoclimate was also called a local climate.
Kayane (1966) and Bjelanović (1967) also used the term "mesoclimate" to
denote the intermediate scale between micro- and macroclimate.

Knoch (1942) has emphasized the relevance of research on a regional climate named *Heimatklimakunde* to a concept of *Geländeklima* (topoclimate),
which will be discussed later.

Mörikofer of Switzerland has also emphasized the importance of local
climate, mainly from the standpoint of bioclimatology. He made the first classification of climatic dimensions with numerical values, as shown in Table 1.2
(Mörikofer, 1947). His classification, however, can also be applied to pure climatology.

1.2.3. Topoclimate and Other Climatic Phenomena on an Intermediate Scale

In 1953, the term "topoclimatology" was introduced by Thornthwaite
(1953) in relation to microclimatology and micrometeorology. He stated that
the climate of a very small space might be called the topoclimate and its study,
topoclimatology. The term "topoclimate" was introduced by analogy from
classical geographical topography, which was a description of places of very
limited area-fields or villages.

On the other hand, the term *Geländeklima* has been used in Germany since
1949 in charting climatic conditions of countries on a scale of 1:25,000 (Knoch,
1949). According to Knoch's definition, *Geländeklima* is a local climate under
the influence of the local relief of the region. Troll (1950) states that *Geländeklima* is equal to the topographical climate. The term "topographical climate"
is therefore different from the topoclimate presented originally by Thornthwaite,
but *Geländeklima* or *Topoklima* in German have both been translated into
English by "topoclimate" and considered synonymous in recent years. The
revised edition of the famous standard textbook on microclimatology by Geiger
(1961) devotes one section to *Geländeklimatologie*, or *Topoklimatologie*, and these

terms have been translated in the English edition as the climatology of a terrain, or topoclimatology (Geiger, 1965). According to his scale division of climatic phenomena, the topoclimate is located between micro- and macroclimate. But the term "local climate" is avoided in his book on the ground that "it is used in so many different senses in the literature." Flemming (1971) has offered the opinion that the scale division can be made from the viewpoints of space units (e.g., local, regional, or global climate) and of space contrast (e.g., micro-, meso-, or macro-climate), and according to Flemming, the lower subdivision of local climate is topoclimate.

From the viewpoint of phytobiology, especially ecological climatography, Boyko (1962) has given the climatic grades microclimate, ecoclimate, and macroclimate. He states that topoclimate may be defined as ecoclimate at the same altitude and with the same exposure. That is, the main factors are generally of the same order.

Many papers have been presented that stress the importance of studies on climatic phenomena between micro- and macro-scales. These include studies on the *Geländeklima* from the meteorological as well as climatological standpoint (Böer, 1959, 1964; Eriksen, 1967), on the topoclimate from the viewpoint of agricultural meteorology (Schnelle, 1968; van Eimern, 1968), on the topoclimate in relation to the microclimate (Mihăilescu et al., 1965; Teodoreanu, 1971; Bogdan, 1972), or on the urban climate defined in terms of climatological concepts (Eriksen, 1964).

1.2.4. Relation to the Meteorological Scale

Since approximately 1950, mesometeorology has made rapid progress. Figure 1.1 gives the ranges of the mesometeorological phenomena observed by some meteorologists. While Hall and Holloway (1955) and Tepper (1959) show similar classifications, Iudin's (1955) are somewhat different, since his viewpoint is not the same as the others. Iudin classified air movement by the ratio of horizontal relative acceleration velocity (r) of the air mass to the Coriolis acceleration (c) and defined a movement where r is a decisive factor as small scale, a movement where c is a decisive factor as large scale, and a movement where c and r are approximately of the same order (several kilometers to scores of kilometers) as meso-scale. On the other hand, Flohn places the *Lokalsynoptik* (e.g., a phenomenon like a rain shower) on the scale of 10^3 to 10^4m, the *Regionalsynoptik* (e.g., a warm front) on $(1–5) \times 10^5$m (Flohn, 1959). Watanabe (1960) classified all the disturbances under 10 km in one group as small scale and disturbances between 10 and 200 km as medium scale. Furthermore, according to a textbook in America (Byers, 1959), mesometeorology is defined as meteorology related to the weather patterns of a small scale, and the area covered ranges from 10 to 100 miles. Especially, standard synoptic meteorology covers a range from 50 to 500 miles. Raethjen (1960) has analyzed the differences of vertical and horizontal air movements between small- and large-scale phenomena and has

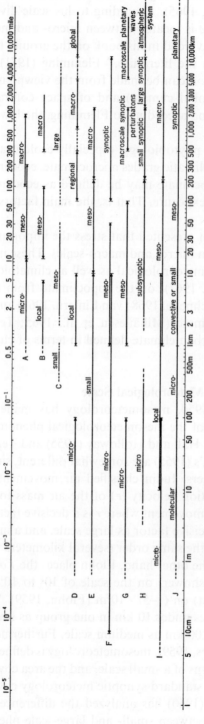

Fig. 1.1 Horizontal range of scales in meteorological and climatological phenomena.
A: Hall and Holloway (1955), B: Tepper (1959), C: Indin (1955), D: Flohn (1959),
E: Watanabe (1960), F: Takahashi (1969), G: Barry (1970), H: Kurashima (1968),
I: Yoshino (1961 and Table 1.1), and J: Mason (1970).

made it clear that small-scale movement appears during a short period and is independent of the Coriolis force. Barry (1970) has concluded that phenomena on the micro-scale in meteorological systems have a horizontal scale smaller than 100 m and those on the macroscale 1–100 km.

Takahashi (1969) divided the horizontal scale of meteorological phenomena: colloid meteorology $10^{-5}-5\times10^{-1}$ cm, micrometeorology $10-10^6$ cm, mesometeorology $10^5-5\times10^6$ cm, and synoptic meteorology 10^7-10^9 cm. Some representative divisions mentioned above are shown in Fig. 1.1. The great differences between the different horizontal scales may be caused by the uncertainty of our knowledge of meteorological phenomena in general and from the discontinuity in the scale of the systems, as pointed out by Barry (1970).

According to a recent study on the spectrum of atmospheric processes, the entire spectrum of the oscillation periods is divided into nine intervals (Monin, 1972). In the smallest micrometeorological oscillations, the energy spectrum $f\,s(f)$ of small-scale turbulence has a maximum value when the period of oscillation τ is of the order of 1 min [f is the frequency, and $s(f)$ the spectral energy density]. This corresponds to a scale of the horizontal-turbulent inhomogeneities $L=u\tau_{max}\sim600$ meters. On the other hand, the mesometeorological oscillations with periods ranging from 1 min to 1 hr are relatively rare and, therefore, there is usually a broad minimum in the spectrum $f\,s(f)$ within this interval. This minimum corresponds to a period τ on the order of 20 min and to a scale $L=u\tau$ on the order of the effective thickness of the atmosphere $H\sim10$ km. The minimum separates the quasi-horizontal synoptic inhomogeneities with scales $L\gg H$ from the quasi-isotropic micrometeorological inhomogeneities with scale $L<H$.

Fiedler and Panofsky (1970) indicate that the kinetic energy of the atmosphere is not spread uniformly over all wavelengths, but has certain preferred scales, with gaps in between. The first group of systems has horizontal wavelengths of the order of 100 m to several kilometers. Barry and Perry (1973) considered the scale problems in synoptic climatology introducing the space and time dimensions presented by Mason (1970), which are shown in Fig. 1.1.

Climate phenomena are the chronological accumulation of meteorological phenomena, but it is difficult to generalize their relationship to the length of time of meteorological phenomena. It might be safe to consider that the length of time of climatic phenomena is of approximately the same order or slightly longer than the life time of meteorological phenomena. Although the classification in Table 1.1 has been obtained after taking all these considerations into account, it still reveals discrepancies, as seen in Fig. 1.1.

1.2.5. Relation to the Concept of Geographical Region

Weischet (1956) has offered a different view from the preceding ones. Based on the concept of geographical regions, he organized climatic phenomena and, paying careful attention to the semantics of the terms, made the following classification. *Regionalklima* was adopted in place of "macroclimate" and *Subregional-*

klima was adopted in place of "orographic climate" and "regional climate". The distribution of subregional climate was considered on a scale corresponding to a map on the scale of 1:100,000–1:1,000,000. He further grouped together mesoclimate, *Kleinklima*, and local climate, and named the group *Lokalklima*, corresponding to a map on a scale of 1:5,000–1:50,000. A few problems remain to be examined concerning those names and scales, but his classification is worthy of attention, since he saw climate and region as one unit.

Stone (1968) has suggested that the subdivision of geographical phenomena may be designated as regional, sectional, local, and individual, where the local scale is represented by a map with a scale of 1:20,000. Scale problems in geography have also been discussed from the viewpoint of agricultural regions and land use in detail (Ukita, 1970, 1972). Here, it is only pointed out that these studies are valuable for consideration of the problems, especially on a local scale.

It is important to consider the scale problems of local climate in relation to the arrangement of *Landschaft*, which means roughly landscape or region. Paffen (1948) has studied the scale of *Landschaft* ecologically and divided it as follows (unites are arranged from smaller to larger): *Landschaft* cell, small *Landschaft*, individual *Landschaft*, large *Landschaft*, *Landschaft* group, *Landschaft* region, etc. According to his division, the scale of local climate defined in Table 1.1 corresponds to small, individual, and large *Landschafts*. The concept of a geographical *Landschaft*, which may be translated as a geographical region, has been reviewed thoroughly by Troll (1950), with special reference to the ecological and climatological conditions. Using vegetation space as an example, he has maintained that a microclimate is composed of the climates of tree crown and trunk layers, an air layer near the earth's surface, and a soil layer and that the microclimatic space corresponds to "ecotope." He has asserted, with the aid of a *Landschaft* profile, that a scale of the climate called local climate, topographical climate, or *Geländeklima* ought to be arranged between the micro- and macroclimates.

In a study on the classification of natural regions, the lower order unit is named *Fliese* (mosaic stone) and *Fliese* group or complex, and the higher order is *Fliese* combination (Schmithüsen, 1948, 1949). Ecological area studies depend on some basic conditions like homogeneity of areas and significance of types and their regional validity. Starting from this concept, Neef (1963, 1964) arranged the different scales of regions by recognizing topological and chorological units. He has asserted that the geographical regions of various grades must be decided through the typical ecotope and other elements. The chorological mosaic (pattern) type, according to his definition, may be affected mainly by local climatic conditions, which are under consideration here. This should be discussed further, applying the concept of order of natural regions from the viewpoint of *Landschaft* ecology (Haase, 1964).

1.2.6. Scale and Concept of Local Climate

The scale divisions on climatic phenomena mentioned above are given in Table 1.2. Except for some cases, respective figures for each scale are left out. Therefore, comparison of the classifications in this table with different researchers cannot always be done accurately.

In Table 1.2, climatic phenomena are classified into (a) three scales, (b) four scales, or (c) five or more.

The consistent use of the terms "macro-", "meso-", and "micro-" certainly provides a basis for systematic classification, but it is doubtful whether the scale of the meteorological and/or climatic phenomena can adequately be determined

Table 1.2 Division of scale of climate according to the research workers (Yoshino).

Research worker (year)	Microclimate (Mikroklima)	Microclimate (Klein- od. Lokalklima)	Mesoclimate	Macroclimate
Geiger, 1927	Mikroklima	Orographisches Mikroklima		Makroklima
Geiger, 1929	Mikroklima	Kleinklima	Landschaftsklima	Makroklima
Geiger and Schmidt, 1934	Mikroklima	Kleinklima		Grossklima
Scaétta, 1935	Microclimat	Mesoclimat		Macroclimat
Fukui, 1938	Microclimate	Local climate		Macroclimate
Landsberg, 1941	Spot climate	Local climate	Regional climate	World climate
Mörikofer, 1947	Mikroklima (Vert. 1m) / Kurortklima (Vert. 2m)	Lokalklima (100–1000m) / Regionalklima (10–20 km)	Landschaftskl. (100 km)	Makrokl. (1000 km) / Klimagürtelkl. (5000 km)
Knoch, 1949	Mikroklima	Geländeklima od. Lokalklima		Makroklima
Troll, 1950	Mikroklima	Lokalklaim		Makroklima
Dammann, 1950	Mikroklima	Landschaftsklima / Lokalklima		Makroklima
Sekiguti, 1950, 1952	Microclimate	Lokal climate		Macroclimate
Sasakura, 1950	Mikroklima	Kleinklima		Grossklima
Keil, 1950	Mikroklima	Mesoklima		Makroklima
Sapozhnikova, 1950	Mikroklimat	Mestnii klimat		Makroklimat
Suzuki, 1951	Mikroklima	Kleinklima		Makroklima
Chromov, 1952	Mikroklimat	Mestnii klimat		Makroklimat
Thornthwaite, 1953	Mikroclimate	Topoclimate	Mesoclimate	Macroclimate
Wagner, 1956	Mikroklima	Örtliches Klima	Mesoklima	Makroklima
Weischet, 1956	Mikroklima	(Meso-, Klein- od.) Lokalklima	(Gelände-, Landschafts- od.) Subregionalklima	(Makro, Gross- od.) Regionalklima
Flohn, 1959	Mikroklima (1–100 m)	Lokal- or. Mesoklima (1–10 km)	Regional- od. Landschaftskl. (100–500 km)	Makrokl. ((1–5)·10³ km) / Globalklima (10⁴ km)
Huschke, 1959	Microclimate	Kleinklima òr Mesoclimate		Macroclimate
Böer, 1959	Mikroklima	Lokal- od. Mesoklima		Makroklima
Okolowicz, 1960	Microclimate	Mesoclimate		Macroclimate
Geiger, 1961	Mikroklima	Geländeklima (Topoclimate)		Makroklima
Yoshino, 1961	Microclimate	Microclimate (Lokal- od. Kleinklima)	Mesoclimate	Macroclimate
Fukui, 1962	Microclimate	Local climate	Mesoclimate	Macroclimate
Boyko, 1962	Microclimate	Ecoclimate		Macroclimate
Mäde, 1964	Mikroklima	Geländeklima	Mesoklima	Makroklima
Mattsson, 1964	Mikroklima	Meso (Lokal) klima		Makroklima
Kayane, 1966	Microclimate	Mesoclimate		Macroclimate
Berényi, 1967	Mikroklima	Mesoklima (Lokalklima)		Makroklima
Eriksen, 1967	Mikroklima	Gelände-, Meso-, od. Lokalklima		Makroklima
Shcherbany, 1968	Mikroklima	Local climate		Makroklima
Barry, 1970	Microclimate (100 m)	Local (Topo) climate (1–10 km)	Regional macroclimate (5×10²–10³ km)	
Flemming, 1971	Mikroklima	Mesoklima		Makroklima
Hess, 1971	Mikroklima	Mesoklima		Makroklima

in accordance with those terms. Therefore, ideally speaking, it would be most fair to prepare the frequency distribution of the size of all the possible climatic phenomena on the earth's surface and determine the scale according to the frequency distribution curve. At present, however, since our knowledge of meteorological and/or climatic phenomena is rather limited, not many scientific methods are available, except classifying these phenomena subjectively, relying either on the scale of the meteorological phenomena or on the concept of geographic regions.

In my opinion, four scales can be systematized: i.e., microclimate, local climate, mesoclimate, and macroclimate. The scales of each climate and its corresponding meteorological phenomena shown in Table 1.1 were obtained from the results given in Table 1.2 and Fig. 1.1 and from what has been discussed in Sections 1.2.1–1.2.5. In this book local climate (*Kleinklima*) will be the major topic. Local climate can also be called "topoclimate (*Geländeklima*)" in mountainous and hilly regions, "forest climate" in forest regions, "shore climate" on the coasts or in the maritime regions, "urban climate" in the city regions and so on.

However, discussion here is intended to be as inclusive as possible of the significant phenomena without being restricted to the tentative classification in Table 1.1 or to the terms microclimate and mesoclimate. The spaces covered by microclimate, local climate, mesoclimate, and macroclimate are shown as an example in Fig. 1.2, which was drawn after referring to the figures by Sapozhnikova (1950) and Wagner (1956).

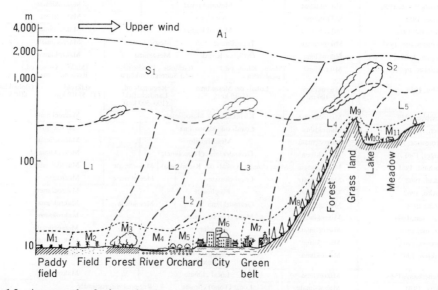

Fig. 1.2 An example of micro-, local, meso-, and macroclimatic phenomena.
M_1–M_{11}: Microclimate, L_1–L_5: Local climate, S_1–S_2: Mesoclimate and A_1: Macroclimate (Yoshino, 1961).

1.3. LOCAL CLIMATOLOGY AND MICROCLIMATOLOGY

The branch of science which is concerned with local climate is called local climatology, and that which is concerned with the microclimate is called micro-climatology. Local climatology is the science of the climate of smaller areas under the influence of vegetation, minor topography, cities, etc. Climatic phenomena in a local climatic region, which will be treated in this book, can be called local climate in a narrow sense but are a part of the microclimate in a broader sense and the bulk of the phenomena overlap with those of the meso-climate. It is, as Sekiguti (1950) says, one of the principal aims of local clima-tology to analyze climatic effects of local surface conditions and to evaluate their effects quantitatively. In microclimatology the vertical gradient near the ground is a major subject, whereas in local climatology the horizontal gradient in addition to the vertical gradient is an important topic. In local climatology specific surveys and observations must be planned and carried out, since the regular, normal observations by the meteorological observatories and weather stations are not sufficient.

In general, the effects of the microclimate die out at a height of four times as high as grass, buildings, microorography, etc. Hence, a survey up to an area of this height becomes indispensable, and the results obtained should be ana-lyzed. The vertical distribution ranges from a height of 10 cm to 1,000 m above the ground surface; and, ideally, observations should be conducted to cover this wide range of space.

The climate of any region is a composite of various local climates. Differ-ences caused by regional variations are ascribable to various physical properties of the ground surface, the location, exposure, and different conditions of the ground surface—such as the color, apparent density, heat capacity, moisture content, and permeability of the soil; the characteristics of the vegetation cover; the albedo and roughness of the ground surface. These are all factors that in-fluence the heat and moisture exchange and the momentum exchange on the ground surface (Brunt, 1953; Thornthwaite, 1953). Accordingly, investigation of local differences in these factors is necessary on level ground. Moreover, in a hilly area we will need to investigate thoroughly the influence of the orography on the air current and to discuss the distribution of the climatic elements brought about by those influences. Endrödi (1965) dealt with the research methods in topoclimatology and the task of the investigation. Holmes and Dingle (1965) stressed the importance of studies on the quantitative relationships between the macro- and microclimate.

The time scale aspect of local climate poses a considerable problem, as discussed by Barry (1970). Most local climates are distinguishable primarily during periods of weak winds and, usually, of clear skies. This means that a synoptic-climatological framework of study is desirable. For instance, in in-vestigating the frost distribution, the occurrence of strong local wind, the existence of city temperature, the precipitation maximum on mountain slopes, etc., the synoptic climatological conditions should be taken into account. In other words, their occurrence frequencies are important for consideration of the phenomena.

A detailed classification of meteorological and geographical factors producing local climates has been given by Stringer (1958). The geographical factors are summarized in Table 1.3. The task of local climatology is to analyze the relationship between the distribution of these factors and the local climatic phenomena.

Table 1.3   Geographical factors producing local climates (Stringer, 1958).

Factor	Items to be mapped
(a) Type of surface	
(i) Rock	Type, color, thermal conductivity.
(ii) Soil	Type, texture, color, air and moisture content, thermal conductivity.
(iii) Water	Surface area, depth, movements.
(iv) Vegetation	Type, height, density, color, seasonal change.
(v) Agricultural	Fallow land; type, height, and color of crops; seasonal change.
(vi) Urban and industrial	Material (concrete, tarmac, wood, metal, etc.) color, thermal conductivity; sources of heat, moisture, pollution, etc.
(b) Properties of surface	
(i) Geometrical shape	Flat, convex, concave, etc.
(ii) Energy supply	Latitude and altitude, degree of screening of natural horizon, aspect, slope, exposure.
(iii) Exposure	Shelter provided by macro- and micro-orographic features; shelter provided by buildings, trees, etc.
(iv) Topographic roughness	Rural areas: extent of woodland, grassland, arable; location of windbreaks and hedges; degree of agglomeration or dispersal of individual buildings. Urban areas: distribution and average height of different types of built-up zones; orientation and exposure of streets, blocks and individual buildings; density of parks, gardens and other open spaces; vertical profiles across area.
(v) Albedo	Type of surface.
(vi) Radiating capacity	Type and maximum temperature of surface; observed earth radiation.

Chapter 2

HISTORY OF RESEARCH

2.1. BEFORE INSTRUMENTAL OBSERVATIONS

The history of research on climate in smaller regions begins before instruments were used for meteorological observations. How did people accept climatic differences in an urban area where a number of people congregated to live? How were climatic differences interpreted in a farming area? How much knowledge was disseminated concerning the climatic differences in the mountain and forest regions? Since instruments for observation were not used in those days, some of the accounts might have been based on purely subjective observation. Yet, these observations based on experience alone gave some results that are no less accurate than those obtained today with the aid of instruments. These observations in the historical age made significant contributions in laying the foundation for the development of climatology in a small area.

2.1.1. Cities

In the Palmyra Ruins of Assilia, 240 km NE of Damascus, it is apparent that climatic conditions were considered in planning the city. The avenue in Palmyra with great colonnades on each side is 11 m wide and exposed to the open air, but the sidewalks are 5 m wide and arcaded to avoid the sun, wind, and rain.

Marcus Vitruvius (B.C. 75–26), a famous Roman engineer and architect, made the following remarks concerning climatic conditions necessary for founding a city based on his own professional experience: Land ideal for the health is slightly elevated, and there should be neither fog nor frost. The direction of the slope and the distance to the swamps, lakes, and beaches must also be considered. The prevailing wind directions observed by a wind tower at the center of the city, like Horologium at Athens, should be taken into consideration in planning a city. The main and narrow streets should be placed in the middle angle of the two prevailing wind directions. Then according to Marcus Vitruvius, the location of the Pantheons and squares should be decided.

In ancient India also the following accounts are given in Mānasāra Śilpaśāstra concerning village planning. The major roads should run from east to west and from south to north according to the sundial, and a square should be located at the center of a village. Roads running from east to west were called Râjapatha (the emperor's street) and were believed to be purified by the sunshine from morning till night. The roads running from south to north (Mahākala or Vamana) were made for ventilation, and they blessed villagers with the breeze.

In Europe from around the 12th or 13th century people became aware of the climate within an urban area, that is, the peculiar climate created by city life. This is due to the fact that the population started growing rapidly in the cities, and the urban areas expanded as well. For instance, the population in London was 40,000–50,000 in the 14th to 15th century, rose to 100,000 in the 16th century, exceeded 200,000 in the early 17th century, and reached 400,000 in the mid-17th century. Such rapid growth inevitably caused problems like epidemics, fires, and air pollution. In London there were 16 epidemics during the 16th century.

Great fires frequently broke out: The great fire in 1666 lasted for four days, reducing 12,300 houses to ashes. Concerning air pollution, Sir John Evelyn (1620–1706), a famous English diarist, wrote in his book "Fumifugium, or the inconvenience of the Aer and Smoak of London dissipated" (1661) that "For when in all other places the Aer is most Serene and Pure, it is here Ecclipsed with such a Cloud of Sulphure, as the Sun itself, which gives day to all the World besides, is hardly able to penetrate and impart it here; and the weary Traveller at many Miles distance, sooner smells, than sees the City to which he repairs." He also stated that in order to clean the air the city of London should order all the industries that used fire to move out of the heart of the city to the area between Greenwich and Woolwich and not to build any more factories near the city in the future. He suggested that sweet-smelling trees should be planted in London to purify the air. A similar situation was seen in Paris. The French House of Councilors made various resolutions to cope with the situation.

Some city planning can also be found in the castle towns in Japan. The city planning by the *samurai* during the first 50 years of the Tokugawa Period included a countermeasure against flood damage, but no consideration seems to have been given to climatic conditions. However, people in the same industry or trade were allocated to the same district considering the prevailing wind conditions to minimize damage by fire. Fish and salt shops, which scarcely used fire, were allocated to the windward parts of a city, and blacksmithes and gun-smithes, which used fire, to the leeward parts. Boulevards were made as fire-prevention zones in Tokyo and Nagoya. The great fire in Edo (old Tokyo) in 1657, however, lasted for two days, and about 108,000 people were burned to death.

As a city grew, the life of the citizens became closely connected with the special climate that city life created—the local climate of the city. At the same time, damage caused by the city's peculiar climate became phenomenal.

2.1.2. Arable Land
When man began to engage actively in production through farming, he

eventually became concerned with crop conditions. Consequently, the concept of the local climate and microclimate in farm land has a longer history, just as farming itself does, than that in urban areas. Herodotus, for example, described the different harvest seasons by location in Libia in his book *Historia*.

Arthastra, a book on politics, written in the 4th century B.C. by Chanakya, a grandfather of King Asoka, is considered to be the earliest literature on Indian meteorology (Sammadar, 1912). There is a chapter on "Superintendent of Agriculture" dealing with local differences in the amount of rainfall, probably measured by a rain gauge, in relation to agriculture. Here, the concept of "local" was on a somewhat broad scale.

In Japan about the time when three volumes of the *Kojiki* (Ancient Chronicle 712), thirty volumes of the *Nihonshoki* (Chronicle of Japan) and one volume of the *Keizu* (Genealogy 720) were completed, the compilation of the *Fudoki* (Regional geography) was undertaken in compliance with an Imperial edict. At the present time the description of only five territories—Hitachi, Harima, Izumo, Hizen, and Bungo—remain intact out of the entire volume, and detailed accounts based on experience are given on the local climatic conditions.

For instance, in the chapter on villages of Harima-Fudoki (regional geography of Okayama), it records that the soil is medium-of-medium. The statement was made in compliance with an Imperial edict; "Make a report on the degree of fertility of the soil." The "soil" here does not mean the soil *per se* but probably means productivity of the land and thus the over-all natural environment suitable for farm products, namely, the soil, water temperature, underground temperature, air temperature, wind, moisture, etc. Although the judgment of

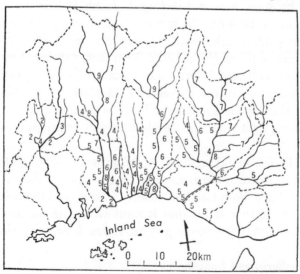

Fig. 2.1 Distribution of grade of productivity by natural environments in Harima, an old province in SW Japan in 8th century (Yoshino, 1961). 1: best of best, 2: medium of best, 3: worst of best, 4: best of medium, 5: medium of medium, 6: worst of medium, 7: best of worst, 8: medium of worst, and 9: worst of worst.

crop productivity seems to have been based on subjective criteria, this record is noteworthy as the first treatise in Japan in which consideration was given to the climatic differences in a small area. Three degrees were given—best, medium, and worst—and each degree was further divided into three grades—best, medium, worst. Therefore, there were nine grades altogether, and the distribution of these is shown in Fig. 2.1. This figure is important not only as a map showing the conditions 1,200 years ago, but also as a map based on observed results.

The *Kikôshinkenroku* (Records of climatic investigation), written in the years 1690–1705 by Gen'an Sato, the results of Sato's trip during which he conducted a survey on longitudes and latitudes of various parts of Japan, and the climate of these places is discussed in relation to the geographical location. He pointed out the possibility that the orography of a certain area could considerably change the climate of that area from the general climate of other regions on the same latitude. Prompted by the request of the lord of the Hoshina family of Aizu who had been exploiting his territory for agriculture, Sato seems to have carried out this survey to aid in more economical production.

Because of the need for a countermeasure against frost damage on farm products, there have long been accounts on forecast and prevention of frost. In the early stages these forecasts and prevention measures were not based on scientific observation. Sir John Evelyn, who had pointed out the air pollution in London in the 17th century, conducted a careful observation also on frost damage to various species of timber and shade trees, shrubs, herbs, and garden plants during the longest, severest winter of the century in 1683–4. He wrote in the Philosophical Transactions of the Royal Society of London in 1684 that plants covered by snow were not damaged. Jacob Bobart noticed also in the same Transactions in 1684 this "surprising" effect of cold snow cover preventing damage from cold, together with the facts that the shrubs and garden plants covered with straw suffered less damage from frost (observed in winter, 1683–1684). W. Derham collected information from his friends in various countries on serious frost damage in the winter of 1708–1709 and found that the wind conditions were the same in England and on the continent during strong winds, but different during light wind and that light wind conditions involved a danger of frost damage.

In Chôshû Fushi, an old region of Aichi, Japan, it is written that in a village called Sotohara in Kasugai Gôri, a mountain breeze from the northeast used to start blowing at dawn from late fall to early winter, and the breeze was called *watakushi-kaze*. It further states that it rains when the mountain breeze dies down. Here, *watakushi-kaze* means private, isolated, or local wind.

2.1.3. Seashores and Lakeshores

Man has always been influenced by the climate of the seashore, which varies greatly from place to place. For instance, sailors used land and sea breezes or lake breezes or the topographically deviated winds along the seashore or lakeshore. People in the regions near coasts could use the wind power only when they had an appropriate knowledge of the local distribution of the wind force. The location of each windmill and the distribution area of the windmills, which we can see today, may be considered as a reflection of this knowledge of the past. Of course, the former concerns micro-topography and wind, and the latter the local distribution of characteristic winds.

The Murakami-suigun, a pirate family with a base of operations on the Inland Sea of Japan, which was very active during the 14th–15th century, had very reliable knowledge on the local difference of wind conditions on the coasts. *Senkô-Yôjutsu* (Handbook of navigation) written by Masahusa Murakami in 1456 provided laws deduced from experience concerning the wind conditions and topography (Fujiwara, 1951). In Section 3 of the chapter on the weather, Vol. III, it says, "It is vitally important to know the nature of the wind at a location. This is the same as learning the inherent characteristics of a country. For instance, the north wind blows considerably in villages facing the north, whereas the south wind blows in villages facing the south." Since knowledge of the characteristics of the local wind was important for coastal navigation in the age of sailing vessels, these accounts based on experience were quite complete and accurate.

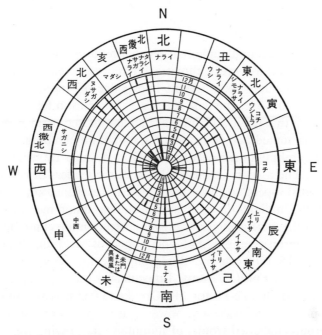

Fig. 2.2 An example of a diagram of coastal winds on the east coast of Izu Peninsula, Shizuoka Prefecture, Japan, given in Shizuoka-ken Suisanshi (1894). Local names of the winds are, for example, as follows: *Narai*=N, *Kochi*=E, *Inasa*=SE, *Minami*=S, *Nishi*=W, *Saganishi*=WNW, *Dashi*=NW, and *Saganarai* or *Dashinarai*=NNW.

In Japan, which is surrounded by the ocean, coastal fishing has developed as well as the art of navigation. An illustration of this tradition can be found in the detailed accounts on the local climate of the wind recorded in Vols. III and IV of the *Shizuoka-ken Suisanshi* or Fisheries in Shizuoka Prefecture (1894). This is the most outstanding climatography of a small area in the age without instrumental observations. In these volumes the coast of Shizuoka Prefecture is divided into 16 zones, and each zone is supplied with a diagram and a detailed description of the wind and fishing conditions. Figure 2.2 is an example. For each wind in the figure there is a description of its daily variation, the wind speed, its relation to the weather, etc. For instance, the *Dashi* (northwest) wind is explained as follows: It rises at 4 a.m. on a calm day and dies down by sunrise. When strong, it will change into the *Narai* (north) by the afternoon and die away by dusk. Even when it continues to blow the following morning, the *Dashi* will change into the *Minami* (south) or the *Kochi* (east) wind under windless conditions. *Dashi* wind mentioned here means a strong mountain breeze or a cold air runoff from inland to the seashore under the condition of fine weather, which caused strong cooling in the inland the night before. If the *Dashi* wind does not occur, the weather becomes bad, and the *Minami* or *Kochi* wind will probably blow, which means the approach of a cyclone. This description presents a vivid description of the local wind conditions in relation to weather changes in the region.

2.1.4. Mountains

It would be natural to assume that much attention was paid to the climate in the mountains because great climatic differences were expected to occur in these regions, but actually our present knowledge on mountain climate is a gradual accumulation over the past 200 years. The reason for this lag could be ascribed to the fact that the local climatic change was so drastic in the mountain regions that it prevented people from going mountain climbing.

Mountain climbing in the past was motivated either by religion or professional necessity. The ascetic devotees who were the leaders of religious groups of mountain worshippers, such as the Fuji-kô, Tateyama-kô, or Ontake-kô, should have been familiar with every inch of the mountain map in question and well informed of the meteorological changes and mountain climbing techniques. Fuji-kô was a fraternity composed of worshippers of Mt. Fuji. From the Muromachi era in the 15–16th century, this group became active, and many worshippers climbed Mt. Fuji (3,776 m). Tateyama-kô was a group for Mt. Tateyama (3,015m), and Ontake-kô for Mt. Ontake (3,063 m). Mountain climbing motivated by professional necessity, however, is more important for discussion here. These groups included troups of mountaineers praying for rain, literary men, scholars, collectors and surveyors for hobbies, local inspection groups of government officials, military activities, etc. The records written by the chief of these troups indicate how enthusiastic people of that time were concerning the local climate of the mountain. *The Komagatake Ichiran-no-Ki* or the Survey on Mt. Komagatake, for example, is a record of a group survey in the summer of 1736 by 114 people of the Takatô Clan in Nagano Prefecture, Central Japan. This record gives full accounts of the deformation of the vegetation near the top of the mountain due to wind and snow. Similar descriptions can be found in the record of the summer of 1811 in the *Naeba-San ni Asobu Ki* (A Trip in the Naeba Mountain) by Bokushi Suzuki as well as in a book on Mt. Hakusan in Central Japan written by Hiroyasu Yamazaki in 1841.

It was after the 18th century that people in Europe started modern mountain climbing. It was rather unusual to go mountain climbing prior to that time; one case is the climbing of Pilatus Peak (2,123 m) by Professor Conrad Gesner of the University of Zürich in 1518 for academic study. In the 18th century mountain climbing by scientists and literary men became rather popular, and Horace Bénédicte de Saussure, James Fobbes, Louis Agazy, and John Tyndal are some of the people who went mountain climbing. Horace Bénédicte de Saussure (1740–1799) was born in Geneva, Switzerland, and tried four times to reach the summit of Mont Blanc between 1760 and 1787 and finally succeeded in 1787. He was the second man in history who got to the summit. He also conducted a meteorological observation in August, 1787. In winter, 1788, he carried out an observation continuously for 14 days in the snow covered Col de Géant (3,405 m).

J. Eoaz, a forest inspector, published a comprehensive work entitled *Die Lauinen der Schweizeralpen* (Snow avalanches in the Swiss alps) in 1881. In this book, many figures and tables were given concerning avalanches in the mountain region. These facts are part of the second stage of the history of climatology in a small area.

2.1.5. Forests

Forests have particular climates. Although this is clearly recognized now, it is not written in old literature. However, forest influences on the climate of nearby regions were mentioned often in the pasts.

Shelterbelts or wind breaks were built gradually in many countries in order to create a favorable climate around settlements and arable land, especially in the 17–18th centuries. Afforestation with pine trees was intended to prevent the cold, severe winter monsoon winds in some parts of the Japan Sea coast; in the Tsugaru provinces in 1681–1683, and in the Izumo provinces in 1673–1680. Wind breaks to prevent strong local wind were also seen. For instance, on the southern foot of Mt. Akagi, Central Japan, thick pine forests planted in 1761–1772 prevented the dry, strong fallwind called *oroshi*.

One of the greatest, although private, afforestation enterprises was done by Zenbei Ishikawa. He began to plant pine trees from the years of Tenmei (1781–1788) on the Japan Sea coast of Yuri-gun near Akita City, NE-Japan. It was said that he planted about 7 million young pine trees before his death, and the forest extended about 20 km. For the establishment of wind break techniques there must have been knowledge of microclimate and local climate. One example was given by Tarôzaemon at Sakanobe-shinden in 1769, when he planned to fix the sand dune in Shônai, Akita Prefecture: As there was no instrument for measuring wind velocity, he tried to observe the local distribution of wind velocity by walking around on the sand dune in summer with his younger brother, Kyûtarô, without any clothes on (an application of the idea of sensible temperature). He concluded that the places with strong wind velocity in summer also had strong velocity in winter and, for this reason, young trees must be planted densely at the site. The distribution pattern of wind velocity on the sand dune during the hot summer, probably under the conditions of a sea breeze like the NW wind, closely resembles that of the winter monsoon, NW prevailing wind in this region. Therefore, his judgement was right in the light of modern local climatology.

The use of shelterbelts may also be seen in other countries. For instance, 700,000 ha of pine forest along the coast near Bordeaux, Southwest France, is very famous as a great work of Napoleon III (1808–1873). Similar forests are seen in Great Britain, Germany, Denmark, Hungary, Switzerland, and Russia. They were afforested not only to reduce high wind velocity or salty wind along the coasts, but also to control snow accumulation or to keep higher air humidity and to protect soil moisture from damage caused by dry wind in the inland regions.

Vegetation studies in small areas have been made intensively since the beginning of the 19th century. O. Heer studied the vegetation in a SE part of Canton Glarus, Anton Kerner von Marilaun the plant life of the Danube Basin in 1863, H. Christ the vegetation in Switzerland in 1879, and R. Gradmann the vegetation in the Schwäbische Alps in 1898.

No record of a survey has been found concerning the study of climatic differences in and out of a forest area during the period prior to the age of apparatus use. Therefore, it was after the second half of the 19th century that the concept of microclimatology in a forest developed, and this period coincides with the age of instrumental observation.

2.2. INSTRUMENTAL OBSERVATION (I)——(Up to the 19th century)

The number of stations at which meteorological observations with modern instruments and methods are conducted continuously has increased since the 18th century. In the 19th century more countries started meteorological observations as a systematized venture, and the number of observation stations increased. Observation records of many years at some stations where observation was started earlier have been accumulated. It was one of the characteristics of climatology of that time to compare the observation records of various stations. This was the awakening of local climatology.

The development of the studies up to the end of the 19th century is summarized below under the individual topics—urban climate, climate of flat open ground, coastal climate, mountain climate, and forest climate. Since this is merely a summary of the historical development of the research and surveys, the description of actual phenomena will be minimized. The important results of the classic studies of this period are discussed in Chapters 3 and 4.

2.2.1. Cities

(A) *City Temperature or Heat Island*

As mentioned above in Section 2.1.1, people in the 17th century were already well aware that the urban area has a unique climate compared to rural districts. Hann (1897, 1908), who noticed that the center of a city has higher temperatures than the suburbs, called this phenomenon city temperature, *Stadttemperatur*, and he confirmed this fact by observation records in the mid-19th century.

Modern meteorological surveys were undertaken in various cities in Europe and America from the 18th to the 19th centuries. In Paris, after reorganizing the observation results, Louis Cotte released in 1774 summarized records of the maximum temperature, minimum temperature, and air pressure for 1699–1770; the weather of 1748–1770; and the amount of precipitation of 1689–1754. In London, L. Howard first compiled in 1818 various records on the climatic elements and released the results in 1833 after reorganizing the observation records for the years 1797–1831. The *Climate of London Deduced from Meteorological Observations* by Howard (1833) is three volumes and discusses such topics as the effect of the ocean and the Thames on the observation records, the climatic differences between the city and the suburban areas, etc. A history of studies

Table 2.1   City temperature in the early period (Yoshino).

Name of city	Researcher	Published year	Observed period	City temperature
London	Howard	1833	1806–1830	1.5–1.8°F
Paris	Renou	1855	1816–1860	0.7°C
München	Wittwer	1860	1820–1850	0.9°C
Many cities in Europe	Hann	1885		0.4–1.1°C
Berlin	Perlewitz	1899	1888–1897	0.2–0.3°C (annual mean)
St. Louis	Hammond	1902	1891	4.6°F (mean daily minimum)
Berlin	Hellmann	1921	1881–1910	1.0°C
Tokyo	Matoda	1922	1913–1916	0.5°C
München	Treibich	1927	1923–1926	0.9°C
Moscow	Bogolepow	1928	1910–1926	0.7°C
Tokyo	Sasakura	1931	Feb. 1931	0.7°C

on city climates was given in detail in the classical works by Fels (1935) and Kratzer (1937).

Later, reports on the city temperatures of Berlin and Vienna were released one after the other. A summary of these records up to the early 20th century is shown in Table 2.1.

(B) *City Fog*

It is the city fog that is first noticed among various phenomena of city climate. City temperature was recorded by a thermometer, but the nature of the fog had to be recorded through visual observation.

Various researchers conducted surveys of fog in various cities, such as London in 1891, Paris in 1893, and later in Vienna. However, the city fog in London was the major interest to researchers. Although the cause for the city temperature and the city fog was thought to be the new environment created by the city's development, the studies of this period were concerned mainly with the recognition of such phenomena and their detailed descriptions. Tables 2.2 and 2.3 show some of the results.

Table 2.2   Increase of number of days with fog in London (Kratzer, 1937 after Brodie).

	1871–1875	1876–1880	1881–1885	1886–1890
Number of days with fog per year	50.8	58.4	62.2	74.2 days
Population (Year)	2,804 (1871)	3,816 (1881)	4,150 (1886)	4,767 × 1,000 (1890)

Table 2.3   Difference of number of days with fog between London and Greenwich in the 19th century (Carpenter, 1904).

	Jan.	Feb.	Mar.	Apr.	May	Jun.	Jul.	Aug.	Sep.	Oct.	Nov.	Dec.	
Difference (L–G)	4.9	2.9	2.3	0.9	0.3	−0.4	0.1	−0.4	1.7	3.5	2.6	4.5	days

(C) *City Wind*

A pressure difference arises when there is a difference in the air temperature of the city and its outskirts. This pressure difference causes a wind system to blow toward the city center. This type of wind is a country wind or *Flurwind*, named by E. Fels, since the wind blows in from the surrounding area.

It was Hann who first noticed the pressure difference within and outside the city boundary, based on the observation results in Paris in 1895. However, it was not until the 1920s that an effort was actually made to confirm by observed results whether or not such a peculiar wind system really existed. A detailed account on this effort will be given in the next chapter.

The most obvious phenomenon is the decrease in the wind velocity in a city. Kremser made the first report on this phenomenon observed in Berlin in 1909. The area to the southwest of Berlin, which was still open in 1894, was completely built up by 1900. Table 2.4 shows the obvious decrease of the wind velocity in that area.

Table 2.4   Decrease of wind velocity in Berlin (Kremser, 1909).

	1884–1893	1894–1903
Mean wind velocity	5.12 m/sec	3.93 m/sec
Max. wind velocity	19.	15.
Diurnal change		
Year	1.49	1.20
Summer	2.09	1.60
Winter	0.94	0.80

(D)   Insolation, Sunshine, and Air Pollution

Polluted air is one of the important factors of city climate, especially influencing the city temperature and city fog. As discussed in Section 2.1.1., Sir John Evelyn's report on London's air pollution in 1661 was a pioneer work in the field of air pollution. It was F. A. Russel, however, who conducted a scientific study on the effect of smoke as a cause of the city fog in London in 1889. In 1893 a report was made on Berlin. These studies indicate that city air contains numerous particles, which act as condensation nuclei. Studies of this sort were conducted in only a few cities; around 1890 there were only a few other cities besides London and Berlin.

The decrease in hours of insolation and sunshine caused by city development was reported around 1890. As seen in Table 2.5, the duration of sunshine within the cities is less than half of that in the outskirts in winter time.

Table 2.5   Sunshine in the city and suburbs of London,
1890–1891 (Russel, 1892).

	Location	Nov.–Feb.	Mar.–Oct.	Year
Woburn	NW	206	1,214	1,420 hr
Kew	W	172	1,233	1,405
City center		96	1,062	1,158
Greenwich	E	150	1,105	1,255
Eastburne	SW	268	1,455	1,724

(E)   Study on the City Climate in the 19th Century: A Summary

The following is a summary of the studies on city climate up until the end of the 19th century. The results indicate the relative values obtained by instrumental observations at certain spots in cities and in the suburbs or outside city boundaries.

The values are calculated as monthly and/or annual mean or summation of the monthly and/or annual values. Such phenomena as the city temperature and the city fog were pointed out and confirmed on the basis of these values.

The fact that the climatic conditions vary generally between a city and the surrounding area has serious implications for the study of macroclimate. For instance, two air temperatures—one observed at a station within a city, the other, outside the city—cannot be used to draw a scientific map of air temperature distribution on the global scale. The 19th century is said "to have been the golden era for climatography," (Leighly, 1949) but in order to keep a more accurate climatography, it was necessary to analyze the environmental effects of a certain location on the survey values. This need seems to have caused the early study of city temperature.

2.2.2. Flat, Open Land

(A) *Air and Ground Temperatures*

In 1878, E. Wollny observed ground temperatures on the slopes facing various directions and found that the air temperature was affected by direction. He built eight slopes inclined at 15°, which faced different directions in his garden and observed air temperatures at the height of 15 cm on each slope three times a day. T. Homén of Finland reported the average air and ground temperatures observed from August 10 to 12, 1893. The highest ground temperature were 35°C on granite fels, 42°C on sandy soil, and 27°C on moorboden. The lowest ground temperatures at night were 6°C on moorboden, 7°C on sandy soil, and 15°C on granite fels. The ground temperature at a depth of 60 cm was the highest on granite fels, and the lowest on moorboden.

R. Rubenson's results in 1876 are considered to be the most important for the study of the conditions of air temperature and humidity in the process of dew formation.

(B) *Heat Balance at the Ground Surface*

T. Homén studied the heat balance on various soils, such as granite fels, sandy soil, and moorboden, for six clear days in 1896. The distribution of insolation energy, reflecting energy to the atmosphere, energy used for evaporation, and energy remaining underground were shown. His results show that a great deal of energy is consumed in evaporation on moorboden. His study on heat balance is noteworthy from a historical standpoint.

The first substantial research on the redistribution of the sun's energy in the atmosphere was undertaken toward the end of the 19th century (Budyko, 1956).

(C) *Vertical Distribution of Wind Velocity*

It was after the 20th century that the vertical distribution of wind velocity was surveyed. Perhaps the only study undertaken in the 19th century was by T. Stevenson, whose results on simultaneous observation of vertical distribution were published in 1880. G. Hellmann discovered the logarithm and power laws after the turn of the century to mark the initial step toward the great development of this field in the first half of the 20th century. The details are described in Section 3.1.4.

(D) *Snow*

A. I. Voeikov was a pioneer researcher on the effect of snow accumulation.

As early as 1871 he discovered that snow cover prevented crops from frost damage, mitigated the rise in the springtime temperature, and prevented floods at the thawing time. He discussed in detail these phenomena by listing various survey results in 1885 and further in two chapters in *Climate of the Earth* (Voeikov, 1887).

Study of snow cover was naturally quite advanced in Russia and Northern European countries. H. F. Abels made a microclimatic observation on snow-drifts in Sveldrovsk in February 1891, and January and February of 1892. He obtained a simple correlation between heat conductivity K and snow density D: The experimental equation was given $K=0.800D$. K varies in a wide range from 0.0010 to 0.3289 cal·min^{-1}·cm^{-2} as D ranges from 0.05 to 0.90.

2.2.3.   Seashores and Lakeshores

(A)   *Air Temperature*

F. Walter observed the meteorological effect of lakes in the Lake Constance (Boden See) area in 1892. An annual mean air temperature difference of 0.4°C was found between the lakeshore temperature (8.6°C) and the inland area temperature (8.2°C). The difference was the largest in January (0.8°C) and was insignificant in March and April.

It was made clear toward the end of the 19th century that the influence of lakes varies from one season to another. A survey on the air temperature distribution in the Great Lakes region showed that the prevailing wind would also affect the width of the lake-effect zone.

Various attempts were made to clarify why heat loss into the atmosphere occurred more slowly on the seashore and islands than in the inland area. The most outstanding study was made by R. Strachey in 1866 on the effect of water vapour on the basis of his observations in Madras from March 4 to 25, 1850. From his results, concerning the relation between the temperature drop at night and temperature rise during the day and water vapour, it can be seen that both rise and drop of temperature are large when the vapour pressure is low. Therefore, the seashore and the islands have a mild climate. A. I. Voeikov observed in 1883 the relation between ground radiation and relative humidity, and J. A. Sutton reported in 1895 on the effect of water vapour (relative humidity) on night cooling based on his observation at Kimberley in South Africa.

(B)   *Land and Sea Breezes: Facts*

The study of land and sea breezes has long been in existence, but it was *Marine Physical Geography* (15th ed.) by M. F. Maury published in London in 1874 that described in detail the time, wind direction, velocity, and effect on human life of breezes in the tropical zone.

At the end of the 19th century, land and sea breezes were surveyed with the appropriate scientific instruments. The vertical distribution of the sea breeze was surveyed with kitoon at Coney Island in New York in August 1879. This showed a circulation system with the sea breeze blowing to a height of 200 m, and the wind direction was reversed above 250 m. An observation at Toulon in mid-October of 1893 by the same observation method disclosed that the sea breeze was blowing to a height of 400 m and the anti-sea breeze was blowing beyond 600 m.

The Meteorological Association of New England conducted surveys on horizontal distribution of the land and sea breezes at 130 different spots on the Massachusetts coast in the summer of 1887. The results were reported by W. M. Davis, L. G. Schultz and R. DcWard during 1890 and 1893. They revealed that the sea breeze began to approach the land at the speed of 1.5–3.6 m/s around 9–10 h; then gradually decreased its speed after reaching land. It reaches as far as 16–30 km from the sea shore late in the afternoon. Then it gradually died out. In addition to these results, the correlation between the sea breeze and the air temperature and humidity was studied through observations along the Senegambia coast and other spots. However, neither study was sufficient to give a full account on the land and sea breezes.

The same is true with lake breezes. A. Forel made observations on Lake Léman, and H. A. Hazen on Lake Michigan in the 1800s. Both of them, however, only contributed to the accumulation of observation records.

(C)   *Land and Sea Breezes: Theory*

The nature of land and sea breezes, along with the föhn in the Alps, was theoretically explained fairly well in the 19th century. At the end of the 19th century land and sea breezes were explained as follows (Hann, 1897):

In the morning the land temperature increases more rapidly than the water temperature; the warmed-up air over the land expands upward. The upper air over the land begins to flow toward the sea at a higher point to decrease the atmospheric pressure over the land and to increase the pressure over the sea. Thus, the sea breeze comes into the land at a lower point. The situation is reversed at night; that is, a land breeze flows to the sea.

On the basis of this theoretical background people tried to confirm that the atmospheric pressure over the land was lower than that over the sea during the daytime and higher at night. They encountered some difficulties, however, because the pressure difference was small. H. F. Blanford pointed out this difficulty in the observation records of the Bengal coast in 1877, and F. Chambers in the records of British Islands in 1880.

2.2.4. Mountains

(A) *Atmospheric Pressure and Air Temperature*

A mercury barometer was invented by Evangelista Torricelli of Italy in 1643. Five years later, in 1648, Blaise Pascal persuaded his brothers to climb Puy-de-Dôme of the Massif Central of France and succeeded in conducting simultaneous observations on the top and foot of the mountain. An aneroid barometer was invented by Vidie in 1848, but 200 years before Vidie. Pascal observed with his own barometer the decrease of atmospheric pressure with increasing altitude.

Behind the development of modern alpinism there was a science-oriented mind. This tendency can easily be understood from the meteorological observations done by numerous alpinists and expedition groups, not to mention H. B. de Saussure and J. Tyndall. A group from Harvard University, for instance, made meteorological observations for eight days at altitudes of 3,960–5,790 m to find the relation between mountain sickness and meteorological elements.

As already mentioned, Saussure made observations on the lapse rate of air temperature in the Alps for two weeks in 1788. The figures registered then were 2.5°C at Col-de-Juan (3,405 m), 17.9°C at Chamonix (1,080 m), and 21.6°C at Geneva (400 m). They were converted to the lapse rates of 0.66°C/100 m and 0.54°C/100 m, respectively. According to A. Kerner, differences between the air and the ground surface temperatures were 1.5°C at 100 m a.s.l., 1.7°C at 1,300 m, and 2.4°C at 1,600 m at the Tirolian Central Alps. Table 2.6 prepared by Ch. Martins shows the results obtained by simultaneous observations at Brussels (50 m) and Faulhorn (2,680 m), Switzerland, at 9 a.m. from August 10 to 18, 1842. In addition, Martins conducted other simultaneous observations at Pic du Midi (2,877 m) and Bagnères (551 m) in 1864. These studies treated the difference between the ground surface and the air temperatures as a function of altitude. Obviously, these points of view are slightly different from the present-day approach, which is concerned with the lapse rate of air temperature in the mountains.

Table 2.6 Decrease of temperature observed by C. Martins,
August 10–18, 1842 (Hann, 1908).

Station	Height a.s.l.	Air temperature	Ground surface temperature
Faulhorn	2,680 m	8.2	16.2°C
Brussels	50 m	21.4	20.1

The change of soil temperature with exposure in the Alps was reported by A. Kerner in 1867, 1868, and 1869 and further by his son, F. Kerner (1891). The three-year average of soil temperatures on the slopes around a hill is given in Table 2.7, which is still valuable today: The highest temperature appeared on the SW slope in winter, but on the SE or S slope in summer.

The present-day knowledge of the lapse rate of temperature in the mountains was perhaps established originally by J. Hann in his *Meteorology* third edition (Hann, 1908). Hann's conclusion was 0.51–0.64°C/100 m around 12 mountainous districts in the nontropical zone. Therefore, 0.55°C/100 m would be the average lapse rate in the mountainous districts between the equator and latitude 60°N. He also stated that the lapse rate would be different on plains and in valleys.

Table 2.7 Soil temperature at 70 cm depth on the eight exposition, average of 3 years' Observation at Inntal 600 m (F. Kerner, 1891).

	N	NE	E	SE	S	SW	W	NW	Amp.
Dec.	5.2	5.4	4.8	5.8	6.2	7.5	6.2	5.6	2.7°C
Jan.	3.9	4.3	3.4	4.3	4.5	5.8	4.7	4.0	2.4
Feb.	3.5	3.6	3.8	5.2	5.3	6.4	5.5	3.8	2.9
Mar.	3.6	4.2	4.8	6.5	6.5	7.5	6.7	4.2	3.9
Apr.	6.3	7.4	8.4	9.4	9.5	10.0	9.4	6.9	3.7
May	11.6	13.3	14.9	16.6	16.0	15.4	15.3	12.3	5.0
Jun.	14.0	15.7	17.3	18.5	18.4	17.4	17.4	14.6	4.5
Jul.	15.8	17.4	19.1	20.2	19.8	18.8	19.0	16.5	4.4
Aug.	16.0	17.9	19.5	20.5	20.1	19.6	19.5	17.0	4.5
Sep.	14.9	16.9	17.9	19.5	20.1	19.4	18.9	16.2	5.2
Oct.	11.4	13.0	13.5	15.3	15.8	15.5	14.7	12.8	4.4
Nov.	7.2	7.8	7.9	9.1	9.8	10.3	9.1	8.0	3.1
Year	9.5	10.6	11.3	12.6	12.6	12.7	12.2	10.2	3.2
Amp.	12.5	14.3	16.1	16.2	15.6	13.3	14.8	13.2	

The temperature inversion phenomenon in the mountains was already observed in the latter half of the 19th century. On valley slopes and mountain slopes it was noticed, by flower blooming and bud-shooting, that the temperature on the middle part of the slopes is slightly higher than that on the lowest spots, and this phenomenon was proved by a series of observations.

(B) *Insolation*

J. Violle, one of the founders of modern actinometry, observed the absolute insolation value on the summit of Mont Blanc and Glands Mulets and simultaneously Margottet at the foot of Bosson Glacier on August 16 and 17, 1875. The results are given in Table 2.8. The insolation was 15% stronger at Mont Blanc than at the Bosson Glacier for a relative height of 3,600 m, 26% stronger at Mont Blanc than in Paris at 60 m a.s.l. J. M. Pernter also made observations of insolation at Monte Rosa and Mont Blanc, as Violle did, and concluded that the absorption coefficient of the vapour was 1,900 times larger than that of the atmosphere.

Detailed survey results were available on the general increase of ultraviolet with altitude. It was also known that the ultraviolet content decreased considerably in the lower layer, 1,500–2,000 m.

Development of the heat budget concept in the 19th century was thoroughly reviewed by Miller (1968).

(C) *Rain and Clouds*

People learned through experience that precipitation increased in moun-

Table 2.8 Results of intensity of insolation on August 16–17, 1875, observed by Violle (Hann, 1908).

	Height a.s.l.	Air pressure	Vapour pressure	Intensity of insolation*
Mont Blanc	4,810 m	430 mm	1.0 mm	2.39
Grands Mulets	3,050	533	4.0	2.26
Bossons Glacier	1,200	661	5.3	2.02

* Energy of radiation·cm^{-2}·min^{-1}. Unit is unknown. Adjusted to the value at noon.

tainous districts. They also noticed through the condition of vegetation in mountainous districts that the side facing a prevailing wind was humid and had more rain, whereas the leeward sides which was called the "rain shadow," was dry. Sir Richard Strachey, who was stationed in India in 1850, is credited with discovering that the heaviest rainfall in mountains occurs on the hillside rather than on the summit. However, it was not until the 1860s that these phenomena were proved by observation results. These phenomena are obvious in the Hawaii Islands, which are located in the NE trade wind zone. The clouds form around the 1,200 to 1,500 m layer, annual precipitation is 4,000 mm on the NE seacoast, 6,000–7,000 mm near the cloud zone, 1,300–1,400 mm on the leeward SW coast. In addition to Hawaii, similar observations were made in various mountains in such places as Java, Burma, the Himalayas, the Alps in Switzerland, Halz, the Pyrenees, and Sierra Nevada in California. These observations were analyzed in terms of the monthly or annual precipitation in relation to the elevation and distance either from the coast or from the mountain foot. These studies made clear that the relation between the precipitation amount and the elevation varies in accordance with the season as well as the direction of the mountain slopes.

Meteorological observations in the mountains started in Japan at the end of the 19th century. The Central Meteorological Observatory made observations for 1–2 months at Mt. Gozaisho-take in Mie Prefecture, 1,200 m, in 1888; at Mt. Fuji, 3,776 m, in 1889; at Mt. Ontake in Nagano Prefecture, 3,069 m in 1891; at Mt. Issaikyô in Fukushima Prefecture, 1,956 m, in 1893; and at Mt. Ishizuchi in Ehime Prefecture, 1,977 m, in 1894.

Elaborate observations on clouds in the mountains were conducted, and several results were published on the cloud formation of Helm wind, one of the local winds in England, from 1884 to the end of the 19th century.

(D) *Snow*

During the second half of the 19th century, studies were directed mainly to the relation between altitude and such phenomena as amount, frequency, and period of snow fall and duration and depth of snow cover.

The study on the snow line developed considerably. After H. B. de Saussure first took up the problem of the snow line, such great geographers as A. von Humboldt and F. Ratzel followed, and the distinction between the climatic snow line and the orographic snow line was made clear.

Conditions of remaining snow are different on the sunny side slopes from on the shadowy side of mountains. Since the remaining snow in spring could become a cause of serious damage in mountainous districts, study in this field also developed. As avalanches were dangerous for the traffic and daily life of the people in the mountainous districts, various measures of prevention were developed.

(E) *Wind*

The fact that local names were used for mountain and valley breezes suggests that there existed a close relation between the breezes and the daily life of the people. However, accurate observations of the breezes were slow because wind observations require more skills than measuring the amount of precipitation. In the first half of the 19th century study of winds in mountainous districts was begun by observing the cloud motion and other phenomena. H. B. de Saussure's observations on cloud motion in the mountains, reported in Leon in 1842, was one example. He pointed out that it was due to the valley breeze that fewer clouds lingered around the summit in the morning but more in the afternoon.

More descriptions are available for föhn than for mountain and valley breezes. Already in 1867, H. W. Dove published a monograph on föhn and *sirocco*. About the same time, many studies were done on föhn in the Swiss and Austrian Alps. J. Hann published a paper on the cause of föhn in 1868.

Toward the end of the 19th century data were collected on the air temperature and humidity when föhn blew and its frequency. Various information on the winds with the föhn-effect was brought from different parts of the world. E. Knipping (1890) introduced, for example, the föhn in Kanazawa on the Japan Sea coast in a meteorological journal in Germany. Although slightly behind the study on föhn around the Alps, knowledge about the local winds in mountain regions was gathered from many parts of the world: chinook in North America, bora on the Adriatic coast, and *mistral* on the Mediterranean coast. C. C. McCaul reported the climatic effects of the chinook wind in South Alberta in 1888–1889.

(F)   *Outline of Studies on Mountain Climate up to the 19th Century*

Mountain climate studies were more or less concerned with comparison of the survey results on the summit, on the slopes and at the foot of the mountain. However, without a mountain weather observatory, the observation at the summit could not be made for an extended period. Generally, results based on observations for a week or month, mostly in summer, were compared. The same techniques as used for urban climatic study were employed in comparing the average or the total values at various stations.

The theory on mountain and valley breezes was developed, and synoptic meteorological study on the föhn, in addition to the theory, was also developed.

2.2.5.   Forests

(A)   *Effect of Forests on Climate*

One of the major arguments in the second half of the 19th century was whether or not a forest actually took part in increasing precipitation amount and mitigating the air temperature change. This problem was first discussed by G. P. Marsh in America in 1864.

S. Aughey of Nebraska reported in 1880 that the precipitation amount increased with the expansion of farm land. According to his theory, more rain would permeate the soil and then vaporize much more to the atmosphere. The humidity thus goes up to increase the precipitation amount.

H. E. Hamberg pointed out a similar phenomenon in Sweden in 1885–1896. In Assam and Malaya, Voeikov (1887) pointed out that the forest would play a role in mitigating the extreme values of air temperature. H. F. Blanford in India in 1882 and B. E. Fernow and others discussed in the Report of the Department of Agriculture in 1893 the effect of forests on climate, particularly on the precipitation and runoff amounts.

E. W. F. Ebermayer in Bavaria, A. Müttrich in Prussia and v. Lorenz-Liburnau in Austria conducted microclimatic observations in forests between 1870 and 1890. Observation techniques and studies advanced considerably during the last ten years of the 19th century. The work by J. Schubert is particularly noteworthy among these studies.

More elaborate observations were conducted on vertical and horizontal climatic conditions within and outside forests during this century. It was concluded in this stage of the studies that forests have a definite influence on climate from the standpoint of microclimate, but no clear influence on local or meso-scale climate.

(B) *Air Temperature, Humidity, and Ground Temperature*

E. W. F. Ebermayer, one of the pioneering scholars in the microclimatology of forests, measured ground temperature, air temperature, and humidity in forests, and evaporation from water surfaces, precipitation amount, and ozone amount within and outside forests. He published a 253-page report in 1873 in which he discussed daily, monthly, and annual variations.

A. I. Voeikov pointed out that a forest crown acted just like the ground on heat balance. He named the upper surface of the forest crown the "outer, active surface." This was already noted by v. Lorenz-Liburnau in 1878–1890.

(C) *Wind and Precipitation*

Almost no observation was made on the wind in the 19th century because it was difficult to conduct special observations within and outside of a forest.

E. Hoppes made observations on the rainfall amount toward the end of the 19th century. He studied carefully the rainfall conditions underneath a crown. The rainfall amount was small near a trunk, but the farther away from the trunk he went, the larger it became. On a 60-year-old spruce, it was 55% at 0–0.5 m from the trunk, 60% at 0.5–1.0 m, 66% beyond 1.5 m, and 76% on the ground beneath the crown rim.

Y. Tanaka (1887) discussed thoroughly the vegetation distribution in Japan in relation to climatic conditions, especially the monsoon wind and precipitation, and the wind breaks on the sand dunes on the Japan Sea coast.

2.3. INSTRUMENTAL OBSERVATION (II)——(20th century)

The detailed study scope and results will subsequently be presented in Chapters 3-6. The developments in various countries are reviewed here with the countries arranged in alphabetical order.

Austria was the center of the world in meteorological study. Besides J. Hann and W. Köppen, valuable contributions were made by V. Conrad, A. Defant, E. Ekhart, F. Lauscher, F. Sauberer, W. Schmidt, F. Steinhauser and A. Wagner. The work of the following researchers is also noteworthy: H. Aulitzky, I. Dirmhirn, O. Eckel, F. Fliri, H. Reuter, H. Tollner, H. Turner, and W. Undt.

The recent studies on the climate of Vienna and other cities by F. Steinhauser and F. Lauscher and on high mountain ecological climate in the Austrian Alps by H. Aulitzky and others are most important contributions.

In Canada, studies on climate in a small area started later, but showed excellent development. I. Y. Ashwell, W. Baier, B. J. Garnier, F. K. Hare, R. W. Longley, R. E. Munn, S. Orvig, M. Sanderson, M. K. Thomas, and E. Vowinckel contributed in this field. *Descriptive Micrometeorology* (Munn, 1966) is a comprehensive work on various aspects of the physics and dynamics of the atmospheric layer up to about 50 m and can be considered a basic textbook dealing with climates in a small area.

In Czechoslovakia, J. Förchtgott's contribution on the air flow over mountains and lee waves was very valuable. M. Konček, M. Nosek, P. Plesnik, and E. Quitt have also been studying the problems concerning local and microclimatology.

In England, C. E. P. Brooks dealt with local climate in his books, *Climate in Everyday Life* (1950) and *the English Climate* (1954). O. G. Sutton's *Micrometeology* which has been used as a standard textbook, is a remarkable work even though it does not contain much discussion of horizontal distribution. Although they did not devote themselves exclusively to the study of local and microclimatology, England's great meteorologists D. Brunt and geographer L. D. Stamp have made various significant discussions.

Climatologist G. Manley conducted several excellent studies on the small-scale climate mainly after World War II. W. G. V. Balchin, T. J. Chandler, J. Glasspool, E. L. Hawke, G. S. P. Heywood, H. H. Lamb, E. N. Lawrence, J. Oliver, M. Parry, N. Pye, R. S. Scorer, L. P. Smith and E. T. Stringer also did valuable work.

In Finland, T. Homén is called the father of microclimatology. His work was introduced in the preceding chapter. V. Rossi has also rigorously engaged in the study of this field.

In France, a number of studies on the city climate in Paris were conducted from the 1920s to the 1930s. After World War II a study on Lyon and its vicinity by Piéry was published in 1946, on Paris C. Maurain in 1947 and on the

Paris Basin by P. Pédrabolde in 1957–1958. N. Gerbier and M. Bérenger made an excellent observation of the lee waves in the French Alps.

Germany, together with Austria, was one of the centers of climatology in the history of modern climatology from the 19th to 20th century, and the same is true in the history of local and microclimatology. It was 1927 when R. Geiger (1927) published his book, *Climate near the Ground* (first edition). In 1930 he presented a short, but excellent, monograph entitled *Microclimate and Plant Climate*. A. Schmauss originally influenced Geiger to study in this field. Geiger says that G. Kraus (1841–1915), botanist in Würzburg, should be called the father of microclimate studies, because Kraus published the pioneer work *Soil and Climate in Small Regions* in 1911. Geiger's book *Climate near the Ground* was translated into English and is widely used throughout the world as a standard textbook.

Besides Geiger, other German scholars who made a profound contribution to the development of local and microclimatology were, G. Hellmann, A. Wagner, K. Knoch, J. Schubert, and A. Mäde, A. Kratzer summarized *City Climate* in 1937.

During World War II, in addition to Geiger's research, M. Woelfle's study in forest meteorology was published. K. Knoch presented in 1942 the paper *World Climate and Local Climate* and proposed topoclimatology. Again in 1949 he emphasized the importance of topoclimatology as a kind of applied local climatology. His valuable contribution on the study methods (Knoch, 1951, 1961) is summarized (Knoch, 1963). As topoclimate plays a very important role in land use, a number of surveys and studies on topoclimate or local climate have been carried out in Germany after World War II. H. Aichele, A. Baumgartner, W. Böer, H. Burckhardt, W. Dammann, J. van Eimern, G. Flemming, E. Franken, E. Frankenberger, J. Grunow, W. Kaempfert, H. Kaiser, E. Kaps, H. G. Koch, W. Kreutz, P. Lehmann, H. Lessmann, A. Morgen, H. Schirmer, M. Schneider, F. Schnelle, V. Schöne, J. Seemann, S. Uhlig, N. Weger, W. Weischet, etc., were also research specialists.

A summary on topoclimatological surveys in Germany was made by Knoch (1963). Schnelle (1967, 1968) reviewed the recent problems in topoclimatology in the Federal Republic of Germany.

In Hungary, microclimatology was started by botanists like R. Soó, I. Máthé, B. Zólyomi and M. Újvárossy. A cooperative work by a botanist and a meteorologist on the plant distribution in the Bükk plateau appeared in Időjárás in 1934. After World War II, Z. Dobosi guided studies on the water and heat balance problems, and R. Wagner studied many problems from a local and microclimatic standpoint. Under their institutes many papers were published; especially in *Acta Climatologica* from Szeged.

D. Berényi, Debrecen, published a book of microclimatology in 1967. It concerns mainly physical processes and the vertical structure of atmospheric layers near the ground. It also treats the problems of the horizontal distribution of climatological elements. Götz (1964) reported some aspects of the studies on local climates.

In Israel, J. Lomas and his collaborators have been studying in this field, especially in relation to agricultural meteorology.

In Japan, *Kishô Shû-shi* (Journal of the Meteorological Society of Japan) was first published in 1882. Research papers in the journal show that it was after 1890 that mountain climatology started. Some excellent accounts are given in *Nihon Kikô-Gaku* (Climatology in Japan) written by Fusakichi Koide in 1897, which dealts with the microclimate of forests and mountains. Studies were extended by Minoru Katsuya and Shigetoshi Takeda in the Forest Meteorology Observatory in the early 20th century.

Some local climatic views were emphasized among the people concerned with industries. For instance, Teizo Namie, a sericulturalist in Saitama Prefecture, discussed in his book, *Meteorology and the Sericultural Industry*, published in 1903, the microclimatic conditions of rooms for silk worms in relation to their production.

Katsue Misawa (1885–1937) of Nagano Prefecture was a founder of local climatology in Japan. K. Sasakura, who started the study with Misawa, was one of the leading specialists during 1930s. He published monograph *Shôkikô Gaku* (Local climatology) in 1950. K. Takasu under the supervision of Tadao Namekawa made an excellent study in this field. H. Hatakeyama published his research on the climatology of Tokyo and of the Toyohara-chô area in Saghalien. E. Fukui studied the city climate, snow cover distribution, etc., from the late 1930s to the early 1940s. Along with Taro Tsujimura, T. Yazawa emphasized the importance of analyzing the climatic landscape, which reflects the climatic conditions in a given small area, and summarized the studies in a book entitled *Kikô Keikan* (Climatic landscape) in 1953. Y. Daigo, T. Kato, M. Nakahara, and T. Ueno also contributed to this study during this time.

After World War II, Y. Daigo, M. Matsuno, N. Nishina, T. Sekiguti, K. Takeda, and T. Yamanaka made a number of studies in local climatology. T. Sekiguti, who published studies in this field since the pre-war days, wrote a number of papers on the air temperature and humidity distributions in a small area, on urban climatology, and on the local rainfall distribution. Y. Tsuboi, Y. Ozawa, and T. Asai reported the air temperature distribution, etc., in connection with agricultural problems. S. Suzuki and M. Nakahara treated local climatological phenomena in their books on agricultural meteorology. On the other hand, in connection with water resource measures and flood prevention devices, surveys on the distribution of precipitation and snow accumulation in the mountainous regions have been regorously conducted. The work by J. Sugaya, under the guidance of U. Nakaya, around 1948–1949, should be listed here as a pioneer work in this field. M. M. Yoshino published a monograph on local and microclimatology in 1961. Y. Ozawa and M. M. Yoshino summarized the observation methods of local climate in 1965. A review on mesoclimatology and microclimatology in Japan in recent years is given by Shitara (1966).

C. L. Godske, a successor of C. G. Rossby in Norwegian meteorology, showed a strong interest in local climatology and published several research results. Under the theme "local climate, microclimate, and bioclimate," a symposium was held in Bergen March 27–April 2, 1944, and Godske presented five papers. F. Spinnangr, G. Spinnangr, and J. Knudsen are also Norwegian scholars in this field.

In Poland, much attention has been given to local and microclimatology for many years. M. Hess, A. Kosiba, W. Okołowicz, and J. Paszyński are recent contributors in this field.

In Rumania, topoclimatological investigations are carried out under the guidance of V. Mihăilescu. O. Bogdan, El. Mihai, Gh. Neamu, and El. Teodoreanu also studied this subjects.

Sweden produced a number of brilliant climatologists, but there were not many studies in local climatology. However, Å. Sundborg's detailed research on the city climate in Uppsula (1950, 1951) is an important work in the history of city climatology. Quite recent contributions from Lund are observable.

In Switzerland, local climatology advanced as in Germany and Austria. There are excellent research results on the climate in the Alps, at health resorts, and in forests and on plant climate from the standpoint of local climatology. R. Billwiller, H. Brockmann-Jerosch, W. Lüdi, W. Mörikofer, W. Nägeli, M. Schüepp, and H. Zoller must be noted in the history of this field.

In the United States of America ecologists have rigorously conducted local and microclimatic studies. H. C. Cowles did research on the development and distribution of vegetation in the Chicago area in 1901, and pointed out the differences caused by local climate. After Cowles the studies on local and microclimate in connection with vegetation developed in the following three categories.

(i)　Research by meteorologists in the Weather Bureau has been done in the field of agricultural meteorology, and some study has been done on meteorological conditions for forest fires. These studies are mainly for practical purposes.

(ii)　Research has been carried on by specialists in botany, ecology, forestry, etc., chiefly from the ecological viewpoint. They concentrated on the analyses of the relationship between the inclination and surface conditions of mountain slopes or soil conditions and the climatic elements. Lately a number of detailed reports on analyses of results obtained through comprehensive observations have been published. The studies by J. E. Cantlon, J. Kittredge, D. B. Lawrence, H. H. Rasche, P. Ross, and J. N. Wolfe are excellent.

One outstanding contribution after World War II was the study on microclimates and macroclimates of the Neotoma valley by J. N. Wolfe, R. T. Wareham, and H. T. Scofield, which appeared in the Ohio Biological Survey Bulletin

No. 41 in 1949. They made clear how great the local and microclimatic differences are in a small valley compared with the macroclimatic differences in Ohio State and showed the year to year variation of microclimatic conditions.

(iii) Studies have been done by meteorologists, climatologists, and geographers who were interested especially in local and microclimatic phenomena. The studies by W. A. Baum, C. F. Brooks, J. E. Church, H. J. Cox, A. J. Henry, E. W. Hewson, R. L. Ives, H. E. Landsberg, D. H. Miller, R. Peattie, J. Leighly, C. W. Thornthwaite, S. S. Visher, and F. D. Young form the principal axes of the research that belongs to this category.

The observation of local and microclimate in the Soviet Union started in the second half of the 19th century. At a signal tower of the Central Physics Observatory in Pulkobo, the vertical distribution of air temperature in the lowest air layer was measured in 1872. On the basis of this data, A. I. Voeikov studied the vertical distribution of the air temperature and humidity near the ground surface and the effect of the vegetation and snow cover on climate. His conclusions were summarized in a monograph, *Global Climate: Particularly of Russia* published in 1884. A. I. Voeikov (1842–1916), born in Moscow, was a brilliant scholar who pursued a number of climatology studies after graduation from Göttingen University. He studied heat balance of the active surface and the local differences in characteristics of climates. His ideas were reflected and developed in the works of an expedition headed by V. V. Dokuchayev. He emphasized the organization of meteorological observation in 1892–1894 for understanding the diversity in climates of the city, forest, steppe, plateau, and valley. As early as 1894, G. N. Vysotskiy, who took part in the Dokuchayev expedition, characterized the basic regularities in the distribution of minimum temperatures with relation to the degree of shielding of a locality provided by the vegetation and also with relation to the peculiarities of the underlying surface.

G. A. Lynboslavsky of the Forestry Institute in Leningrad studied the thermal cycle of the soil and the effect of the vegetation cover in the beginning of the 20th century. P. I. Koloskov published a monograph *Relief as a Climatic Factor* in 1915. In the early 1930s, research in local and microclimate was promoted for application to the needs of agricultural production, health resort construction, transportation, national defense, and other purposes. Under the direction of G. T. Selyaninov, the Agrohydrometeorological Institute, agroclimatic zonation of the subtropical belt in the U.S.S.R. was investigated. The chair of climatology of the Geographic Faculty of Leningrad State University, under A. V. Voznesenskiy and A. A. Kaminskiy, and later under their followers, also conducted important work on microclimate and local climate.

Since World War II many researchers have been studying the problems physically and geographically. They are B. P. Alisov, T. G. Berlyant, M. I. Budyko, S. P. Chromov, F. F. Davitaya, O. A. Drozdov, E. E. Federov, I. A.

Gol'tsberg, D. L. Laykhtman, E. S. Rubinstein, S. A. Sapozhnikova, G. T. Selyaninov, V. A. Smirnov, and P. A. Vorontsov.

Sapozhnikova (1950) wrote a comprehensive monograph, *Microclimate and Local Climate*, admitted by the Ministry of Higher Education of the U.S.S.R. as a textbook for use in hydrometeorological institutes and universities. This monograph reports many results of the earlier studies in the U.S.S.R. In a book edited by Gol'tsberg (1969), there are the results and methods of research carried out by the Main Geophysical Observatory, which has been in the center of this field for many years. A short summary of the last 50 years' progress in this observatory was given also by Gol'tsberg (1970).

2.4. DEVELOPMENT OF RESEARCH METHODS

The development of modern research methods up to the present time might be chronologically classified into the following three periods:

(1) The period when analyses were made on how the local conditions were reflected in the results observed at a certain station. This is the period from the 19th century to the beginning of the 20th century when the observed results at a certain spot were analyzed and compared with those at several other adjacent stations for the determination of the effect of their local conditions. For such comparison the mean values were mainly used.

(2) The period when the study was conducted as much in detail as possible on the horizontal distribution of the climatic elements in small areas. On May 12, 1927, W. Schmidt conducted a survey in Vienna by a car-mounted thermometer and drew detailed temperature distribution maps. After this experiment, cars came to be indispensable as "a moving laboratory." Thus the local climate can be recorded on a distribution map based on the observed values at many points in a small area.

(3) The period when the climatic phenomena in small regions are examined by experiments. Local climatology is an experimental science. Hence, the examination through repeated experiments is one of the study methods. The experimental methods vary from indoor methods, such as wind tunnel experiments, to outdoor ones. The outdoor experiments here are different from conventional observations: The conditions must be controlled so that the phenomena can be analyzed one after the other under certain set conditions. For example, an experiment is conducted under a certain wind direction or on clear days only. In other words, synoptic conditions are controlled. Further, numerical experiments are a powerful method in approaching the problems.

The foregoing chronological classification is not necessarily applied to the individual research of local climatology. In today's study the research methods of the first stage are still significant and can teach us how to generalize the results of the second stage.

The indicators of the local and microclimatological phenomena can be used in a survey at the second and third stages. Furthermore, according to the results of the quantitative analysis, we can try to make assessment of the local and microclimatic environments quantitatively. As for another applied purpose, it must be pointed out that our interest in local climatology and microclimatology is linked to the possibilities existing for its modification.

Chapter 3

GROUND SURFACE AND CLIMATE IN A SMALL AREA

3.1. FLAT, OPEN LAND

In a flat, open area where there is no remarkable effect of orography, the climatic conditions of the area depend on the surface cover. The climate differs according to whether the ground surface is sandy, bare, or covered with vegetation, snow, or water.

3.1.1. Radiation Balance

(A) *Heat Balance Equation*

The total heat energy flux of a unit surface of the ground and of the surrounding space within a given time is represented by the formula:

$$R = LE + P + A \tag{3.1}$$

where R, the radiation flux of heat, is positive when the surface receives heat but negative when the surface emits heat (Budyko, 1956).

R is called the net radiation; L is the latent heat of evaporation; E is the velocity of evaporation or of condensation. Thus, LE stands for the heat emission caused by evaporation, or the heat supply caused by condensation. P is the heat flux caused by the turbulence between the surface and the atmosphere; A is the heat flux between the surface and the deeper layer of the earth. It is the sum of the horizontal heat exchange (F) and the variation of heat storage in soil (B) within a given period. F is in general very small and may be thought to be null. Both A and B are equal to zero if averaged over a year. The values of E and P are determined by direct observation, as is shown in Eqs. (3.28) and (3.29).

In an arid region, P is large and LE is small. In contrast to this, LE is large and P is small in high latitudes. In an arid region, because of the small amount of water consumption, LE is small, and because of the higher surface temperature of the ground, P becomes large. On the other hand, in higher latitudes, where water is abundant, LE takes large value. Geiger (1961, 1965) gives many examples of these values.

The net radiation (R), i.e., the difference between the solar radiation absorbed by the surface of the earth and the effective radiation from it, can be obtained from the following equation:

$$R = (Q + q)(1 - \alpha) - I \tag{3.2}$$

41

where Q is the total amount of direct solar radiation; q is the total amount of diffused sky radiation; α is the albedo, that is, the reflective power of the active surface in relation to the short-wave radiation; I is the effective out going radiation; that is, the difference between I_T and I_A, where I_T is the long-wave part of the radiation of the active surface and I_A the atmospheric back radiation.

According to A. Ångström, the total radiation, i.e., the amount of direct radiation and scattering radiation is obtained from the following equation:

$$(Q+q)=(Q+q)_0 \cdot [k+(1-k)s] \tag{3.3}$$

where $(Q+q)$ is the actual total solar radiation; $(Q+q)_0$ is the total radiation when the sky is completely clear; k is a constant determining the potential radiation when the sky is completely overcast; s is the rate of sunshine during the period under consideration.

On the other hand, an alternate equation is given by S. I. Savinov:

$$(Q+q)=(Q+q)_0 \cdot (1-c\bar{n}) \tag{3.4}$$

where c is the constant determined by the influence of the clouds on radiation; $\bar{n}$ is $\left(1-\dfrac{1-n+s}{2}\right)$; and n is cloud amount in unit 0.1.

The Main Geophysical Observatory (GGO in Leningrad), U.S.S.R., uses the following formula:

$$(Q+q)=(Q+q)_0 \cdot [1-(1-k)n] \tag{3.5}$$

The total radiation amount $(Q+q)_0$ on a completely clear day for individual months at respective latitudes is presented in Table 3.1. The value of k is also shown in the right-hand column. From this table the total radiation of a given latitude is obtained. It is possible of course to get an exact measurement of the total amount of short-wave radiation using an actinometer. The short-

Table 3.1   Total radiation $(Q+q)_0$ kcal·cm^{-2}·month^{-1} under completely clear conditions and k in Eq. (3.3) (Budyko, 1956).

	Jan.	Feb.	Mar.	Apr.	May	Jun.	Jul.	Aug.	Sep.	Oct.	Nov.	Dec.	k
80°N	0.0	0.0	2.5	9.6	17.9	20.3	18.9	10.8	3.6	0.4	0.0	0.0	
75	0.1	0.6	4.0	11.2	18.7	20.9	19.7	12.3	5.3	1.7	0.2	0.0	0.55
70	0.2	1.4	5.8	12.7	19.4	21.4	20.3	13.7	7.0	3.0	0.7	0.1	0.50
65	0.8	2.5	7.6	14.1	20.1	21.9	21.0	15.1	8.8	4.5	1.5	0.4	0.45
60	1.7	3.9	9.6	15.4	20.8	22.3	21.6	16.4	10.5	6.1	2.6	1.2	0.40
55	3.0	5.6	11.5	16.6	21.5	22.7	22.1	17.7	12.3	7.7	4.1	2.3	0.38
50	4.7	7.5	13.5	17.8	22.1	23.0	22.5	18.8	14.2	9.6	5.8	3.8	0.36
45	6.6	9.4	15.4	19.0	22.6	23.3	22.9	20.1	16.0	11.6	7.7	5.7	0.34
40	8.7	11.5	17.0	20.0	22.9	23.5	23.2	21.1	17.6	13.4	9.7	7.7	0.33
35	10.8	13.6	18.5	21.0	23.0	23.5	23.3	21.8	18.8	15.1	11.8	9.6	0.32
30	12.7	15.2	19.5	21.6	23.0	23.5	23.3	22.2	19.8	16.5	13.6	11.4	0.32
25	14.3	16.5	20.3	21.8	22.9	23.4	23.1	22.3	20.5	17.6	15.0	13.1	0.32
20	15.5	17.5	20.8	21.8	22.6	22.9	22.7	22.2	21.0	18.5	16.3	14.5	0.33
15	16.6	18.3	21.0	21.6	22.0	22.2	22.1	21.8	21.1	19.2	17.3	15.7	0.33
10	17.4	19.0	21.0	21.3	21.2	21.2	21.2	21.2	21.1	19.6	18.0	16.0	0.34
5	18.0	19.5	20.8	20.8	20.4	19.8	20.1	20.5	20.8	19.9	18.6	17.3	0.34
0	18.5	19.8	20.4	20.2	19.2	18.0	18.7	19.6	20.4	20.0	19.0	18.0	0.35

wave radiation is equal to the sum of the solar radiation and scattering radiation. On the other hand, the long-wave radiation includes the radiation from the surface and the atmosphere.

The difference of albedo (α) exerts the most important influence on the local differences of the radiation balance in a small area. For example, when R is obtained by substituting ($Q+q$) from Eq. (3.5) into Eq. (3.2), the value of α varies according to the difference of the surface conditions, so that the local value of R is determined by the actual value of α.

(B) *Albedo*

The albedo is defined here as the ratio of the amount of reflected radiation to the solar radiation. For example, an albedo 0.4 (or 40%) means that 40% of the solar radiation is reflected on the surface of the earth, and 60% of the solar radiation is absorbed. The albedo has different values for three different wavelengths (λ) in the solar spectrum: the ultraviolet portion of the solar spectrum ($\lambda \leq 0.36\mu$), the visible portion of the spectrum ($0.36\mu \leq \lambda \leq 0.76\mu$), and the infrared portion of the spectrum ($0.76\mu \leq \lambda \leq 100\mu$). In the ultraviolet portion of the spectrum, the value of the albedo is 0.80–0.85 for a surface with snow cover, 0.20–0.25 for gravel, granite rock, and chalk rock, 0.17 for dry sand, 0.06 for soil, and 0.02 for a desert (Geiger, 1950).

Many observations have been made concerning the visible portion of the spectrum: 0.80–0.85 for a surface with snow cover, 0.18 for dry sand, 0.10–0.20 for bare ground, 0.09 for wet sand, and 0.03–0.10 for a green forest.

As for the infrared portion of the spectrum, the following results were obtained: 0.11 for sandy soil of light color, 0.08–0.09 for light gray limestone rock, 0.08–0.09 for rough gravel, and 0.05 for snow cover.

Table 3.2 shows the values of albedo obtained by observations in U.S.S.R. for short-wave radiation (Budyko, 1956). The average value of albedo in broad region, such as a climatic zone, is slightly different from that in Table 3.2.

The albedo shows diurnal variations, which arise mainly from the diurnal variations of the height of the sun and the cloud amount. With an increase of cloud amount, the direct solar radiation decreases, and the scattering radiation is more and more absorbed. As a result, the relation between albedo and the

Table 3.2   Albedo of the natural ground surface (Budyko, 1956).

Surface	Albedo	Surface	Albedo
Snow and ice		Arable land, grass land,	
dry, new snow	0.80–0.95	Tundra	
slightly wet snow	0.60–0.70	Rye and wheat field	0.10–0.25
old snow	0.40–0.50	Potato field	0.15–0.25
sea ice	0.30–0.40	Cotton field	0.20–0.25
		Grassland	0.15–0.25
Bare ground		Dry grassland	0.20–0.30
black soil	0.05–0.15	Tundra	0.15–0.20
wet, gray soil	0.10–0.20	Forest	
dry clay, gray soil	0.20–0.35	Needle trees	0.10–0.15
dry, sandy soil	0.20–0.45	Broad-leaved trees	0.15–0.20

height of the sun becomes weak. When there is snow cover, the daily variation of albedo can be ignored.

The albedo on a water surface is generally lower than that on a ground surface. The albedo on a water surface for direct solar radiation is closely related to the height of the sun. It decreases to a very small percent when the sun is high, but increases almost to 100 percent when the sun reaches the horizon. On the other hand, the albedo on a water surface for scattering radiation is not directly related to the height of the sun. The albedo on a water surface for total radiation, at latitudes 20°–40°, is 0.06–0.07. It is 0.10–0.12 from November to January around latitude 40° and is 0.08–0.09 around latitude 30°.

The estimated mean monthly albedos as a function of month for the area around stations in the tundra, forest-tundra, open woodland, and closed forest zones in Canada show a very clear contrast (Hare and Ritchie, 1972). Midsummer albedos are similar in all zones, about 0.2, but the large differences in winter, 0.35–0.80, continue well into the high radiation period of spring, except in the closed forest zone. The vegetation structure determines whether snow cover will be effective as a reflector. This effect is of great importance in spring, when solar radiation is strong and the ground is still covered with snow.

Monthly changes of albedo for a meadow and for the water surface of the River Danube near Vienna, for radiation from the sun and the sky, were measured by Sauberer (1952) and Dirmhirn (1953), as shown in Table 3.3. It is interesting to note that the seasonal change of surface cover conditions is the most effective for causing a greater range of annual variation of the albedo.

Chia (1967) estimated the albedo of the natural surface in Barbados, which is the eastern-most island of the Caribbean chain (13° 10′N, 59° 35′W) to be 0.17 for sugar cane fields, 0.155 for vegetable crops, 0.11 for bare soil surfaces, 0.20 for sour grass (*Panicum conjugatum*) pastures and wasteland, and 0.20 for built-up areas and roads. The percentages are 47, 8, 5, 15, and 25, respectively.

The reflection coefficient—the ratio of the radiant energy reflected along the geometrical reflection path to the total that is incident up on the surface—was

Table 3.3   Monthly average albedo (percent) for a meadow and water surface near Vienna (Sauberer, 1952; Dirmhirn, 1953).

Month	Jan.	Feb.	Mar.	Apr.	May	Jun.	Jul.	Aug.	Sep.	Oct.	Nov.	Dec.
Meadowland, depending on the development of vegetation	13	13	16	20	20	20	20	20	19	18	15	13
Meadowland, taking into account winter snow cover	44	39	27	20	20	20	20	20	19	18	21	36
Water surface of the River Danube	11.2	11.4	10.7	10.0	9.0	8.6	8.6	9.7	9.5	11.7	11.8	11.9

measured over natural and arable land in Nigeria (Oguntoyinbo, 1970). The reflection coefficient is inversely related to the vegetation height. Distinct seasonal differences were found with unexpectedly low values in harmattan haze during the dry season. The coefficients in an urban area were lower than those in the rural settlements.

(C) *Effective Radiation*

We will be concerned here with the effective outgoing radiation (I) in Eq. (3.2). According to the Stefan-Boltzmann's law, the surface radiation of a substance is $s\sigma T_n{}^4$ cal·cm^{-2}·min^{-1}, where s is a coefficient representing the ratio of the actual surface radiation to the black body radiation and its usual value being 0.85–1.00; σ is Stefan-Boltzmann's constant ($=8.14 \times 10^{-11}$ cal·cm^{-2}·min^{-1}· deg^{-4}); and T is the surface temperature in degrees Kelvin (°K$=273+$°C). On a fine day, according to Ångström, the relation between the effective radiation amount I_0 and the atmospheric temperature and moisture is formulated as

$$I_0 = \sigma T^4(0.194 + 0.236 \times 10^{-0.069e}) \tag{3.6}$$

or, according to Brunt (1932),

$$I_0 = s\sigma T^4(1 - a - b\sqrt{e}) \tag{3.7}$$

where $a=0.256$ and $b=0.065$. The medians of twenty-two evaluations are $a=0.605$ and $b=0.048$ (Sellers, 1965). According to T. G. Berlyant (1956),

$$I_0 = s\sigma T^4(0.39 - 0.058\sqrt{e}) \tag{3.8}$$

where T is the temperature in °K and e the vapour pressure in mm. I_0 is represented in units of cal·cm^{-2}·min^{-1}. Budyko (1956) took 0.050 instead of 0.058 in Eq. (3.8).

The effective radiation (I) when it is cloudy is formulated as follows:

$$I = I_0(1 - cn) \tag{3.9}$$

where c is the coefficient, its values being 0.15–0.20 when the cloud is high, 0.5–0.6 when the cloud is not very high, and 0.7–0.8 when the cloud is low; n is the cloud amount expressed in unit 0.1. In recent research, the effect of the difference between the surface temperature (T_w) and the air temperature (T) on the effective radiation has been taken into consideration. When there is a difference in temperature between them, the effective radiation can be represented as $s\sigma T_w{}^4 - s\sigma T^4$, which in turn may be represented as $4s\sigma T^3 (T_w - T)$, which yields the following equation:

$$I = I_0(1 - cn^2) + 4s\sigma T^3(T_w - T) \tag{3.10}$$

In Eq. (3.10), 0.9 may be used as an average value of s. The value of I_0 is obtained in tables calculated by Eq. (3.8) (Budyko, 1956). The monthly value of I at Pleven (43°25′N, 24°36′E) in Bulgaria varies from 2.4 in December and January to 6.0 in August, and the annual total amounts to 52.4 kcal·cm^{-2} (Lingova, 1965).

In a small area, the distribution of I can be obtained by observing the distributions of T, T_w, and e and making use of Eq. (3.10) and the table of I_o for each point.

(D) *Distribution of Heat Balance and Radiation Balance in a Small Area*

The heat balance conditions on the earth's surface during the day and at night have been studied for many years. Good reviews of this problem have been given by Budyko (1956), Geiger (1961, 1965), Baumgartner (1963), Miller (1965), and Sellers (1965). There is fairly useful information about the heat balance and the radiation balance at a certain point under various ground conditions. However, there are very few studies on the local differences of the heat and radiation balance or of the respective balance terms in a small area. As stated earlier by Thornthwaite (1953), "the ultimate objective of climatology may be to make maps of the heat budget and the moisture budget of the earth on a topoclimatological scale." This is, unfortunately, still true today.

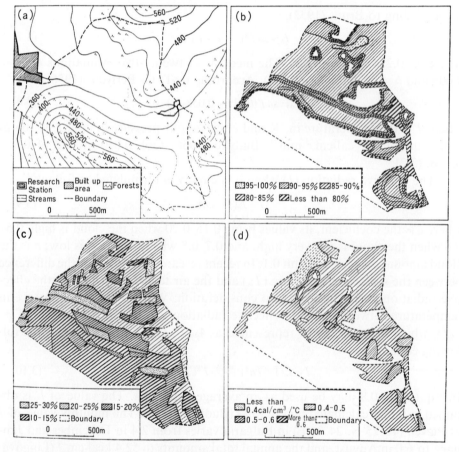

Fig. 3.1 Distribution of factors in relation to the heat balance terms (Paszyński, 1964b). a: Topography, b: Albedo, c: Index of horizon screening, and d: Heat capacity of soil.

Paszyński (1964a, b) studied the heat balance distribution in a small area, 5 km², in Wojcieszów, Poland, illustrating the topography (inclination of the slope), albedo, horizontal vegetation coefficient, heat conductivity, heat capacity of soil, roughness of the ground surface, area percentage of water surface, etc. Some of his figures are shown in Fig. 3.1. The intervals of the isolines in these maps must be carefully determined taking into account the degree of the contribution to the heat balance equation and the accuracy of observation and calculation. A similar representation was made with a colored map showing the topoclimate in an area of Pińczów (Paszyński, 1970).

The monthly and annual values of radiation balance (R) were determined at about 60 meteorological stations in Poland, during 1951–1960 (Paszyński, 1966). The distribution of short-wave net radiation in Poland was also given in figures by Paszyński (1965). The mean annual sums of R cover a wide range from 28 to 40 kcal·cm⁻², as shown in Fig. 3.2. The maximum occurs in the Krakow area, and the lowest values in the Upper Silesia industrial area. This map corresponds to meso-scale distribution, but it may serve for the first approach to the problems in a small area.

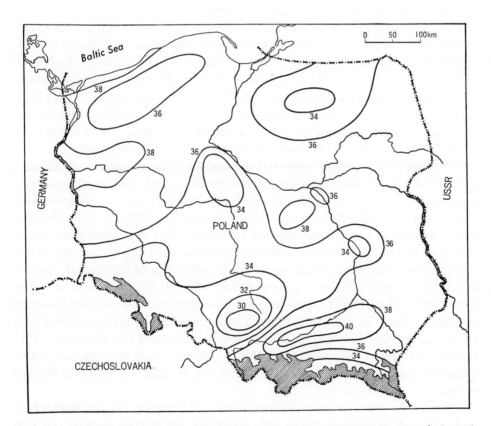

Fig. 3.2   Mean annual sum of radiation balance (R) for 1951–1960 (kcal·cm⁻²) (Paszyński, 1966).

Further results can be found in the studies on the heat balance and evaporation in Poland (Paszyński, 1972). The distribution of the ratio of the heat used for evaporation to the net radiation in Poland, i.e., E/Q, varies from 50% to 80%. This means a large share of evaporation in the heat balance, which is especially important in SW-Poland, Silesia, and the region bordering the chain of Sudety mountains. Detailed observations have been carried out at Wojcieszów in the Sudety mountains in the Kaczawa Hills (Kluge and Krawczyk, 1964; Miara, 1969; Skoczek, 1970) to find the local factors influencing the heat balance of the regions.

The incoming short-wave radiation and the effective long-wave radiation at about 80 stations in Japan were calculated by Kondo (1967) and the sensible heat transfer coefficient by Nishizawa (1966). Laitinen (1970) presented a paper dealing with the distribution of monthly mean values of the components of the energy balance of the earth's surface in Finland, based on seventeen grid points. His results indicate that the annual variation of the radiation balance is locally very regular, and there are two types of annual variation of turbulent heat exchange, north and south Finland types. In north Finland, melting and evaporation become intensified later, which is why the sensible heat flux from the earth's surface to the atmosphere is greater in spring.

A good general view of heat budget together with water budget—this is the most important approach toward local climate—was given by Miller (1965). He compiled the results of Aslyng (1960), Kristensen (1959), and Aslyng-Nielsen (1960) as shown in Table 3.4. In this table, the unit one langley (ly) is equal to 1 small calorie of heat per unit area of 1 cm^2. There is a large annual range (372—24=348 ly/day) of short-wave radiation absorbed by the earth's surface, as seen in item f in Table 3.4. The annual range of the net whole-spectrum radiation (item h) varies from -63 ly/day in winter to $+234$ ly/day in summer.

Terjung et al. (1970b) observed the budget of energy and moisture in a desert in Death Valley, California, in mid-summer, and came to the following conclusions: (i) In spite of the observed results under hazy conditions, solar radiation at normal incidence reached 1.43 $ly \cdot min^{-1}$ by 1000 hr, while the same radiation on a horizontal surface was 1.20 $ly \cdot min^{-1}$ at that same time. About a half of the latter amount was diffuse radiation. (ii) The surface temperature of the desert vegetation, which covers about one-fourth of the ground—mainly bushes of mesquite (*Prosopis* sp.) and desert holly (*Atriplex hymenelytra*)—was lower during the day and higher at night in comparison with the conditions of the bare surface. (iii) Because of the very steady loss of infrared radiation during the night, net radiation was negative at a rather constant rate of about -0.13 $ly \cdot min^{-1}$. (iv) The results of the field observation agree well with the values generated by the theoretical model of net radiation by Sellers (1965) and Budyko (1956).

Monteith (1961) estimated annual net (total) radiation for the British Isles and obtained 29 $kcal \cdot cm^{-2}$ with little variation. He further pointed out that

Table 3.4 Fluxes in heat and water balances, with auxiliary data, from Copenhagen, obtained by Aslyng et al. (1960) (Miller, 1965).

Flux	Range of Monthly means		Yearly mean or total
	Lowest month	Highest month	
a. Noon sun height	11°	58°	35°
b. Day length	7 hr	17 hr	12 hr
c. Downward short-wave radiation	30 ly/day	468 ly/day	85,300 ly
d. Upward short-wave radiation			−19,100 ly
e. Albedo	0.19	0.50	0.22
f. Net difference between downward and upward short-wave radiation	24 ly/day	372 ly/day	66,200 ly
g. Net difference between downward and upward long-wave radiation	−64 ly/day	−143 ly/day	−40,700 ly
h. Net whole-spectrum radiation	−63 ly/day	+234 ly/day	+25,500 ly
i. Heat exchange with the soil			Approx. 0
(From the soil)	+22 ly/day		
(Into the soil)		−29 ly/day	
j. Sensible-heat flux			−2,200 ly
From the air	+72 ly/day		
(Into the air)		−78 ly/day	
k. Latent-heat flux	zero	−133 ly/day	−23,300 ly
l. Evapotranspiration	zero	70 mm/month	392 mm
m. Precipitation	21 mm/month	104 mm/month	585 mm
n. Water exchange with the soil			Approx. 0
(Into soil storage)	−44 mm/month		
(Out of storage)		+21 mm/month	
o. Water surplus	zero	42 mm	193 mm
p. Air temperature (Averages 2°C lower than ocean; varies from 6°C lower to 2°C higher)	−2°C	17°C	7°C
q. Humidity (vapour pressure)	5 mb	14 mb	9 mb
r. Soil temperature, at −2.5 cm depth	0°C	23°C	10°C

radiation totals at Aberporth on the Welsh coast are anomalously high (Monteith, 1966). The annual mean intensity of direct radiation at Aberporth is 31 m w·cm^{-2} compared with 25 m w·m^{-2} elsewhere. This local difference might be caused by air pollution.

Miller (1965) discussed further the influences of surface cover on the balances of heat and water under various conditions such as grassland, low vegetation, bare ground, mulches, irrigated land, tall crop plants, orchards, and forests. Some examples are compiled in Table 3.5, adapted from his tables. The radiation surplus is caused in any case by the downward short-wave radiation. Most of the surplus is spent in steaming water out of the grass, rather than in heating the air, as shown on the meadow near Hamburg (Frankenberger, 1962). The energy balance over prairie grass was computed by Nkemdirim and Yamashita (1972). The patterns of the soil heat flux were fairly steady from day to day in summer, and the relation between hourly flux of sensible heat and soil heat flux was linear on a daily basis.

Table 3.5　Radiation and heat balances of various places in langleys/day after the tables compiled by Miller (1965).

Land surface		Short-wave radiation		Long-wave radiation	Whole-spectrum radiation	Disposition of heat	Place	Researcher
		Direct	Diffuse					
Meadow	Downward	+371		+620	+991	−53, Sensible	Hamburg	Franken-berger (1962)
	Upward	−56		−720	−776	−150, Latent		
	Net	+315		−100	+215	−12, Soil		
Open desert	Downward	+76	+10	+39	+125	−36, Sensible	Bukhara	Aizenshtat (1958)
	Upward	−22		−57	−79	0, Latent		
	Net	+64		−18	+46	−10, Soil		
Saxaul foliage	Downward	+51	+7	+39	+97	−43, Sensible	Bukhara	Aizenshtat (1958)
	Upward	+5		−38	−33	−21, Latent		
	Net	+63		+1	+64			
Ground beneath saxaul	Downward	+13	+9	+41	+63	−3, Sensible	Bukhara	Aizenshtat (1958)
	Upward	−6		−49	−55	0, Latent		
	Net	+16		−8	+8	−5, Soil		
Cotton field	Downward	+78	+8	+38	+124	−3, Sensible	Bukhara	Aizenshtat (1958)
	Upward	−16		−41	−57	−60, Latent		
	Net	+70		−3	+67	−4, Soil		
Sugar cane	Downward	+600			+405	−75, Sensible heat + photosynthesis	Hawaii	Chang (1961)
	Upward	−95		−100		−330, Evapotranspiration		
	Net	+505				0, Soil		

Aizenshtat (1958) measured the radiation budgets and heat balances at a point in the desert, Saxaul brush (*Haloxylon* sp.) land, irrigated cotton fields, etc. The absolute values of the radiation surplus at these places are lower than else where, but the circumstances of the net radiation are also true at these places.

Peculiarities of the radiation and heat balance of the icing were studied in the Ulakhan-Taryn valley in Central Yakutia of Siberia (Gavrilova, 1972). The icing develops from the freezing of underground fresh waters and begins to form in October–November and to melt in the second half of April and ends at the end of June. The amount of heat released in the process of the icing formation is equal to the annual total of the radiation balance, 19.6 kcal·cm^{-2}, in the Ulakhan-Taryn valley. When the icing melts away, the greater part of the thermal energy is used for evaporation and heating of the soil. In the Lena River valley, where the icing is lacking, the annual total of the radiation balance was 29.7 kcal·cm^{-2}.

In regions heavily with snow covered, it is very important to melt snow accumulation as early as possible in spring for agricultural use of the land. A comprehensive study was made on the heat balance at the snow surface in the thawing season in Japan (Nakamura, 1964). The absorbed short-wave radiation ranged from 30 to 500 ly/day, and the effective long-wave radiation from -50 to -250 ly/day. The ratio of these amounts to the expended heat of snow-melting is given in Table 3.6. This table shows that the short-wave radiation absorbed on the snow surface contributes to snow-melting, particularly in the first half of the snow-melting period, much more than long-wave radiation. This implies that artificial methods such as spreading carbon-black powder on a snow surface may be effective to promote melting in the early stages.

Table 3.6 The ratio of the amounts of that balance items to the expended heat snow melting (Nakamura, 1964).

	First half of snow-melting period	Second half of snow-melting period
Absorbed short-wave radiation	1– 2.5	0.5– 2.0
Effective long-wave radiation	−0.5–−1.9	−0.2–−1.6
Sensible heat flux	0.3– 0.4	0.4– 0.6
Latent heat flux	<0.0	0.0– 0.2

Grunow (1961) pointed out that, in determining a local condition of radiation, the use of the relative values of incoming radiation from the sun and sky to the maximum values on days with clear sky or to the values of extraterrestrial solar energy is by no means suitable and that the values relative to the Rayleigh atmosphere, that is, atmosphere without dust and water vapour, should be used.

3.1.2. Ground Temperature

The ground temperature determines the air temperature within a small area when the wind velocity is low and the sky is clear. The ground temperature distribution has the most intimate relation to the above-mentioned local distribution of heat balance.

The major factors of soil climate were listed in 16 items by Taylor (1962). Among them, the followings are important from the standpoint of local climatology: (i) the nature and depth of the parent material in relation to its thermal conductivity and water-holding capacity, (ii) the depth of the soil itself and of the weathered material, if any, below it, (iii) the texture and the structure of the soil itself and of its different horizons, (iv) the degree of roughness of the ground surface, (v) the color of the ground surface, (vi) the degree of wetness or dryness of the soil, (vii) the nature, morphology, and degree of wetness of vegetation, if any, (viii) the level and fluctuation of the water table, and (ix) the presence or absence and disposition of organic horizons and organic matter within the soil profile. Some of these factors are discussed below.

The radiation and the heat flux that reach the earth raise the ground temperature only after their absorption in the soil. When the earth's surface is covered by vegetation, the increase of the ground temperature in the daytime and cooling at night will be hindered. According to Brunt (1945, 1953), the ratio of the temperature variation of the ground surface is proportional to

$$\frac{1}{\sqrt{ks}} \tag{3.11}$$

where k is the thermal conductivity, and s the heat capacity of the soil. However, the soil is composed of soil particles with air or water filling the spaces. Consequently, the values of k and s both depend on whether the space is occupied by air or by water. Since air has little thermal conductivity and heat capacity, soil containing much air has a large value of $1/\sqrt{ks}$. In light soil, such as sandy soil, containing much air, surface temperature increases very rapidly in the daytime, and decreases very quickly at night. If, on the other hand, the soil contains much water, the variation of temperature will be slow, and the diurnal range will be small, because the thermal conductivity and the heat capacity are large. Because the water content becomes higher in the following order, the diurnal and annual range of temperature also become smaller in the following order: humus, sandy soil, light loam, chalky soil, heavy loam, and clay soil. The observed results of the soil temperature at Giessen, W-Germany, from 1939 to

Table 3.7 Soil temperature of humus, sand, and loam at Giessen, W-Germany (Kreutz, 1943).

Depth	Annual mean			Annual range		
	Humus	Sand	Loam	Humus	Sand	Loam
0 cm	10.1°C	9.3	9.1	23.0	21.6	20.0
10	10.3	9.5	9.3	22.0	20.8	19.6
20	10.4	9.7	9.5	21.0	19.8	18.8
50	10.8	10.3	9.9	18.2	17.4	17.0
100	11.3	11.3	10.7	14.4	14.2	14.2

1941 for humus, sand, and loam are given in Table 3.7 (Kreutz, 1943). The value at 0 cm in this table is obtained by an extrapolation. The annual means of the ground temperature at a depth of 10 cm have differences of about 1°C, and an annual range of about 2–3°C. Even when the moisture content is large, however, the ground temperature remains low at marshy or ill-drained places where the heat is lost by the evaporation from the ground surface.

The heat absorption rate changes according to the color of the ground surface. When the solar radiation is relatively constant, soil with smaller specific heat shows a stronger temperature dependence on its color. The surface temperature dependence on color is stronger in dry soil with small specific heat than in moist soil. Since dry soil has low thermal conductivity, the ground temperature cannot be generalized at a certain depth under the ground. The results of varying surface color of moist soil are as follows: 35.1°C for white, 36.8°C for yellow, 37.5°C for gray, 39.8°C for black. The results for dry soil are: 42.6°C for white, 44.1°C for yellow, 45.3°C for gray, 47.4°C for black.

World distribution of ground temperature was studied thoroughly by Chang (1957a, b, 1958). In general, the mean ground temperatures practically do not vary in depth, usually the variation being less than 1–2°C. The difference may be caused by a flux of latent heat. In most parts of the world the mean annual ground temperature at a depth where the annual temperature wave is imperceptible exceeds the air temperature by 0.5–3.0°C. Exceptions are: (i) In areas where there is a prolonged snow cover, the mean annual ground temperature may exceed the air temperature by more than 3°C, due to the higher temperature under the snow cover. (ii) The same is true in regions with volcanic activity. (iii) Because of the increasing solar radiation and the consequent rise of the daytime surface temperature, the amount by which the ground temperature exceeds the air temperature increases with altitude. (iv) Desert ground temperature appears to have a somewhat higher maximum, a markedly higher minimum and, consequently, a higher annual mean. The mean annual ground temperature is 3.6°C warmer than that of the air at Tucson, Arizona (Turnage, 1939). In Japan, the mean annual ground temperature at 1 m depth ($\bar{T}s$) is expressed in linear relation to the annual mean air temperature ($\bar{T}a$) by the equation (Nishizawa and Hasegawa, 1969):

$$\bar{T}s = 2.5 + 0.96\,\bar{T}a. \tag{3.12}$$

The correlation coefficient is $+0.96$, and the standard deviation of the calculated and observed $\bar{T}s$ is ± 0.51. This equation means that in general $\bar{T}s$ is 2.5°C warmer than $\bar{T}a$.

The mean monthly ground temperatures at Khartoum (15°37′N, 32°33′E) at 08h, 14h, and 20h from 1958 (1959) to 1962 are shown in Table 3.8. It is thought that the factors that affect the daily heat balance of an area will reduce the uniformity of ground temperature conditions from day to day and place to place (Oliver, 1966). The mean monthly values ranging from 65.3°C in April to

Table 3.8 The mean monthly ground temperature and extreme values (°C) at Khartoum, 1958–1962, compiled from the tables by Oliver (1966).

Local time	08h00m				14h00m				20h00m			
Depth (cm)	1	10	50	100	1	10	50	100	1	10	50	100
Jan.	20.7	22.7	29.1	30.2	50.6	33.7	28.8	30.2	21.6	30.5	29.0	30.1
Apr.	33.8	30.6	35.4	34.0	65.3	44.9	35.0	34.1	30.5	40.4	35.3	34.0
Jun.	36.4	34.2	38.5	37.1	63.2	46.6	38.2	37.2	34.1	42.7	38.4	37.1
Jul.	31.8	31.5	36.6	36.5	54.4	42.1	36.3	36.3	32.2	38.3	36.6	36.2
Oct.	33.2	31.7	36.2	35.9	59.6	43.5	35.8	35.8	30.4	37.6	36.3	35.7
Year	29.7	29.1	34.1	33.9	57.2	40.8	33.9	33.9	28.6	36.5	34.1	33.8
Extreme min.	—	17.2	—	27.9	—	—	25.8	—	12.0	—	—	27.9
Extreme max.	—	—	—	—	73.0	50.5	39.8	38.5	—	—	—	—

20.7°C in January at 1 cm depth as well as the extreme values of 73.0° to 12.0°C given in Table 3.8 must mean an extremely severe environment for the flora and fauna in arid tropical regions. Infrequent rain showers are readily evaporated by the heat of the surface and by heat stored in the soil below. However, the local character of the shower plays some role in the ground temperature regime in such regions. Temperature measurements have been carried out for the various kind of soils in East Europe (Junghaus, 1961; Berényi, 1965; Wagner, 1966) and in North Europe (Oliver, 1962) in relation to heat balance. These studies are an appropriate approach to the problems.

The effect of vegetation on the ground temperature cannot be neglected. For example, the ground temperature of a meadow was higher than that of a hilltop in Jávorkút in the Bükk mountains, Hungary, where there is dense grassy vegetation (Boros, 1966). The effect of forests is discussed in Section 3.3 below. Ground temperature conditions together with the light, wind and humidity over the crop fields were measured during the vegetation periods at the outside of Malmö in southernmost Sweden (Mattsson, 1961, 1966a, b). Of in-

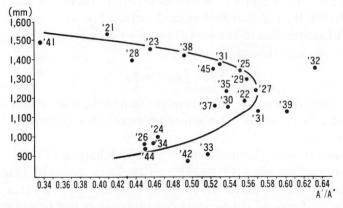

Fig. 3.3 Relationship between annual total precipitation and the amplitude ratio A′/A″ at Maebashi, Japan ('x means year 1900+x) (Fukuoka, 1966).

terest are his descriptions of the change of conditions at the various stages of crop growth, in the day and nighttime, and according to cloudiness.

The ratio of the annual range of ground temperature 3 m under the surface, A′, to that of 1 m, A″, is proportional to the square of thermal diffusivity in the soil (Fukuoka, 1966). The relationship between annual total precipitation and the ratio A′/A″ at Maebashi, Japan, is given in Fig. 3.3. It is clear that the ratio, thermal diffusivity, increases with increasing precipitation, but after a certain value of precipitation, the ratio decreases. Such a relationship must exist in the horizontal distribution in an area, but there is little knowledge about this. Only the effect of sea or river water has been analyzed using the date from 34 stations in Japan (Fukuoka, 1965): The ratio A′/A″ has its maximum at a point about 300 m from the shore line. Inside of this point, the thermal diffusivity is smaller due to the heat capacity augmented by the soil moisture near the shore. On the other hand, outside of that point, the thermal diffusivity becomes small due to the lower thermal conductivity. This is shown in Fig. 3.4.

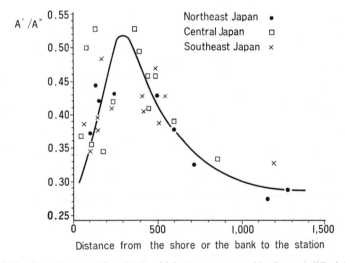

Fig. 3.4 Relation between the ratio A′/A″, which represents roughly thermal diffusivity in the soil, and the distance from the shoreline of a sea or river (Fukuoka, 1965).

As stated by Sarson (1962), the ideal observation of ground temperature must be made several times a day over a long period. This is almost impossible for many stations distributed in a small area. Therefore, the problems mentioned above still remain to be solved. Much more information on the soil microclimate is available from the descriptions by Slatyer and McIlroy (1961), Geiger (1961), van Wijk and de Wilde (1962), and van Wijk (1965).

3.1.3. Air Temperature

The climatological characteristics of the atmospheric layer up to altitudes of 800–1,000 m were dealt with in detail by Devyatova (1957). She summarized the results observed by the aerostatic lifts at 2, 50, 100, 150, 200, 300, 400, 500, 600 and 700 m during the years 1950–1954. In solving the problem of the daily variations in the air temperature, we start with the equation of the influx of heat, in the form

$$C_p \rho \frac{dT}{dt} = \varepsilon_1 + \varepsilon_2 \tag{3.13}$$

where dT/dt is the change in the air temperature with time, ρ is the density of the air, C_p the specific heat at constant pressure, ε_1 the influx of heat from turbulent heat conduction, and ε_2 the influx of heat due to radiation of the air. The diurnal variation in the air temperature is determined essentially by ε_1, but the nocturnal variation principally by ε_2. The results are given in Fig. 3.5. From this figure, the following facts are noted: (i) Daily variation is large in spring and

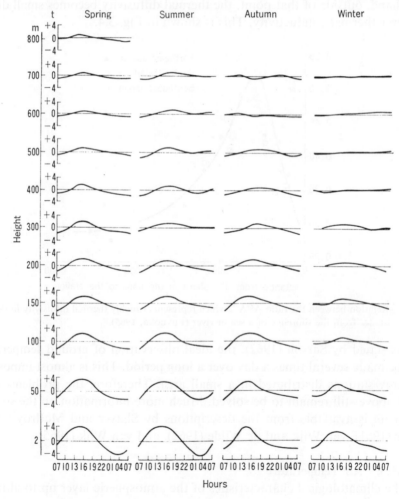

Fig. 3.5 Daily variation of air temperature (°C) at various altitudes (Devyatova, 1957).

summer and small in winter; in winter it is 2/3 of that of the summer. (ii) Daily variation becomes small with increasing altitude. The range of variation at 2 m decreases to 1/2 at a height of 200 m in spring and summer, but at 100 m in winter. (iii) In the layer up to 200 m during the time 1–7 h in the morning, the temperature remains almost constant, and at high altitudes a smooth course starts from 19 h. (iv) Above 400 m, one observes only a weak increase from 7 to 10 h. (v) Of great interest are the two maximums at altitudes higher than 500–600 m in summer, autumn, and winter. (vi) In winter the variation in the two maximums appears from 300–400 m, even though it is not very obvious. This is caused by the decrease in temperature due to the destruction of the inversion layer during the sunrise hours and then the temperature rises increasing ε_1 and ε_2 in association with the increase of sun height.

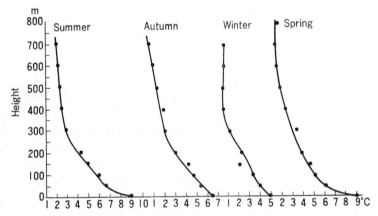

Fig. 3.6 Changes in the average daily amplitude of the air temperature with respect to altitude (Devyatova, 1957).

The amplitude of the daily variation, in other words, daily range of air temperature decreases rapidly with increasing altitude, as shown in Fig. 3.6. The maximum average amplitudes of $\Delta T°C$ are observed during the summer months. Close to these are the amplitudes during the spring months. Both are quite different from those in winter. The relation between ΔT and height z is expressed in the following form:

$$\Delta T = \Delta T_o e^{-pz} \tag{3.14}$$

where ΔT is the daily amplitude of the air temperature at a height of z m, ΔT_o those on the earth, p a coefficient that depends on the time of the year. According to Devyatova (1957), p is 257 ($\times 10^5 m^{-1}$) in autumn, 268 in winter, 227 in spring, and 218 in summer.

The daily variations of the vertical temperature gradient are shown in Fig. 3.7. In summer the variation up to 100 m is quite clearly pronounced. From 10 h to 13 h, the maximum appears in the layer 0–50 m, reaching superadiabatic

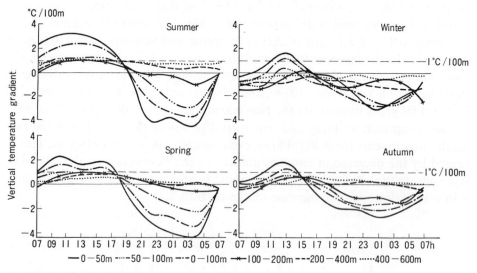

Fig. 3.7 Daily variation of the vertical temperature gradient (Devyatova, 1957).

values (it exceeds 3°C/100 m), and on some days it exceeds the autoconvection gradient. The negative values, which means inversion, appear at approximately 18–19 h and continue until around 7 h. The average maximum value of the negative gradient is −4.6°C/100 m, reached at 4 h in the morning. In winter and autumn, the durations of the negative gradient are longer because the daytime hours are short. The maximum of the negative gradient occurs at midnight.

The details of such microaerological conditions of the boundary layer of the atmosphere have also been described in monographs by Vorontsov (1960, 1961), which are summaries of his earlier works. The state of the inversion layer is discussed again in Chapter 5 (Section 5.5).

The air temperature on grassland is lower at night and higher in the daytime than on bare soil. This is due in part to: (i) There is much space with air between the grass; (ii) the roots of the grass are expanded in the soil; and (iii) the evaporation from the grass surface is effective for cooling the air temperature at night. Accordingly, the cooling rate depends on the height and thickness of the grass. Even if the above conditions are equal, the rate of cooling varies slightly in grass with spreading or creeping roots and other plants with straight roots.

A comparison of the minimum air temperatures about 1 m above the ground with varying vegetation shows that bare land has 9.7°C, raspberry bushes 9.0°C, thick grass 7.6°C, tall grass 6.1°C, and grass and clover 6.4°C. The difference of temperature between bare land and grass-covered land was no less than 2.1°C.

Small variation of air temperature within a short period of time reflects the turbulence in the atmospheric layer. It changes with the cover of the earth's surface. The results obtained with a thermister thermometer, at noon in summer, are given in Table 3.9. The observation was done through for 12 min at intervals of 15 sec. The results are shown in the standard deviation (Kawamura and Mizukoshi, 1957). This table shows that the small variation is largest at 20 cm. However, except for clover fields, the variation at 120 cm is smaller than that at 180 cm; the reasons await further study. By means of a frequency analysis of the observed data at 120 cm, it became clear that, above a lawn in the daytime, cycles with the periods of 4 or 8 min dominate, and the period becomes longer in the morning and at night. It becomes shorter on bare soil in grass-covered areas (Kawamura and Mizukoshi, 1959). Mattsson (1965) made a similar observation in and over a potato field and over fallow ground in Sweden.

Table 3.9   Micro-variation of air temperature over different vegetation (Kawamura and Mizukoshi, 1957).

Vegetation	Note	20 cm	120 cm	180 cm
Potato field	Almost dead. Ridges in field	0.96	0.71	0.85°C
Clover field	Grass height 10 cm	0.69	0.51	0.45
Larch forest	Forest width 10 m	0.22	0.16	0.33
Bare land	Diameter 200 m	0.80	0.51	0.58

The effect of the color of the ground surface on the amount of absorption and reflection of radiation is remarkable. The air temperatures near the ground and of the underground vary according to the color of the soil. Temperatures measured with a thermocouple, even though in the partly cloudy afternoon showed a difference of more than 20°C between soils with different colors, sunshine conditions, and water contents (Daubenmire, 1950). Schöne (1962) observed the various meteorological elements in the Havelländisches Luch, middle Germany, and evaluated the effects of surface conditions, especially sand areas, on them.

3.1.4. Wind Velocity

There are three stages in the history of investigation into the vertical distribution of wind velocity in the atmosphere near the ground: (i) The period of power law (from the beginning of the twentieth century to the end of the 1930s. (ii) The period of conflict between the power law and the logarithmic law, from the early 1930s to the end of 1940s. (iii) The period since the end of 1940s during which the stability effect was taken into consideration. A historical sketch (1943–1962) of micrometeorology in Japan was given by Inoue (1963) and 30 years' development of micrometeorology of cultivated fields in Japan by Uchijima (1974).

For an account of the subjects in Sections 3.1.4, 3.1.5, and 3.1.6, the readers are referred to the excellent textbooks *Micrometeorology* by O. G. Sutton (1953), and *Turbulent Transfer in the Lower Atmosphere* by C. H. B. Priestley (1959).

Exploring the Atmosphere's First Mile by H. H. Lettau and B. Davidson (1957) and *Aerial Microclimate* by E. K. Webb (1965) are also useful.

(A) *Power Law*

The study of vertical wind velocity distribution of the air layer near the ground has a rather long history. Hellmann made observations on a farm near Potsdam for 1,488 hr at five definite heights, viz., 5, 25, 50, 100, and 200 cm above the ground. Analyzing these data, he has obtained the following formula (Hellmann, 1915):

$$\bar{u} = \bar{u}_1 z^{\alpha} \tag{3.15}$$

where $\bar{u}_1$ is the mean wind velocity at 1 m above the ground, and $\bar{u}$ the mean wind velocity at height z. The value of α in Eq. (3.15) is small when it is windy, and its minimum is 0.14. The average value of α is 0.3. The value α depends on the height; it becomes smaller with an increase in height, and has diurnal variation.

If the values of α are classified by stability or the sign of temperature lapse rate (Best, 1935; Flower, 1937), under the stable conditions, the wind velocity difference between two different heights increases quickly in accordance with the wind velocity, and under unstable conditions, the wind velocity has no remarkable increase, and thus the velocity difference is small.

The value of α depends also on the ground surface conditions, i.e., the roughness of the surface (Paeschke, 1937a, b). Observation results are summarized in Table 3.10. In this table, z_0, the roughness parameter of the ground surface, stands for the critical height above which Eq. (3.15) is applicable. The value z_0 is discussed in detail below.

Table 3.10   The value α of Eq. (3.15) according to the roughness
of the ground (Paeschke, 1937a, b).

Surface condition	z_0	α of Eq. (3.15)
Smooth snow surface	3 cm	0.20
Göttingen airport	10	0.23
Bracken	10	0.25
Low grass	20	0.26
Tall grass	30	0.28
Beet field	45	0.33
Wheat field	130	0.29

(B) *Power Law and Logarithmic Law*

Hellmann (1917) made it clear that the vertical distribution of wind velocity could also be represented as the following logarithmic law:

$$\bar{u} = a \log z + b \tag{3.16}$$

where $\bar{u}$ is the mean wind velocity at a height z. The discussions as to whether the power law or the logarithmic law were to be used continued from 1936 to 1948. Thornthwaite and others, by precise observations from 1939 to 1942, found the following equation:

$$\bar{u} = \left(\frac{\log z - \log z_0}{\log \alpha} \right)^{\frac{1}{p}} \tag{3.17}$$

where $p=2$ for unstable atmosphere and $p \leqq 1$ for stable atmosphere. After World War II, Laykhtman and Chudnovskii (1949) showed the following equation:

$$\frac{\bar{u}_2}{\bar{u}_1} = \frac{z_2^{1-n} - z_0^{1-n}}{z_1^{1-n} - z_0^{1-n}} \tag{3.18}$$

where $\bar{u}_2$ and $\bar{u}_1$ are the wind velocity at heights z_2 and z_1, respectively, and z_0 is the roughness parameter of the surface. Also, $(1-n)$ changes according to the weather type; $n>1$ for unstable air, $n<1$ for stable air, and $n=1$ for adiabatic air. This means that the following logarithmic law can be obtained for adiabatic atmosphere:

$$\frac{\bar{u}_2}{\bar{u}_1} = \frac{\ln \dfrac{z_2}{z_0}}{\ln \dfrac{z_1}{z_0}} \tag{3.19}$$

Equation (3.19) suggests that the logarithmic law is applicable only to adiabatic atmosphere.

Equation (3.18) is most convenient for interpolation and extrapolation of the results of direct observation at any two points to the entire air which lies near the ground surface up to 20–30 m (Sapozhnikova, 1950).

To facilitate computations, the curve linking the rate of wind velocities at two levels, R, and α at a given value of z_0, is preliminarily computed and plotted. When any value of R is obtained by observation, the value of α is determined by the curve, after which the wind velocities at any level are computed. The same curves can be used to solve the reverse problem: that is, for the determination of the velocity ratios and, consequently, of the velocity itself if the other velocity is known.

One example might be as follows; If the observed wind velocities at heights of 5 and 2 m were 2.8 and 2.0 m/sec, respectively, the ratio of the velocities is given by $R = \bar{u}_5/\bar{u}_2 = 1.40$. Figure 3.8, which is constructed for a case of $z_0 = 3$ cm, shows that α is 0.25 when $R_{5:2}$ is 1.40.

From the same graph in Fig. 3.8: $R_{10:2} = \bar{u}_{10}/\bar{u}_2 = 1.77$. Since $\bar{u}_2$ is 2.0 m/sec, we can estimate $\bar{u}_{10}$ to be 3.5 m/sec.

Deacon (1949) introduced the quantity β, which varies according to the stability, and presented the following formula for the vertical distribution of wind velocity:

$$\frac{d\bar{u}}{dz} = \alpha z^{-\beta} \tag{3.20}$$

In this formula, $\beta > 1$ for an unstable state; $\beta = 1$ for a neutral state; and $\beta < 1$ for a stable state. By integrating Eq. (3.20) and rearranging the terms, the following equating is obtained:

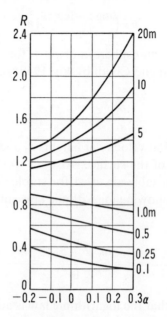

Fig. 3.8 "Laykhtman" graph for the estimation of wind velocity. Wind velocity ratio, R, at indicated heights and the wind velocity at the height of 2 m, with different values of α. This graph is applicable only in the case where $z_0 = 3$cm (Sapozhnikowa, 1950).

$$\bar{u} = \frac{u_*}{k\,(1-\beta)}\left[\left(\frac{z}{z_0}\right)^{1-\beta} - 1\right] \tag{3.21}$$

where u_* is the friction velocity. Equation (3.21) is reducible to the logarithmic law, when $\beta = 1$ (adiabatic atmosphere). Equation (3.21) is the same expression as Laykhtman's Eq. (3.18) when $\beta \neq 1$, and is reduced to Eq. (3.19) when $\beta = 1$. Equation (3.21) is called a generalized power law.

(C) *Logarithmic Law Taking Stability into Consideration*

The vertical distribution of wind velocity is empirically given by

$$\bar{u} = \frac{u_*}{k} \ln \frac{z-d}{z_0} \tag{3.22}$$

or

$$\bar{u} = \frac{u_*}{k} \ln \frac{z+d}{z_0} \tag{3.23}$$

where $z > z_0 + d$, z_0 is the roughness parameter, and d is the zero-plane displacement.

Ogura has introduced a new term by considering the stability of the atmosphere (Ogura, 1952). That is:

$$\bar{u} = \frac{W_*}{k} \ln \frac{z+z_0}{z_0} - \frac{\alpha\,gLz}{2C_p\rho\theta W_*{}^2} \tag{3.24}$$

where W_* is the maximum turbulence velocity under adiabatic conditions in units of cm/sec (equivalent to $\sqrt{\tau/\rho}$); α is the empirical constant, about 6; g is the acceleration of gravity; L is the heat flux in units of $cal \cdot cm^{-2} \cdot sec^{-1}$; C_p is the specific heat at constant pressure; and θ is the average temperature of the atmospheric layer in units of °K.

Monin and Obuhkov (1954) obtained the following expression:

$$\bar{u} = \frac{u_*}{k} \ln \left(\frac{z + z_0}{z_0} + \beta \frac{z}{L} \right) \tag{3.25}$$

where β is the empirical constant (about 0.6). L is defined as:

$$L = -u_*^3 \bigg/ \left(k \cdot \frac{g}{\theta} \cdot \frac{H}{C_p \rho} \right) \tag{3.26}$$

where H is the vertical heat flux. The values L, u_*, z_0, etc., were obtained for the stable, neutral, and unstable conditions from profiles up to the height of 300 m (Volkovitskaya and Mashkova, 1963). These results show that u_* is about 10% of $\bar{u}$.

It has been noted that the values of z_0 and d in Eq. (3.22) change with the wind velocity, even at the same observation point. Recent results are summarized in Table 3.11. From this table it is clear that they vary according to the wind velocity. An experiment on this problem was done by Bradley (1968).

Table 3.11   Relation between wind velocity and z_0 or d (Yoshino).

Researcher		z_0	d	Condition of surface	Note
Deacon	1949	Negative*	Const.	Tall grass	—
Rider	1954	Const.	Positive	Oat field	—
Tani·Inoue· Imai }	1955	{ Negative Positive	Positive Negative	Rice field, ear period	$\bar{u}_{200} < 5$ m/sec $\bar{u}_{200} > 5$ m/sec
Tani·Inoue· Imai }	1956	Positive	Negative	Rice field	$\bar{u}_{200} > 5$ m/sec
Uchijima	1957	Positive	—	Water surface of pond	Sharp increase after $\bar{u} = 2$–3 m/sec
Yoshino	1958	Positive	Negative	Grass with 50–70 cm height	In some cases, z_0 increases when $\bar{u}_{490}$ <2 m/sec
Tani	1963	(Positive Negative (Positive	Negative Positive Negative	{Paddy field 90 cm height	$\bar{u}_{150} < 2$ m/sec 2 m/sec $< \bar{u}_{150}$ <4 m/sec $\bar{u}_{150} > 4$ m/sec
Long et al.	1964	Negative	—	Barley crop 35 cm and 110 cm	$\bar{u}_{110} < 2.5$ m/sec

* Negative means reverse relation. $\bar{u}_{150}$ means mean wind velocity at a level of 150 cm above the ground.

In Fig. 3.9, the seasonal variation of z_0 of a barley crop is shown. The relation between wind velocity and z_0 is negative. In this figure, the abscissa is the wind speed at 110 cm above the top of the crop. In mid-June the height of the crop was 35 cm and at the end of July 105 cm (Long et al., 1964). The greater the roughness—other things being equal—the greater the rate of evaporation, transpiration, and transfer of pollutants. For a water surface as well as a ground surface with vegetation, Schmitz (1962) discussed theoretically the change of z_0

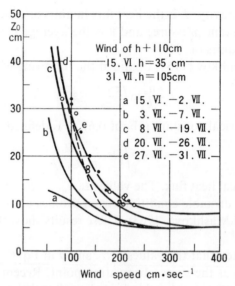

Fig. 3.9 Seasonal change in aerodynamic roughness parameter, z_0, of barley crop. Full points show hourly mean values on July 23, 1963 (Long et al., 1964).

and other terms with increasing wind velocity. If the surface is very smooth, z_0 becomes very small. For instance, when the surface consisted of fine-grained frozen snow at Ice Cap Station in spring, z_0 amounted to the order of 10^{-4} m on the average (Holmgren, 1971).

Friction velocity u_* is one of the most important micrometeorological elements, but no systematic observation has been undertaken in relation to the surface conditions. It is usually said that in case of tall grass, u_* is about 10% of u_{500}. According to indirect calculation it was 4 to 13% of $\bar{u}_{490}$ over meadow grass at Sugadaira, Nagano Prefecture, Japan (Yoshino, 1961a). The relation $u_* = 3.9 \times 10^{-3} \times \bar{u}_{75}^{1.5}$ was obtained over the water surface of a pond (Uchijima, 1959). According to the observations on rice fields by means of Robinson's cup anemometer, u_* was 22.4% of $\bar{u}_{200}$, and by means of a hot-wire anemometer it was 12.4–12.9% of $\bar{u}_{180}$ (Tani et al., 1955).

An "effective" roughness length was defined for use over heterogeneous terrain as the roughness length that homogeneous terrain would have in order to produce the correct space-average downward flux of momentum near the ground, with a given wind near the ground. With wind speed 15 m/sec and a height of 75 m, the "effective" roughness lengths derived are given in Table 3.12 (Fiedler et al., 1972). Thom (1972) treated the vegetation as a complex surface roughness to which the transfer of mass or heat encounters greater aerodynamic resistance, r_p, than the transfer of momentum, r_d. The excess resistance $(r_p - r_d)$ is equated to B_p^{-1}/u_*, where B^{-1} is the nondimensional bulk parameter. B^{-1} is expressed in terms of the exchange characteristics of the individual elements of a vegetation canopy, and this expression does not contain z_0.

Table 3.12 Effective roughness lengths in the case of wind speed 15 m/sec and a height of 75 m (Fiedler and Panofsky, 1972).

Region	σ_w/u	Effective roughness length
Plains	0.089	0.42m
Low mountains	0.107	0.99
High mountains	0.117	1.42

σ_w is standard deviations of vertical velocities.

The logarithmic spectral density for the longitudinal wind component, $nS(n)$, is defined by $\int_0^\infty nS(n)d(\ln n) =$ total variance, where n is the frequency and is expressed by $nS(n) = u_*^2\ F(f, \mathrm{Ri})$, where u_* is the friction velocity, $f = nz/\bar{u}$ and a nondimensional frequency, $\bar{u}$ is the mean wind speed at height z, and Ri is Richardson number (Berman, 1965). Despite variations in detail, the following characteristics can be noted: (i) At all frequencies the ratio $nS(n)/u_*^2$ varies little for $\mathrm{Ri} < -0.4$. (ii) At low frequencies the ratio tends to decrease from about $\mathrm{Ri} = -0.4$ through the stable portion of the graph. (iii) At high frequencies the ratio remains fairly constant for about $\mathrm{Ri} < -0.1$, but then definitely increases past $\mathrm{Ri} = -0.1$ into the stable region [*Q.J.R.M.S.* **91** (1965) 302–317].

According to a recent measurement of spectra of atmospheric turbulence at various heights and sites (Busch and Panofsky, 1968), it was shown that longitudinal spectra do not obey the similarity theory at low frequencies. A comparison of the longitudinal spectra over land and water also shows that the spectra over water contain significantly more energy than the spectra over land at low frequencies.

Due to the difficulty of wind velocity measurement at the layers up to 100–300 m, there are very few results concerning the velocity profiles of these layers. The mean wind velocity in the bottom 300 m atmospheric layer was measured on the upper-air mast by means of a photoelectronic anemograph during 1961–1962 (Klinov and Poltavskii, 1963). They noted four groups of profiles: (a) the normal profile, an increase of wind velocity with height, (b) the maximum appears between 100 m and 200 m, (c) the minimum appears at about 200 m, and (d) the velocity profiles are constant up to 170–190 m and vary appreciably from 190 m to 300 m. The (c) and (d) profiles are apparently caused by inhomogeneities in the atmosphere.

The normal wind profile in the bottom of the boundary layer between 25–30 m and 300–500 m is expressed approximately by the power law:

$$\frac{\bar{u}}{\bar{u}_1} = \left(\frac{z}{z_1}\right)^n \qquad (3.27)$$

in which the value n depends on the state of stratification of the layer (Sutton, 1953; Vorontsov, 1960) and the surface conditions. For instance, n takes 7.5–8.5 over an open land or water surface, about 5.5 over a grassland with scattered trees or houses, 3–3.5 over forests or suburban areas and 1.5–2.5 over built-up areas in a great city.

The daily variations of the wind velocity were more clearly pronounced in the analysis of averaged data up to the 700 m layer (Devyatova, 1957). Figure 3.10 shows curves for the daily variation of the wind velocity during the warm and cold half years. During the warm half year, the 100 m level is a transitional region, and above 150 m the variation type is opposite that on the surface. The minimum appears approximately at 10 h in the morning, and the maximum at 22 h in the late evening above the 150 m level. During the cold half year, the transitional region occurs at the 50 m level.

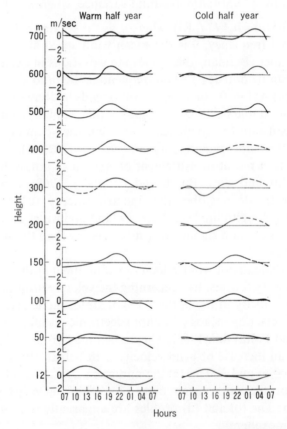

Fig. 3.10 Daily variation of wind velocity during the warm and cold half-years (Devyatova, 1957).

3.1.5. Exchange Coefficient and Some Other Characteristic Numbers or Factors

The temperature on the ground or the water surface is generally different from that in the lower layer of the atmosphere. Accordingly, a vertical flux of heat is generated between the surface and the atmosphere. This flux P is controlled by turbulent thermal conductivity in the air layer near the ground.

The process of thermal diffusion between the ground surface and the atmosphere caused by turbulence is given in the following equation, inferred from the process of molecular diffusion:

$$P = -\rho K C_p \frac{\partial \theta}{\partial z} \tag{3.28}$$

where ρ stands for the density of air; K is the exchange coefficient or Austausch coefficient; C_p is the specific heat at constant pressure of air, and $\partial \theta / \partial z$ the vertical temperature gradient.

By using this Eq. (3.28) for P, P can be obtained from the observed value of air temperature. To get R by Eq. (3.1) from the observed values, P obtained from Eq. (3.28) is used. K in Eq. (3.28) is obtained, however, by a method mentioned below.

Secondly E, the speed of evaporation, is shown in the following equation, similar to the form of Eq. (3.28):

$$E = -\rho K \frac{\partial q}{\partial z} = -\rho K \frac{0.622}{p} \frac{\partial e}{\partial z} \tag{3.29}$$

where $\partial q / \partial z$ stands for the vertical gradient of specific humidity, and $\partial e / \partial z$ is that of water vapour pressure. Baumgartner (1967c) gave various formulas for meteorological evaluation of evaporation from natural surface covers using the water, water vapour, and energy exchange terms and compared them with the observed results.

There are various methods to get K (m²/sec), the exchange coefficient; three formulae are as follows (Alissow, Drosdow and Rubinstein, 1956). According to M. I. Budyko, it can be calculated by the equation shown below:

$$K = \frac{0.144\Delta u}{\ln \frac{z_2}{z_1}} \left[1 + \frac{\Delta \theta}{(\Delta u)^2} \ln \frac{z_2}{z_1} \right] \tag{3.30}$$

where Δu is the difference of the wind velocity between the altitudes z_1 and z_2, and $\Delta \theta$ is that of the air temperature between the heights z_1 and z_2. In most cases the heights z_1 and z_2 are taken as 0.5 m and 2 m for K at a height of 1 m. Since the observation error on the difference of the wind velocity between two heights is apt to be great, another way to get K from the observed values at one height is expressed according to M. P. Timofejev:

$$K = \frac{0.16u}{\ln \left(\frac{1}{z_0}\right)} \left(1 + 7.5 \frac{\Delta \theta}{u^2} \right) \tag{3.31}$$

in which z_0 is the roughness parameter of the ground surface (m), u the wind velocity at a height of 1 m above the ground surface, and $\Delta \theta$ the difference of the air temperatures between 0.5 and 2 m. The equation by D. L. Laykhtman is:

$$K = 0.16\,(1 + 1.8\varepsilon)^2 \frac{\varepsilon z_0 u}{\left(\dfrac{z_1}{z_0}\right)^2 - 1}\left(\frac{z_1}{z_0}\right)^{1-\varepsilon} \tag{3.32}$$

where ε is a parameter smaller than 0 under lapse conditions, and greater than 0 under inversion conditions. K, obtained from the above three formulae, is almost the same in each equation.

K in the air layer near the ground increases according to the altitude. This is expressed by:

$$K = K_1 \cdot z^{1-\delta} \tag{3.33}$$

where δ is a constant ($0 < \delta \leq 1$), and K_1 is shown by K at a certain height above the ground. If Sutton's hypothesis (Sutton, 1953) is accepted here, K_1 and δ in Eq. (3.33) can be determined from the vertical distribution of the wind velocity. From the results measured up to 100 cm on Hamasaka Dune, Tottori Pref., on the coast of the Sea of Japan, the following values were obtained: $\delta = 0.135$, and $K_1 = 2.68$ cm²/sec. Another result obtained in the same way on the snow surface in Hiruno, Gifu Prefecture, showed $\delta = 0.260$, and $K_1 = 1.77$ cm²/sec (Azuma, 1958).

Figure 3.11 shows the vertical distributions of K up to 700 m with the conditions on the surface. In this case, the exchange coefficient K_z has been obtained in the following way:

$$K_z = \frac{w'^2\,TV}{2\Delta V} \tag{3.34}$$

where w' and ΔV are fluctuations of the vertical and horizontal components of wind velocity; T is time of persistency of the same symbol of vertical fluctuation, and V is wind velocity. Balloons with the measuring apparatus were used to record each value. From Fig. 3.11 the maximum can be seen generally at 100 m above the surface. In a cotton field or irrigated oasis, however, the maximum appears at 200 m above the surface (Vorontsov, 1958b).

Takasu (1957) obtained the exchange coefficient in the following way: Com-

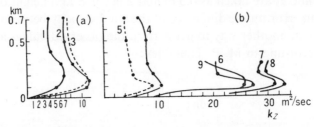

Fig. 3.11 Vertical distribution of coefficient of turbulent exchange (Vorontsov, 1958b).
(a) Pahta-Arale district: 1. cotton field; 2. semi-desert; 3. desert (July, 1952).
(b) Voeikov district: 4. forest; 5. grassland (March and April, 1953); 6. summer; 7. spring; 8. winter; 9. autumn.

bining H. Ertel's intermediate formula with T. Hesselberg's relation, he proposed the formula below:

$$K_e = \frac{A_e}{\rho} = \frac{4 \overline{(s-\bar{s})^2}}{p \, (\partial \bar{s}/\partial z)^2} \tag{3.35}$$

where s is the mean value of air temperature over 20 sec, and $\bar{s}$ is the mean value of s over 10 min. The value p is the mean period given by $(10 \times 60)/n$ sec if n is the number of the maximum or the minimum point in the variation curve of $s-\bar{s}$. And then he proposed the use of the formula below:

$$K_s = \frac{A_s}{\rho} = \frac{1}{40} \frac{R^2}{(\partial \bar{s}/\partial z)^2} \tag{3.36}$$

where R is the mean value over 10 min of the range in 20-sec intervals. If $A_e \gg A_s$, then A_e can be regarded as the exchange coefficient. Table 3.13 was compiled from his results by the present writer only for K_e at a height of 50 cm above the ground surface.

As seen in Table 3.13, K_e not only varies according to the condition of the surface, but has diurnal variation. When the weather is fine, it becomes 10 to 100 times as great in the daytime as at night. Over land, K_e at 1 m above the surface is 500 to 2,000 cm²/sec in the daytime in summer. Uchijima (1959) obtained the value of K on a pond. At 300 cm above water in the daytime in summer the figures were 200 to 1,400 cm²/sec.

Table 3.13   The values of K_e(cm²/sec) at 50 cm above the surface in various conditions compiled by Yoshino after Takasu (1957).

Time	Surface condition	ice	water	field	sand
	Place	Lake Suwa	Lake Biwa	Foot of Mt. Akagi	—
Morning (7–11 h)	Range	—	169.7–191.2	38.6–644	16.1–56.9
	Average (Frequency)	—	180.5 (2)	229.2 (5)	38.8 (3)
Afternoon (12–17 h)	Range	32.7–1858	—	15.8–72.5	12.4–51.8
	Average (Frequency)	440.9 (9)	—	46.4 (5)	31.9 (6)
Night (18–24 h)	Range	20.7–483	—	8–707	—
	Average (Frequency)	131.2 (5)	—	248.7 (3)	—
Night (1–6 h)	Range	2.3–77.7	45.8	10	—
	Average (Frequency)	30.0 (3)	45.8 (1)	10 (1)	—

On water, the vertical temperature gradient is not large in general, so that the exchange coefficient in the layer of the atmosphere near the water depends mainly on wind velocity.

The observations of K_m, exchange coefficient of momentum or eddy viscosity, and K_h, exchange coefficient of heat or eddy conductivity, were made at a spot 8 m inlandwards from the coast by Cramer and Record (1953). Transports of momentum and heat flux in the layer near the ground are expressed, respectively, by:

$$\tau = \rho K_m \frac{\partial \bar{u}}{z} \tag{3.37}$$

$$p = -\rho K_h C_p \frac{\partial \theta}{\partial z} \qquad (3.38)$$

The main results were as follows: (a) K_m and K_h increase sharply with an increase of height in the air layer near the ground. (b) The ranges of K_m and K_h depend on the thermal stratification. (c) The values were $K_m = 1.4—53.0 \times 10^3$ cm²/sec and $K_h = 0.8—81.1 \times 10^3$ cm²/sec. (d) $K_h > K_m$ under the sharp lapse condition, and $K_h < K_m$ under the inversion condition. (e) In the case of winds from the sea, $K_h < K_m$, and $K_h \doteqdot K_m$ with winds from the land. These are the results observed about 16 h in August, and so it is naturally supposed that the results are influenced by the difference between the thermal conditions of the land and the sea.

Blackadar (1962) studied the hodograph in the Ekman layer, which is the layer of transition between the surface boundary layer and the free atmosphere, supposing a vertical change of mixing length, l. That is:

$$l = \frac{kz}{1 + kz/\lambda} \qquad (3.39)$$

where, k is Kármán constant, $\lambda = 0.00027 \cdot G/f$, G geostrophic wind $= \sqrt{U_g^2 + V_g^2}$, f Coriolis parameter. According to his results, wind velocity reaches the maximum at a level of 600 m above the ground in the Eckman layer.

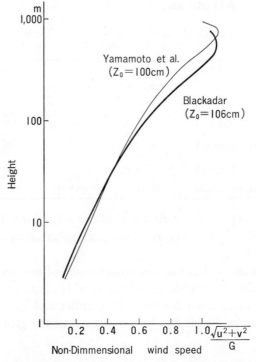

Fig. 3.12 Vertical distribution of wind velocity in the Ekman layer for the neutral state (Yamamoto et al., 1968).

This study by Blackadar was in the case of neutral conditions. Yamamoto et al. (1968) inquired further into this problem considering the effect of thermal stratification. Their results as well as some by Blackadar are given in Fig. 3.12.

In the formula concerning the logarithmic velocity profile, von Kármán's constant, k, is involved. The value of k is approximately equal to 0.4, but $k>0.4$ under the lapse condition, and $k<0.4$ under the inversion condition. As the thermal stratification varies greatly from place to place in a small area, the vertical wind profile would also show a local peculiarity.

A non-dimensional number, Richardson number, which represents the rate of growth of turbulence is derived by:

$$\mathrm{Ri}=\frac{g\beta}{\left(\frac{\partial u}{\partial z}\right)^2}=\frac{g}{\theta}\frac{\frac{\partial \theta}{\partial z}}{\left(\frac{\partial u}{\partial z}\right)^2} \tag{3.40}$$

where g is the acceleration of gravity; β is a representative vertical stability, commonly $\partial\theta/\theta\partial z$; θ is potential temperature; and $\partial u/\partial z$ is a characteristic vertical wind shear. A negative Ri means lapse conditions, and a positive Ri inversion. The ratio of mechanical production of turbulence from the mean wind shear $\partial u/\partial z$ to the thermal buoyancy effect is a measure of the relative importance of thermal effects and is represented by the flux Richardson number, $\mathrm{R_f}$, defined by

$$\mathrm{R_f}=-\frac{gH/C_p\theta}{\tau\frac{\partial u}{\partial z}} \tag{3.41}$$

According to Deacon (1953a) and Lettau (1957), the Ri given by Eq. (3.40) is called "local Richardson number," whereas "bulk Richardson number" can be defined as:

$$\mathrm{Ri}_{(2.1)}=\mathrm{const.}\frac{(T_2-T_1)}{V_2{}^2} \tag{3.42}$$

where (T_2-T_1) is the temperature difference at heights z_2 and z_1, and V_2 is the wind velocity at height z_2.

The values of Ri were calculated by Devyatova (1957) for the 5 layers up to 600 m above the ground. The results revealed that the layer with developed turbulence reaches the maximum altitude (In summer and spring to an altitude of 600 m, Ri<1).

For a lapse condition, wind speed can be read as functions of height, z_m, or $|L|$, defined in Eq. (3.26), and roughness parameter, z_0, as given in Fig. 3.13. (Webb, 1965). This figure is based on a heat flux H of 25 m W/cm² which is typical in middle latitudes on a clear summer day. For other values of H, one must read the chart at the correct z_0 and at wind $(25/H)^{\frac{1}{3}}\cdot u_2$. Alternatively, as a good practical approximation of $z/|L|$ is less than 0.2 or so, simply adjust $|L|$ in proportion to H^{-1}.

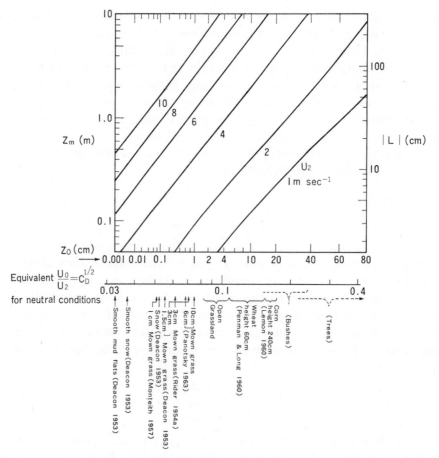

Fig. 3.13 Lower scales: drag coefficient C_D at 2m in neutral conditions, represented by $C_D{}^{0.5}$, related to roughness parameter, z_0. Upper section: dependence of 2m and $|L|$ on surface roughness and on wind speed U_2 (Webb, 1965).

Gust is a sudden brief increase of wind speed. It is expressed by the gust factor:

$$G_tF = \frac{u_{\max}}{\bar{u}} \qquad (3.43)$$

where $u_{\max}$ is the instantaneous maximum wind speed, and $\bar{u}$ is the mean wind speed (commonly during 10 min). G_tF takes the values ranging between 1.2–1.8, and depends on the stability of the air layer, the surface conditions, surrounding topography, and varies with height above the ground. It is an important measure not only from the standpoint of climatology in a small area, but also from that of applied problems, such as architecture and bridge design. A similar measure of the intensity of gusts is the gustiness factor, which is expressed as

$$G_sF = \frac{u_{\max} - u_{\min}}{\bar{u}} \qquad (3.44)$$

Intensity of turbulence is defined as the ratios of the root-mean-squares of the eddy velocities to the mean wind speed for the longitudinal component:

$$g_x = \frac{\sqrt{\overline{u'^2}}}{\bar{u}} \tag{3.45}$$

Similarly, it is obtained for the lateral and vertical components. In some cases, intensity of turbulence is expressed simply by standard deviation of the eddy velocities. This intensity of turbulence is related to height, wind speed, heat flux, and stability. The relation in air layers over the sea was measured by Warner (1972) and on a tower up to 313 m by Yokoyama (1971). Its horizontal distribution in a valley is mentioned below.

3.1.6.   Some Other Phenomena

(A)   *Thermic*

The local convection air current that ascends from near the surface in bubbles is called "thermic" or "thermion." Meteorologically, it comes out when there is an unstable air layer and the temperature is heterogeneous both horizontally and vertically. It looks like bubbles that rise in boiling water. Even at night, so long as the conditions are sufficient, it takes place. The height of the ascent of the thermics is limited by the level at which the temperature (more correctly, density) of the rising air and that of the surrounding air become equal. The speed of the ascent, W, is given by the formula (Vorontsov, 1956):

$$W = \sqrt{2gz\left(\frac{\Delta T}{T_1} + \frac{(\gamma - \gamma_a)\,z}{T_1}\right)} \tag{3.46}$$

where T_1 is the mean temperature of the air layer, ΔT is the difference of the air temperature between the rising and the surrounding air at the initial level, γ and γ_a are the actual and adiabatic gradients of the temperature, z is the altitude of ascent, and g is the acceleration of gravity. Equation (3.46) means that the speed of the ascent of the thermics will increase with an increase of the initial overheating ΔT and the thermal instability of the atmosphere $(\gamma - \gamma_a)$.

The altitude of ascent is greatest at about 12 noon on a fine day. It is 200 to 300 m on the average, and lower in the morning and in the evening. James (1953) separated the zone with a developed convection from the ground to the base of the cumulus cloud into three layers: (a) the first layer from the ground surface up to 250 m, which corresponds roughly to the layer of thermics mentioned above——here, the temperature pulsations reach maximums at 0.3–0.5°C, (b) the second layer from 250 to 900 m, and (c) the third layer from 900 m to the base of the cloud.

The results observed in the Soviet Union are summarized in Table 3.14 (Vorontsov, 1956). The vertical speed shown in Table 3.14 is 3 or 4 times lower than W calculated from Eq. (3.46).

The local distribution of thermic is important for the flight of a glider; this makes it possible for a glider to continue flying for hours.

Table 3.14 Average conditions of the movement of thermics from the earth's surface (Vorontsov, 1956).

Underlying Surface	Laps rate		Temperature at height of 2 m	Height of ascent of thermics		Vertical speed		Extent of over-heating of the stream			Number of cases	
	in layer of ascent	in layer above ascent		Average during the day	during 14 hours	Average during the day	during 14 hours	maximum	average	minimum	without over-heating	with over-heating
Semidesert	1.40	0.40°C/100m	32.4°C	145	300m	51	106cm/sec	3.1	0.9	0.2°C	7	5
Slope of hilly valley	1.34	0.65	15.6	430	850	108	176	2.1	1.2	0.8	16	4
On water surface during winter	1.40	−0.42	−14.8	85	101	41	48	1.3	0.8	0.4	23	5
Irrigated oasis	0.55	−0.10	31.7	92	240	22	43	3.6	1.1	0.2	17	18
Massif of forest belts	2.10	0.95	23.0	60	180	39	115	−	0.1	−	10	1

(B) *Diffusion*

According to Sutton (1953), for a continuous point source at a height h above the ground, concentration χ at a position, x, y, z, is expressed by:

$$\chi=\frac{Q}{u}\frac{e^{-\frac{y^2}{A}}}{\sqrt{A\pi}}(e^{-\frac{(h+z)^2}{B}}+e^{-\frac{(h-z)^2}{B}}) \tag{3.47}$$

where $A=C_y^2 x^{2-n}$, $B=C_z^2 x^{2-n}$; Q is the source intensity; x is leeward distance from the source; y is cross wind distance from the mean wind direction; z is height from the ground; u is mean wind speed; and C_y and C_z are Sutton's diffusion parameters.

Sakagami (1960) proposed the following diffusion equation:

$$\chi=\frac{Q}{u}e^{-\frac{y^2}{A}}e^{-\frac{h+z}{B}}J_0\left(i\frac{2\sqrt{hz}}{B}\right)$$

$$A=a(x), \qquad B=b(x), \qquad x=ut \tag{3.48}$$

where J_0 is the zero order Bessel function of the 1st kind, and i is the imaginary unit.

As a problem of climatology in a small area, horizontal concentrations are of special interest. The horizontal profiles at a point 50 or 100 m leeward from the source always show a form of normal distribution, but at further distances, they gradually attain plateau-type distribution. With standard deviation $\overline{y^2}$ of profiles or plume widths, L, is expressed by:

$$L=k\overline{y^2}=q_A(\varphi_A x+e^{-\varphi A^x}-1) \tag{3.49}$$

where k is a proportionality constant, and q_A and φ_A are diffusion parameters.

The volcanic ashes that accumulated at distances ranging from 40 to 200 km from Mt. Asama were also analyzed and compiled by Sakagami. In Fig. 3.14, the relation between φ_A or $\sqrt{q_A}$ and distance from the source, x, is shown. This figure indicates that the magnitude of diffusion depends on the scale of the phenomena, expressed by the simple relations:

$$\varphi_A=2.88\times10^3 x_m^{-1.73} \tag{3.50}$$

and

$$\sqrt{q_A}=4.57\times10^4 x_m^{1.75} \tag{3.51}$$

Equations (3.50) and (3.51) are applicable in the range 1–400 km.

The transverse distribution of aerosol concentration in the stream is usually assumed to conform to a normal law with a dispersion, σ^2, that is a function of the distance, x, from the source epicenter (Byzova, 1963):

$$\sigma=ax^{\alpha} \tag{3.52}$$

where $a=0.118$ and $\alpha=0.89$ were obtained in the range from a distance of some hundred meters with a source height 25 m to about 10 km with a source height of 100–300 m.

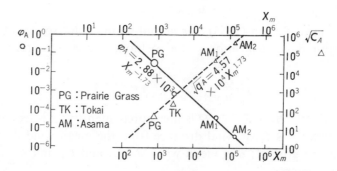

Fig. 3.14 Relationship between the diffusion parameters, φ_A or q_A, and the distance from the source, x (Sakagami, 1960).

(C)   *The Temperature on Snow Cover*

The temperature profile of the air layer on snow cover shows an inversion state. The results observed in Uppsala in a layer as thin as 25 mm over a snow cover are shown in Table 3.15 (Nyberg, 1938). The less the wind velocity is, the more the degree of inversion is, and there is a variation of about 2°C within the 25 mm when it is calm. The variation of the temperature according to the height is similar to the logarithmic law. It is also clear that the temperature of the whole air layer becomes higher as the wind velocity becomes stronger. Air temperature profiles up to 15 cm over the snow were classified into three types (Heigel, 1964): (a) radiation type with strong inversion up to 5 cm, (b) exchange type, which shows a nearly neutral profile and the air temperature is relatively high, and (c) mixed (radiation and exchange) type with inversion up to 5 cm and the air temperature is relatively low.

Table 3.15   Vertical distribution of air temperature up to 25 mm
over snow cover (Nyberg, 1938).

Wind velocity	1	5	10	15	20	25 mm	140 cm
Calm	−17.6	−17.0	−16.4	−16.1	−15.9	−15.7	−12.1°C
0.3–0.6 m/sec	−11.5	−10.7	−10.1	−9.8	−9.4	−9.2	−6.7
0.9–1.2	−9.3	−8.7	−8.4	−8.2	−8.1	−8.0	−6.4
1.8	−4.1	−3.7	−3.5	−3.4	−3.3	−3.3	−2.7

How does the low temperature of the stratum over the snow cover surface occur? Voeikov thought that the four characteristics of snow cover have an important influence on the climate, as discussed below. But most of the explanations for this low temperature from the heat balance near the snow cover surface have been qualitative. It is ture that snow radiates at a value approximate to the rate of a black body, but the value is not great at all. The emissivity is 0.95 to 0.96 over vegetation, and 0.995 over snow, which is not much greater. Besides, as the conductivity of snow is low, the temperature over that surface becomes lower than, for instance, on the bare ground. Therefore, at night the surface of snow cover must radiate less than the radiation in other conditions. On a cloudy night in the Sierra Nevada, the flux of radiation on the grassland was -2.2 cal·cm^{-2}·hr^{-1}, and it was -1.0 cal·cm^{-2}·hr^{-1} over snow, with a difference of 15 cal or so during the whole night. At dawn of a fine night, the difference of input and output of heat by long-wave radiation is very small. According to the calculation by Ångström, the incoming flux is 21 cal·cm^{-2}·hr^{-1}, when the water vapour pressure in the layer near the ground is 6 mb. The outgoing flux is 24 cal·cm^{-2}·hr^{-1} when it is $-10°$C on the surface of snow cover. The balance is the loss of only 3 cal·cm^{-2}·hr^{-1}. The importance of the balance of long-wave radiation in the layer of the moist air near the surface has recently been discussed, and it is said to be 1/3 to 1/2 of the whole flux of incoming radiation.

Table 3.16 Albedo of snow cover in winter (Miller, 1955).

	Mean value albedo	Albedo adjusted to fresh snow cover	Researcher (Year)	Observed region
High latitude	0.80[a]	0.82	Kalitin (1930)	U.S.S.R.
	0.85	0.85	Kalitin (1930)	U.S.S.R.
	0.92[b]	0.87	Kalitin (1930)	U.S.S.R.
	0.87	0.87	Chernigovskii (1936, 1939)	The Arctic Ocean
	0.81	0.81	Ångström (1925)	Sweden
	0.81[a]	0.81	Lunelund (1926)	Finland
	0.81[a]	0.81	Olsson (1936)	Spitzbergen
Middle latitude	0.89	0.85	Dorno (1918)	The Alps
	0.85	0.85	Nutting, Jones, Elliott (1914)	North America
	0.85[a]	0.85	Miller (1950)	Sierra (California)
	0.77–0.88[a]	0.83	Eckel, Thams (1939)	The Alps
	0.80[b]	0.80	Sauberer (1938)	The Alps
	0.93	—	Iaroslavtsev (1948)	Turkistan
	0.75[b]	0.80	Kimball, Hand (1930)	North America
	0.70	0.75	Devaux (1935)	The Alps, Pyrenees
Summary	0.89	0.89	Büttner (1938)	
	0.81–0.85	0.83	Conrad (1936)	
	0.75–0.90	0.83	Geiger (1950)	

a: the mean value of series of detailed observations.
b: the albedo to the radiation of whole spectrum observed optically.

Voeikov (1885, 1889) said that the influence of snow cover on the climate consists of albedo, long-wave radiation, fusion, and conductivity. Further the water content of snow cover, and the cycle of evaporation and condensation must also be concerned.

First, the results of the observation so far made on albedo are shown in Table 3.16. The mean value of each experiment is shown in the first column, and in the second column is given the adjusted value of the first to the condition of fresh snow, so that the values may be compared with each other in case the experiment was not made immediately after the snowfall (Miller, 1955). Gavrilova (1966) used the mean values of albedo for calculating the absorbed radiation: 0.85 on stable snow cover for latitudes above 70°, 0.80 for latitudes 65–70°, and 0.75 for latitudes 60–65°.

Table 3.17 Interdiurnal change of albedo of snow cover in the Sierra Nevada (Miller, 1955).

| | Days after new snow accumulation | | | | | | | |
	0	2	4	6	8	10	12	14 days
Albedo in winter (Mean of Dec., Jan., and Feb.)	0.83	0.76	0.71	0.67	0.65	0.64	—	—
Albedo in spring (Mean of April and May)	0.80	0.66	0.58	0.53	0.50	0.47	0.45	0.43

The albedo of snow cover becomes small as the days pass; the tendency is clearer in spring than in winter. The interdiurnal change of albedo in the Sierra Nevada is shown in Table 3.17, in which it decreases considerably. The experimental formula to show the decrease of albedo in relation to the days passing is:

$$\log A = m - nX \qquad (3.53)$$

where A is albedo, X is the accumulated value of the highest temperature of the day, and m and n are constants determined by the area or the season (Miller, 1955). As shown in Table 3.17, the value of n in the same area becomes greater towards late spring, the end of the period of snow cover. When it is fine in April, the mean albedo $\bar{A}$ in the Sierra Mountains may also be expressed in the following empirical formula:

$$\bar{A} = 0.68 - 0.8N + 0.016(31 - T_x) \qquad (3.54)$$

where N is the number of fine days in April, and T_x is the mean value of the daily maximum temperature (°F) in the period of snowfall of the preceding winter (December, January, and February). These conditions contribute much more to the heat balance on the surface of snow cover and accordingly, the local distribution of air temperature. Miller (1956b) also analyzed the influence of snow cover on local climate in Greenland and concluded that the classic theory does not accurately describe the local climate there, even when, as in Greenland, modifying factors are much less important than in other snow-covered regions.

To accelerate snow melting, the idea was to blacken the surface of snow accumulation in order to reduce the albedo. On this problem, Azuma (1956) carried out a micrometeorological experiment. He found that the increasing (20–55% of incident solar radiation) of absorbed solar radiation occurred on the artificially blackened snow surface and this amount was consumed to melt the snow.

(D) *CO_2 Concentration*

There are few observations on the horizontal distribution of CO_2 concentration in flat, open land, because the studies have been made in relation to vegetation or air pollution problems. Takasu and Kimura (1970) observed the vertical distribution of CO_2 concentration at 10, 50 and 100 cm above a sweet potato field and compared the results with those at a standard meteorological observation field. As shown in Fig. 3.15 (a), CO_2 concentration is minimum in the daytime and maximum at night, due to the diurnal change of photosynthesis and aspiration of plants. Reverse diurnal change between the CO_2 concentration and solar radiation is apparent. The difference of CO_2 concentration at three heights above the fields is interesting, as shown in Fig. 3.15 (b). At 10 cm above the field, CO_2 concentration is higher in the sweet potato field in the daytime, but at 50 cm above the field, the relation is reversed. This is thought to be because the CO_2 concentration is the lowest just above the active surface of the sweet potato field, in which the leaves spreading horizontally. Such a condition

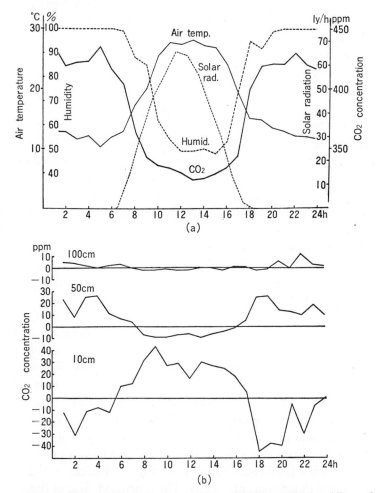

Fig. 3.15 (a) Diurnal change of air temperature, solar radiation, relative humidity and CO_2 concentration at a height of 50 cm above a sweet potato field. (b) Difference of CO_2 concentration (sweet potato field minus standard meteorological observation field). (Takasu et al., 1970).

differs from that in vegetation with relatively tall plants, such as corn (Uchijima et al., 1967) and soybeans (Takasu et al., 1971). Within corn stands, CO_2 decreased with height from the soil surface, and the lowest value appears in the lower part of the canopy. Within soybean stands, the lowest value occurs in the part where the leaves develop most thickly.

3.2. CITY

If the whole earth is taken into consideration, the area covered by cities is limited, but, since they contain a large population, they are important to human life. Many outstanding and comprehensive studies have been made on the climate of cities. As mentioned in Section 2.2.1 of Chapter 2, Luke Howard (1772–1864) was a pioneer of urban climate studies. Since then a great deal has been made clear by a series of research, which has been summarized by Kratzer (1956) and in the review articles by Landsberg (1956, 1957). Peterson (1969), Chandler (1970), and Nishizawa (1973) wrote comprehensive reviews on city climatology. These reviews threw new light on the various faces of urban climates at present. The important problems in these reviews are discussed in this section below.

Some of the studies on city climate since 1950 include: Uppsala (Sundborg, 1951), Bonn (Emonds, 1954), Vienna (Steinhauser et al., 1955, 1957, 1959; Undt et al., 1970), Linz (Lauscher et al., 1959), Köln (Kalb, 1962), Kiel (Eriksen, 1964), London (Chandler, 1965), Stuttgart (Hamm, 1969). All of these studies is that they are climatographical studies of a particular city. Besides these monographical works, there are many excellent studies concerning individual climatic elements, which are cited below.

Research on urban climate in Japan are reviewed by Kawamura (1966) and summarized by Daigo and Nagao (1972).

3.2.1. Air Temperature
(A) *Distribution of Air Temperature*
The most remarkable phenomenon in city climate is the high air temperature in cities; this is called *city temperature*, or *urban temperature*. As mentioned in Section 2.2.1, this was already pointed out in the 19th century. In the decade of the 1930's, detailed observations by means of automobiles were made in Vienna, Berlin, Karlsruhe, and other European cities, and also in several Japanese cities, such as Suwa, Shinano-Ōmachi, Tokyo, Osaka, and Nagoya. In each of these observations a high temperature was observed in midtown areas.

The results obtained by Takahashi (1959) in Ogaki City are as follows: On calm and fine nights, the distribution of air temperature in the city shows that the temperature gradient is steep around the city areas, which are divided into the inner and outer areas of the city. The diurnal change of the air temperature in the city areas is later than in the suburbs. The air temperature at any place in the city has a linear relation to the house density: As the house density increases by 10%, the air temperature on a calm night rises by 0.23°C. However, around the time of the maximum temperature the rise of the air temperature is comparatively slight, i.e., 0.10°C as the house density increases by 10%. When it is cloudy, it rises still more slightly, i.e., 0.06°C. This value may be considered to be common to all medium-sized or small cities in Japan. The distribution of air temperature observed by means of an automobile mounted thermistor in Ogaki City in the evening around 20 h, September 24, 1956, is given in Fig. 3.16.

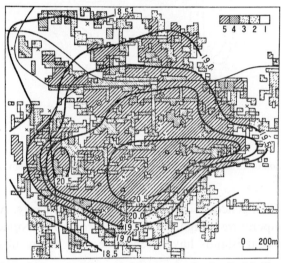

Fig. 3.16 Distribution of air temperature and house density in Ogaki City (Takahashi, 1959).
Note: House density is $1 < 5\% < 2 < 20\% < 3 < 40\% < 4 < 60\% < 5$.

A similar distribution map for Yonezawa City is given in Fig. 3.17. In this case, Sekiguti (1960) succeeded in drawing isotherms at 0.2°C intervals by careful observation with automobile-mounted thermistor.

The maximum horizontal gradient of the air temperature between the central part of the city and the suburbs, which is measured on the distribution maps, is on the whole 1–2°C/1 km in small or medium-sized cities, but it reaches 5°C/1 km or more in extreme cases. For instance, it was 2.0°C/1 km in Lund, Sweden, at 18 h, March 22, 1965 (Lindqvist, 1968), and in Ashdod, Israel, at 01 h 30 m, April 10, 1970 (Sharon et al., 1972). On the other hand, it reached 2.0°C/0.4 km (5°C/1 km) in Ogaki, Gifu Prefecture, Japan, as given in Fig. 3.16 (Takahashi, 1959) and 3.5°C/1.3 km (2.7°C/1 km) in Ise City, Japan, in the case of the minimum temperature on August 21, 1964 (Mizukoshi et al., 1968). In a study on air temperature distribution in Pretoria in winter, it was shown that the city area is up to 4°C warmer than adjacent urban areas on July 1, 1965 (Louw and Meyer, 1965). These figures, of course, depend on the synoptic conditions and diminish to zero under unfavorable conditions, discussed below. The

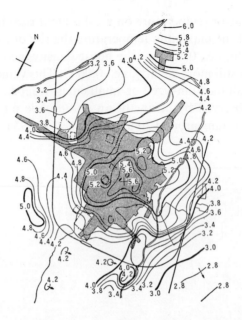

Fig. 3.17 Distribution of air temperature in Yonezawa City between 21h 30m and 23h(JST), May 3, 1957 (Sekiguti, 1960).

air temperature is in general high in a built-up area of a city. An area in a city, where the air temperature is high, is called the *heat island*. Numerical and model experiments for heat islands have been done, and the results are mentioned below.

The characteristics of city temperature become clearer with the development of cities. Several studies have been done on the tendency of air tempera-

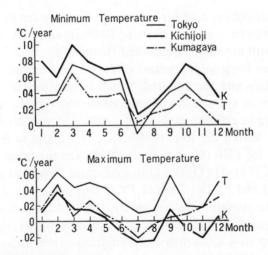

Fig. 3.18 Rate of secular change in monthly mean maximum and minimum temperatures during 1926–1956 (Kayane, 1960a).

ture to rise in several cities in Japan (Arakawa, 1937; Fukui, 1943). The rate of rise of the monthly mean maximum and minimum temperatures for 31 years from 1926 to 1956 is determined by the least square method, as shown in Fig. 3.18 (Kayane, 1960a). From this figure it will be seen that in an extreme case the rate for a month is 0.1°C/year. The influence of the climatic variation on a hemispheric scale during those years might be added to the value. The absolute value of the rate of rise, therefore, cannot be said to depend solely on the growth of the city. A local comparison shows that there is a tendency to rise, and the following two facts become clear: (a) in Tokyo, both the maximum and minimum temperatures rise a little, and (b) at Kichijôji (western suburbs of Tokyo) the minimum temperature rises remarkably, and the maximum temperature decreases slightly in some months.

The rate of secular change in the monthly mean temperature in large cities in Japan during the years 1941–1970 is shown in Fig. 3.19. In Osaka and Fukuoka, both SW-Japan, it is always positive. That is, the temperature is increasing in every month, but in Sapporo, N-Japan, it is negative or zero from July to

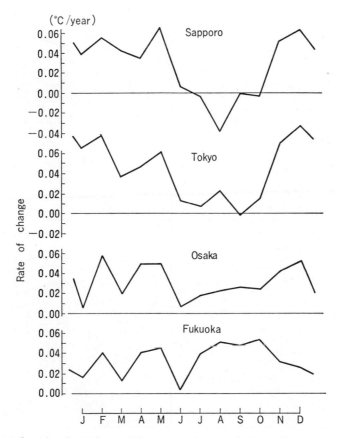

Fig. 3.19 Rate of secular change in monthly mean temperature in large cities in Japan during 1941–1970 (Yoshino).

October. Such regionality has a close relationship to the secular change of appearing frequency of some synoptic situations in East Asia (Yoshino and Kai, 1973).

The area of the heat island becomes larger year after year in accordance with development of the city. An example with Tokyo arranged for 5-years' average shows striking intensification of the heat island since 1951 (Kawamura, 1974). As shown in Fig. 3.20, the average of the minimum air temperature in January showed a 2.5°C difference between urban and suburban areas in 1951–1955, but 4°C in 1966–1970. The effect of the expansion of the city area is clearly seen.

Garnett and Bach (1967) presented a map of the heat island in Sheffield, England, by the distribution of potential temperature to eliminate the effect of relief in the urban area. In the city center, temperatures reduced to potential temperature were −5.0°C, and −8.0−−9.0°C in the surrounding district just before dawn under midwinter anticyclonic conditions on January 17, 1963.

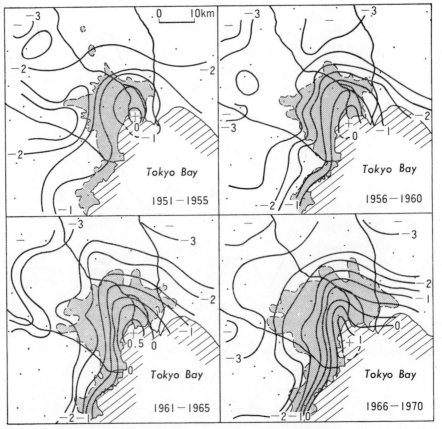

Fig. 3.20 Distribution of monthly mean minimum temperature in January for Tokyo urban area (Kawamura, 1974).

(B) *Three Dimensional Structure of Heat Island*

Horizontal distribution of the heat island has been observed for many cities of the world. Now the problems on the vertical structure remain to be solved.

Kratzer (1956) described the smog or haze layer above cities, calling it haze-cap (*Dunsthaube*), which can be observed from an airplane. Because haze is visible, it is noticed, but it might be only one of many urban climatic phenomena.

Duckworth et al. (1954) observed vertical temperature distribution with wiresonde over the built-up and adjacent open areas in San Francisco and Palo Alto, California. The results showed that the built-up areas frequently caused instability up to about 3 times roof height in otherwise stable air and that a *cross-over* point sometimes existed above which the air over the urban center was cooler than that over the surrounding country. This cross-over effect is also influenced by the prevailing wind velocity.

In Japan, Sekiguti (1963b) observed the horizontal distribution of air temperature in Ogaki City (population in the city area is about 70,000), Central Japan, at the surface 10, 20, and 30 m levels, as shown in Fig. 3.21. According

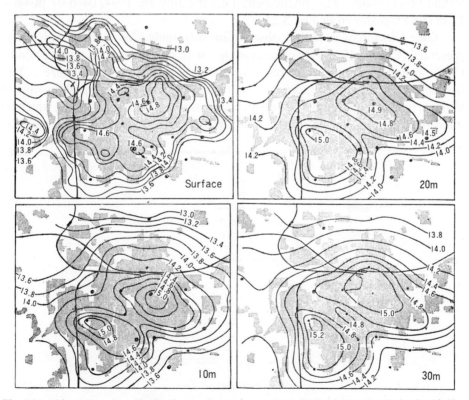

Fig. 3.21 Air temperature distribution at the surface and 10, 20, and 30 m levels in Ogaki City, Central Japan, on October 28, 1957 (Sekiguti, 1963b).

to these results, the heat island was detected up to 30 m, as far as the observations were made, and had two warm centers. The difference between inside and outside the city becomes small with increasing height and the distribution patterns of heat islands change slightly with height.

According to the observations at the tower, the surface inversion at night occurs more frequently and is stronger over the city than in the surrounding suburban area. This was ascertained at the tower in Louisville (DeMarrais, 1961) and Montreal (Munn and Stewart, 1967).

In Tokyo, there is a layer between 60–100 m, where the correlation coefficient of the microfluctuation of air temperature between the various levels becomes small. From this fact, it can be said that the effects of buildings in the cities reach up to 3–5 times their average height.

Bornstein (1968) reported the heat island effect in New York City, based on 42 observations from July 1964 to December 1966. It was confirmed that the surface temperature inversion is weaker in the city than in the suburbs, and its occurrence frequency is less in the city. Temperature excess (city temperature) is maximum about 1.6°C, at ground level and becomes 0 at 300 m as shown Fig. 3.22. The cross-over phenomenon was observed in about two-thirds of all the observations. This phenomenon on the July 16, 1964, 04 h 07 m–06 h 12 m (EST), is shown in Fig. 3.23.

Over Cincinnati, a similar observation was made, and a cross section was presented (Clarke, 1969). On clear evenings, with light consistent surface winds, Clarke found a strong surface-based inversion upwind of the urban area. Over the build-up area, lapse conditions occurred in the lowest 70 m, and downwind of the urban area he found a strong inversion or weak lapse conditions, which he interpreted as the downwind effect of the city. This "urban heat plume" was

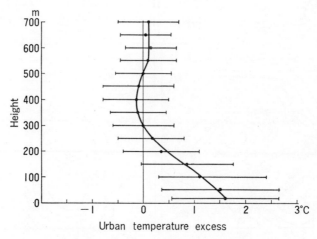

Fig. 3.22 Height variation of the magnitude and its range (standard deviation) of the urban heat island of New York City (Bornstein, 1968).

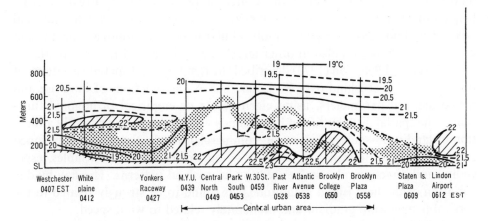

Fig. 3.23 Vertical and horizontal distribution of air temperature over the New York City area during the morning of July 16, 1964 (Bornstein, 1968).

detectable by vertical temperature measurements for several kilometers downwind of the city.

In a review paper on the recent topics in urban meteorology (Munn, 1973), research for universal models is stressed. For heat islands, Hanna (1969) presented the following equation:

$$\Delta\theta = \left(HW \frac{\partial\theta}{\partial z} \Big/ C_p \rho u \right)^{\frac{1}{2}} \tag{3.55}$$

where $\Delta\theta$ is the difference between the air temperature at the center of the urban heat island and a representative air temperature in the rural areas; H is the anthropogenic heat input; W is the city width; $\frac{\partial\theta}{\partial z}$ is the rural vertical gradient of potential temperature; C_p is specific heat of air at constant pressure; ρ is air density, and u is a representative rural wind speed. Padmanabhamurty et al. (Munn, 1973) obtained the following linear regression for clear and partly cloudy conditions:

$$\Delta\theta = 3.81 + 0.88 \left(\frac{T_2 - T_1}{u} \right)^{\frac{1}{2}} \times 10^3 \tag{3.56}$$

where $T_2 - T_1$ is the vertical temperature difference between 48.8 and 6.1 m at a representative rural station. The constant in the second term on the right-hand side of Eq. (3.56), compared with Eq. (3.55), implies an anthropogenic urban heat input of approximately 0.3 ly/min. Equation (3.56) means that, if the temperature profile is neutral, the heat island in Toronto is about 3.8°C.

Oke and East (1971) studied the horizontal and vertical structure of the urban boundary layer in Montreal and showed that, under conditions of strong rural stability, the lowest layers of the urban atmosphere become progressively modified as air moves toward the center of the city. In Montreal differing heights of heat and SO_2 emission appear to produce more than one internal layer. It was pointed out that the multicellular nature of Montreal's heat island center probably occurs also in other large cities, where land uses is varied.

Clarke and Peterson (1973) described the temporal and spatial variations of the nocturnal heat island over St. Louis, U.S.A., as functions of land use and meteorological parameters, using eigenvectors or empirical orthogonal functions. The spatial distribution of heat islands was shown to be primarily a function of land use—such as industrial, commercial, residential, and undeveloped areas—population density, elevation etc., but meteorological variables, especially wind direction, also influenced the heat island patterns. The day by day variation of the heat island magnitude was found to be strongly related to the vertical temperature gradient and moderately related to wind speed.

A model experiment of thermal convective circulation over the heat island under the condition of inversion was made in a stratified wind tunnel (Tsuchiya et al., 1973). The results coincided quite well with the pattern obtained by numerical simulation by Delage et al. (1970): The convective currents do not thrust through the upper light layer, but spread out horizontally.

A numerical model of a heat island was studied by Myrup (1969). It was suggested that such a model could be used in engineering calculations to improve the climate of existing and future cities.

(C)   *The Cause of City Temperature*

Various causes of city temperature, heat islands, have been analyzed. The following four seem to be of special importance.

(i)   *Heat of Combustion in the City*

Formerly, the heat used in factories or in houses was thought to have a great influence on the rise of city temperature; since the sources of heat are concentrated in the city, the temperature becomes higher than the outer areas.

Supposing that the heat released by factories, houses, and human activities directly warms the city air, Kratzer (1956) has summarized the following: A warming of 1.4°C for the 30 m thick air layer near the ground for London, energy of 8.1 kcal·cm^{-2}·year^{-1} for Vienna, and 16.8 kcal·cm^{-2}·year^{-1} for Berlin were estimated for the conditions at the beginning of the 20th century. In 1939, the calculated value on the heat released artificially in Tokyo was 3.1 kcal·cm^{-2}·year^{-1}, which was equal to about 1/50 of the insolation intensity received on a horizontal surface per year (Fukui and Wada, 1941). Limiting this to winter, insolation decreases and calorific value increases, so the ratio is 1/17, and artificial calorific value is found to be quite large. Artificial heat generated in the built-up areas of Sheffield from multiple sources has been calculated to be about 1/5 of the direct solar radiation received (Garnett and Bach, 1965).

Recently Miller (1971) has made an estimate for the coastal plain of South California, where the low-density agglomeration of Los Angeles is predominant (see Table 3.18). Total energy conversion is approximately 50 cal·cm^{-2}·day^{-1},

Table 3.18 The daily energy output in the coastal plain of S-California (Miller, 1971).

Process or operation	Source of energy	Daily rate of energy conversion $cal \cdot cm^{-2} \cdot day^{-1}$
Human metabolism	Food	1
Space heating of buildings	Gas	7
Manufacturing operations	Gas, oil	8
Electric motors, light, and other uses	Hydropower	1
	Oil—low in sulfur	7
	Gas—no sulfur	0
Waste heat from generation of electric power		15
Solid-waste disposal	Paper, trash	2
Urban circulation	Gasoline engines	9
Total		50

which is 10–15% of the rate at which solar energy is received by this surface. This amount is probably of about the same magnitude as the natural flux of sensible heat from the surface to the atmosphere.

An extreme example is also given by Miller (1971): In the great steel-producing areas of southeast Chicago, the energy input is estimated at 6×10^{14} $cal \cdot day^{-1}$ to an area of 50 km². The daily output is thus about 1,200 $cal \cdot cm^{-2} \cdot day^{-1}$, which is about 3.4 times the incoming solar energy averaged over the year.

Thus energy produced artificially cannot be disregarded in city areas. However, ventilation of the city air must be taken into consideration. How fast or at what rate is the city air ventilated? Further study is needed on the effect of heat released in city areas, including the ventilation rate over the city.

(ii)   *Haze or Smog Layer*

As shown below, cities contain a large quantity of fine dust, CO_2, SO_2, etc. In many cases the development of city areas is closely connected with that of the surrounding industrial areas. Consequently, dust and other pollutants from the factory areas cover the city areas, and exhaust gas of cars is no less influential. These effects are remarkable especially when the wind is low and the inversion layer fully developed. The smog layer absorbs long-wave radiation, and so it decreases the incoming radiation by day, and prevents outgoing radiation by night. In consequence, the maximum temperature becomes a little lower, and the minimum temperature fairly higher in cities. But city temperature can be observed not only in the nighttime, but also in the daytime. Therefore, this layer may play a greater role in the nighttime.

(iii)   *Exchange of Heat due to the Turbulence Increased by Buildings*

In cities crowded with building, the turbulence of so-called mechanical origin increases. Consequently, the vertical exchange of heat becomes more active in a city than in outer areas. This causes a comparatively low temperature in the daytime, and a comparatively high temperature at night because of ground inversion.

However this only applies when wind velocity is assumed to be the same both in and outside the city. In general, mean wind velocity itself is weakened by the buildings in the city. There is a correlation between mean wind velocity and turbulence as has been discussed above. If the streets are narrow and the buildings are tall, the air between the buildings cannot move under weak upper wind conditions. It can be said that the roughness parameter z_0 is very high in such cases, which means that the ventilation in these spaces within the height of z_0 is very weak. Accordingly, the high temperature remains till late in the evening. In this sense, the effects of heat exchange due to the turbulence is very important, but quantitative examination awaits further study. As has been pointed out by Nishizawa (1958a), eddy conductivity is sometimes very large and eddy conduction shows a complicated feature in a city. Apparently these factors influence the city temperature.

(iv)   *Materials and Shapes of Buildings*

A built-up area of a city, unlike the rural areas around it, is covered with cement, asphalt, gravel, stone, brick, etc. In the daytime these materials are warmed, though more gradually than the bare ground or surfaces covered with vegetation. The air temperature does not decrease quickly until night, because these materials cool slowly. The phase of the diurnal variation of the air temperature is slower in the urban center than in the suburbs, as has been confirmed by the Sendai data (Sasaki et al., 1969).

Nocturnal cooling of the surface temperature of the earth at the time, t, after sunset is represented by Brunt's formula:

$$T = T_0 - \frac{2R}{\sqrt{\pi} c \rho \sqrt{\kappa}} \sqrt{t} \qquad (3.57)$$

where $T(°C)$ is the temperature of the ground surface; $T_0(°C)$ is the temperature of the ground surface at sunset; R is the nocturnal radiation from the ground surface; ρ is the density of the ground; c is the specific heat of the ground; and κ is the specific heat conductivity of the ground. Kawamura (1964b) applied this formula to estimate the distribution of nocturnal cooling in Kumagaya, a city of about 50,000 population. The value of $1/c\rho\sqrt{\kappa}$ in the city areas with asphalt is 34.2 cm²·°C·cal⁻¹·sec⁻¹/², concrete 33.3, roofing tile 28.4, soil 50.2, and sand 50.5. The results obtained are presented in Fig. 3.24, the values being for each 50-m meshed square on a map of 1/3,000 scale. The relation between the air temperature and the physical properties of the construction materials is clear.

The shape of buildings also plays a significant role. As has been discussed originally by Lauscher (1934) and introduced later by Geiger (1961, 1965), the rate of the effective outgoing radiation from a street surrounded by the rows of houses to the radiation from a completely open horizontal surface is below 50% in some cases. The vertical walls of the houses bordering on the street also radiate, but it amounts only to about 40% of that at the horizontal surface, because each side of the street presents a barrier to the radiation of the other.

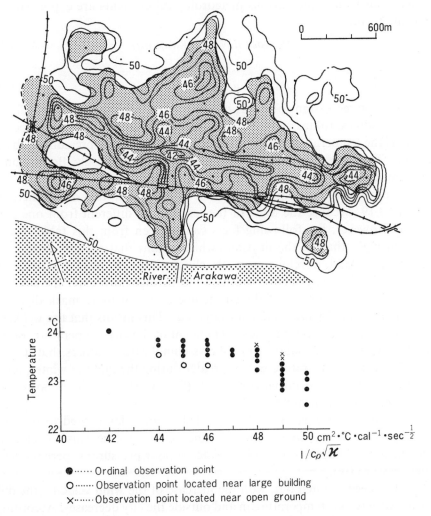

Fig. 3.24 (Upper part) Distribution of values $1/c\rho\sqrt{\kappa}$ in Kumagaya City. Shaded area is the city region with the rate that the area occupied by houses is more than 10%. (Lower part) Relation between mean air temperature in summer and $1/c\rho\sqrt{\kappa}$ (Kawamura, 1964b).

Even though there are few studies on the quantitative analysis of radiation of vertical as well as horizontal surfaces of buildings in cities, it is certain that the outgoing radiation is reduced in city areas as far as the horizontal surface is concerned.

(D) *Meteorological Elements Related with City Temperature, Heat Islands*
Concerning the causes of (i) to (iv) mentioned in the preceding Section no quantitative study has yet been undertaken. Only a few meteorological elements with indirect influence on those effects have been investigated.

For instance, the relationships between the temperature difference in a city and its suburbs, D_n, and various meteorological elements are expressed in the following form:

$$D_n = a_0 + a_1 N + a_2 u + a_3 t + a_4 t_R \qquad (3.58)$$

or

$$D_n = a_0 + a_1 N + a_2 u + a_3 t + a_5 e \qquad (3.59)$$

where a_0 is the regression constant; $a_1, a_2, \ldots a_5$ are the regression coefficients; N is the cloudiness (oktas); u is the wind velocity (m/sec); t is the air temperature (°C); t_R is the temperature range, and e is the water vapour pressure or absolute humidity. The regression constants and coefficients for Uppsala, Bonn, London, and Kumagaya are given in Table 3.19.

It is purposeless to compare the value of a city with that of another, given in Table 3.19, because the units of the terms e are different from one city to another. However, the following facts can be seen from this table: (a) a_0 is greater at night than in the daytime, which means that the city temperature, i.e., the heat island, develops more clearly at night; (b) a_1 is negative and its absolute value is greater at night, which means that the clearer the sky is, the greater is the development of the city temperature, or more markedly, of the heat island; (c) a_2 has the same tendency as a_1. This means that the weaker the wind velocity is, the greater is the development of the city temperature, or more markedly, of the heat island. (d) a_3 shows complicated features, but it can be said that the city temperature will be greater during the night in winter. (e) a_4 is concerned only with nocturnal conditions. (f) a_5 works in quite opposite directions in the day and at night.

The amount of cloud is to be regarded as an index related to the causes (ii) and (iv) mentioned above, and the wind velocity to the causes (ii) and (iii), while the temperature and humidity (water vapour pressure) is perhaps related to causes (i) to (iv).

As has been pointed out, with the increase of the wind velocity, the difference between the air temperature in and outside the city decreases. According to the observations made in Palo Alto, U. S. A., (population 33,000), the critical wind velocity was about 3 m/sec on the ground surface and 6–7 m/sec on the towers at a height of 7–8 m (Duckworth et al., 1954). It has also been ascertained in medium-sized cities in Japan that there is no difference of air temperature at 6 m/sec (Kurashige, 1943).

Observations made in and around the city of Reading, England, on the minimum temperature difference relative to cloud amount and wind scale (Parry, 1956a), have shown the frequency distribution of temperature difference centering between −1 to −2°C and 0°C at a cloud amount of more than 7–8 and a wind force of more than 3.

When the wind is low, the temperature distribution shifts slightly leewards. In Fig. 3.25 as an example of Kumagaya City, Japan, is shown. The observations made 36 times from 1956 to 1957 in Kumagaya show that (a) in nine cases

Table 3.19 The regression constant, a_0, and regression coefficients, a_1–a_5, in Eqs. (3.58 or 3.59) for Uppsala, Bonn, London, and Kumagaya (compiled by Yoshino).

		a_0 Constant	a_1 Cloudiness (Oktas)	a_2 Wind velocity (m/s)	a_3 Air temp. (°C)	a_4 Air temp. range (°C)	a_5 Water vapour pressure	Researcher (Year)
Uppsala	Daytime	1.4	−0.01	−0.09	−0.01	—	−0.04[*]	Sundborg (1951)
	Nighttime	2.8	−0.10	−0.38	−0.02	—	+0.03[*]	"
Bonn	Daytime	1.72	−0.03	−0.13	+0.03	—	−0.10[2*]	Emonds (1954)
	Nighttime	3.05	−0.18	−0.18	+0.18	—	+0.17[2*]	"
London (Summer)	Daytime	0.83	+0.03	−0.00	+0.06	+0.00	—	Chandler (1965)
	Nighttime	1.72	−0.12	−0.17	+0.01	+0.15	—	"
London (Winter)	Daytime	0.75	−0.03	+0.01	−0.00	+0.00	—	"
	Nighttime	1.69	−0.13	−0.10	+0.04	+0.08	—	"
Kumagaya	Nighttime	4.21	−0.08[3*] −0.12[4*]	−0.52	−0.04	—	+0.01[5*]	Kawamura (1964a)

[*]   Absolute humidity (g/kg).
[2*] Water vapour pressure (mmHg).
[3*] Cloud amount at 21h00m.
[4*] Cloud amount during the preceding daytime.
[5*] Water vapour pressure (mb).

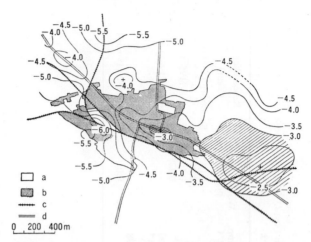

Fig. 3.25 Air temperature distribution under the WNW prevailing wind in Kumagaya City at 5h(JST), January 26, 1957. a: high temperature area, b: city region, c: railroad, and d: road (Mizukoshi, 1965).

the high temperature area are moved 300–400 m leeward, (b) in six cases there is hardly any move, and (c) in three cases a move of several hundred meters is found, sometimes even of 1 km (Mizukoshi, 1965). The NW prevailing wind is weakest in (b), and strongest in (c). On the whole the W component of the wind is 1.18 m/sec in (a), 0.46 m/sec in (b), and 1.53 m/sec in (c). In the Durban Area, South Africa, the center of the heat island in midday is displaced away from the central business district by the sea breeze (Preston-Whyte, 1970b).

The phenomenon of the shift of high temperature areas may be treated in the same way as the problem of pollutant distribution through diffusion.

Nishizawa (1958b) has investigated statistically the relationships of the difference of temperature, wind direction, wind velocity, or cloud amount between the areas in and outside a city. In November, if the wind direction is fairly constant and the wind velocity is about 2 m/sec, the temperature difference varies according to the cloud amount: The temperature difference at cloud amount 0 is about twice as large as that at cloud amount 10. In general, if cloud amount is equal, wind velocity is the decisive factor: at 3–4 m/sec the difference of temperature is about one-third of that at 1–2 m/sec. Munn et al. (1969) pointed out that the change of position of the heat island was not simply due to the influence of prevailing wind speed and direction in Toronto, Canada. Because of the heterogenous horizontal structure of the city and its suburbs, although the position of the heat island shifts in accordance with wind velocity, the range of shifting is different from direction to direction. This was also shown in Kumagaya (Mizukoshi, 1965).

The daily minimum temperature distribution in winter observed at 28 places in and around Tokyo can be classified into five patterns; typical examples are shown in Fig. 3.26. Their relations to various meteorological conditions——

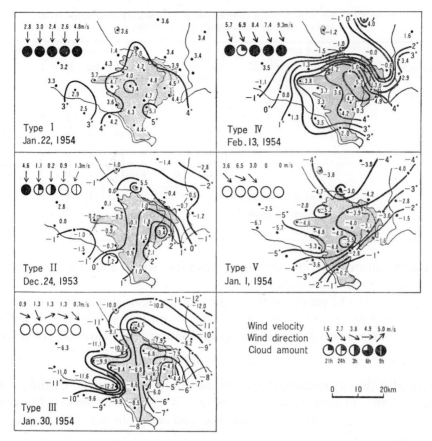

Fig. 3.26 Five types of daily minimum temperature in and around Tokyo (Kayane, 1960b).

cloud amount, precipitation, wind velocity, and temperature——are as follows (Kayane, 1960b): In type I, the temperature difference is less than 2°C; in Type II, the isotherms describe nearly concentric circles around the high temperature areas in the city, with a temperature difference of 2–4°C; Type III, with a difference of more than 4°C, shows most remarkably the city temperature, i.e., the heat island, Type IV occurs when there is a considerable difference on the outskirts of the city; different from any of the above four is Type V, where, with a temperature difference greater than 2°C, the isotherms run fairly independently of the city areas. Weather conditions can be classified according to various combinations of meteorological elements representing them synoptically. Table 3.20 shows the frequency of appearance of the five types, observed during the three winter seasons from 1952 to 1955. From this table it should be clear that the city temperature develops most conspicuously when it is very fine and the wind is weak (i.e., less than 5 m/sec at the tower of the Meteorological Observatory in Tokyo), and, in addition, the temperature is low. On the other hand, the

Table 3.20 Relations between daily minimum temperature distribution types and the weather situations (Kayane, 1960b).

Weather conditions				Types shown in Fig. 3.26					
Clondiness	Precipitation	Wind velo-city (m/sec)	Air temp. (t°C)	I	II	III	IV	V	Total
8–10	with rain	—	—	26	2	0	1	0	29
—	no rain	$\leqq 5$	—	10	18	0	0	3	31
		> 5	—	2	0	0	2	0	4
3–7	—	—	—	1	15	3	0	2	21
0–2	—	$\leqq 5$	$0 < t$	0	24	8	0	2	34
			$2.0 < t \leqq 0$	0	16	16	0	2	34
			$t < -2.0$	1	13	24	0	5	43
		> 5	—	3	0	0	3	0	6
	Change in wind direction $> 90°$			0	4	2	0	6	12
	Irregular diurnal change			—	—	—	—	—	56
	Total			43	92	53	6	20	214

city temperature is hardly observable when it is windy and cloudy, especially after precipitation during the night.

Ludwig (1970) presented the relation of the heat island, urban-rural temperature difference, to the lapse rate in the lowest layers. For the smallest five cities; Winnepeg, Albuquerque, Leicester, Denton, and Palo Alto, (14 cases):

$$\Delta T = 1.3 - 6.7\gamma \tag{3.60}$$

where ΔT is the city temperature minus the rural temperature in °C, and γ is the rate of change of temperature with pressure in °C/mb. For the next five cities; Minneapolis, Dallas, Ft Worth, San Jose, and San Francisco, (52 cases):

$$\Delta T = 1.7 - 7.2\gamma \tag{3.61}$$

ΔT and γ are the same as in Eq. (3.60). For the two largest cities; London and St. Louis, (12 cases):

$$\Delta T = 2.6 - 14.8\gamma \tag{3.62}$$

ΔT and γ are the same as in Eq. (3.60). It is, therefore, clear that the constant and coefficient depend on the city size.

One explanation for heat islands is that aerosols increase atmospheric counter radiation (long-wave radiation emitted by the atmosphere) over the cities. Evidence from Montreal, Canada, and Johannesburg, South Africa, indicates that the observed increase in counter radiation in urban areas does not result from an increase in the emissivity of polluted air and does not present a source of energy preferentially found in urban areas (Fuggle, 1972). He concluded that the observed increase in counter radiation is an effect rather than a cause of heat islands.

Eriksen (1964b), arranging the temperature differences between Kiel City and its suburbs in accordance with the weather situation, has obtained similar results.

The magnitude of the heat island shows a weekly cycle as has long been known. This is of course attributed to the weekly cycle of human activity, including air pollution exhaust. The mean weekly amplitude of the difference in daily maximum temperature is about 0.5°C, whereas the lowest mean difference of daily maximum temperature (London minus its surroundings) occurs on Thursday and the highest on Sunday (Lawrence, 1971).

(E) *Relation to the Population*

City growth can be expressed by the expansion of build-up areas or the increase in population. Mitchell (1953) has assumed as the first approximation that the area of a typical city is essentially proportional to its population and the change in area is proportional to the change in its population.

Duckworth et al. (1954) have analyzed the temperature difference in and outside a city in relation to its population: Introducing an index R, which is the shortest distance from the urban center to the outward region whose temperature is lower than the center by 1°F, they have shown that R is roughly proportional to the square root of the population. Fukui (1957), comparing the values in 14 cities in Japan has come to the following conclusion: The temperature difference, Δt, between the years 1900 and 1940 is proportional to the value $\sqrt{p}$, where $\sqrt{p}$ is $\sqrt{n_2} - \sqrt{n_1}$ and n_2 and n_1 are the population in 1900 and 1940, respectively. Following this line of investigation, he has further shown that the relation applies to cities in Japan with a population of less than 500,000 (Fukui, 1969). As for the large cities, such as Tokyo and Osaka, the rate of temperature change is not influenced by rise in population, where the rate remains uniform irrespective of the increase in population. This means that the assumption by Mitchell (1953) does not apply to these cities.

Yoshino and Kai (1973) made clear that the temperature rise, Δt, has a linear relation to its square root of population density, $\sqrt{d}$, in Japanese cities. That is, Δt increases sharply in the cities in Hokkaido and Kyushu, but slowly in central Japan, in accordance with increasing $\sqrt{d}$. Differences in the occurrence frequency of synoptic patterns may cause such a regionally different relationship.

Oke and Hannell (1970) studied the heat island in Hamilton (43°N, 79°W), Ontario, Canada, and compared the results with those in other cities. Although there are some differences in observed anemometer height and character of the city, there is evidence to suggest a relationship between city size and the wind speed necessary to obliterate the heat island effect. If city size is expressed by city population, p,

$$U = 3.4 \log p - 11.6 \qquad (3.63)$$

where U is the critical wind speed in m/sec. When $U=0$, Eq. (3.63) yields a population of about 2,500. Tamiya (1968) made clear the heat island in a small new town with a size of 400×700 m and population of 10,000 in Hibariga-oka, northwest of Tokyo. It would be interesting to ascertain whether other cities and towns fit the relation given in Eq. (3.63) or not.

3.2.2. Humidity

Because the air temperature is high in the central part of a city, when the absolute amount of vapour is equal, the relative humidity becomes lower inside than outside the city. Also, differences in the surface conditions for evaporation must be taken into consideration. In city areas, roads are usually paved, roofs drain well and sewerage is good, so that rain flows out quickly without being retained on the surface of the ground. This brings about a change in hydrological input and output to and from city areas. Consequently, the absolute humidity is smaller in a city than in its suburbs.

In European cities, the difference of humidity inside and outside a city is generally greater in summer than in winter. The difference of relative humidity in and outside of Stuttgart is, according to Hamm (1969),—5.7% for the winter average (December, January, February) and —6.9% for the summer average. Monthly averages of relative humidity in Tokyo show that the humidity is lower in the city by 1–3% in January and by 6–10% in June; it is lower by 4–6% averaged annually. Research in Tokyo has shown annual averages of humidity in 1935 of 63–65% in the city and 67–68% in the suburbs—about 4% lower in the city (Sasakura, 1950). An observable difference has been reported for Brno, a Czechoslovak city, and its surrounding regions (Quitt, 1964): The relative humidity at the densely built-up center of the city is 15% lower during the day and 20–30% lower during the night.

Eriksen (1964a) has made a minute observation of the relative humidity distribution as well as the air temperature in Kiel. An example of the distribution maps he made is given in Fig. 3.27. The central areas of the city show a very low value, 45%, in contrast to the northern coastal region, where the value is more than 65%, because the effect of humid air from the sea is apparent in this case. Between the city and the forest and the park, the difference is more than 5%. He summarizes the distribution of relative humidity as follows: (a) The distribution of relative humidity depends essentially on the air temperature distribution in the daytime. (b) A pure arid condition appears only in the thickly built-up areas. (c) The green area is humid and, it affects the surrounding arid areas.

Humidity decreases with the growth of a city. Tokyo, for example, has a 30-year average from 1921 to 1950 of an annual mean relative humidity of 72.1%. If the data since the beginning of observation is examined, however, it is clear that the humidity gradually diminishes, although there are some fluctuations as shown in Table 3.21.

As for absolute humidity, it is observed in European cities to be 0.2–0.5 mm Hg lower in the city than in the suburbs. The amount of vapour pressure is a direct measure of the water content in the city air. The distribution of water

Fig. 3.27 Distribution of relative humidity in Kiel at 14h15m—15h45m, on July 2, 1964 (Eriksen, 1964a).

Table 3.21 Secular change of annual mean relative humidity in Tokyo (Yoshino).

Year	5-year mean	Year	5-year mean
1876–1880	77.4%	1926–1930	72.9%
1881–1885	76.9	1931–1935	72.1
1886–1890	75.4	1936–1940	70.5
1891–1895	72.8	1941–1945	70.8
1896–1900	75.2	1946–1950	72.4
1901–1905	75.7	1951–1955	71.4
1906–1910	73.5	1956–1960	69.0
1911–1915	74.0	1961–1965	64.6
1916–1920	74.1	1966–1970	62.2
1921–1925	73.7		

vapour in Ogaki City, Japan, was presented by Sekiguti (1963a). It is reported that it is greater in the suburbs in the daytime, and the amount of daily change is smaller in the city than in the suburbs.

Kopec (1973) observed the humidity distribution in Chapel Hill, North Carolina, U. S. A., and provided that, under the optimum conditions of sky cover and wind velocity, city vapour pressures were found to be higher at night and lower in the morning and afternoon in comparison with suburban and rural areas. Although relationships between the surface character and micro-patterns of absolute humidity are not clear, basic vapour pressure patterns associated with city and noncity surfaces are remarkably consistent and persistent on a meso-scale level. Kopec suggested use of the expression *humidity island*.

3.2.3. Precipitation

Many reports show that the cloud amount is greater in a city than in its suburbs. In the city, where soot and dust discharged by factories, houses, and various means of transportation supplies more condensation nuclei, condensation occurs easily in spite of the low absolute humidity. In addition, as the central parts of a city are a little warmer, the up-currents must be slightly stronger there. Moreover, greater friction with buildings in the city causes greater up-glide of air. In other words, turbulence resulting from increased surface roughness cannot be disregarded. For these three reasons the cloud amount is greater in the city, and as a result, the precipitation amount becomes greater.

This conclusion has been reached by studying annual or monthly total precipitation in Bremen, Moscow, Stockholm, Helsinki, Cologne, Munich, Budapest, Chicago, St. Louis, and many other cities. According to Changnon (1961b), the average annual precipitation during the years between 1945 and 1956, and the average winter and summer precipitation shows a clear effect in Chicago. Eriksen (1964a) pointed out that no city influence can be recognized on the map showing the distribution of the average annual total precipitation in Kiel and its surrounding areas during the years 1891–1950, but a clear influence appears on the maps from December 1960 to November 1961 showing annual or seasonal total precipitation. There are many reports, however, that deny the difference of toal precipitation amount in and outside the city for any period of the year. In Japanese cities, no difference is observed, as shown below.

Generally speaking, this fact may be explained in this way: The phenomenon of city temperature, or heat island, is clearly observed only in the layer near the earth's surface, and it seldom reaches the condensation levels and, in addition, the phenomenon itself disappears when there is rainfall.

In Munich the amount of rain is greater in the city areas, irrespective of rainfall intensity: For the 30-year average of annual rainfall (1901–1930), there were 10 to 20 more days with drizzle (less than 1 mm/day) and about one more day with heavy rain (30–40 mm/day) in the city areas than in the suburbs (Schneider and Sonntag, 1936). Thus, the difference of precipitation between inside and outside a city should not be as clear in a climatic zone where heavy rainfall is frequent. In every one of the above-mentioned European and American cities, the annual total of precipitation is less than 1,000 mm. In Japanese cities, however, where rainfall is plentiful——up to 1,000–2,000 mm or even more ——the city-suburb difference is by no means clear.

However, a recent study of the urban industrial complex at Bombay, India, revealed that the region downwind of the urban industrial complex received about 15% more rainfall from 1941 to 1969, which is the period of increased industrialization (Khemani et al., 1973). Annual precipitation at Bombay is 2,078 mm. Further study is needed to confirm the conditions in every climatic zone.

The frequency of rainy days, excessive rainfall, thunderstorms, hail, and snowfall in the city regions in the U.S.A. were examined in relation to urban effect (Changnon, 1970). Similar studies must be done for cities in the various climatic regions.

Even though the difference between the annual or monthly totals of precipitation is not clear, a contrast in the number of rainy days inside and outside a city is often clear. Changnon (1961a) points out that the difference in the number of days with light rain and thunderstorms is 11–18% in the European and North American cities, and the precipitation amount is 5–10% higher in the built-up areas.

Figure 3.28 shows the distribution of annual total precipitation and the number of drizzly days in the urban areas of Tokyo in 1943. Here a drizzly day means a day with precipitation of 0.1–1.0 mm/day. Whereas the number of drizzly days is concentrated in the city areas, the distribution of the total annual rainfall shows no influence of the city areas in Tokyo. The total amount of drizzle is greater by 50–100 mm in the city than in the suburbs (Yoshino, 1957a). Lawrence (1971) called it a *rain island* and elucidated its weekly cycle which is typical in London. The distribution of drizzly days is of course influenced by the amount of condensation nuclei, which must be due to the polluted air in the city.

Many reports have made it clear that the number of drizzly days increases

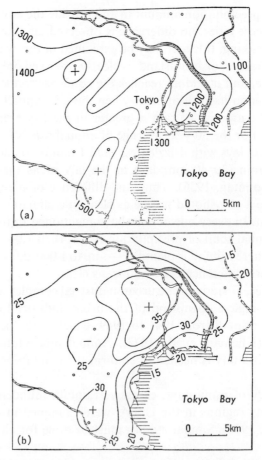

Fig. 3.28   (a) The annual total amount of precipitation in the Tokyo area for 1943. (b) Distribution of annual total number of drizzle days for the same area for 1943 (Yoshino, 1957a).

in accordance with the development of cities. The marked increase of precipitation days with 0.0 mm/day in Tokyo is given in Fig. 3.29. The same tendency may be seen in most cities in the world.

Whether or not the thunderstorms increase in the city has been a matter of interest to many meteorologists and climatologists. In England, for instance, it has been reported that the areas with heavy thunderstorms are roughly coincident with the built-up areas of London (Chandler, 1965), Reading (Parry, 1956b), and certain Midlandtowns (Barnes, 1960). Atkinson (1968, 1969) evaluated the effect of London's urban areas on the distribution of thunderstorm rainfall in SE-England. He has presented a statistically significant maximum of thunderstorm rainfall over the central part of London, which is due to about ten summer storms out of an overall total of over 600 that occurred in SE-England during the period. This study shows that urban-induced thunder clouds are capable of precipitating over London only in summer. The analysis suggests

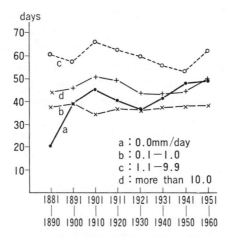

Fig. 3.29   Number of days with respective intervals of precipitation amount (mm/day) in Tokyo (Yoshino).

that the urban effect is strongly operative in only six synoptic types and that, of these types, fairly heavy warm frontal falls (more than 12 mm/day) are the most important. Atkinson (1971) made detailed case study of thunderstorm clouds over London on September 9, 1955. As the storm passed over the city, there was rapid cloud growth due to the high values of potential and wet-bulb potential temperatures in the urban areas, and precipitation was heavy. He added, however, that "the urban effect is real, but is reflected in daily precipitation patterns only when all conditions are just right," because the cases under consideration are very rare. He analyzed another case of thunderstorms over London on August 21, 1959, and came to the conclusion that the storms were triggered by the high urban temperatures and that turbulence and potential condensation and ice nuclei in the urban area played a negligible role in their initiation (Atkinson, 1970). Annual precipitation averages in the Detroit urban area did not appear to differ from regional values (Sanderson et al., 1973). However, seasonal precipitation amounts seem to be affected by the presence of the city: In autumn and winter, Detroit receives less precipitation than the surrounding rural areas and in summer about 20% more. The reason for this might be that there is more warm weather thunderstorm precipitation in the city than in the rural areas.

Changnon (1968) showed an increase of precipitation over La Porte, Indiana, by presenting the climatological values. He attributed this increase to the release of heat, water vapour and nuclei by the Chicago steel complex. Huff and Changnon (1972) presented a map showing the average annual pattern of the number of days with thunder in the Chicago-La Porte area from 1942 to 1968. This map shows that the maximum value of 50 days appears near the southeast corner of Lake Michigan, i.e., ESE of Chicago. Outside this relatively concentrated area the values are 36–38 days. Weickmann (1972) states that La Porte must be located in an area where the natural convergences in the low-level

develop due to the different configurations in the geographic boundaries between water and land in the Great Lakes Basin. Ogden (1969) studied the effect
of large steelworks on rainfall and came to the following conclusion: When
other extraneous effects are allowed for, the records show no influence of the
steelworks greater than 5% on total rainfall, summer rain, or light rain. He
wrote that, the effect of the Chicago industrial complex reported by Changnon
(1968) is unlikely.

A satisfactory answer to these problems must await the results of further
studies. If the climatological maximum of days with thunder or that of thunderstorm rainfall in a certain area is not caused by the city industrial areas, but by
geographical, orographical, or topographical factors, this will involve another
aspect of small-scale climatology.

The difference of snowfall inside and outside a city is not clear in respect to
the amount of snow and the number of snowfall days. The city temperature, i.e.,
heat island, does not develop under the conditions that cause snowfall, so that
no distribution of snowfall peculiar to cities is observed. For instance, Tokyo
has a distribution of snow accumulation clearly in proportion to the distance
from Tokyo Bay. The boundary lines between rain and sleet and between sleet
and snow are most influenced by Tokyo Bay. There is, however, a report that
reveals clearly the effect of the heat island on snowfall (Grillo et al., 1971):
The probability of snow on a precipitation day in the New York metropolitan
area in winter is 30%, and that in the western island suburbs is more than 40%.

However, snow melts more quickly inside a city than in its suburbs. Lindqvist (1968) measured and evaluated the changes in the snow depth on the
ground in aerial photographs of Lund and its environs. In his evaluation, a
grid, 300 m × 300 m, is employed and the snow-covered area on the house roofs
in each square is estimated in percent. The results are shown by the isarithms
for equal snow frequency: The city center is 30% and the suburbs 70–80%.

3.2.4. Fog
London is perhaps the best-known for city fog, but now no less famous are
the other cities. Since the beginning of this century, foggy days have been ever
on the increase (twenty to thirty days a month) in these cities. Table 3.22 shows
the change in the number of foggy days in Tokyo.

Table 3.22   Secular change of annual total fog days in Tokyo (Yoshino).

Year	5-year mean	Year	5-year mean
1896–1900	14.2 days	1936–1940	52.4 days
1901–1905	10.2	1941–1945	32.2
1906–1910	12.4	1946–1950	38.2
1911–1915	11.4	1951–1955	24.8
1916–1920	24.4	1956–1960	21.0
1921–1925	33.8	1961–1965	26.0
1926–1930	52.4	1966–1970	20.6
1931–1935	44.6		

No detailed observations on distribution of fog in cities have ever been made, but fog is known to form under these conditions: ground inversion in temperature, very weak wind, ample water vapour, and abundant condensation nuclei. In Tokyo dense fog often covers the eastern low-level areas near the Sumida River, the Arakawa Canal, the Naka River, and the Edo River.

City fog occurs in winter and the city-suburb difference in the number of foggy days is great in winter and small in summer. In London, city fog reduces solar radiation to 1/2 in winter and to 1/6 in summer. Fog often forms at night in areas other than cities, but in cities it forms mostly in the morning, the frequency from 6 to 10 h being 60–70% of the total. Five-year average mean hourly frequencies of fog and thick fog at London Airport are shown by Davis (1951). There are two maxima of occurrence in winter: one about one hour after sunrise and the other about midnight. The diurnal change has a weekly cycle and a marked seasonal change caused by the invading air mass characteristics.

Dense fog in cities hinders not only various means of ground transportation but also the arrival and departure of airplanes. Fog in London has a bearing even on the death rate. When unusually dense fog covered the city in 1880, 1892, 1948 and 1952 the death rate was higher by as much as 40% than the average death rate in winter.

3.2.5. Frost

Frost is related to the minimum temperature, and accordingly, its distribution inside and outside a city is comparable to that of the minimum temperature. As an average from 15-winter from 1926 to 1941, the western suburbs of Tokyo show first frost 10 days earlier and last frost 12 days later. The distribution of frost days in the Tokyo area is shown in Fig. 3.30, where only 30 frost days are recorded (the fewest in Tokyo) around Honjo and Fukagawa, the eastern parts

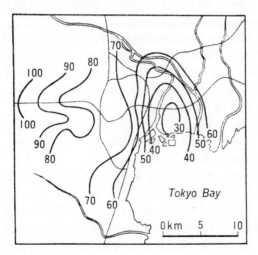

Fig. 3.30 Distribution of frost days in the Tokyo area (by Yoshino).

of Tokyo, but 90–100 near Kichijoji, the western suburbs. The distribution pattern closely corresponds to that of mean monthly minimum temperature or of monthly mean temperature in winter.

3.2.6. Wind

There are two important points to be considered about wind in cities: (a) development of peculiar wind systems in and around a city and (b) change, normally reduction, of wind velocity blowing under the general pressure field over a city.

The former, i.e., wind peculiar to a city, arises from the difference of pressure caused by the higher temperature in the city and is called *Flurwind*, or *country breeze*, which means wind from the surrounding districts toward the city. Existence of this wind was reported from the end of the 19th to the first half of the 20th century through statistical research of difference of pressure, wind direction, etc., between inside and outside the city. These factors were indentified with this kind of wind in many European cities. It should be noted, as pointed out by Bach (1970), that the country breeze blows at night as well as in the daytime, if there is country breeze circulation. Pooler (1963) reported that the urban area of Louisville, Kentucky, U.S.A., acts as a heat island and tends to cause an inflow of air near the ground surface under stable conditions.

However, various questions about country breeze remain. First, the difference of wind direction between inside and outside a city is not very large. Second, the influence of factors other than temperature or pressure difference, e.g., local orography, may be recognized in certain cities. Third, air currents by-passing the city are sometimes observed even when the wind is weak.

In Japanese cities, the influence of land and sea breezes is great on the coast, and in mountain districts, mountain and valley winds have a stronger influence, so that there is no report on the existence of country breeze in Japan, excepting for Asahikawa (Okita, 1960).

As for Tokyo, there is comparatively detailed research of wind distribution in this city. As early as 1926 it was reported that the wind direction at the Tokyo Weather Observatory was influenced by a small valley, about 1 km in width, in the low upland in the Tokyo area. Then, in the 1930s, the local distribution was observed several times by many researchers.

After these studies, detailed observations of the wind direction and wind force were made in Tokyo during World War II, by means of the wind suck. The whole metropolitan area was divided into squares 1.5 km × 1.5 km and one point, a primary school, was chosen within each of the squares. At about 200 points chosen in this way, observations were made 4 times a week—at 10 h and 14 h (JST) on Tuesday and Friday, from 1943 to 1945. The results are as follows (Yoshino, 1955, 1957c): (a) Although the wind direction with the annual maximum frequency in the whole area is almost north at 10 h, it changes to south or east at some stations at 14 h. That is, the wind directions from south to east are

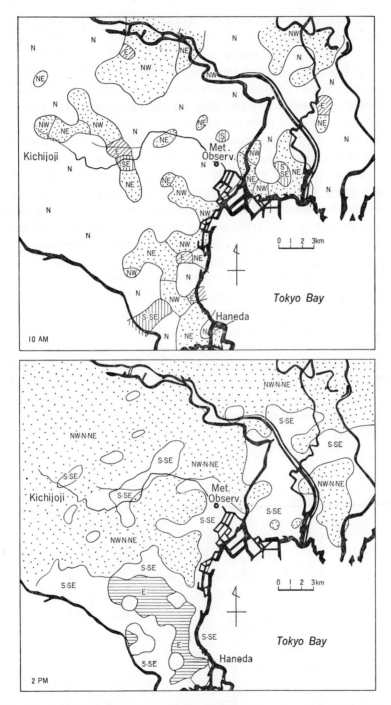

Fig. 3.31 Distribution of prevailing wind direction observed at about 200 points
in the city area of Tokyo, 1943–1945, at 10h and 14h (Yoshino, 1955).

dominant in the afternoon along the coastal region of the lowland area and also in the area along the rivers, both of which are 8–10 km away from the coast. These conditions are given in Fig. 3.31. (b) Throughout the year, the difference in the wind direction between the observation points is greater at 10 h than at 14 h. Since this originates from the development of the sea breeze, it is the clearest in summer. (c) The stronger the prevailing wind velocity is, the smaller is the deviation of the wind direction between stations. This is mentioned below in Section 5.1.1. (d) A closer examination of the cases of weak wind will make it clear that the frequency of occurrence of winds from the sea blowing opposite to the prevailing northerly wind at angles of 180°—90° decreases in proportion to the logarithm of distance from the sea. (e) The areas where the sea breeze invades are shown schematically in Fig. 3.32. In the eastern part of Tokyo, the prevailing wind from the north blows along the rivers, and the sea breeze blowing against it invades as far as 4–5 km from the coast. (f) On the other hand, in the areas where sea winds blow from the southeast or east, their invasion reaches as far as 6–7 km. Beyond this is a calm zone. (g) The situation of this windless zone varies a little according to direction and velocity of the prevailing wind. But it plays an important role in the local distribution of polluted air, which is mentioned below. (h) In conclusion, a peculiar wind system like the country breeze is not detected in the Tokyo area.

Using the constant-volume balloon, tetroon, Angell et al. (1966, 1971, 1973), Druyan et al. (1966, 1968), and Pack et al. (1963) observed the low-level trajectories over an urban atmosphere. In these studies, the influences of the urban area both on the wind direction and wind speed are clearly detected, although they are not strong. For instance, the results observed at Columbus,

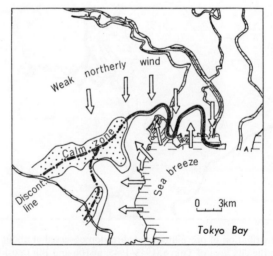

Fig. 3.32 A schematic illustration of the calm zone and the discontinuous line between sea breezes and weak northerly winds (Yoshino, 1957c).

Ohio, U.S.A., in March 1969, revealed the effect of the city on the nighttime airflow at heights of 100–200 m: The effects on wind direction are small at a height of 100 m, but an anticyclonic turning of 10° was observed at 200 m. The turning is greater under inversion conditions than under lapse conditions.

Concerning the latter of the two problems mentioned at the very start of this section the change of wind velocity in the city has been studied chiefly in connection with the growth of the city. Yokoi (1953) has shown the influence of the city on the wind velocity by extrapolating the results of vertical distribution observed at 8 points in Shizuoka City. The wind velocity at a height of 7 m over the city is 60–80% of that of the suburbs. This rate is low in the central part of the city and high in the windward part.

Munn (1970) reviewed thoroughly the wind conditions in urban areas. The vertical distribution of the wind speed is expressed by the following equation, the same to Eq. (3.27):

$$\bar{u}/u_1 = (z/z_1)^p \tag{3.64}$$

where u_1 is the wind speed at heights z_1. Exponent p for urban areas may range between 0.15 in the daytime to 0.5 at night. If we take the gradient wind speed, V_g at heights z_1, p is 0.28 in the city suburbs taking $z_1 = 390$ m, 0.40 in the built-up urban centers taking $z_1 = 420$ m.

There are many studies that deal with the vertical distribution of gusts over the city area. The vertical profile depends on the type of anemometer, the roughness of the site, the duration under consideration, and the stability of the air layer.

Arakawa and Tsutsumi (1967), analyzing the data observed at the Tokyo Tower up to 250 m above the ground, have pointed out that the strongest gust 52.1 m/sec was observed at the 253 m point on September 25, 1966, when a typhoon passed nearby. The vertical profile of the mean wind speed was expressed by Eq. (3.64), but the maximum wind gust appears frequently at a height of 67 m above the ground, and accordingly, the vertical profile of the wind gust cannot be expressed by Eq. (3.64). According to their opinion, this is attributed to the fact that there are many buildings about 30 m high around this tower.

Sekine (1956) has observed the vertical profile of the mean wind velocity at the various points in Tokyo up to 45 m and determined the term zero plane displacement, d, and the roughness parameter, z_0, of the following equation:

$$\bar{u} = \frac{\bar{u}_*}{k} \ln \frac{z-d}{z_0} \quad z > z_0 + d \tag{3.65}$$

This is in form the same as Eq. (3.22). According to Sekine, the d values range 0–6 m, and the z_0 values 0.45–2.70 m. On the other hand, Jones et al. (1971) have confirmed, observing the profile up to 330 m above the ground, that the rate of growth of the urban boundary layer is slow.

3.2.7. Air Pollution

Besides the city temperature, the most important cause for the special climate in cities is air pollution. Pollution is the main source of condensation nuclei, which form city fog and increase drizzly days in the city, the source of substances for the green-house effect and the source of lowering intensity of solar radiation. The effects of air pollution on urban climates have been discussed by Landsberg (1967) and Georgii (1970). Moreover, concentration of dust, CO_2, SO_2, etc., itself is a serious problem in the cities. Space does not permit thorough discussion of air pollution. In this Section, the horizontal distribution in the city area will be treated only from the viewpoint of small-scale climatology.

It is said that in London an average of 12 g/m^2 per day of dust is accumulated. Generally speaking, the air in the country contains about 10 times, in small towns about 35 times, and in large cities about 150 times as much dust as it does on the ocean. Especially, in such areas under the influence of ill-conditioned factories the amount of dust often reaches 4,000 times as much as that contained in the air on the ocean. According to a recent report, the amount of smoke is proportional to the square root of population.

Research into vertical distribution of dust made in Leipzig has made it clear that there are three strata of dust in the city: (a) the lower layer situated between roofs and the ground and consisting chiefly of smoke discharged by vehicles and of dust blown up from the ground, (b) the middle layer situated higher than roofs, that is, about 20 m above the ground surface, and consisting chiefly of smoke from chimneys of houses, and (c) the upper layer about 50–60 m above the ground and formed by smoke released from the chimneys of factories (Löbner, 1935). The altitudes of these layers may differ from city to city, but it has been ascertained in many cities that dust does form two or three layers.

A comparison of dust distribution was made in the city areas of Leipzig for 1935 (Löbner, 1935) and 1950 (Schmidt, 1952). An exact comparison between them is impossible, because of the different instruments and methods. It can be said, however, that (a) the distribution pattern did not change essentially, (b) the maximum in the city center and the minimum in the suburbs were 50 ($\times$ 100 parts/1 air) and 3 in 1935, but 75 and 40 in 1952, and (c) there are many closed areas, subcenters, with maximum more than 70. According to the figure showing the increase of annual values of dust in an industrial area in Halle-Leipzig (Rassow, 1959), the values in the beginning 1950s were estimated at about 1.5–2 times the values in the middle 1930s. Taking into consideration these circumstances, concerning the dust distribution change in accordance with the development of cities, (a) the pattern does not change, (b) the maximum values in the city center change by almost the same rate as industrial areas, (c) there are many subcenters of dust maximums in accordance with the city development, and (d) there is a remarkable increase in the values in the suburbs.

In London, the annual average smoke concentration, April 1957–March

1958, was 30 mg/100 m³ in the center; a tendency to higher concentrations than elsewhere was observed in the northeast suburbs in contrast to the relatively clean air in the higher, more open areas of S-, W-, and N-London (Chandler, 1965).

There are few studies on the horizontal distribution of submicron particles. Concerning the vertical distribution, Asakuno et al. (1970) reported the values of Mn and Br in air measured at the Tokyo Tower. The results obtained for Mn show the same distribution characteristics as SO_2 and Br the same as NO. The particles 0.01–0.05 μm occupy 70% of the total number; particles 0.5–0.8μm are 20–40%, and particles smaller than 0.5μm are 80% of the total weight except Br at 125 m, as given in Table 3.23.

Table 3.23 Vertical distribution of submicron particles at the Tokyo Tower (Asakuno et al., 1970).

Diameter	25 m		125 m	
	Mn	Br	Mn	Br
0.80–2.0 μ	0%	3%	0%	6%
0.50–0.80	18	29	38	80
<0.50	82	68	62	14

% is calculated by weight.

It has been observed in a typical large city in the U.S.A. with a population of more than a million that the total amount of dust falling in a year is 2,000 t/mile² in the center, 1,000 t/mile² 3 miles from the center, and 500 t/mile² 6 miles from the center. The amount of dust, though depending also on topography in and around the city, direction of the prevailing wind, and location of the sources of dust, decreases by geometrical series every 3 miles from the center of the city (Landsberg, 1951).

On a fine and calm night, an inversion layer develops. Therefore, early in the morning, the amount of dust is great, about 4 times greater than in the afternoon. In summer, the amount of dust is naturally smaller than in winter, because heating apparatus is not used, and the stronger solar radiation causes an active vertical exchange of atmosphere.

Brooks (1954) has pointed out the following facts concerning the distribution of dust in England: (a) the amount of coal burnt in a year for industrial and domestic purposes reaches 4,000 t/km² or more in the densely populated industrial areas, (b) the amount of dust accumulated in deposit gauges in the country with 40 t/km²/year appears in the east or southeast of large cities; (c) the annual number of days with low visibility, less than 1 km at noon, amounts to 150 days in areas roughly coinciding with the areas in (b).

The well-known London smog of December 5–9, 1952, was the most serious in that country for at least 90 years. Average daily smoke concentrations were 200–400 mg/100m³ in the central areas on December 6 and 7, and the average SO_2 concentration had increased by 50% or more at most locations, as shown in

Fig. 3.33. There were some relatively clean areas in the more elevated southern parts of London. Smoke concentrations of 300–400 mg/100 m³ occurred again exactly ten years in December 1962. In this case, the heavy concentration areas had a ring form. The zonal area with 200–300 mg/100 m³ was roughly coincident with the inner suburban area of 19th century London. The outer part of this area was 100–200 mg/100 m³.

In Tokyo, an exact measurement of SO₂ distribution was first made at seven stations in January 1955 (Itoo et al., 1955). As shown in Fig. 3. 34, SO₂ concentration of more than 0.10 ppm appeared in the southern and eastern parts of Tokyo, where industry was then developing. The SO₂ concentration in Tokyo,

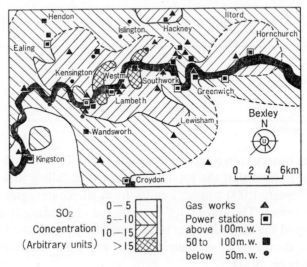

Fig. 3.33 Average sulphur dioxide concentration in London, December 5–9, 1952 (Wilkins, 1954).

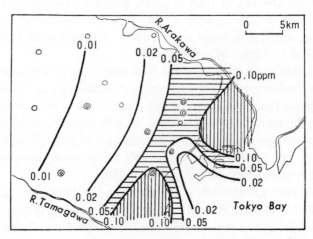

Fig. 3.34 Distribution of SO₂ concentration in Tokyo, January 18–24, 1955 (Itoo et al., 1955).

registered by means of the solution electro-conductometric method, were 0.04–0.06 ppm for 1962–1964 and 0.07–0.08 for 1965–1967. The distribution of SO_2 concentrations for the heavy industry regions, the "*Keihin* industrial region" and "*Keiyô* industrial region", around Tokyo Bay are given in Fig. 3.35. It is worth noting that the SO_2 distribution has a closer relationship to the industrial region and not to the city areas (Fukuoka et al., 1970). However, the pollutants produced in the industrial areas around Tokyo Bay are brought by the sea breezes from the coast to the inland area, where the city areas are developing. Such situations are also found in Osaka and other regions in Japan.

The vertical distribution of SO_2 concentration is also of interest. The results of observations at a height of 25, 125, and 225 m above the ground at Tokyo Tower during March to May, 1969, indicate that SO_2 concentration has a maxi-

Fig. 3.35 Distribution of SO_2 concentration in the industrial regions around Tokyo Bay in winter, 1967. Unit is mg $SO_3 \cdot 100cm^{-2} \cdot day^{-1}$ [PbO_2] (Fukuoka et al., 1970).

Table 3.24 Vertical distribution of SO_2 concentration at Tokyo Tower in March, April, and May, 1969 (Ôdaira et al., 1970).

	Average value			Occurrence freq. of high concentr.					
	March	April	May	Total hours	0.2–0.29	0.3–0.39	0.4–0.49	0.5–0.69	0.7 <
225 m	0.083	0.076	0.111 ppm	2,134	138	54	30	19	6
125 m	0.118	0.102	0.120	2,195	180	70	29	10	1
25 m	0.074	0.034	0.049	2,158	16	1	0	0	0

mum at 125 m, and the frequency of high values is shown in Table 3.24. The maximum at 125 m (and sometimes 225 m) appears under S-SSW winds conditions, which bring the air with high SO_2 concentration from the *Keihin* industrial region south of Tokyo Tower.

Hamm (1969) made a minute observation of the SO_2 concentration as well as dust distribution in Stuttgart from the standpoint of local climatology. Höschele (1966) also presented the SO_2 concentration values in the Karlsruhe region in relation to temperature and wind velocity, which is expressed by the following equation:

$$\log S = -1.9 + 1.0 \log(25 - t) - 0.45 \log v \qquad (3.66)$$

where S is the maximum 30-min mean value of SO_2 concentration in mg/m^3; t is daily mean temperature (°C), and v is wind velocity (m/sec). Equation (3.66) means that the lower the temperature is and the weaker the wind velocity is, the greater the SO_2 concentration value becomes.

There are still many problems to be described here. McCormick (1969) made a brief, but thorough review on urban air pollution. Further information on the problems together with a list of references will be found in his paper.

Currently, the urgent problem, among others, is diffusion modeling. Since the pioneer work by F. N. Frenkiel in 1956, mathematical models of urban air pollution have been developed. Turner (1964) was the first to develop a multi-source diffusion model for an urban area. Shieh et al. (1972) studied a model incorporating point and area sources, time and space dependence of source strengths, and time and space dependence of meteorological variables. It was shown that the numerical simulation of the SO_2 concentration distribution for New York City on January 11, 1971, agreed favorably with the observed values.

Hanna (1971) suggested the equation of air pollution concentration:

$$X = cQ/u \qquad (3.67)$$

where X is the volume concentration of a pollutant emitted with area source strength; Q and u are the average wind speed. Equation (3.67) follows from the simple urban diffusion model by Gifford and Hanna, if the source strength is assumed to be constant. The parameter c is a weak function of city size and should be approximately constant (Gifford and Hanna, 1973); c is found to equal about 225 for particles, and 50 for SO_2.

Turner (1970) reviewed the studies on the estimation methods of atmospheric dispersion. In his well-written book first published in 1957, there is a complete list of references on the problem. Frimescu et al. (1973) applied one of the estimation methods to an industrial region. 400 km^2, with complex topography in Rumania, and the values calculated coincide fairly well with those estimated by the model.

The town of Reading, England, has been subjected to air pollution studies in recent years (Parry, 1967, 1970; Marsh et al., 1967a, b). Reading, situated

64 km west of the center of London, is of moderate size (about 150,000 population) and mainly residential and commercial with only light industry. However, more than one-third of Reading suffers from pollution originating from external sources. The results have indicated that location near a major pollution source, the London conurbation, has particular consequences. There are many spots in the distribution pattern caused by road traffic, low industrial chimneys, parks, and "Smoke Control Areas." The positions of the spots with maximum concentration depend on the prevailing wind direction.

As mentioned below, the smog distribution is strongly affected by the sea breeze front on the coastal area. However, the distribution of oxidant concentration in the greater Tokyo region does not show a simple relation to the sea breeze front, as shown in Fig. 3.36 (Fukuoka et al., 1972). It seems that oxidant pollution occurring due to photochemical reaction is generated in the air layer with 200–500 m above the ground and comes down to the surface level by local down currents.

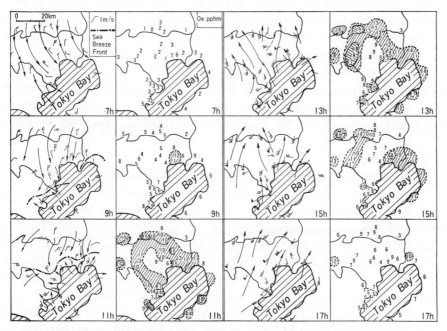

Fig. 3.36 Distribution of oxidant concentration (pphm) and wind in Tokyo and its surrounding areas on June 26, 1971 (Fukuoka et al., 1972).

Yoshino (1968) estimated the possibility of air pollution (H) from the standpoint of climatology. The experimental regression equation is,

$$H = h \{1 + (G + 0.7)\} \tag{3.68}$$

where h is the annual total of monthly hours with the condition of relative humidity above 60% and simultaneously, wind velocity below 3 m/s and G is the monthly average lapse rate of air temperature (°C) (plus means inversion). For instance, Tokyo in the coastal region is 388, but Matsumoto in a basin in Central Japan, where wind is weak and inversion takes place frequently, is 740.

The air pollution causes lichen to develop on tree trunks and injure the leaves of the trees in the city. Using the relation between the air pollution and the vegetation characteristics, a detailed observation of the distribution of air pollution affecting the vegetation at many points in the city can be made. As indicators of air pollution, lichens have been used in the Tyne Valley, England (Gilbert, 1965), and in the Ruhr District, Germany (Domrös, 1966). Domrös used a scale of lichen coverage as follows:

Coverage grade 4:   100–50% of the area is covered by lichen
Coverage grade 3:   50–25%   "
Coverage grade 2:   25–10%   "
Coverage grade 1:   lower than 10%   "
Coverage grade 0:   no   "

Applying this scale in the central part of the Ruhr District, he has drawn a detailed map. It has been confirmed that the distribution of measured air pollution, especially the SO_2 concentration, shows a good coincidence with his map. Recently, Thiele (1974) applied similar method of study to Munich.

An observation of the contamination of rain was made in Uppsala and its surroundings with an area of about 400 m² (Andersson, 1969). The contaminants, potassium, calcium, chlorine, and sulphur, have a maximum in and around the city, because of the wash-out in the subcloud layer over the city.

3.2.8. Insolation

So far, only a few careful studies have been made of the distribution of the amount of insolation in city areas. An investigation concerning the Tokyo area was conducted from March 8 to 25, 1958. The distribution on March 8 is shown in Fig. 3.37; the sky was entirely clear in the daytime on these days (Sekiguti et al., 1960; Kawamura, 1963). According to this figure, the amount of insolation is as small as 90% in the industrial areas in the eastern and southwestern parts of Tokyo. The area of zonal shape below 70% is particularly remarkable. North of this zone there is a steep gradient. The distribution of the insolation amount on March 13 and 17 was similar to that on March 8, but a little different from that on March 15 and 24. In every case, the amount of insolation reduction within the Tokyo area amounted to 40 to 50%.

The reason why the zonal area with little insolation lies in a WSW direction is thought to be that, although simultaneous observations of the wind were not made on March 8, this zonal area corresponds to the calm zone, an inland boundary of the sea breeze front, as schematically illustrated in Fig. 3.32. In this area, the air is filled with pollutants coming from the Kawasaki district of the *Keihin* industrial region and accordingly, the insolation is weakened. The evidence that the coastal side of this weak insolation area has a gentle gradient,

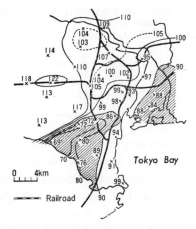

Fig. 3.37 Distribution of insolation in Tokyo on March 8, 1958. Unit is %. (Sekiguti et al., 1960).

but the northwest side is steep, favors our explication introducing the sea breeze front.

Mizukoshi (1968) also made clear the influence of the city and industrial areas of Yokkaichi City on the decrease of insolation. The weak insolation areas receive 80–85% of uninfluenced areas. Besides these, there is Paszyński's report on the distribution of transparence amount of atmosphere in the industrial region of Upper Silesia (Paszyński, 1960). Hufty (1970) dealt with the radiation conditions in the Liège region, Belgium, and Olecki (1973) the radiation balance in Cracow. Poland.

According to a study by Terjung et al. (1970a) on the urban energy balance of the city and man in the Los Angeles basin on a cloudy day, "Catalina eddy", the energy parameters have considerable areal variations in intensity and trend. In the daytime, 70% of the solar radiation is absorbed. It is asserted that a man living in the city receives only 37% and 70% of the values available at the horizontal pavement for solar radiation and net radiation, respectively.

Yamashita (1973) studied the reduction of solar radiation in relation to air pollution concentration in Toronto. He found that the reduction shows a pronounced seasonal variation, although SO_2 concentration has no seasonal variation. This is because of the seasonal change of the path length of the solar beam through the urban atmosphere. He computed also the values of Linke's turbidity factor in Toronto and nearby rural areas and found that the turbidity factor was larger in the city both seasonally and daily (Yamashita, 1974).

An investigation on the local differences of ozone amount was made on fine days in Bad Kissingen, West Germany (Zimmermann, 1953). The amount of ozone on the streets in the city was about 30 to 40 γ/m^3 less than at the other places, as shown in Table 3.25. This difference is said to be due to the automobile traffic in the city. But, as the amount of ozone on the surface generally bears

Table 3.25 Ozone concentration in Bad Kissingen, 1950
(Zimmermann, 1953).

Observed point	October 6			October 11		
Park, I	14 h	29 m	49 γ/m^3	14 h	33 m	49 γ/m^3
Park, II	14	48	43	14	50	59
Bismark Ave.	15	03	10	15	05	24
Park, III	15	18	52	15	19	59
Park, IV	15	38	49	16	10	52

a close relationship to wind velocity and relative humidity, the difference may be explained also from this point. On September 6, 1951, when there was no great difference of wind velocity and relative humidity between the investigation spots, no great difference in the amount of ozone was observed between the streets, stations, and parks.

Because insolation is weak in cities, UV radiation is also weak there, especially in the industrial areas. Fukui (1939) measured the UV radiation in and around several large cities in Japan. According to his measurement, the intensity in Tokyo was greater in the city center and outer part of the suburbs, while in the intervening ring-like region, the intensity was weaker. This was considered to be caused by the smoke-dust from factories and sandy dust carried from the country fields.

Yamamoto (1957) has investigated the heat budget of large cities in Japan. Because of a large extinction coefficient for visible light and a small absorption coefficient for infrared radiation of aerosols, the net effect of aerosols on solar and terrestrial radiation is to diminish the amount of radiation received at the surface of large cities. The mean daily amounts of insolation were 20–30% less in Osaka and Tokyo from February through August, 1953.

Along this line of investigation, Kondo (1967) has pointed out the existence of another factor, f_A, examining the additional aerosol particles over the cities. The factor is defined by

$$I'_s = f_A \times I_s \tag{3.69}$$

where I'_s is the corrected monthly mean value of the solar radiation at a point in the cities, and I_s is the monthly mean value of the solar radiation on a cloudy day at the point. I_s is obtained by

$$\frac{I_s}{I^*} = \frac{1}{N} \left\{ N_1 \left(\frac{I_s}{I^*} \right)_1 + (N - N_1) \left(\frac{I_s}{I^*} \right)_2 \right\} \tag{3.70}$$

where $\left(\dfrac{I_s}{I^*} \right)_1 = 1.02(0.3 + 0.7 \times 10^{-0.055\bar{x}})$

$\left(\dfrac{I_s}{I^*} \right)_2 = 0.34 + 0.5\bar{s}_2 - 0.1\bar{n}_2$

N_1 = the number of clear days

N = the number of days in a month

$\bar{x} = (1 + 0.04\bar{e})(\sec \bar{Z}_0 + 0.8)$

$$\bar{s}_2 = \frac{\bar{s}N - \bar{s}_1 N_1}{N - N_1} \doteq \frac{\bar{s}N - 0.9N_1}{N - N_1}$$ (s is the ratio of the number of hours of sunshine to the max. number of sunshine hours)

$$\bar{n}_2 = \frac{\bar{n}N - \bar{n}_1 N_1}{N - N_1} \doteq \frac{\bar{n}N - 0.1N_1}{N - N_1}$$ (n is the daily mean amount of cloudiness)

$\bar{e}$ = daily mean value of vapour pressure at the surface(mb)

$\bar{Z}_0$ = the solar zenith angle at noon

I^* = the monthly mean value of the solar radiation incident on a horizontal surface at the top of the atmosphere.

(An upper bar in the above equations means the monthly mean of the respective values.)

By comparing the calculated I_s with the observed I'_s values, the value of factor f_A is obtained (see Table 3.26). The relationship between f_A and the population is detectable.

Table 3.26 Factor f_A defined by Eq. (3.69) (Kondo, 1967).

Station	Factor f_A
Sapporo	0.94
Sendai	0.94
Wajima*	0.95
Nagoya	0.89
Tokyo	0.82 and 0.77 (June–Sept.)
Yokohama	0.91
Kyoto	1.00 and 0.92 (April–Sept.)
Shimonoseki	0.92 and 0.86 (May–Sept.)
Hiroshima	0.95 and 0.88 (April–Sept.)
Kobe	0.95 and 0.89 (April–Sept.)
Osaka	0.83 and 0.76 (April–Sept.)
Fukuoka	0.91

* Inferred from the transmission of the direct solar radiation observed in 1953–1958.

3.3. FOREST

The climate in forests has so far been studied as a branch of forest meteorology. The problems of forest climate in a small area are roughly classified into the following three: (a) the influence of the forest upon the climate in the neighboring areas; (b) climatic difference between within and outside a forest and (c) the relation between the climate and the local difference of the growth of seedlings in a nursery or younger trees in an afforested area. The third problem is not discussed here, because a field with small seedlings can be treated as identical with a crop field from the climatological point of view. According to Flemming (1968a), the first problem belongs to silvimeteorology (*Waldmeteorologie* in German) and the second to forest meteorology (*Forstmeteorologie* in German).

Baumgartner has made a thorough review on forest meteorology under the titles *Forstmeteorologie, Forstliche Meteorologie und Hydrologie* or *Forstliche Meteorologie und Klimatologie* in the series of *Schriftenreihe des AID* (e.g. Nos.

115, 127, 138). In these review, he summarized the recent development of forest meteorology.

There has been much project research on the subalpine forests ranging over many years. Among them, the most important, synthetic studies at the subalpine zone in Ober Gurgl, Ötz-Valley, Austria, made by the research group in the Forschungsstelle für Lawinenvorbeugung Innsbruck. The results have been published in the *Mitteilungen der Forstlichen Bundes-Versuchsanstalt Mariabrunn* Nos. 59 (1961) and 61 (1963). Comprehensive studies on forest meteorology from the standpoint of small-scale climatology have been done in many parts of the world: in the Wermsdorfer forest near Leipzig, Germany (Koch, 1934); in the spruce forest at Os, Norway (Paulsen, 1948); in the Neotoma Valley, Ohio, U.S.A. (Wolfe et al., 1943, 1949); in Muttenz near Basel, Switzerland (Lüdi and Zoller, 1949); on the Cushetunk Mountain, New Jersey, U.S.A. (Cantlon, 1953); in Gr. Falkenstein and Fichtelgebirge, Germany (Baumgartner, 1956; Baumgartner et al., 1958); in the Black Rock Forest, New York, U.S.A. (Ross, 1958); on the Staufenberg mountain, Harz, Germany (Hartmann et al., 1959); in the Green Ridge and Montgomery Ridge, Ontario, Canada (MacHattie et al., 1961). Recent investigations on the climatic conditions of the natural forests in Mittelgebirge, Germany (Hartmann and Schnelle, 1970) were done to get informative results, of which are shown below. The micrometeorological and biological experiments in Thetford Forest, Norfolk, England, have produced outstanding results (Stewart, 1971; Oliver, 1971; Stewart and Thom, 1973). The results of these studies are mentioned below.

3.3.1. Radiation, Heat, and Energy Balance
(A) *Solar Radiation*

About 80% of the insolation is absorbed by forest canopies and about 5% reaches the forest floor. If the total radiation is considered, the percentage ranges from 100% at the tree tops to only 5% within the forest on a fine summer day in a forest of young fir (Baumgartner, 1956). This percentage greatly differs according to the conditions of the tree crowns and also according to whether it is clear or cloudy. That is the percentage of insolation absorbed by tree crowns is great on fine days, and the difference between the absolute values on fine days and on cloudy days is small at the forest floor.

A summary of observations that have so far been conducted is shown in Fig. 3.38, which indicates the relation between the number of trees and the transmission of insolation to the unit area in the case of coniferous trees (Miller, 1955). There are considerable variations because the method of observation and the conditions of the forest differ from investigator to investigator, but the general relation can be grasped: A large amount of interception by open-crowned trees, even in sparse stands, is observable. The tree-trunk density in this figure is expressed originally by the total of the diameter measured in inches of every tree within one acre of the forest, but is rearranged in units of meters and hectares.

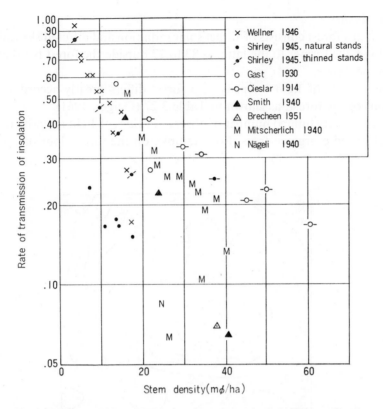

Fig. 3.38 Transmission of insolation through stands of chnifers as a function of stand density, expressed as sum of diameters of stems of all trees. Observations, except those by Mitscherlich and Nägeli, were made in stands of intolerant tree (Miller, 1955).

Measurements of direct and diffuse solar radiation were made beneath a pine canopy and a hardwood canopy near New Haven, Connecticut, U.S.A. (Reifsnyder et al., 1971). It was suggested that the pine (*Pinus resinosa*) canopy acted as a uniformly scattering medium, whereas the hardwood canopy acted much more like a layer of horizontally-oriented flat leaves.

Table 3.27 Light intensity in forests. Unit is % of the value outside of the forest (Geiger, 1961).

Kind of tree	Without leaves	With leaves
Deciduous trees		
Red beech	26–66	2–40%
Oak	43–69	3–35
Ash	39–80	8–60
Birch	—	20–30
Needle trees		
Silver fir	—	2–20
Spruce	—	4–40
Pine	—	22–40

(B) *Light*

Light in a forest consists of direct solar radiation and diffused sky radiation. The ratio of the former to the latter is 83 to 17 outside the forest between noon and 14h local time.

The amount of light intensity in a forest differs greatly according to the conditions of the forest, as shown in Table 3.27 (Geiger, 1961).

There is a very limited number of observed values of light intensity in forests of broad-leaved evergreen trees. For example, on the forest floor of a tropical

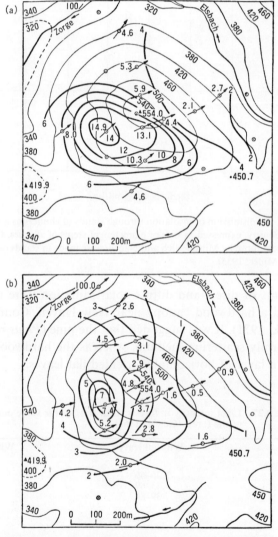

Fig. 3.39 Light intensity (% of outside of the forest) in a beech forest near Staufenberg, Germany (Hartmann et al., 1959) a: September 16, 1953; b: June 20, 1954.

rainforest, the light reaches only 0.5% or less of that outside the forest as mentioned below (Dirmhirn, 1961).

The average percentage of light measured on a fine day in August in a forest in Mt. Kirishima, Kyushu, Japan, where the trees in the upper layer, 15–20 high, are Japanese red pines (*Pinus densiflora*) and firs (*Abies firma*), and the trees in the middle layer, 10 m, are broad-leaved evergreen trees, was 1–3% on the forest floor (Yoshino, 1961b). In a forest of evergreen trees (mostly *Distylium racemosum*) in the southern part of the Ôsumi Peninsula, Japan, the average was 25% (Kitazawa et al., 1960).

As the distribution of light in a forest has a great influence on vegetation on the forest floor, various minute examinations are currently being carried on. An observation in a forest of beech trees (340–350 m above sea level) near Staufenberg, Germany, is shown in Fig. 3.39 (Hartmann, Eimern and Jahn, 1959). These maps are drawn by the values measured by mobile observation spending 2 hr each on cloudless days. The values of the light intensity measured outside of the forest is given here as 100%. The arrows show the directions of light. The absolute values are a little greater in June than in September. Differences caused by the slopes, i.e., influences of the topography are remarkable, the greatest being on the SW-facing slope. These differences are more obvious in the afternoon than in the morning or around noon.

Light has a vertical distribution in the forest. The results of observation of light up to 2 m above the forest floor in a forest of white firs (*Abies sachalinensis*) in Hokkaido, Japan, were 220 luxes on the forest floor, 280 luxes at a height of 1 m, and 220 luxes at a height of 2 m; light intensity was slightly weak at the

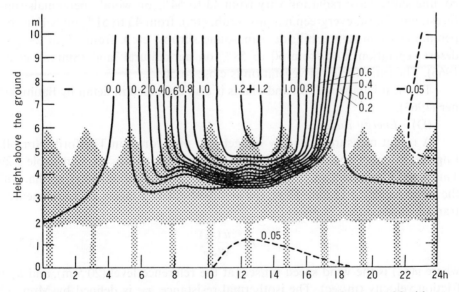

Fig. 3.40 Vertical distribution of net radiation in a young spruce forest near Munich on a hot summer day (Baumgartner, 1967a).

height of the lowest branches. The vertical profile of the light intensity in Mpanga Forest shows that the decline down to about 33 m above the forest floor is slow and below this, the intensity decreases rapidly in the zone where the dense crowns of trees develop (Dirmhirn, 1961). Between the top of the undergrowth layer and 15 m height, it is 15–20%. Below the undergrowth, it becomes 0.3–0.5%

   (C)   *Radiation Balance*

The vertical distribution of the radiation budget on a hot summer day as given by Baumgartner (1956, 1967a) is shown in Fig. 3.40. This figure shows how sharp the gradient in the upper layer of the canopy of a spruce forest is.

The main radiation balance of various long- and short-wave radiation follows the upper layer condition in a natural closed forest. The radiation balance in this forest can be expressed by the equation, z/H, as in Table 3.28, where z is the height above the forest floor and H the mean height of the forest. Together with the state shown in Fig. 3.40, the values given in Table 3.28 indicate that the radiation balance increases sharply in the upper layer of the forest.

Table 3.28   Relative distribution of radiation budget in a spruce forest (Baumgartner, 1967a).

z/H	0.04	0.59	0.73	0.89	1.78
Daytime	6	6	39	98	100%
Night	5	9	18	96	100

Daily totals of net radiation balance expressed as percentages of total incoming short-wave radiation vary from 43 to 54% on woody perennials (pine forest, oak forest, evergreen maquis scrub, etc.), from 43 to 51% on herbaceous perennials (semi-steppe hillside, sand-dune shrubs, etc.), from 25 to 29% on desert vegetation, and from 50 to 58% on agricultural land (Stanhill et al., 1966). The ratio for most vegetation is close to 50%.

The heat balance of soil and plants is also discussed in detail by Baumgartner (1963).

   (D)   *Energy Budget*

A recent measure of energy budget in a pine forest in Thetford, Norfolk, England (Stewart and Thom, 1973) has obtained the values given in Table 3.29. This pine forest extends uniformly over 70 km² and has about 800 trees/ha, with the mean height of 15.8 m. Here, the aerodynamic resistance, r_D, to momentum transfer to the forest is given by

$$r_D = \frac{u(z_R)}{(u_*)_0^2}$$
(3.71)

where $u(z_R)$ is the wind speed (m/sec) at the reference level (20.5 m), and u_* is friction velocity (m/sec). The isothermal resistance, r_I, is defined by Monteith (1965) as

$$r_I = \frac{\rho C_p}{\gamma} \frac{\{e_w(T) - e\}}{A} \tag{3.72}$$

where, e is vapour pressure (mb), e_w is saturation vapour pressure, T is air temperature, and γ is psychrometric constant; $C_p p / (\frac{5}{8}\lambda) \simeq 0.66 \text{mb}/°\text{C}$. A is the available energy given by

$$A = R_n - G - S - P = H + \lambda E \tag{3.73}$$

where R_n is net radiation, G is energy flux into the ground; S is total flux of energy into storage between a chosen height in the forest and forest floor, P is net rate of energy absorption by photosynthesis and respiration (assumed to be 1% of R_n), H is sensible heat flux (+upwards), and λE is latent heat flux. The value r_I will be referred to as the "climatological resistance," although it must be understood that this is not a proper resistance. r_I is a climatological parameter or measure of the overhead conditions. The surface resistance, r_S, is given by

$$r_S = \left(\frac{\Delta}{\gamma}\beta - 1\right) r_D + (1 + \beta) r_I \tag{3.74}$$

where Δ is de_w/dT (mb/°C) taken at the air temperature at a height of 10 m within the canopy (approximation to the mean of air and needle temperatures), and β is Bowen ratio ($H/\lambda E$). Similar to Eq. (3.74), the actual bulk physiological resistance, r_{ST}, defined by Thom (1972) can be expressed as

$$r_{ST} = \left(\frac{\Delta}{\gamma}\beta r_H - r_V\right) + (1 + \beta) r_I \tag{3.75}$$

The bulk physiological resistance r_{ST} of the forest as given in Table 3.29 ranges from 1.12 to 1.62 s/cm in the forenoon. It becomes 4 s/cm in the late afternoon, which is consistent with independent biological measurements. In contrast, the bulk aerodynamic resistance generally takes values between 0.05

Table 3.29 Some values of aerodynamic and bulk physiological resistances and associated parameters at 11–12 h, during May–September (Compiled by Yoshino after Stewart and Thom, 1973).

	Variables		Units	Range
Aerodynamic	u,	wind velocity at 20.5 m height	m/sec	1.7–5.2
	u_*,	friction velocity	m/sec	0.43–1.21
	F,	stability factor	—	2.9–7.0
	Ri,	Richardson number	—	−0.025–−0.36
	r_D,	aerodynamic resistance to momentum transfer to the forest shown in Eq. (3.71)	sec/cm	0.034–0.089
	r_H,	aerodynamic resistance to heat exchange between the forest and a chosen level, higher than h	sec/cm	0.056–0.091
	r_V,	aerodynamic resistance to transpiration flux from forest to a chosen level	sec/cm	0.049–0.078
Bulk physiological	r_I,	isothermal resistance given by Eq. (3.72)	sec/cm	0.23–0.47
	r_S,	surface resistance given by Eq. (3.73)	sec/cm	1.08–1.51
	r_{ST},	bulk physiological resistance given by Eq. (3.75)	sec/cm	1.12–1.62
	β,	Bowen ratio	—	1.38–3.91

and 0.10 s/cm. The ratio r_{ST}/r is of the order 20:1. This implies that the transpiration from the forest must occur at rates much less dependent on net radiation R_n than on ambient vapour pressure deficit and that the evaporation of intercepted rainfall from the trees must occur at about 5 times as fast as the corresponding transpiration rate under the same meteorological conditions. These conclusion, from a pine forest, must be enlarged to apply to other types of forests in the future.

The influence of an open pine forest on daytime temperature in the Sierra Nevada during the snow-melting season was studied in relation to the heat balance by Miller (1956a). His evaluation of the heat balance terms is given in Table 3.30. This table gives his estimation of the heat transfer to the air. In Table 3.31, the areal total of heat transfer to the air are given. On a typical spring day, the heat of about 120 cal·cm^{-2} is supplied to the air of the region, most of the heat is from needles of the pine forest. He also considers the effect of the local air layer on the heat balance in this region. It is concluded that the daytime temperature is not always cold in the snow-covered region, contrary to what generally believed, if the forest is open.

Table 3.30 Heat balance of pine needles and crowns during a midday hour in spring (Miller, 1956a).

Component	Needles		Crowns	
			cal·cm^{-2}·min^{-1}	
Solar radiation				
Reflected to sky	13		7	
Transmitted to ground	0		15	
Absorbed		77		154
Long-wave radiation				
Net loss upward	20		18	
Net loss downward	9		11	
Photosynthesis	1		3	
Stored in wood	0		2	
Transferred to air				
By convection	32		80	
By transpiration	15	77	40	154

Table 3.31 Heat transfer to the air in the Sierra Nevada (Miller, 1956a).

	From needles	From crowns	From forest stands	Average over crest region
As sensible heat	32	80	24	10 cal·cm^{-2}·min^{-1}
As latent heat	15	40	13	4

3.3.2. Temperature

(A) *Air Temperature Inside and Outside a Forest*

The maximum temperature within a forest is low, but the minimum temperature is comparatively high. Because the tree crowns prevent solar radiation in the daytime, the specific heat of trees is greater than that of the soil, and the transpiration from the surface of leaves requires heat for evaporation, so the temperature of trees does not rise rapidly. Thus, the air temperature in a forest is prevented from quick rising and is lower in the daytime. The occurrence of the maximum air temperature is markedly retarded. At night, however, as the cooling by the long-wave radiation from the surface of the forest floor is prevented by the tree canopies, the air temperature above the floor stays higher than that outside the forest.

The daily mean temperature is a little lower in a forest than outside the forest because the difference in air temperature is greater in the daytime than at night. The difference in air temperature between inside and outside a forest is usually great when the sky is clear and in the growing seasons, but it differs also according to kind, height, and age of the trees composing the forest. An example observed in Japan is presented in Table 3.32. This table shows that, in general, the difference of the monthly mean temperature between inside and outside the forest is great in July, when the tree leaves are dense, and this tendency is naturally great in the case of deciduous trees. As shown in Table 3.33, the difference between the daily range of air temperature inside and outside of the forest is great in summer, for the same reason.

Table 3.32 Difference of monthly mean air temperature between the inside and outside of the forest. Plus means the outside is warmer (Yoshino).

Region	Height a.s.l.	Forest	Jan.	Apr.	July	Oct.	Year
Nikkô	1,270 m	Deciduous broad-leaved trees	0.07	0.00	0.80	0.52	0.31°C
Myôgi	427	Needle trees	0.67	0.73	0.78	1.02	0.85
Tokyo	60	Japanese cedar trees	0.16	0.11	0.38	0.16	0.23

Table 3.33 Difference of mean daily range of air temperature between the inside and outside of a Japanese cedar forest in Tokyo (Yoshino).

	Jan.	Apr.	July	Oct.	Year
In the forest	11.47	10.36	7.42	8.05	9.81°C
Outside of forest	9.82	9.06	5.55	6.22	8.23
Difference	1.65	1.30	1.87	1.83	1.58

Koch (1934) made a comprehensive study on the temperature and wind in Wermsdorfer forest and the surrounding region near Leipzig. He drew the isopleths of the air temperature during the night hours. An example of the results for July 8–9, 1933, under summer conditions, is shown in Fig. 195 of the book by Geiger (1961, 1965) and another one, under winter conditions, is shown in Fig. 3.41. About 1 hr before sunset, the air temperature is almost the same, but about 2 hr after sunset the air temperature becomes lower in the hollows occupied by the younger trees, where it is −4 to −5°C. Before sunrise, the temperature becomes lower again. About 4 hr after sunrise, it becomes almost

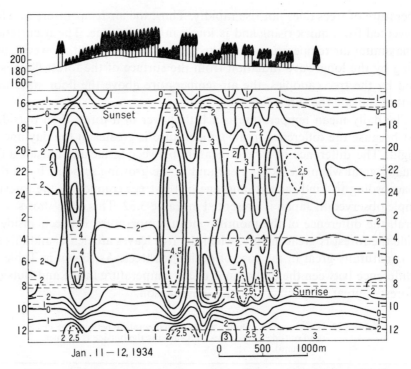

Fig. 3.41 Isopleth of air temperature in a closed forest region near Leipzig on January 11–12, 1934 (Koch, 1934).

stable. The low temperatures between the forests are especially important from the standpoint of afforestation. Konda (1962) observed the minimum temperatures at many points under various conditions in a spruce forest in order to determine the frost height above the ground. Burgos et al. (1951) investigated the frost occurrence in the forest region of NW-Argentina in relation to crop planting such as coffee, tea or cocoa.

(B) *Vertical Distribution of Air Temperature within a Forest*

In general, the vertical distribution of the air temperature in a forest differs greatly according to the kind and height of the trees and the trunk density. The maximum air temperature in the daytime appears near the upper level of the forest canopies, and the air temperature profile between the lower part of the forest canopies and the surface of the forest floor shows inversion or neutral, if the canopy is thick enough. If the canopy is thin, the air temperature near the forest floor rises high, and in the extreme cases where insolation fully reaches the forest floor, the air temperature 120 cm above the earth in the forest is sometimes higher than that at the same height outside the forest.

At night the air temperature in a forest shows almost no difference in the trunk space, that is, between the forest floor and the forest canopy, though a slightly lower temperature is observable in the canopy. There is an obvious inversion in the upper part of the canopy layer. If the forest canopy is not dense enough, then the cold air falls down from the forest canopy towards the forest floor, and the minimum air temperature is observed near the forest floor.

Thus the upper surface of forest canopies acts as the *active surface*, as Voeikov pointed out in the 19th century. Heckert (1959) made two-year observation of oak forests in Potsdam, whose canopies form a layer from 10 to 15 m high above the forest floor. He measured air temperature at eight different heights—0.5, 2, 10, 12, 14, 15, 20, and 25 m above the forest floor—with instruments attached to a cable on a mast. An example of his results from a calm summer day is given in Fig. 3.42. According to these results, the inversion of the air temperature is seen all through the day in the trunk space in the forest, and even above the forest, lapse conditions are observed only between 8 h and 15 h 30 m local time. Thus, compared with the diurnal variations of the vertical distribution of air temperature on the earth's surface, the duration of lapse conditions above the forest canopy is short. This is understood to be due to the fact that there is a large amount of air and heat transferred vertically through the layer of forest canopies. To examine the distribution of water vapour and air temperature above the forest in the afternoon, it is necessary to pay special attention to the layer of the forest canopy, which plays an important role in the heat and energy budget.

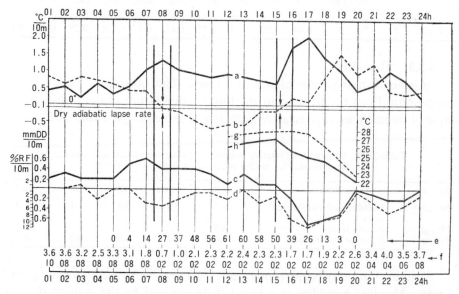

Fig. 3.42 Vertical gradient of meteorological elements in a deciduous broad-leaved (oak) forest at Potsdam on August 21, 1955 (Heckert, 1959). a: Air temperature gradient between the canopy and forest floor (°C/10m); b: That between the canopy and 25 m height (°C/10m); c: Water vapour pressure gradient between the canopy and forest floor (mm/10m); d: Relative humidity gradient between the same space (%/10m); e: Hourly total radiation (g cal·cm⁻²); f: upper values are wind velocity (m/sec) and lower values wind direction, 32=N; g: Air temperature at a height of 14 m above the ground. h: That at a height of 0.5 m. Arrow shows that the dry adiabatic lapse rate over the forest.

According to a study on the temperature profile in and above a tropical forest in NW-Colombia, it was shown that the lapse rate above the forest canopy is a direct function of the increase in temperature between the ground surface and the canopy (Baynton et al., 1965).

(C) *Ground Temperature*

Ground temperature in a forest in mountainous areas decreases in accordance with the altitude. Table 3.34 shows the results of an investigation carried on in 1950 at ten spots between 450 m and 1,800 m above sea level in the Great Smoky Mountains in North Carolina and Tennessee (Shanks, 1956). The lapse rate of ground temperature is greater in spring (May and June) and autumn (October and November) but smaller in the other seasons than that of air temperature. This depends to some extent upon the kind of trees in the forest and upon the seasonal change of air masses, but in general it is due to the fact that in summer at low altitudes above sea level the rate of rise in air temperature is greater than the rise in ground temperature in the forest, and in winter at high altitudes above sea level the rate of decrease in air temperature is greater than that in ground temperature in the forest.

Table 3.34 Lapse rate of ground temperature (15cm depth) and air temperature in a forest on the mountain slope (°C/100m) (Shanks, 1956).

	Jan.	Feb.	Mar.	Apr.	May	June	July	Aug.	Sep.	Oct.	Nov.	Dec.
Ground Temp.	0.43	0.46	0.55	0.58	0.58	0.59	0.55	0.51	0.48	0.43	0.41	—
Air Temp.	0.50	0.49	0.60	0.59	0.54	0.55	0.61	0.52	0.51	0.29	0.36	0.34

3.3.3. Humidity

In general, humidity is higher in a forest than outside it. This is due to transpiration from the leaves and the lower temperature in the forest. An example of monthly average relative humidity is presented in Table 3.35.

Table 3.35 Difference of relative humidity (%) between inside and outside of a forest. Plus means the inside is higher (Yoshino).

Region	Forest	Jan.	Apr.	July	Oct.	Year
Nikkô	Deciduous broad-leaved trees	3.4	3.2	−0.8	1.1	2.2
Myôgi	Needle trees	4.8	4.8	6.5	9.5	6.8
Tokyo	Japanese cedar trees	1.6	−1.1	1.5	0.5	0.8

Absolute humidity is more important than relative humidity for the growth of trees. Absolute humidity in a forest is determined by the content of water vapour from the ground and by the content of transpiration from the plants. Saturation deficit is determined by absolute humidity and air temperature. Accordingly, while saturation deficit influences evaporation and transpiration, the evaporation and transpiration influences the saturation deficit by causing changes in absolute humidity. Thus the problem of humidity in a forest is fairly complicated. Some examples of transpiration mechanisms in a pine forest were mentioned in Section 3.3.1 (D).

According to one observation, the difference in saturation deficit between inside and outside of a forest is comparatively great on fine days, and slight on cloudy or rainy days.

Vertical distribution of humidity is as follows: Humidity is great at the canopy and a little less below. It increases again near the ground surface and is the greatest on the forest floor. Above the forest canopy it decreases in proportion to height. The difference in humidity between the forest canopy and on the forest floor is 1–2% at night but begins to increase after sunrise, amounting to 7–8%. Between 11 h and 15 h (local time), the difference becomes a little lower, 5–7%, but it again increases from about 15 h and reaches 20–30% about 17 h.

This change of humidity is explained as follows: The maximum difference in the morning occurs after sunrise because then evaporation from the ground is greater than transpiration in the canopy. The maximum difference in the afternoon occurs because, in spite of increasing humidity with the fall of air temperature, exchange of the air at the canopy for free air is still going on, and thus humidity in the canopy is not as high as within the forest.

This diurnal variation is greatly affected by the general wind conditions, though it has some relation to the density of the tree crowns. Great areal variations are closely related to what extent the exchange of air takes place between the forest and outside. For instance, in mountainous areas, the influence of the slopes cannot be disregarded: In the case of the observation of Staufenberg (Hartmann et al., 1959), the differences between the driest and the most humid spots was from 8 to 10% in the morning, 14% at noon, and 16% in the afternoon.

Sekiguti (1951) observed the distribution of relative humidity in the Uchihara region with an area of 2 km × 2 km and found that it is 52.0% in forest and 51.0% in the bush land, in contract to 49.1% at paddy fields, 43.5% at grassland and 39.7% at bareland with clear weather on an autumn day.

3.3.4. Evaporation

Evaporation in a forest is related to air temperature, humidity, winds, and solar radiation. The discussion of transpiration from a pine forest in Section 3.3.1 (D) included the role of vapour pressure deficit, humidity.

The amount of evaporation on the surface of water in a forest, though it differs greatly according to the state of the forest, is 25–90%, but in most cases 40–50%, of the amount outside the forest. The difference depends on the kind of tree. An example obtained in Germany shows that the ratio of evaporation in forests of beeches, firs, and pines to the evaporation outside the forest is 55%, 40%, and 33% respectively.

Evaporation in the forest is an important factor in the renewal of a forest, because it is related to the budding and growth of seedlings through the soil moisture. The percentages above indicate the amount of evaporation from the surface of water, but not from the actual soil surface. Unfortunately, there is no detailed study of evaporation from the actual soil surface in a forest. It was observed that the ratio of the annual total amount of evaporation from the soil surface to the annual total precipitation is generally 30–50% outside of a forest, but it is 8% in a forest where the soil is covered with fallen leaves and 20% in a forest where the soil is not covered with fallen leaves. These values are in contrast to 15% and 4% on the same kinds of soil outside the forest, respectively.

If the amount of evaporation from the soil in a forest is compared with that outside the forest by a simple method (Lowry, 1956), the latter is four times as much as the former (Selleck and Schuppert, 1957).

The annual variation of evaporation from the water surface together with the relative humidity and mean air temperature in a mixed forest composed of deciduous broad-leaved trees (10–15 years old) is shown in Fig. 3.43 (Oosawa,

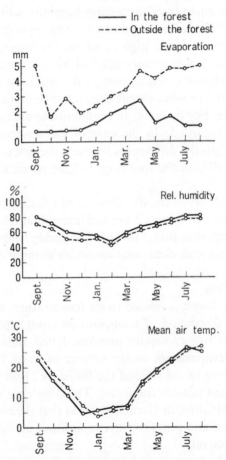

Fig. 3.43 Annual variation of evaporation, relative humidity, and mean air temperature in a mixed forest and outside the forest (Oosawa, 1950).

1950). The smaller differences in evaporation in winter and spring are found in association with the reverse relation of the mean air temperature from January to July.

Total evaporation from a forest is composed of (a) evaporation from the soil, (b) evaporation of precipitation intercepted by the canopy and (c) transpiration of the trees. The total evaporation from a fir forest differs from year to year, according to the wetness of the year, as shown in Table 3.36. (Fedorov, 1965). It is important that the evaporation of the intercepted precipitation is much greater in a wet year than in a dry year.

Table 3.36 Total evaporation (E), transpiration (E_{tr}), evaporation under the forest canopy (E_n), and evaporation of the intercepted part of precipitation (E_z) in a fir forest for May–September in years of different wetness (mm) (Fedorov, 1965).

Year	Index of aridity May–Sept.	Precipitation	E	E_{tr}	E_n	E_z
1953	1.02	620 mm	419	163	110	146 mm
1957	1.66	496	412	165	112	135
1962	1.00	527	414	131	115	168
Average		548	415	153	112	150
	%	—	100	37	27	36
1951	2.18	255	368	164	118	86
1959	2.92	223	334	124	120	90
1960	2.33	299	369	166	120	83
Average		259	357	151	119	87
	%	—	100	43	33	24

3.3.5. Precipitation

(A) *Interception of Precipitation by a Forest*

The rain that has fallen on a forest is intercepted by the forest canopy. Some of it evaporates there, and some of it reaches the ground either down the tree trunks or through the air as waterdrops.

The amount of precipitation differs greatly at different points in a forest, but is far smaller than outside of the forest. This is due to the intercepting action of the tree crowns. As a rule, the amount of precipitation intercepted by the tree crowns increases together with an increase of total precipitation, and the relation can be represented as a linear function. Therefore, the ratio of the amount of interception (I) to the amount of rainfall (R) is represented by a hyperbolic curve:

$$\frac{I}{R} = a + \frac{b}{R} \qquad (3.76)$$

where a and b are constants determined by the kind of the trees composing the forest. An actual observation shows that when the amount of rainfall is very small, I/R is approximately 100%, but when it amounts to 5–6 mm, I/R rapidly decreases, and when it is more than 10 mm, I/R is equal to 20–30% and almost invariable. Of the total rainfall (5–10 mm), 2–5% goes down the trunks of needle trees, whereas 5–20% of the total rainfall (1–5 mm) goes down the trunks of deciduous broad-leaved trees.

The amount of interception (I) is also related to the intensity and duration of the rainfall. When tree crowns are artificially watered, the amount of interception becomes almost constant for the first 5 min in the case of pine trees, and for the first 25 min in the case of broad-leaved (*Baccharis pilularis*) trees (Fig. 3.44). This is the maximum amount of water that can be held in the tree crowns (A+B+C) in calm conditions. The amount of interception after the rain clears away and the waterdrops fall off under calm conditions is (B+C). When more of the waterdrops have been shaken off by a strong wind, the amount held in the tree crowns decreases to (C). Grah and Wilson (1944) observed that in a forest the amount of interception that is held in the crowns is something between B and C.

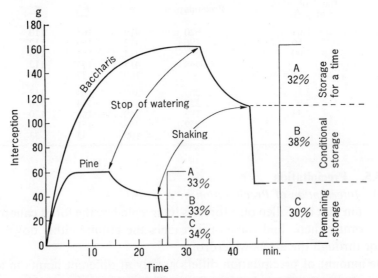

Fig. 3.44 An experiment on the interception of crowns of Baccharis and pine trees (Grah and Wilson, 1944).

The relationship between the amount of interception and the number of leaves is expressed generally by a linear function. This has been confirmed by the experiment by Ishihara et al. (1970) using pine (*Pinus thumbergii*), Japanese cypress (*Chamaecyparis obtusa*), and camellia (*Camellia japonica*). Furthermore, they have shown that, in the unsteady state of the interception process, the interception during the rainfall is expressed by the following equation:

$$\frac{dS}{dt} = A_t \cdot r + \frac{A_t}{\alpha} \ln \left(1 - \frac{S}{S_m}\right) \qquad (3.77)$$

where S is the amount of interception; A_t is the equivalent area of tree crowns that concerns the interception process; S_m is the maximum amount of interception; r is rainfall intensity, and α is constant.

In foggy areas, as the tree crowns catch fog particles by the phenomenon called fog precipitation, discussed below, and let the waterdrops fall to the ground, the rain gauge near the tree receives more rainfall than at a spot far from the tree. In calculating the difference of rainfall inside and outside of the forest, this phenomenon must be considered. The results of several observations are presented in Table 3.37. The amount of interception is greater in Japan than in Moscow because Japan has a greater amount of rainfall. It is greatest in a subtropical forest, probably because the rainfall is heaviest there.

Table 3.37   Amount of interception of rainfall in various forests.
(Alissov et al., 1956; Yoshino, 1961a).

Observed place	Observation period	Kind of forest	Age (Years)	Interception amount (%)
Moscow (Timirya-zev-Akad. Agr.)	1909–1929	Spruce	32–60	12
"	"	"	80–110	14
"	"	Fir	40–60	36
"	1907–1920	Birch	80–100	9
Brasil (22°59′S, 44°40′W, 622mH)	1908–1935	Sub-tropical forest	—	68
Japan (Many stations)		Japanese cedar* and cypress**	—	35
		Pine	—	18
		Deciduous broad-leaved trees	—	24

* : *Cryptomeria japonica,* ** : *Chamaecyparis obtusa*

Since the amount of precipitation intercepted by the tree crowns differs considerably from snow to rain, it is also influenced by the percentage of snow within the total amount of precipitation. The following paragraphs outline the relation between the forest and snow cover. The amount of snow caught in the tree crowns has been observed at Tôkamachi, Niigata Prefecture, Japan, which has the heaviest snowfall in Japan every year. The results of measurements of snow-load on trees were reported by the Laboratory of Snow Damage (1952) and Watanabe et al. (1964).

(B) *Forest and Snow Cover*

The influence of a forest on the distribution of snow cover arises directly from the prevention of snowfall by the forest canopy and indirectly by the facts that forests weaken the wind velocity, change the state of radiation balance and, accordingly, change the distribution of air temperature.

Near a tree the amount of snowfall begins to decrease inside the projection area of the crown of the tree and decreases further toward the tree trunk. The rate of decrease is large for coniferous trees like cedars, firs, and spruce and small for broad-leaved trees like beeches and oaks.

The amount of snowfall in a forest in comparison with that outside the forest is governed by the kind of the trees and density of the tree crowns. The amount of snowfall in the forest becomes scarce in proportion to density, and in the case of cedars, which have dense crowns, it is 20–50% less than outside the forest. On the other hand, in a forest of deciduous broadleaved trees, the amount of snowfall is often greater than outside the forest. Sometimes it is 1.5–2 times as great as that outside the forest. This is due to the influence of winds, discussed below.

In Japan, snow prevention forests and fences for railways and roads and snow shelter fences for houses are an indispensable means of protection against the winter monsoon carrying snow. Although they differ a little from each other in function, their purpose is to slow down the wind velocity and to decrease the amount of leeward snow accumulation.

When a wind blows at right angles to the running direction of forests or fences, in the case of a snow shelter forest, for example, a snowdrift (hill of snow) is formed there. This phenomenon should be clear from Fig. 3.45. The shape and the position of the snowdrift change according to the density of trees in the forest. On the lee side of the snowdrift, the depth of snow accumulation is small within and as well as outside the forest, as a result of the influence of the forest. On the lee side, at a distance of eight times the height of trees, the depth of snow accumulation is 50%, and at a distance of 8–20 times the height of trees it is the same as on the fields.

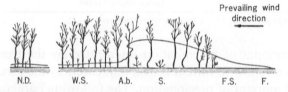

Fig. 3.45 Snow accumulation at the fringe of forests (Alissow et al., 1956). F: Field (Steppe). F.S.: Foot of snow drift on a field near the forest fringe. S: Snow drift. A.b.: Steep cliff leeward of snow drift. W.S.: Snow drift in the forest. N.D.: Snow accumulation in a forest of broad-leaved trees or in a forest parallel to the prevailing wind direction.

Considering the total water equivalent of snow accumulation in a drainage area, the influence of a forest is not significant; the forest only changes its distribution pattern within a limited area. However, the thawing season is retarded 15–25 days and 5–10 days in a forest of needle trees and deciduous broad-leaved trees, respectively, which are both 30–50 years old.

In the steppe in the shelterbelt of the U.S.S.R., this phenomenon of retardation is to retain moisture in the farm soil and protect the crops from severe cold by keeping them covered with snow for a longer time by changing the distribution of snowfall on the fields.

The influence of a shelterbelt on the snow accumulation differs according to the width of the shelter belt, the number of tree rows, and the distance between trees. In the case of a densely planted forest that hardly allows the wind to pass through, the wind blows only over the forest, and there are eddies, and snow heaps up where the wind is weakened. In the case of a sparsely planted narrow shelterbelt, through which the wind may blow, the depth of snow accumulation increases nearly constantly on the lee side. In the case of a pruned shelterbelt with intermediate width, the accumulation of snow intermediate between the two cases mentioned above.

Djunin (1963) expressed the snow transport, Q, by the following experimental equation:

$$Q = 0.0774 \times \bar{u}^3 \qquad (3.78)$$

where $\bar{u}$ is wind velocity (m/sec) at a height of 11 m. Flemming (1969) argues that the movement of snow cover by wind around the forest, snow drift, can be observed not only by the depth of snow accumulation, but also by snow density. The relationship between the depth of snow accumulation, the density of snow, and the density of trees in a pine-dominated forest was studied in Finland (Seppänen, 1961). The experimental equation obtained in that study is

$$w = -0.6n + 167 \qquad (3.79)$$

where w is water equivalent of snow in mm, (depth of snow) $\times$ (density of snow); n is the number of pine trees in an area of 400 m². This means that the greater the number of trees, the smaller the water equivalent of snow is. The coefficient and constant of Eq. (3.79) are different in the case of an area of 100 or 25m².

Photo 3.1 Lower part of trunk of Japanese cedar trees (*Cryptomeria japonica*) on a slope with heavy snow accumulation in Niigata Prefecture, Japan (February, 1972, by M. M. Yoshino).

There is a report by Kataoka et al. (1959, 1964) on bending of the lower part of afforested Japanese cedars (*Cryptomeria japonica*) as shown in Photo. 3.1, caused by creeping pressure and settling force of snow accumulation. Occurrence of bending depends on the inclination of the slope. If the slope angle is smaller than 5°, the percentage of bent trees is low. Further, it depends on the micro-topographical situation of the slope, convex, concave, or terrace.

Drooping branches are usually observed in snowy regions. Ishikawa et al. (1970) has investigated the relationship between the depth of snow accumulation and the shape of Japanese cedar and larch trees. By using this relationship reversely, they have shown that the depth of snow accumulation can be estimated for the mountain region. This will be discussed again in Sec. 6.1.4(E).

(C) *Fog Precipitation or Kisame*

When fog is thick, large water drops fall from the leaves of trees to which fog particles have adhered. This is called fog precipitation, or *Kisame* in Japanese. *Kisame* means "tree-rain."

A fog prevention forest is a forest used to catch fog by the application of this phenomenon. The structure of fog prevention forests is explained below. It has been explained that tree crowns intercept rainfall, but this is limited to the case when the rain is heavy. When the rain is light or under dense fog conditions, there is less interception, and in a dense fog the amount of precipitation in a forest increases because of fog precipitation.

The fog precipitation measured at Mt. Oodaigahara in Nara Prefecture, Japan, from April to October in 1922 was as follows (Mayama, 1923): Fog precipitation was remarkable in a forest of 40-year-old firs, including beeches and maples, with a density of 90%, a breast height diameter from 12 to 25 cm, and heights of 7–8 m. The amount of precipitation in July is shown in Table 3.38. From this table it should be clear that the amount of precipitation is about 10% greater in a dense fog in the forest than outside the forest. The total amount inside the forest from April to September in a dense fog is 2% greater than outside the forest. Similarly, as shown in a table by Rubner (1932), the amount of fog precipitation was 13% of the total precipitation outside the forest for the

Table 3.38 Difference of rainfall between the inside and the outside of the fir forests on Mt. Oodaigahara (Mayama, 1923) and of the spruce forests on Mt. Erz (Rubner, 1932).

Observation period	Oodaigahara		Mt. Erz	
	With dense fog	Without dense fog	With fog	Without fog
	July 1–31, 1922		May 7–Oct. 27, 1931	
Precipitation outside the forest	469.4 mm	26.3 mm	263.8 mm	286.2 mm
Precipitation in the forest	517.1	18.1	298.3	218.6
Difference	+47.7	−7.8	+34.5	−67.6
% Compared to the precipitation outside the forest	+10%	−29%	+13%	−24%

period from May 7 through Oct. 27, 1931, in a spruce forests of Mt. Erz, Germany, 450 m a.s.l. The figures are given in Table 3.38.

As in the case of a fog prevention forest, the catching amount by the forest increases with increasing the wind force. The relation between the wind force and the amount of precipitation in a dense fog both in and out of a forest is shown in Table 3.39. When the wind force is four or more, the amount of precipitation is greater in the forest; that is, it is found that the increase in precipitation caused by fog precipitation becomes greater than the amount of precipitation intercepted by the tree crowns.

Table 3.39 Differences of rainfall amount in relation to wind force change between inside and outside of a forest under the thick fog conditions (Mayama, 1923).

Wind force	Rainfall in the forest	Rainfall outside the forest	Difference	Percentage compared to the amount outside the forest
5	306.9 mm	275.3	31.6	+11%
4	681.2	608.2	73.0	+12
3	195.8	249.2	−53.4	−21
2	135.2	159.0	−23.8	−15

Grunow (1965) also reports on the statistical results of precipitation in a forest and the fog precipitation. According to this report, the amount of fog precipitation is greater in seasons when the fog occurs frequently.

3.3.6. Fog Prevention Forest

Fog prevention forests work by turbulent diffusion, by warming up the temperature, and by catching the fog particles. Concerning these functions, numerous studies were made in Hokkaido, Japan, in 1950–1953 (Forest Section, Hokkaido Government Office 1951, 1954). Following is a summary of these studies.

The three functions mentioned above are thought to take place simultaneously. The function of turbulence diffusion is effective in the case of a comparatively low fog. The thermal effect is seen when the temperature inversion is destroyed (Imahori, 1952). The collection efficiency (=collection amount per 1 g/m³ of liquid water content of fog) is different on a forest from that before and behind the forest, according to an observation in the Ochiishi area made in June, 1951 (Ooura, 1952). On the forest, collection amount was 0.37 $g \cdot m^{-2} \cdot sec^{-1}$ at most. But, as the amount varies widely from place to place, 0.04 $g \cdot m^{-2} \cdot sec^{-1}$ was observed at another point at the same time. This is caused by the fact that the size distribution of fog particles differs greatly and the wind velocity also varies markedly from place to place. Before and behind the forest, when the wind velocity is about 6 m/sec and the fog water content is 200–250 mg/m³, the collection amount is 5 kg/hr in the case of the forest 1 m in width.

Because of this effect, the fog water content becomes smaller at the back of

Table 3.40   Fog water content at the leeward side of the fog prevention
forest at Ochiishi (Fukutomi et al., 1951).

	Front (windward fringe) of the forest	Lee of the forest (from the fringe)		
		90 m	170 m	340 m
Ratio of fog water content	1.00	0.46	0.60	0.53

the fog prevention forest than in front of it. The results of an observation in Ochiishi is presented in Table 3.40, from which it will be clear that the catch amount decreases by 50% behind the fog prevention forest compared with that in front (Fukutomi et al., 1951). Of course, the rate of decrease is related to the wind velocity and the absolute amount of fog water at that time.

Tabata et al. (1952) made clear that the fog water content decreases only at the canopy to about 80%, but at 3 m above the canopy they remain unchanged compared with stations in front of the forest. In the canopy, fog water content was not only decreased, but the phase of the temporal variation was found to be lagging behind that in the free atmosphere.

Small unsheltered opennings in a fog prevention forest have special climatic relationships with low wind velocity, higher air temperature in the daytime, and accordingly, low fog water content (Onodera et al., 1954). On the morning of July 22, 1953, fog quantities outside the forest were at 145 mg/m³, whereas the average fog quantity in several openings (about 0.52 ha) was 48 mg/m³.

3.3.7.   Wind

As pointed out by Flemming (1967), the three important and basic climatic functions in the canopy layers of a forest are radiation, wind, and precipitation. Quantitative effect of the wind is the weakening or braking action of the forest. Only in some special cases, such as in a small space between two narrow shelter-belts with parallel wind direction or in a gap in the shelterbelt with the wind direction at right-angles to the running direction, does the wind become strong. As a quantitative effect, the transformation of the turbulence spectrum is important in connection with the energy budget of the forest.

(A)   *Wind Velocity in a Forest*

Woelfle (1939, 1942, 1950) deals very extensively with the wind conditions in and above a forest. Summarizing the results, he states that the wind velocity near the fringe of a forest with thick middle and lower layer vegetation is 60–80% of that outside of the forest. He also deals with the gustiness in and above the forest, which causes the vertical exchange of air. Baumgartner (1961) studied the wind field in relation to the forest.

Inside a forest the wind velocity decreases exponentially down to zero with increasing distance from the fringe of the forest. According to an observation in a spruce forest, the rate of the wind velocity in the forest to that at the fringe of the forest is from 55 to 75%, from 24 to 27%, 7%, and only 2% at spots 50, 70,

100, and 200 m from the fringe of the forest, respectively. Needless to say, the percentage differs according to the kind of trees, the density of crowns, and the height of the trees.

The vertical distribution of the wind velocity above the forest also differs greatly according to the conditions of the forest. Detailed structures have recently been studied in relation to the energy budget, as mentioned above. Roughly speaking, the forest raises the wind field of the open land surface up to an altitude of two-thirds the height of the trees.

Figure 3.46 shows the vertical distribution of the wind velocity in and above a forest of oaks in Potsdam (Heckert, 1959). When the oaks have leaves on, the rate of weakening is 10–20% greater, but even then just above the layer of the tree crowns the wind velocity suddenly increases.

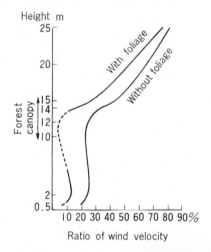

Fig. 3.46 Vertical distribution of wind velocity in and above an oak forest. Wind velocity is expressed by ratio to the tower values (Heckert, 1959).

Kalma and Stanhill (1972) measured aerodynamic characteristics of an orange orchard of 35-year-old *Citrus sinensis* (Average tree height was 4.25 m and planting distance 4×4 m). The average values of z_0 and d of Eq. (3.22) were 40.4 cm and 286 cm, respectively.

Martin (1971) observed the winds above and in a pine forest. It was proposed that the ratio of the wind speed in the trunk space to the wind speed above the canopy is a measure of the magnitude of momentum transfer through the canopy. This ratio reaches a maximum at noon and drops at night to 75% its daytime maximum value.

Since the forest is never horizontally even, the velocity in or above the forest differs greatly according to the wind direction. The difference of the velocity is greater especially above the forest.

(B) *Shelterbelts and Wind Breaks*

There are many comprehensive monographs on shelterbelts and wind breaks. The followings listed in chronological order, are the important studies done after 1945: monographs by Nägeli (1946), Woelfle (1950) and Kreutz (1952), experimental studies of many functions of shelterbelts (Iizuka, 1952a, b, c); a study of wind break effects on various meteorological conditions by van Eimern (1957) and by Eskuche (1957); a thorough study on the problems of shelterbelts and microclimate by Caborn (1957); a wind tunnel test of the model shelter-hedge by Tani (1958); a report prepared by the Working Group of WMO on "Windbreaks and Shelterbelts" (van Eimern et al., 1964); and a paper on wind conditions in the range of staggered shelterbelts by Nägeli (1965). However, discussion here is limited to the phenomena that are important for understanding the horizontal distribution of climatological variables.

The effect of a shelter belt, that is, the influence of the forest on wind velocity is greater at night than in the daytime. According to the results of numerous studies, the scope of the influence is 2–10 times the tree height on the windward side, and 8–60 times on the leeward side. The reasons for such great differences in its influence may be that the extent actually influenced fluctuates, and that the criteria of the observations themselves are not always the same. The tree height, the permeability (also called penetrability or density of trees), and the width of the shelterbelt differ from one observation to another, and the

Photo 3.2   Wind break of poplar trees on the lower part of Rhône river in Switzerland (June 9, 1963, by M. M. Yoshino).

height where anemometers are placed is not uniform. Besides, definitions of the weakening degree of the wind velocity very often differ.

The weakening degree of the wind velocity by the forest is usually represented by a ratio to the wind velocity on open fields. A suitable height for the anemometer is said to be 1.4 or 1.5 m above the ground. At this altitude, the distance of the point where wind velocity decreases by 10% is 5 times the tree height on the windward side and the 25 times on the leeward side. The permeability is related to the rate of the weakening of the wind, but not strongly related to the limit distance of influence; that is, behind of a dense shelterbelt, the wind velocity diminishes greatly, but it soon regains its velocity. The effect of wind prevention decreases when the wind has a high velocity. According to a study by Caborn (1957), the ratio of the width (thickness) to the height in the wind break has a significant effect on determining the extent and nature of the area to be sheltered on the leeward zone. This may become apparent only when the degree of permeability to the wind falls below a critical value, estimated to be 20%. Wide belts appear to lead the wind parallel to their upper surfaces with consequent rapid downward transfer of energy after leaving the leeward edges and restriction of the leeward eddy zone, giving rise to early resumption of the unobstructed with velocity and a reduction of the protection distance.

Iizuka (1952a) states that the resistance offered by the shelterbelt has greater dependence on the sum of the tree diameters at breast height than on the average diameter. However, the ratio width/height may be a better index for practical purposes.

The plantation of shelterbelts was scheduled in the U.S.S.R. to prevent dry wind. Here the results of A. P. Konstantinov's study are introduced after Sapozhnikova (1950) and Alissow et al. (1956). The effect of the shelterbelts is summarized in Figs. 3.47 and 3.48. Figure 3.47 presents a case where the wind blows

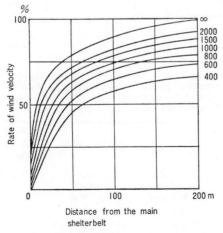

Fig. 3.47 Change of wind velocity according to the intervals of cross-shelterbelts. The sign ∞ means no cross-shelterbelts (Alissow et al., 1956 after Konstantinov).

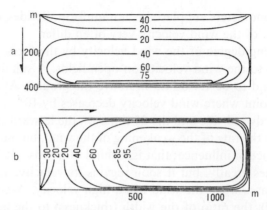

Fig. 3.48   Change of wind velocity in an area with mesh size (400 m × 1,200 m) in the case of tree height of 16 m (Alissow et al., 1956 after Konstantinov). a: In the case of cross-winds; b: In the case of parallel winds.

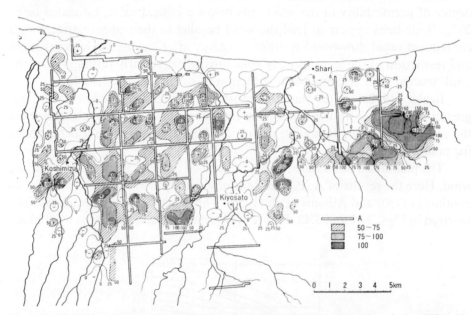

Fig. 3.49   Density (m/ha) of shelterbelts in the Shari-Abashiri region (Yoshino et al., 1972a). For instance, Density 25 means the total length of the shelterbelts of Grade 2 is 25m/ha. Grades 1–6 are given in the following table. Criteria of shelterbelt width on the airphoto in scale of 1:25,000. *A* denotes national shelterbelt wider than 60 m.

Grade	Width on the airphoto	Actual width	Weight in the calculation of the total length
1	A line of dispersed points	One row dispersed	×0.5
2	0.2mm	ca. 5 m	×1.0
3	0.4	10	×1.5
4	1.0	25	×2.0
5	1.5	35	×2.5
6	2.0	50	×3.0

parallel to the main shelterbelts with an interval of 400 m. The abscissa designates the distance from the main shelterbelts, and the axis of the ordinate
designates the wind velocity (the ratio of the wind velocity to a place free from
the effect of the belts), according to the intervals of subshelterbelts, which run
at right angles to the main shelterbelts. The sign infinitive (∞) stands for the
case where there are only main shelterbelts. For example, at a spot 50 m away
from a main shelterbelt with no cross-shelterbelts, the wind velocity weakens to
80%. In the case of the standard shelterbelts (400 m $\times$ 1,200 m), that is, with an
interval of 1,200 m between the cross-shelterbelts, the wind velocity weakens to
67%. Figure 3.48 summarizes these figures, showing the distribution of the wind
velocity in the standard area surrounded by the shelterbelts at intervals of
400 m $\times$ 1,200 m. A problem similar to the wind velocity distributions in clear-
ings surrounded by forest is treated by Flemming (1968b).

When it is impossible to make straight shelterbelts because of land owner-
ship or topographical conditions, shelterbelts must be built along the borderlines
of farmland or along roads. In Germany, a plan of shelterbelts and wind breaks
was drawn up in many areas, based on detailed observations of the wind and air
temperature. One example is discussed in the report for Ohmenheim and the
surrounding region, S-Germany, by Aichele (1957).

Where stronger winds blow very frequently in all seasons, an impressive
landscape such as *Heckenlandshaft* or *Knicklandschaft* in Schleswig-Holstein,
N-Germany, develops. In the Shari-Abashiri region, Hokkaido, Japan, there are

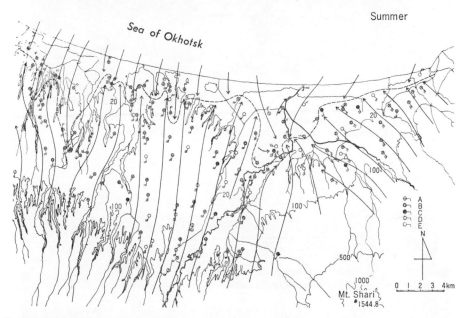

Fig. 3.50 Prevailing wind distribution in summer as revealed by windshaped trees in the Shari-
Abashiri region. A: *Larix leptolepis*, B: *Fraxinus mandshurica*, C: *Populus sp.*, D: *Betula platyphylla*,
and E: others. (Yoshino et al., 1972a).

dense shelterbelt nets, which were studied by means of airphoto interpretation (Yoshino et al., 1972). In this study, the wind distribution was observed in detail using wind-shaped larch trees as an index of wind velocity (see Section 6.1.3), and the wind distribution was compared with the distribution of shelterbelt density. The results are given in Figs. 3.49 and 3.50. From these figures, the following conclusion can be made: The Shari-Abashiri region can be divided into three parts from the viewpoint of wind conditions: (a) the northern foot region of Mt. Shari, (b) the coastal region and the marshy region along the Shari River, and (c) the hilly region between the cities Koshimizu and Kiyosato. The characteristics of these regions are: In the first region, the density of shelterbelts is 100 m/ha on the average with a maximum of 150–200 m/ha. Up to 100 m above sea level at the foot of Mt. Shari, a southeasterly wind prevails.

Photo 3.3 Shelter of beech (*Fagus silvatica*) trees around a farmer's house near Monschau, W- Germany (June, 1963, by M. M. Yoshino).

Photo 3.4 Shelter of pine trees (*Pinus thumbergii*) around a farmer's house in the Izumo plain, Shimane Prefecture, Japan (October, 1968, by M. Ishii).

This wind, which blow especially in spring, is called *Sharidake-oroshi* (fallwind from Mt. Shari). The direction of the shelterbelts is mainly from SW to NE. In the second region, the density of shelterbelts is less than 25 m/ha. In this coastal region, lower than 20 m above sea level and 2–3 km wide, northerly winds from the Sea of Okhotsk prevail. Where the marshy lowland extends 8–10 km inland from the coast, the northerly winds invade into this region. In the third region, the density of shelterbelts is mainly 50–75 m/ha. Almost all wind breaks run in a W-E direction. Shelters on the different micro-topographic conditions show a sharp contrast as shown in Photo 3.5.

Mäde (1970) proposes an agrometeorological research method, and among his "climatological" problems, of special interest are these two questions: "What is the best angle between the running direction of the wind breaks and the prevailing wind directions?" and "What is the most suitable length of wind breaks for practical purposes?"

Shelterbelts and wind breaks exercise a great effect on other climatological elements, such as insolation, radiation, air and ground temperature, humidity, dew, fog, and evaporation. However, the effect of wind breaks on crop yields differs from year to year. For instance, according to the yield data of an irrigated sugar beet crop with two rows of corn windbreaks spaced at 15 m, yield increases of as much as 25% were possible during years when yields were low. During the years of high yield, however, the windbreaks did not increase sugar beet yield (Brown et al., 1972).

3.3.8. The Forest and Run-off

The influence of forests on run-off is seen more clearly in the retardation in time of run-off rather than in the amount of run-off. A heavy rainfall within a short period is prevented to a large extent from permeating into the soil by the interception of the forest canopy. Depth of snow accumulation is relatively small in a forest, but the melting of snow is retarded. For example, according to observations by Rutkovskii (1948) at 2,000 spots in European Russia, snow melted 5 mm/day in spruce forests, 10 mm/day in bushes, and 16 mm/day on farmland. Accordingly, the run-off of melted snow is also retarded in this relation.

Table 3.41   Monthly change of specific run-off in different vegetation regions. Average of 1948–1953 (Maruyama et al., 1952).

Area	Vegetation	Area	Dec.	Jan.	Feb.	Mar.	Apr.	May	Mean of winter half-year
Forest	Japanese cedar 40 years' old and broad-leaved trees	3.2ha	62%	38	63	175	237	48	95%
Without forest	Upper-layer trees were cut down and lower-layer vegetation was cut twice a year	2.5ha	63	40	72	202	225	47	99

Photo 3.5 Contrast of citrus orchards between the east-facing slopes (left hand) without shelters and the west-facing slopes (right hand) with shelters against ↗

Table 3.41 shows the monthly variation in the rate of run-off in areas with and without forests in Kamabuchi, Yamagata Prefecture, Japan. The rates are greater up to March in the forest area, but in April and May, the area without forests has greater rates. But attention should be paid to the fact that the average rate of run-off during the six winter months is greater in the area without forests.

In a small watershed area (572.9 ha) in Hokkaido, Japan, the annual run-off (R) and precipitation (P) were measured before and after cutting down the forests (Katsumi, 1956). The experimental equation for the relation before the cutting is

$$R = 0.704P - 297 \qquad\qquad (3.80)$$

and the relation after the cutting is

$$R = 1.105P - 740 \qquad\qquad (3.81)$$

where R and P are expressed in mm. If Eq. (3.80) is compared with Eq. (3.81), a great deal of run-off is expected after the cutting. As a matter of fact, the run-off increased year by year after the cutting during the years 1951–1953. The rate of increase is greater in summer than in winter.

In general, the ground water outflow is difficult to measure in a drainage area. Ishihara et al. (1971, 1972) measured the outflow from a joint of exposed rock on the slope of the Ara Experimental Basin, Japan, in order to estimate the ground water outflow, while calculating the water balance in the basin.

The question of whether or not the runoff from a drainage area with a forest is more than the run-off from a drainage area without a forest remains. Some reports—Keller (1968) among them—conclude that there is an increase of the run-off in forest areas, but many other investigations suggest the opposite conclusion. Here I would like to point out the necessity for comparative studies: (a) the influences should be analyzed by considering not only the absolute amount of run-off, but also the specific run-off; (b) in this analysis, the homogeneity of geology, topographical configuration, climate, vegetation and, of course, the forest structure must be taken into consideration; (c) this analysis must be made

↗ strong W-winds in winter in Nishiura near Numazu
City, Shizuoka Prefecture, Japan. Shelters are pine trees
(*Pinus thumbergii*). (March, 1952, by M. M. Yoshino).

according to the class of precipitation intensity for every month or season, because the climatic conditions, especially the frequency distributions of precipitation intensity, and the forest structures differ greatly from month to month; (d)
the run-off phenomena before and after afforestation or cutting in the same
drainage area must be compared using standard (unchanged) conditions in another area, because of the broad climatic changes or anomalies in different
years; (e) the topographical conditions together with the drainage structures
must be expressed quantitatively in the comparison (this must be done not only
for the drainage area under consideration but also for the surrounding regions);
(f) this problem must be solved in relation to the water balance in the whole
drainage area for many years. The results measured according to the same run-
off model only can be compared.

3.3.9.   Influence of Forests on Climate

The influence of forests on climate has been of great climatological interest
since the earliest days of its study. However, there are still many problems to be
solved. The excellent paper by Turner (1968) is a thorough review of the literature on this subject. He asserts that the urgent problems are the assessment of
forest influences and topoclimatological influences on the actual forest climates.

Areas near the fringe of a forest, where trees are scattered, do not present an
intermediate climate between the climate in a forest and that on open land far
from the forest, but a considerably different climate from either of them. What
is characteristic of the climate of the areas near the fringe of a forest is that it is
far milder than that of other areas. According to observations, the air temperature is 3°C higher there than inside and outside the forest. In addition, the air
temperature at night in an area surrounded by forests with sparsely planted low
trees, whose angle of elevation is lower than 50°, is about 4°C higher than on
treeless land. These phenomena suggest that, because the forest canopies are not
dense, the solar radiation reaches down to the lower part of the crown, a half
to two-thirds of the solar radiation is absorbed there, and the amount of long-

wave radiation, which goes back to the sky, is lessened. This fact is more significant when the sun height is low. The crown of a single tree is also significant in
its effect on heat balance on a vertical surface of tree crowns. This is the case, as
has been pointed out by Egli (1951), we are to locate high buildings with broad,
vertical space in a city in high altitudes for the purpose of utilizing heat from the
sun.

The Sierra Nevada, California, including the Western mountainous areas
around it, is covered with deep snow in winter and spring. But, as skiers and new
settlers who come to this region point out, in spite of a deep, persistent snow
pack, nights are moderately warm and days are warm in this region. The conditions are the same in the Apalachian Mountains. A comprehensive study has
been made of this phenomenon in the Sierra Nevada (Miller, 1955). The distribution of trees is sparse, and 40–45% of the total solar radiation on the tree
crowns is permeated, 8% is reflected into the sky, and 50% is absorbed. The
tree crowns, moreover, absorb part of the radiation that penetrates into them
and part of the radiation that is reflected from the surface of the snow beneath
the crown. As a result, it is indicated that in this area, the albedo has 0.45–0.50
on the surface of spring snow, 18% of the total solar radiation is reflected, 25%
is absorbed by the snow and 57% is absorbed by the tree crowns. Of the heat
from the absorbed insolation: 4% is used for the growth of the trees, 25% for
the latent heat in the process of transpiration, 47% for the convection of the air,
14% as a loss in the process of long-wave radiation into the sky, 8% for the
exchange of long-wave radiation with the snow, and 2% is held over from daytime to nighttime within the branches or trunks. In the daytime on fine days in
winter and spring, 15–25 cal·cm^{-2}·h^{-1} is transported into the air by convection
and 5–15 cal·cm^{-2}·h^{-1} by transpiration in one of the standard forest stands in
the area. The heat balance within 1 hour around noon in a forest in this area is
given in Table 3.42. The high air temperature of this area in the daytime is
thought to be caused by the great quantity of heat that is carried away into the
air.

Table 3.42   Heat balance in the whole forest stands in Sierra Nevada
around noon in the case of fine weather (Miller, 1955).

	Winter	Spring
Solar radiation	50 cal·cm^{-2}·hr^{-1}	90 cal·cm^{-2}·hr^{-1}
Distribution of solar radiation		
Reflection from trees	2	4
Reflection from snow surface	11	12
Absorption by snow surface	6	23
Absorption by trees	31	51
Loss of heat absorbed by trees	31	51
Into air by convection	17	24
Latent heat by transpiration	5	13
Long-wave radiation into sky	5	7
Long-wave radiation to snow cover	3	4
Photosynthesis	1	2
Amount reserved within branches or trunks	0	1

Hare and Ritchie (1972) present a profile from Dawson Creek, B.C., to Cambridge Bay, N.W.T., Canada, showing the thermal parameters such as mean annual net radiation, mean net radiation during the seasons with air temperature above 0°C, Thornthwaite's mean annual P-E function, and the duration of the seasons with air temperature above 0°C. Among these parameters, mean annual net radiation has a strong poleward gradient in the open woodland zone and near the border between the open woodland and tundra zone. Mean net radiation for the growing season shows, on the one hand, a sharp decrease in the open woodland zone. This is because of the low albedo of the dense forest zone, which extends from the open woodland zone on the lower latitude side. This results in the rapid warming in spring. On the other hand, in the tundra zone the snow remains intensely reflective and hence, cold. As a result, a sharp gradient is seen in the open woodland zone.

The influences of forest on the water balance have been studied since the second half of the 19th century as mentioned earlier. Recent development of the studies concerning this problems, as well as on run-off, snow cover, snowmelt, interception, underground water, coastal vegetation, water quality, modeling, etc., was summarized in great detail by Keller (1968).

Holmes (1970) measured the *oasis-effects* of forests on the Cypress Hills situated in SE-Alberta and SW-Saskatchewan, Canada. The oasis-effects produced by the relatively more moist and forested hills are apparent when compared to the semi-arid surrounding prairie: Air flow over the forests, air temperature, dew point temperature, and ground temperature showed conspicuous effects.

3.4. SEASHORES, LAKESHORES, AND RIVERBANKS

3.4.1. Air Temperature

As land and water have physically different characters, seashores, lakeshores, and riverbanks, which are the borders between the two, show peculiar climatic conditions. The lakeshores and riverbanks that are discussed here are, of course, those of great lakes and rivers.

The characteristic feature of air temperature in these areas is that the maximum temperature in the daytime is relatively lower and the minimum at night is warmer than in the inland areas. This means that they have *coastal climate*. This phenomenon is explained by the fact that, since the thermal rise caused by insolation in the daytime is smaller above the water surface than on land, the maximum air temperature above water is low and also by the fact that because the cooling by long-wave radiation above the water at night is smaller than on land, the minimum air temperature is high.

However, windless conditions are seldom in actual cases. From the standpoint of local climatology, winds in those areas play an important role. That is to say, the lowering of the maximum air temperature is related to the fact that the breezes on oceans, lakes, and rivers blow inland in the daytime and lower the air temperature. This point is discussed in Section 3.4.2 below. The following section is concerned with the high minimum air temperature in these areas.

(A) *Air Temperature on the Seashore*

The air temperature on the seashore at night is characterized by the wind that blows from land toward the sea; thus the surrounding topography is important. For example, on a seashore with steep slopes facing the sea, the influence of the sea is great, but in an area where a river running from inland facilitates the blowing of mountain or land breezes from inland, the influence of the land is great. In Hiroshima, situated 20 km away from Kure, cold air flows down along the Ota river, which valley has 1–2 km wide, and the minimum air temperature falls. The mean monthly minimum air temperature in April, 1952–1954, was 1.2°C lower, and the mean minimum air temperature on clear nights in April was 1.74°C lower than in Kure, where the surrounding mountain slopes are steeper (Shitara, 1955).

When there is a counter air current over the land breeze, a local line of discontinuity is frequently formed. For instance, in Fukuoka, Kyushu, Japan, air temperature sometimes suddenly rises about 5°C in the early morning in winter (Okanoue, 1950). This is caused by a weak discontinuity line between the cold land breeze (below 1.8 m/sec) and the comparatively warm winter monsoon (0.7–2 m/sec), passing from north to south over Fukuoka, when the monsoon becomes strong (sometimes more than 5 m/sec) or when the land breeze be-

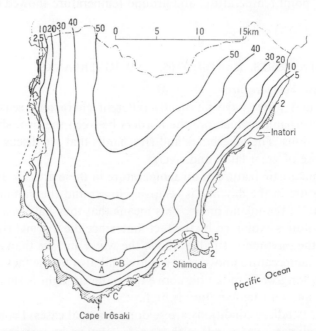

Fig. 3.51 The frostless zone in the southern part of Izu Peninsula, Central Japan (Sasakura et al., 1956).

comes weak. The influence of this land breeze explains the fact that as fishermen in these areas say, in Hakata Bay, north of Fukuoka, the farther they are from Fukuoka, the warmer it becomes and the loquat trees on Shiga Island, north of Fukuoka, bear better fruit in the northern part of the island, which is protected from the cold land breeze.

The seashore area is comparatively warm at night and early in the morning, because of the weaker influence of cold air currents from inland, as mentioned above, and because of the weaker development of the inversion layer. Because the friction on the sea surface is not as great as on the land surface, winds of the general pressure field are usually stronger on the sea than on land. For this reason the inversion layer does not develop enough on the sea, and therefore, the decrease of air temperature in the air layer near the ground is weakened. The sourthern part of Izu Peninsula is an exceptional area in Japan in that it has very little frost, as shown in Fig. 3.51 (Sasakura et al., 1956). This is thought to be due to the fact that the inversion layer can hardly develop there because the winter monsoon, a relatively warm, wet wind from the W or WSW, is especially strong.

Characteristics of the distribution of air temperature are clearly reflected in the distributions of frost on the seashore. The distribution of air temperature shows a great difference between inland and the seashore, especially when there is little frost. According to an investigation in the Fenland region, England, the frequency of frost after a certain day, F, is given in the following experimental equation (Lawrence, 1952):

$$\log F = -5.55 + 0.18T + 0.015x - 0.15t \qquad (3.82)$$

where T stands for the degree of frost, x the distance in miles from the coast, and t the number of days (counted in weeks) since April 1. The value -5.55 is determined by the condition of the land. T is expressed by values such as 32°, 30°, and 28°F, and naturally the frequency decreases in the case of lower temperature frost.

Neuber (1970) reports on the influence of the Baltic Sea on the annual mean air temperature and frostless period by statistical tables.

Even in a small peninsula, air temperature varies considerably. The Dale Peninsula of England, situated at approximately 5°10'W, and 51°42'N, has an almost uninterrupted exposure to oceanic influences from the south and west. A comparison of air temperature between two spots on this peninsula—Dale Fort (33 m a.s.l.) and St. Ann's Head (43 m a.s.l.)—is given in Table 3.43 (Oliver,

Table 3.43 Deviation of air temperature of Dale Fort readings from St. Ann's Head for 1950 (Oliver, 1959).

	Jan.	Feb.	Mar.	Apr.	May	June	July	Aug.	Sep.	Oct.	Nov.	Dec.
Mean min.	−1.2	−0.6	−0.2	−0.8	−0.5	−0.1	−0.2	−0.2	−0.7	−0.3	−0.8	−0.3°C
Mean max.	—	0.6	1.2	1.2	1.9	1.7	1.2	1.3	0.7	0.8	0.9	0.9
At 9 h	0.1	0.1	−0.4	0.3	−0.4	0.3	−0.2	0.1	−0.1	−0.3	0.1	0.1

1959). The two spots are 2.7 km away from each other: St. Ann's Head is at the southern end of the peninsula, and Dale Fort at the east extending small cape. Table 3.43 shows, in Dale Fort the minimum air temperature is lower and the maximum air temperature is higher than in St. Ann's Head. This is because St. Ann's Head is more exposed to the westerlies; in other words, it undergoes more oceanic influences, and has a shorter duration of sunshine. The deviation of air temperature is obvious, though slight. In addition, this deviation varies from year to year.

The diurnal change and the sea-land profile of air temperature near Zingst in June, 1966, is given in Fig 3.52 (Nitzschke, 1970). The sharp increase of temperature inside the shore occurs from 10 m to 2 km. Such phenomena are mentioned again in Section 3.4.2(c).

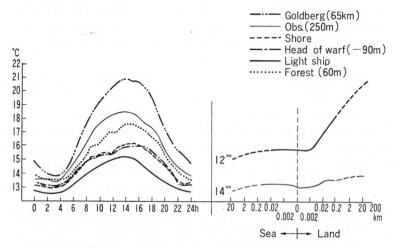

Fig. 3.52 Monthly mean diurnal change and sea-land profile of air temperature near Zingst, Germany, in June 1966 (Nitzschke, 1970).

(B) *Air Temperature on Lakeshores and Riverbanks*

In a region near a great lake, there is a phenomenon similar to that near the sea. The influence of the Great Lakes appears in the whole surrounding areas. For example, in January, the average air temperature is 2.8°C and the average minimum 5.6°C warmer on the lakeshore, and in July the average minimum is 1.7°C lower on the lake shore than in the area far from the lakes. The period with no frost is 30 to 40 days longer in the areas near the lakes than in areas far from it (Visher, 1943). The result of observations of temperature at various stations around Lake Erie is presented in Table 3.44. The farther the station is from the lakeshore, the greater the range of air temperature becomes and the shorter the frostless period becomes (Verber, 1955).

Investigations on the effects of the Great Lakes' thermal influence on freeze-free dates in spring and autumn and on seasonal temperature anomalies

Table 3.44 Climatic variations within 80 km from the Lake Erie shore (Verber, 1955).

	Distance from L. Erie	Average monthly range for 1950	Frost-free season
Put-in-Bay	0 km	7.6 °C	205 days
Sandusky	1.6	8.8	194
Toledo	24	10.4	185
Tiffin	48	10.8	162
Bucyrus	80	11.9	154

were made by Kopec (1966, 1967). Almost everywhere along the shores of the Great Lakes the freeze-free season is advanced and the pattern of isochrones runs parallel to the coastlines. However, the wind-facing shores along some lakes register a greater advance in the start of the freeze-free season than the wind-shadow shores. Based on Hopkins' isophanic model (Hopkins, 1938) and the continentality distribution in the Great Lakes Region (Kopec, 1965), the following three influence regions are defined by Kopec (1967): (a) the region of maximal lakes' effect: the influence of the lakes on air temperature element is seen at all times, (b) the region of mean lakes' effect: the average spread of the lakes' thermal influence, and (c) the region of minimal effect: the influence of the lakes on any air temperature element is measurable at some time during the year but not continuously.

In the case of s smaller lake, the effect is, of course, not so obvious through the year. Shitara (1967) shows results observed in the area around Lake Inawashiro (104 km^2), Fukushima Prefecture, Japan. It is cooler on the wind-facing shore (wind direction from lake to land) than on the wind-shadow shore (wind direction from land to lake) by 2.8–4.5°C on a clear day in summer. The air temperature difference between the wind-facing shore and the wind-shadow shore, Td, is expressed by the following equation:

$$Td = a \log x^2 + b \tag{3.83}$$

where x is the distance from the wind-facing shoreline at the inland station in the range <1.0 km. It is clear, also, that the lake breeze becomes warmer with the magnitude of 2.5°C within 3.5 km from the shore.

The horizontal profile of air temperature at 1.5 m level from the lakeshore of the wind-facing side to the point 800 m inside is shown in Fig. 3.53. The sharp increase near the shore can be observed when the wind blows from the shore toward the land (Shitara, 1969).

Gregory et al. (1967) report a difference of air temperature between two stations located on the opposite sides of small lakes and a reservoir. The temperature observed at the leeward station is never more than 2.8°C warmer or cooler than that at the windward station.

The local climate at 37 spots on South Bass Island—A narrow island about 5 km long located about 10 km away from the shore of Lake Erie—(Verber, 1955). The distribution of air temperature on this island in February of 1945 is

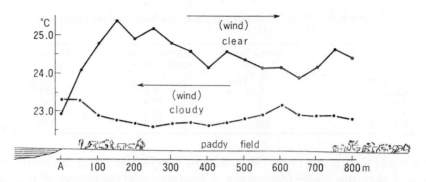

Fig. 3.53 Air temperature profile at 1.5 m level from the south shore of Lake Inawashiro on July 10, 1968 (Shitara, 1969).

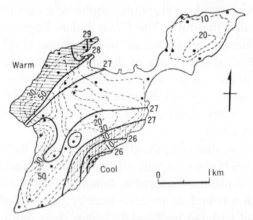

Fig. 3.54 Air temperature distribution on the South Bass Island in Lake Erie for February 1945. Black point: observation point. Contour intervals: 10 feet. (Verber, 1955; revised by personal communication).

given in Fig. 3.54 was observed from November, 1944, to August, 1948. There is only a small difference because the influence of the lake water is great and the comparatively strong southwestern wind (about 6 m/sec average) blows over the island. The local difference of the monthly mean values on the island is greater in late autumn, winter, and early spring. The main factors that control the local difference of the distribution of air temperature, are the wind direction and the wind velocity. If the wind directions are the same, there are similar distribution patterns. If the wind velocity is great, the local difference is slight, and vice versa. For example, when the wind is stronger than 5 m/sec, the local difference is 1–2°C, and when the wind velocity is 5 m/sec, the difference is 2–3°C. This island can be divided into seven climatic areas: the windward area, the leeward area, the central part of the island (each of these three are further subdivided into high and low areas, and the forest areas).

On the shore of a large river one can observe phenomena similar to lake-shores. That is, the minimum air temperature becomes higher, the maximum air temperature lower, and the frostless period longer. These phenomena are especially remarkable in the temperate zones and the subpolar or polar regions. The frostless periods observed on the shore along the upper course of the Angala River in East Siberia and at a spot 7 km away from the shore are given in Table 3.45. According to this table, the average difference between the two observation spots in 34 days. It also changes from year to year; in 1945 the difference reached 69 days—more than 2 months (Galahkov, 1955). As for the influence of rivers, there are several factors—the width of the river, the running direction of the river, the prevailing wind direction and velocity—all of which decide the sphere to be influenced. Kaps (1953/1954) gives an example of the Elbe River in Germany. The influence on the air temperature at 50 cm above the ground is observable in an area 150 m from the shore when the wind blows from the shore toward the land and 70 m in the case of weak winds. Furthermore, the air temperature becomes 1–3°C higher at night and 2°C lower in the daytime.

Table 3.45   Influence of a large river, the River Angala, on the frostless period (Galahkov, 1955).

Obs. station	1939	1940	1941	1942	1943	1944	1945
A: riverbank	111	123	122	100	114	110	134 days
B: 7 km from the riverbank	106	71	97	80	75	81	65

In Japan, an investigation has been made in the neighborhood of Maebashi in the northwestern part of the Kanto Plain along the Tone River (Kubo, 1943). When the temperature on the water surface is higher than the air temperature, there are ascending air currents over the river that spread horizontally at right angles to the river at the height of the inversion layer and form small circulation systems near the river. These small circulation systems develop more conspicuously if there is a great difference between the temperatures of the ground surface and of the water surface and if the general winds at higher levels are weak. Calculated from the results of the observation on the ground, the height of these circulation systems is 65 m, and their horizontal extent is 213 m.

Table 3.46   Heat exchange between the air and water near the waterfall, Shiraito, on the foot of Mt. Fuji (Asai, 1964).

	Total*	Sensible heat	Latent heat	Total**
Daytime	3.33	1.16	1.83	2.99 $\times 10^6$cal/sec
Evening and morning	2.17	0.67	1.56	2.23

   * Calculated from the water temperature rise.
  ** Calculated from the air temperature and its vapour pressure decrease.

Near a waterfall, air temperature is relatively low, and the inversion layer si formed in the valley in the daytime. The Shiraito Waterfall on the SW-foot of Mt. Fuji, from which 2 ton/sec of water with a temperature 12.6°C falls, has a

height difference of 20 m. Asai (1964), based on an observation of the microclimate near this waterfall, has come to the conclusion that the heat exchanges
between the air and water are like those in Table 3.46. The amount of heat exchange is unexpectedly small within the space in the valley under the influence
of the waterfall, but the inversion of air temperature (4–6°C) and of vapour
pressure (3–6 mb) from the valley bottom up to a height of 25 m are observed.

3.4.2. Winds
(A) *Land and Sea Breezes: Facts*

Land and sea breezes are caused by thermal and physical differences between land and water. Water has a greater thermal capacity; the specific heat of
water to the unit volume is 40% greater than that of soil, but the thermal conductivity of water is very small. Hence, the range of temperature change on the
water surface is about the same as that on the ground surface. In the daytime,
however, the temperature on the ground surface is higher than that on the water
surface, because the heat on the water surface is transmitted to the lower layer
by the turbulence caused by the waves.

The air in contact with the ground surface has a high temperature in the
daytime, and therefore, the thickness of the lower atmosphere is greater than on
the sea, but at night the air temperature falls on the ground and the lower atmosphere becomes thinner. Accordingly, the wind blows landward in the daytime and seaward at night. However, in compensation for this wind in the lower
layer, the wind blows in the opposite direction in the upper atmospheric layer:
this current flows seaward in the daytime and landward at night (the current in
the upper layer in the daytime and at night is called the anti-sea breeze and at
night, the anti-land breeze).

In general, the land and sea breezes develop on fine days (when the percentage of possible sunshine is more than 50%) with a weak-gradient wind (when the
velocity is less than 7–8 m/sec). Accordingly, the annual variation in the frequency of the occurrences of the land and sea breezes corresponds to the occurrence of such synoptic conditions. In Japan, land and sea breezes develop when
the NW-Pacific anticyclone develops in summer and when the migratory anticyclone develops in spring and autumn. The frequency of occurrence in several
points on the Japan Sea coast is given in Table 3.47.

Since the strength of the circulation systems of the land and sea breezes is
related to the difference of the air temperature between the surface of the water
and the surface of the land, it differs according to the latitude and, at the same
spot, according to the season and the weather. But even in closely adjacent areas,
the degree of development of the land and sea breezes varies considerably according to the topography of the sea coast and its neighborhood, the form and
size of the plains, the state of vegetation in the area, and the relation between the
direction of the land and sea breezes and that of the gradient wind under synoptic conditions favorable to the development of the breezes. For instance, a

Table 3.47 Occurrence frequency of land and sea breezes at several points on the Japan Sea coast (Yoshino et al., 1973).

	Tsuruga	Toyama	Niigata
January	0	0	0
February	1	2	0
March	0	7	3
April	3	15	8
May	5	11	9
June	6	14	11
July	5	11	15
August	14	19	14
September	8	15	17
October	3	13	6
November	1	11	3
December	0	1	1
Period	1943–1947	1959	1956

statistical investigation on the sea breeze frequency at Athens, Greece (37°58'N, 23°43'E, 107 m), shows that it develops from March to October with a mean wind velocity of 3.1 m/sec and has a steadiness of 87% (Karapiperis, 1953). In contrast, the *Etesians* have a mean wind velocity of 6.2 m/sec and a steadiness of 92%.

Generally speaking, sea breeze layer is thick and velocity is great in regions of low latitudes. On the other hand, it is thin and velocity is small in regions of high latitudes. In most cases, sea breezes develop up to altitudes of 1,300–1,400 m in Batavia (6°S) and 500–700 m in Japan, ranging from 300–900 m. At high latitudes, for example, in Ilmala, Finland (60°12'N, 24°55'E), the altitude of the sea breeze was 700–900 m according to observations made from 1920 to 1934.

As distinguished from sea breezes, land breezes are thin and their velocity is small at low latitudes, and they are thick and their velocity is great at high latitudes. According to some observations in Batavia (Bemmelen, 1922) and in Japan, the thickness of the land breeze is half or less than half of the thickness of the sea breeze. The thickness of the land breeze observed in Kôbe and Tsu-

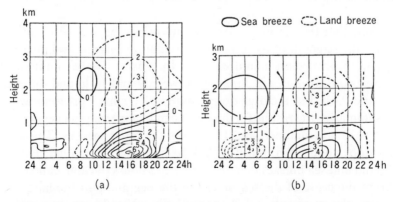

Fig. 3.55 Diurnal change of the vertical structure of land and sea breeze over (a) Batavia (6°S) (Bemmelen, 1922) and (b) Ilmala (60°N) (Rossi, 1957).

ruga is mostly 200–300 m, which is equal to 1/3–1/2 of that of the sea breeze. In Ilmala, Finland, however, the land breeze is about 700–800 m thick which is almost equal to the sea breeze, and the velocities are also nearly equal (Rossi, 1957). The results observed in Batavia and Ilmala are presented in Fig. 3.55. A comparison clearly shows the difference of high and low latitudes. Results from the coast of Danzig Bay show that the height of the sea breeze might be 600–800 m on the coast and 500 m at a point 15 km inland (Kohlbach, 1942).

The extent of sea breezes is 20–30 km on the sea and about 20 km on land. The extent of land breezes is narrower than sea breezes in middle latitudes. The maximum seaward and landward extents of the sea breeze are both 20 km in Ilmala (Rossi, 1957). It is most strongly influenced by the gradient wind. In the northern part of Tokyo Bay, when the direction of the sea breeze is opposite the direction of the gradient wind, the sea breeze goes 4–5 km inland, and when the two directions meet at right angles to each other, it goes 6–7 km inland (Yoshino, 1957c).

When a sea breeze is connected to a valley wind or an upslope wind in a mountain region behind a coastal plain, the sea breeze is strengthened and reaches farther inland. In the Toyama Plains, Japan, surrounded by the mountains 1,500–2,000 m high, the sea breeze, starting on the coast in the morning, invades with a speed of 15 km/hr to 30 km inland in the afternoon. The height of the sea breeze is 1,100 m at 16–17 h (Toyama Local Met. Obs., 1967). The height of land breeze is 700 m at 3–4 h.

In the same way, if a land breeze is joined by mountain winds or downslope winds in the mountains behind the plain, it becomes stronger. For instance, there are reports from the Durban area, South Africa that although the depth of land breezes are of the order of 300 m on the coast (Preston-Whyte, 1968), they induce an uplift of the undercutting moist sea air. The land breeze is initially independent of the mountain wind and, during this period, low intensity rainfall is most frequent. But, when the mountain winds are joined by the land breeze, the rainfall is inhibited because of the strengthened uplift (Preston-Whyte, 1970a).

The time that the sea breeze or the land breeze begins or subsides is connected with the strength of its respective circulation systems and greatly differs according to locality, season, and weather. In general, the beginning of the sea breezes in various localities in Japan is between 9 and 11 h, and the end is between 19 and 23 h, but a closer observation at Toyama Port is shown in Fig. 3.56. To summarize the results for Japan, it can be said that the beginning of the sea breeze is 10 h in summer, and 12–13 h in the other seasons. The sea breeze generally stops about 18 h, but it is 23 h to midnight in summer. Between the hours of land and sea breezes there is mostly 1–2 hours of calm, called *nagi*. The period from the cease of the sea breeze to the beginning of the land breeze, is very muggy, almost windless (0.2–0.3 m/sec) evening hours, especially in the Inland Sea Region in SW-Japan, where people call it *yû-nagi*.

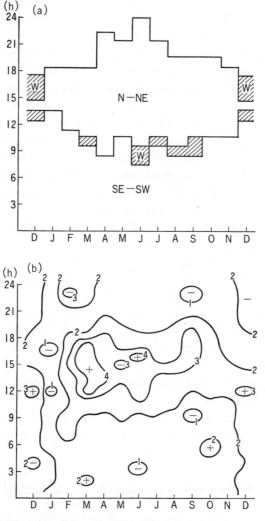

Fig. 3.56 (a) Diurnal and monthly change of the sea breeze (N–NE) hours and the land breeze (SE–SW) hours at Toyama Port, Japan.
(b) The wind velocity (m/sec) (Yoshino et al., 1973).

The wind direction changes from the sea to the land breeze or from the land to the sea breeze clockwise.

The maximum sea breeze velocity appears at 13–16 h, but the land breeze has no definite time with maximum velocity.

(B) *Theories of Land and Sea Breezes*

Defant (1951) has postulated theory of land and sea breezes, and Takahashi (1955) has reviewed the theories of such authors as Kobayashi, Sasaki, Arakawa, and Sakuraba.

All the models in these studies are linearized, but linear models are not capable of treating many of the interesting characteristics that are obviously brought about by non linear processes. Pearce (1955) has used nonlinear models, introducing a different heating mechanism.

The intensity, duration, and dimensions of sea breezes must be governed largely by the amount of heat supplied by the ground to the air layer and also of the prevailing synoptic conditions (Estoque, 1961). Estoque has formulated a primitive equation model, which is an extension of an atmospheric boundary layer model previously developed for a homogeneous terrain and presented the distributions of the wind component normal to the shoreline and the vertical velocities. Estoque clarified the following points; the initial formation of the circulation over the coast, its intensification, the movement of the whole pattern inland, and its subsequent decay. In one of his later studies, Estoque (1962) applied his model to various synoptic situations.

Fisher (1960) observed the sea breeze on the coast of New England, U.S.A., by means of balloons and aircraft and later (Fisher, 1961), numerically analyzed the results, formulating a nonlinear model. In the latter work he assumes that the coefficient of diffusion is simply a function of height. Pearce (1962) applied numerical integration techniques to the solution of the differential equations: The equations of heating and motion are averaged vertically, assuming vertical distributions of velocity and temperature. A nonlinear partical differential equation of a hyperbolic type is found to be satisfied by the surface velocity of the sea breeze.

Studies on the numerical computation of sea breezes have been done by Magata (1965) and Geisler et al. (1969).

Nordlund (1971) made a study of the sea breeze under conditions typical of the south coast of Finland, improving the numerical prognostic model, based on the following equation system:

$$\frac{\partial V_H}{\partial t} = -(V_H \cdot) \nabla V_H - w \frac{\partial V_H}{\partial z} - \frac{RT}{p} \nabla p - f(k \times V_H) + \frac{\partial}{\partial z}\left(K \frac{\partial V_H}{\partial z}\right) \quad (3.84)$$

$$\frac{\partial \Theta}{\partial t} = -V_H \cdot \nabla \Theta - w \frac{\partial \Theta}{\partial z} + \frac{\partial}{\partial z}\left(K \frac{\partial \Theta}{\partial z}\right) \quad (3.85)$$

$$\frac{1}{\rho}\frac{d\rho}{dt} = -\cdot \nabla V_H - \frac{\partial w}{\partial z} \quad (3.86)$$

$$\partial p = -g\rho\partial z \quad (3.87)$$

$$\Theta = T\left(\frac{p_o}{p}\right)^{R/c_p} \quad (3.88)$$

$$p = \rho RT \quad (3.89)$$

In the above equations V_H denotes the horizontal wind vector, ∇ the Hamilton operator, w the vertical wind, T the temperature, p the pressure, f the coriolis parameter, k the unit vector in the vertical direction, K the diffusion coefficient, Θ the potential temperature, ρ the density, and p_o the surface pressure. R represents the specific gas constant for dry air, c_p the specific heat of dry air, and g gravitational acceleration.

In the vertical direction the atmosphere is divided into the friction sublayer nearest to the surface, $0 \leq z \leq 50$ m, and an upper layer, $z > 50$ m, in which the actual dynamic process is assumed to occur. In the friction sublayer the assumption of the vertical flux of heat and of momentum is a usual one.

Space does not allow discussion of the details of the method of integration. The results after the sea breeze have reached its maximum are given in Fig. 3.57. The results obtained in this study agree fairly well with the statistical results of the 15 years' observation at Ilmala (Rossi, 1957).

(C) *Wind on the Coast*

The wind direction tends to be at right angles to the running direction of the coastline. Futi (1933) did theoretical research on this phenomena and it is proved at many points also statistically. At some points, however, the winds frequently blow parallel to the coastline. A survey on the coastal region of

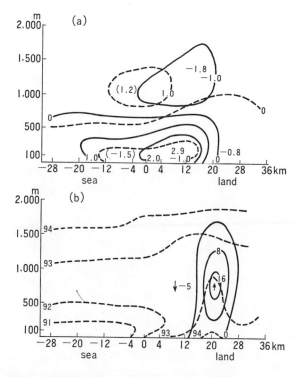

Fig. 3.57 Predicted sea breeze at 17h (Nordlund, 1971).

(a) Horizontal wind components in m/sec normal, u, (solid lines) and parallel, v, (dashed lines) to the shore. The numbers denote maximum and minimum values of u and v (in parentheses). u is positive landward and v is positive to the left of the positive u-direction.

(b) Vertical winds (solid lines) in cm/sec and potential temperature (dashed lines) in °K without the number denoting hundreds.

Klützer Winkel between Wismar and Lübeck, Germany, shows that the prevailing wind direction is at right angles or parallel to the running direction of the coast (Schöne, 1964).

The wind profiles in the layer near the surface present different characteristics according to topographical situations. According to an investigation on the 125 m tower at Risø, Denmark, the roughness lengths (z_0) vary from 1 cm for the water surface sector to 30 cm for the inland sector (300 m to water and 6.5 km to far shore) (Panofsky et al., 1972).

Blom and Wartena (1969) studied the development of a turbulent boundary layer in a neutral atmosphere downwind of an abrupt change of surface roughness. It was indicated that, for two successive changes of roughness, the turbulent boundary layer is devided into the following four regions, from the surface upward: (a) A layer, adjacent to the surface, with heights h'_2 adapted to the underlying surface roughness z_2; (b) A layer, $h'_2 < z < h_2$, where the velocity profile determined by the roughness heights z_1 and z_2; (c) A layer, $h_2 < z < h_1$, where the velocity profile is fully determined by the influence of the first abrupt change in roughness on the original profile for $x < 0$, i.e., it depends on the roughness height z_0 and z_1; (d) A layer, $z > h_1$, where the velocity profile equals the original velocity profile for $x < 0$. In this description, they assumed for $x < 0$ a horizontal surface with a uniform roughness height z_0. At $x = 0$ an abrupt change of surface roughness occurs such that for $0 < x < L$ the roughness height is z_1. At $x = L$ there is another abrupt change of surface roughness such that for $x > L$ the roughness height is z_2.

Photo 3.6 Fishermen's houses with stonewall shelters against the winter monsoon at Sotodomari, Nishiumi, W-Shikoku, Japan (November, 1973, by M. M. Yoshino).

Knudsen (1941) thoroughly discussed the wind conditions under the influences of micro-topography. He analyzed the topographical effects according to direction at the station Kråkenes Fyr (62°2′N, 4°59′E, 80 m) at an island, Vågsøy, and at another station, Sula Fyr (63°56′N, 8°27′E, 30 m), at Sula Island, Norway. Knudsen (1963) studied the wind conditions in the regions at the port of Bergen and its surroundings in detail and made clear striking differences in the distribution patterns of wind directions and velocity. Namely, the strongest wind velocity with great deviation in direction appears in different places according to the prevailing wind direction over the region.

(D)   *Lake Breezes: Facts*

Around a great lake, as at the seashore, winds blow landward in the daytime and toward the lake at night. The former is called lake breeze and the latter, land breeze.

The lake breeze varies in its degree of development according to the conditions of the land around the lake, i.e. whether it is a plain or mountainous region as well as the size of the lake. Around Lake Constance (540 km^2) on the border

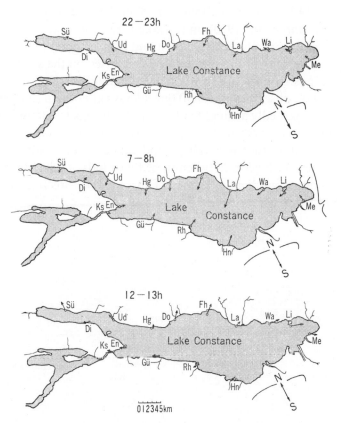

Fig. 3.58   Distribution of vectors of land and lake breezes around Lake Constance (Huss and Stranz, 1970).

of Switzerland, Austria, and Germany, the lake breeze reaches as high as 150 m (Peppler, 1936). A recent "Lake Constance Project" has drawn well defined pictures of lake breeze around the lake (Huss and Stranz, 1970). Among the results, the distribution of the vector of the lake breeze and land breeze is of great interest (See Fig. 3.58). Their direction and velocity vary greatly from place to place.

On the northern coast of Lake Leman in Switzerland, the land breeze is called *morget* and this north wind blows roughly from 17 h in the evening to about 8 h the next morning. In autumn, as the thermal difference between the cooled land and the warm lakewater increases, the land breeze blows over the whole lakeside area continuously through the night. This land breeze is easily distinguished from the mountain breeze; the latter, called *joran*, blows strongly here. The lake breeze, which is called *le rebat* blows approximately from 10 h to 16 h. It rises on the lake and gradually blows toward the shore.

According to an investigation of the lake breeze of Lake Suwa, Japan, which has an area of 14.5 km² and an 18 km shoreline (Yamashita, 1953), the lake breeze with a wind force of 1–2 begins to blow at 9 h 30 m and blows most strongly about 14 h 30 m, and about 16 h 00 m the land breeze begins to blow. A paper by Yoshino et al. (1970) on the lake breeze of Lake Suwa reports that the frequency of occurrence is 14% at 9 h and 24% at noon in summer, and its maximum wind velocity at the surface level is 2 m/sec on the shore. Figure 3.59 shows the composite maps of the wind velocity and direction and the air temperature, based on observations for 10 typical days at noon in summer. The lake breeze develops markedly around the lake, and it brings low temperature to the area 0.5–1 km from the lakeshore. The SW-wind coming up the Ina Valley along the western mountain foot is a valley wind called *inakaze*, which is relatively warmer than the lake breeze. Between these two winds, a lake breeze front with a sharp temperature gradient is formed. Thus, the lake breeze at Lake Suwa is an example of from a small lakes.

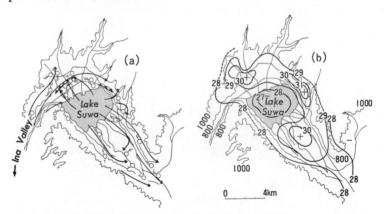

Fig. 3.59 Composite maps of (A) lake breezes and (B) air temperature around Lake Suwa, Central Japan, at noon in summer (Yoshino et al., 1970).

Figure 3.59 shows the low temperature in the center of the lake. However, this low temperature is not always the case: in some cases, the temperature in the central part of the lake is the same as that on the coast, because of the anticyclonic subsidence there. A similar case is observed at Lake Inawashiro (104 km²), NE-Japan (Shitara, 1964b).

In the case of lakes larger than those described above, the lake breeze tends to assume the nature of a sea breeze. An investigation was done in July and August, 1955, around Lake Ladoga, U.S.S.R., the greatest lake in Europe with an area of 18,000 km² (Vorontsov, 1958a). The results follow: on the western coast, when the lake breeze begins to blow, the air temperature falls 1°C abruptly, the humidity increases about 9%, and the breeze comes toward the coast like a cold front from the lake. The lake breeze begins to blow between 7 h to 14 h, and the earlier in the morning it begins to blow, the greater its velocity becomes in the daytime. If the time between sunrise and the beginning of the breeze is represented by τ in hours and the degree of its counter development of the horizontal pressure gradient dp/dx, the following equation is obtained:

$$\tau = 5\left(\frac{dp}{dx}\right) + 4.5 \tag{3.90}$$

The height of the lake breeze differs according to the weather. Between the altitudes 200 and 900 m, the maximum velocity is observed at an altitude 0.15–0.20 times as high as the height of the lake breeze. If the height of the lake breeze is represented by H and the horizontal gradient of air temperature by dt/dx, H is given by the equation below:

$$H = a\left[\left(\frac{dt}{dx}\right) - 5\right] \tag{3.91}$$

The value of a is 0.71 for the Novaia Ladoga on the SE-coast. Figure 3.60 shows the results of an observation on August 1, 1955. The lake breeze rises on the lake about 30 km away from the coast and blows toward the coast and 10 km inland. The thermal disparity between the area covered by the lake breeze and inland is about 4°C, and there is cumulus above the front of the breeze.

Studies on the lake breezes in the Great Lakes region have been done intensively in the U.S.A. since the beginning of the sixties. Following are some of their results.

The lake breeze occurrence on the shore of Lake Michigan is 36% on the Chicago shore, 25% on the E-shore, 14% on both shores, and 49% on either of the shores (Lyons, 1972). Lyons further reports that the time of onset is most frequently 8–9 h, and that the lake breeze front, defined as a narrow convergence band 1–2 km wide, separating airflows with overland and overwater trajectories, is seen to pass inland anywhere from 1 block to over 40 km. On some occasions, however, a rapidly increasing gradient during the afternoon can result in the convergence zone being pushed back offshore. These lake breeze "breakdowns" result in a sharp temperature rise near the shore.

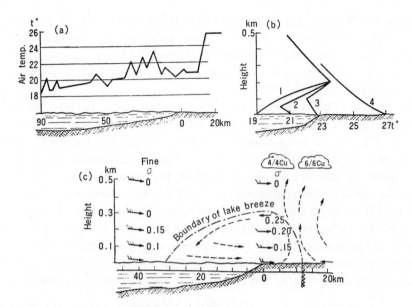

Fig. 3.60 Vertical structure of lake breezes at Ladoga Lake on Aug. 1, 1955 (Vorontsov, 1958a).
    (a) Air temperature change at 60 m level.
    (b) Vertical profiles of air temperature.
        1. Above the water surface at 90 km offshore.
        2. Above the water surface at 40 km offshore.
        3. Above the lakeshore.
        4. Above the land 20 km inland.
    (c) Vertical structure of lake breeze. σ: Turbulence.

    As pointed out by Moroz and Hewson (1966), the outflow from a well-developed thunderstorm can eradicate a lake breeze. In this case, the measured outflow occurs through a depth of approximately 1,000 m at the surface with a total speed of 6 m/sec. The maximum velocity is observed at the 300-m level near the edge of the storm. Therefore, the low-level outflow from the thunderstorm displaces the lake breeze with 1,000 m depth and yields 5 m/sec of onshore component of the wind velocity.

    On the E-shore of Lake Michigan, the lake breeze development and its inland progress were observed and a numerical modeling of them was done by Moroz (1967). Over the land the depth of the layer of onshore flow is approximately 750 m, and a maximum velocity of 5–7 m/sec is observed within 250 m of the surface directly over the lakeshore. Above the lake breeze, a well-defined return current, about twice as high as the lake breeze, is observed in the mid-afternoon. The lake breeze front invades more than 16 km inland. His model is a symmetric lake breeze on one side of a large lake based on the relations developed by Estoque (1962) with appropriate modifications and boundary conditions and on the observed temperature variation over the land. In Fig. 3.61 (A) and (B), 9 hr after time zero, the lake breeze maximum onshore shows a maximum of 3 m/sec at a point about 5 km inland and the vertical wind component

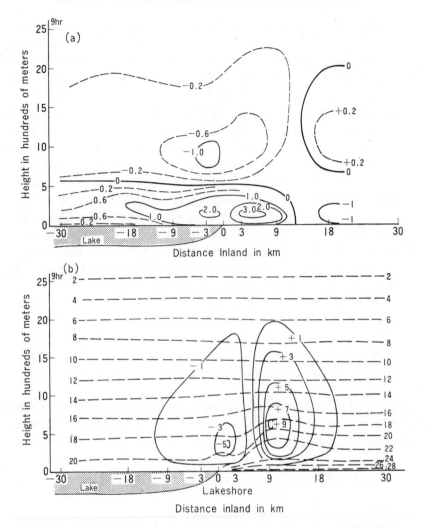

Fig. 3.61 (a) The across-shore wind component u(m/sec), and (b) the vertical wind component ω (cm/sec) and dry bulb temperature (°C), dashed line in a model plane 9 hr after time zero (Moroz, 1967).

ω is 9 cm/sec at a height of 500 m at a point 9 km inland. The return flow at a height of 1,000 m amounts to 1 m/sec.

Biggs and Graves (1962) developed a technique for predicting lake breezes on the western shore of Lake Erie. Their study shows that two dimensionless parameters describe the balance of forces that distinguish between lake breeze days and non-lake preeze days. A lake breeze index is established by the following equation:

$$[L-B]\text{ - Index} = \frac{V_z}{C_p \varDelta T} \qquad (3.92)$$

where V_z is the average wind velocity (m/sec) between 10 h and 16 h (LST) at an inland site; $\varDelta T$ is the temperature difference ($^\circ$K) between the lake surface water temperature (T_w) and the maximum air temperatures reached inland during the day (T_i); and C_p is the specific heat of dry air at constant pressure. It was found that this L-B Index has an accuracy of 97% at a site on the shore of Lake Erie.

(E)   *The Sea Breezes and the Air Temperature on Coastal Plains*

As a sea breeze blows onto the land, its air temperature rises gradually. According to an observation at Tottori sand dune in summer (Asai, 1940), the air temperature of the breeze rises 2–3°C on the average in the daytime during its course from the shoreline to a spot 192 m inland. The thermal change in the sea breeze as it blows from the mouth of the Oirase River to a spot 11 km up the river in the Tohoku district is presented in Table 3.48. When the sea breeze is blowing at an altitude of 150 cm above the ground surface, at 9 h or 13 h for example, its temperature becomes 1–2°C higher at 11.4 km inland (Asai, 1952). The different results from these two regions is thought to be due to the conditions of the ground surface. While the Tottori sand dune is a bare sand surface, the area along the Oirase River is surrounded by paddy fields. The temperature rise depends also on the solar radiation (Sasaki et al., 1970, 1972). The relation between the rate of temperature rise and the solar radiation is: 1.5°C/10 km in the case of 0.2 cal·cm^{-2}·min^{-1}, 2.0°C/10 km in the case of 0.5 cal·cm^{-2}·min^{-1}, and 3.0°C/10 km in the case of 1.0 cal·cm^{-2}·min^{-1}. Thus, the air temperature of sea breeze rises in accordance with the distance from the coastline as it blows onto the land.

At any spot on the coast, the air temperature is naturally lower when the sea breeze is blowing than when it is not. As mentioned above, the maximum air temperature in the daytime is lower when the sea breeze is blowing.

Table 3.48   Change of vertical distribution of air temperature ($^\circ$C) of the sea breeze along the Oirase River on July 7, 1949 (Asai, 1952).

Time	Height above the ground (cm)	A	B	C	D	E
		Distance from shoreline (km)				
		0.4	0.8	3.3	7.2	11.4
5 h	150	11.9 $^\circ$C	11.8	11.6	10.5	10.4
	50	12.1	12.0	11.0	10.0	9.9
	5	11.6	11.7	10.7	13.9	10.0
	0	13.5	14.6	—	—	—
9 h	150	20.9	21.2	22.4	21.7	23.1
	50	21.5	21.2	22.3	22.3	23.7
	5	23.6	22.1	20.8	23.3	25.1
	0	33.6	30.9	—	—	—
13 h	150	22.4	22.8	24.4	24.3	23.4
	50	23.7	24.6	25.2	25.1	24.1
	5	26.8	24.8	26.0	24.2	25.1
	0	34.1	25.7	—	—	—
21 h	150	19.5	18.6	16.4	15.6	16.9
	50	19.2	18.5	15.1	15.5	16.5
	5	18.6	18.3	14.5	14.8	17.7
	0	19.3	18.7	—	—	—

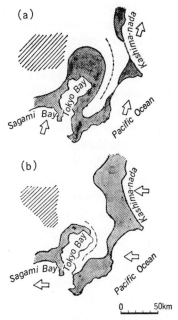

Fig. 3.62 The areas (shaded) on the coastal plains near Tokyo under the influence of sea breezes in the case of (a) SSW gradient wind and (b) E gradient wind (Kayane, 1961).

On the coastal plains of the Kanto district, the sea breeze has an obvious influence on the distribution of the maximum air temperature (Kayane, 1961). Because the direction of the gradient wind is closely connected with the area covered by the sea breeze, the limit of the sea breeze estimated from the distribution of the maximum air temperature is about 40 km under the condition of S-SSW winds, and about 10–15 km when it is an E-ESE wind. On the coastal area facing the sea of Kashima-nada, the distance from the coast covered by the sea breeze is about 10 km in the case of S gradient wind, and the sea breeze does not reach the area in the case of SSW gradient wind (Fig. 3.62). When the pressure gradient is small and the prevailing wind is weak, the decrease of the maximum air temperature caused by the invasion of the sea breeze is observed within about 10–20 km of the coast. Kayane (1966) pursues these problems in connection with the heat budget. His results can be summarized as follows: (a) advective effect of the sea breeze invasion plays a decisive role in forming daily maximum temperature distribution patterns in warmer seasons. The temperature difference between inland and the coast is about 5°C, but in the extreme case it amounts to 10°C. When the general wind is weak, the advective effect appears 15 km inland. (b) Typical sea breezes develop in May, but no marked mesoclimatic temperature distribution pattern is observable in September. (c) On a calm, clear day in winter, a bay breeze develops along the coast of Tokyo Bay, and it may cause a decrease of about 2°C in the maximum temperature along the coast. (d) On a cloudy day in a colder season, Tokyo Bay appears to have no effect on the daily maximum temperature distribution, and a higher temperature zone appears along the coast of the open sea. (e) Several discontinuity lines of temperature distribution are found. The existence of these discontinuity lines suggests that mesoclimatological consideration is very important even in microclimatological study. (f) The above facts are confirmed by the results of heat budget calculation. An evaporation process occupies a greater part of the heat exchange in warmer seasons.

In the coastal plain in SE-Hokkaido, the temperature rise is very sharp (3.5–4.5°C/20 km) from the coast to 20 km inland under the conditions that the wind direction is not from the sea, but is gradual (3.0–3.5°C/50 km) under the conditions that the wind direction is from sea (Ozawa, 1966).

Though it is not really a sea breeze, there is a cold NE-air current called *yamase*, which flows from the anticyclone lying east of N-Japan in the NW-Pacific. When this current flows into the Tohoku district, a low temperature appears in the coastal region. For example, the air temperature in Hachinohe in summer is remarkably low when a NE-wind is blowing, in comparison with winds blowing from other directions. The distribution of air temperature under the *yamase* condition in the Sambongi plains in Aomori Prefecture has been studied by Shitara (1952, 1957, 1963, 1964a). The temperature is 3–4°C lower on the coast when an E-wind *yamase* is blowing, than in the inland area, but the difference is comparatively small when a W-wind is blowing (see Fig. 3.63.)

In the Sambongi plain, the discontinuity mentioned above is seen as a sharp gradient of air temperature, that coincides with the boundary between two wind systems—the warm air current from inland and the cool air current from the sea. The temperature difference between Hachinohe and Gonohe under the

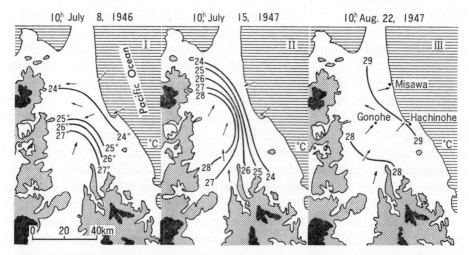

Fig. 3.63 Distribution of air temperature (°C) and wind in the Sanbongi Plain, NE-Honshu, Japan, under the *yamase* conditions, I, II and the W-wind condition (Shitara, 1964a).

yamase condition is about 2°C (temperature increase of *yamase*), but it reaches 3–4°C, when a local front exists between them.

Numerical experiments on humidity change of the air mass moving from the seashore to the land were done by Inoue (1972). He states that the horizontal gradient of specific humidity between 0–5 km is 0.34 $g \cdot kg^{-1} \cdot km^{-1}$ and between 5–10 km, it is 0.22 $g \cdot kg^{-1} \cdot km^{-1}$ under weak wind conditions and 0.23 $g \cdot kg^{-1} \cdot km^{-1}$ and 0.11 $g \cdot kg^{-1} \cdot km^{-1}$ under strong wind conditions. Inoue also showed that the diurual change of specific humidity as well as of temperature and diffusion coefficient is sensitive to the variation in thermal stability near the ground. At an inland station, the sensible heat flux decreases and soil heat flux increases under wet surface conditions, compared with a station on the coast.

(F) *The Sea Breeze Front*

When the sea breeze blows onto the land, a phenomenon similar to a cold front movement is observed, as shown below. In the early morning, the difference of air temperature between the land and the sea is very small, but it gradually increases; the surface of discontinuity formed between the sea and the land becomes clear, and finally the air on the sea blows onto the land like a cold front invasion.

In most cases, the velocity of this sea breeze front does not coincide with the sea breeze velocity on the ground level. This is due to the fact that the velocity of the sea breeze front near the earth's surface slows down through friction, having a shape like the nose of a cold front. The invasion of the sea breeze front is complicated, however, because the inversion layer, the shape of the coastline, and the topography all influence its movement.

The details of the invasion of the sea breeze front moving toward the land have not yet been made clear. Recently, radar investigation has been done, (Atlas, 1960; Simpson, 1967), which can afford much finer details of pulses and wind reversals than the usual network of observations. Kauper (1960) investigated the sea breeze front in the Los Angeles Basin. The air pollution becomes worse with the invasion of the sea breeze, and this is called a *smog front*. With the passing of this smog front, visibility weakens and the wind velocity becomes stronger. The smog front fails to show a simple correspondence to a sea breeze front. Concerning the Carifornia sea breeze, it has recently been pointed out that there are a number of convergence zones in southern California with different causes and characteristics (Aldrich, 1970).

Hourly patterns of streamlines of land and sea breezes for Tokyo from 11 h to 20 h on March 2, 1961, drawn by Ohta (1965) are shown in Fig. 3.64. The sea breeze front starts from the coast line about 10 h in the morning and gradually invades inland, reaching the innermost part of the land, about 10 km from the coast, about 17–18 h. From 19 h the land breeze front begins to move toward the coast, and at 1–2 h, it reaches the coast line. The movement of the land breeze front, i.e., retreat of the sea breeze front, is slightly faster than the invasion of the sea breeze front. The movement velocity of the sea breeze front is

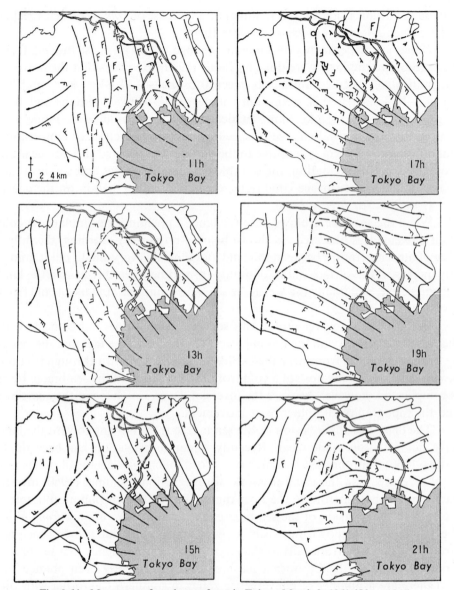

Fig. 3.64 Movement of sea breeze front in Tokyo, March 2, 1961 (Ohta, 1965).

fairly slow in comparison with the wind velocity in the sea breeze region. On the whole, it can be said that the movement velocity of the sea breeze front is 2–4 km/hr (0.5–1.0 m/sec), one-tenth to one-fifth of the velocity of the sea breeze.

In the early morning, a disturbance is generated on the coast and propagates both inland and toward the sea. On the land it produces a wind from the direction of the sea, and it seems to come prior to the arrival of the main nonlinear overturning, which is the sea breeze proper. For this reason it is called the "sea-breeze forerunner" (Geisler et al., 1969). Linear numerical model research has made clear that the structure of the forerunner depends on the large-scale field of stratification. The effects of internal viscosity and of the surface drag are known to limit the amplitude of the forerunner at moderate distances inland to a few m/sec and to limit the inland penetration distance to about 60 km.

An observation of the sea breeze front on the central Oregon coast, U.S.A., made clear the following facts: (a) A sea breeze front is distinguishable in the zonal wind field, which penetrated more than 60 km inland. (b) A distinct wind maximum followed the front inland (Johnson et al., 1973). This study also described the vertical structure of the front.

If the sea breeze or the land breeze is joined by winds caused by a synoptic-scale situation, a monsoon or trade wind for instance, a local line of discontinuity is formed there, as mentioned above for the Sambongi plains in Aomori and other regions. Similar phenomena occur in various regions of the world, though sometimes the formation of the line of discontinuity is not be obvious.

A careful study has been made concerning the conditions of the line of discontinuity or local front formed between the trade wind and the sea breeze in the Hawaii Islands (Leopold, 1949). According to this study, four main types are recognizable, as shown in Fig. 3.65. In the *Lanai* type, the summit is low and the trade wind blows over it; in the *Maui* type, the summit is high and the trade wind does not blow over it, but it is divided into two currents and blows around the foot of the mountain; in the *Mauna Kea* and *Cona* types, there is an inclined slope facing the trade wind and at the back of it, respectively. On the surface of the discontinuity formed by the conflict of two air currents, a cloud comes into existence. This cloud, called the sea breeze cloud, is formed by the sea breeze front, and it also develops vertically in the afternoon when the sea breeze blows

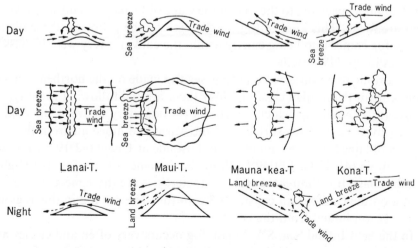

Fig. 3.65 Local front and clouds between the land and sea breezes and the trade winds (Leopold, 1949)

hardest and brings a shower. Accordingly, the local distribution of this cloud determines the local distribution of rainfall on the islands. The local circulation associated with the sea breezes and local circulations in Hawaii will be mentioned again in Secs. 4.1.7 and 5.1.2.(E).

Schroeder et al. (1967), based on observation of the sea breeze circulation, classify the sea breeze fronts into three types: (a) a classical air mass front, that is, a sea breeze front that is characterized by a sharp temperature fall, humidity rise, and sharp wind velocity change; (b) the windshift line, which is a thermally modified air mass type, but not kinematically altered; and (c) the cool change, which is characterized by a sustained cooling and humidity rise and no wind-shear line. Examples of the first type are the cases reported by Leopold (1949), Simpson (1965), and Frizzola and Fisher (1963); those of the second type by Reid (1957), and those of the third type by Berson (1958). For further information, see Schroeder et al. (1967). Hirt and Shaw (1973) estimated the thickness and slope of the lake breeze front from observations of temperature difference between upper and lower levels on a tower on the lakeshore of Toronto. Estimates of 1:11–1:61 were obtained for the slopes and those of thickness varied from 140–4,000 m.

3.4.3. Fog

(A) *Sea Fog*

The distinction between sea fog and land fog is not always clear. Usually, fog during the period from spring to summer, when the water temperature is lower than air temperature, is called sea fog, and fog formed from autumn to winter on the coast or inland is called land fog.

In Japan, fog frequently occurs in early summer along the Pacific coast of the Tohoku district and of Hokkaido and along the coast of the Sea of Okhotsk of Hokkaido. Especially, to the east of the Cape of Erimo as many as 80–90 days a year are foggy. This kind of fog is called *gas* by the people in these districts. This sea fog comes into being chiefly within the cold and very wet air mass that originates in the NW-Pacific.

According to an observation of the sea fog made on the northeastern coast of the Korean Peninsula (Utsumi, 1944), there are the following three types of sea fog: (a) Frontal fog, which occurs on the boundary between the cold and moist air over the cold sea surface and the warm and moist air from the south. The data obtained with the aid of the radiosonde at Unggi (42°19′N, 130°24′E) and Chongjin (41°47′N, 129°49′E) show the existence of this fog. (b) High fog along the coast. It is thought that its characteristics are the same as those of the high fog on the Californian coast. (c) Steam fog. The sea fog in this region invades 8–12 km inland from the coast.

In the Seto Inland Sea, SW-Japan, fog occurs very often and causes many ship accidents. A meteorological investigation into the mechanism of its origin is, however, still to be done.

Table 3.49   Amount of sea fog according to the distance from the coast
in the SE-part of Hokkaido (Morita et al., 1952).

Obs. station	Distance from coast	Amount of fog water	
		18 h–6 h (Average of 9 nights)	6 h–10 h (Average of 3 days)
Kushiro Met. Obs. (10 m Tower)	1 km	>0.106 g/m	0.077 g/m
Kushiro (2 m above the ground)	1	>0.096	0.010
Onehira	9	0.062	0.007
Tsurui	24	0.004	0.000
Touro	22	0.071	0.013
Shibecha	34	0.007	0.000

The condition of the fog ingress along the seacoast of the SE-districts of Hokkaido was observed with a fog gauge from the end of July to the end of August in 1951. The results are presented in Table 3.49 (Morita et al., 1952). There is a rapid decrease in the amount of fog water moving from Kushiro, 1 km from the coast, inland, so that the amount is very small at stations about 9 and 24 km away from the coast. But at Touro, into which the wind blows along the Kushiro River, much of the fog flows in. The values in the table are those measured at an altitude of 2 m above the ground, and the fog stays at the higher level forming a stratus and weakens solar radiation. Thus it has a great influence on sunshine and the air temperature and restricts the growing of crops in the area. The Kushiro area has the shortest duration of sunshine in Japan during the farming season from May to September. This is due to the fog.

Fukaishi (1971) studied statistically the fog in the SE-part of Hokkaido. He points out that the sea fog appears mostly under the conditions of wind velocity of 2 m/sec on the coast, and the fog duration is slightly shorter in cities, such as Kushiro, than the suburbs; probably due to the effect of the urban climate. His recent study (Fukaishi, 1973) reported: (a) The fog invades from the sea at night and dissipates in the morning. The fog frequency at Kushiro reaches over 40% from 5 to 6 h. (b) The fog dissipation begins in most cases between 7 and 8 h. The speed of the fog front retreat is about 2.5 km/hr, but is faster on the coast and slower inland. (c) The fog invasion occurs mostly between 19 and 22 h, with an average speed of 2.0 km/hr. It is slower on the coast and faster inland. (d) In the daytime, the shallow advection fog layer makes the air temperature lower by 1.5–2.3°C and the relative humidity higher by 5–12% for 30–70 min. (e) During the fog invasion in the daytime, the average wind speed is 4.5 m/sec on the coast and 3.0 m/sec inland.

Patton (1956) investigated the summer fog in the San Francisco Bay area, where parallel coast ranges break sufficiently to allow marine air to penetrate inland. Dynamic subsidence in the N-Pacific anticyclone results in a low-level inversion, which is strengthened by cooling from the cold water off the California coast. This inversion concentrates the conflict of the sea and land air masses across the coastline into a shallow layer. As the cold marine air flows inland through gaps in the coast ranges or along the valleys, its streamlines define a wellmarked pattern of temperature and humidity. Fog formed in the cold air over the sea lifts into low stratus (high fog) between 150 and 600 m in elevation as it flows coastward, arriving at the innermost part of the land by the early morning and dissipating during the next day. On the sea near the coast, stratus and stratocumulus appear frequently (more than 30% of the observations) and the height of inversion base is mostly 200–400 m (Neiburger et al., 1961).

Saito (1963) reports on the sea fog distribution in the Yûfutsu plain near Tomakomai, Hokkaido. He concludes that the advection fog from the sea appears over the sea except the area near the coast, where the sea water is cold as shown in Fig. 3.66.

In the middle part of the Sanriku coast of the Tohoku district (about 39°55'N, 141°31E), people call the cool NE-wind with sea fog *okiage* (which means the up-offing wind or the landwards flowing wind in the offing) instead of *yamase*. On the coast the wind blows more than 30 days, at 15 km inland, it blows only 10 days. The sea fog itself, however, reaches approximately 8 km inland from the coast.

Along the Sanriku coast in the Tohoku district, the *yamase* wind accompanied by a cold sea fog blows inland in summer. If this cold fog is thick and lasts long, people in these districts will suffer from a poor crop that year. We

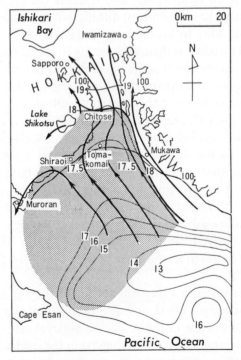

Fig. 3.66 Distribution of sea fog, sea surface water temperature, air temperature, and streamlines in the Yufutsu plain near Tomakomai, Hokkaido, on July 21, 1961 (Saito, 1963). Fog area is shaded.

Photo 3.7   Lake fog about 1 m thick on Myôjin-ike near Kamikôchi, Central Japan (July, 1960, by Y. Tateishi).

have already seen in Sec. 3.4.2 (E) how the air temperature rises when the *yamase* blows inland and what relationship this wind has to the inland air currents.

(B)   *River and Lake Fogs*

When the temperature of the water is much higher than that of the air, and when the water vapour near the water surface reaches saturation, a steam fog develops. This kind of fog is called river fog when it arises on river water, and lake fog on lake water. In mountainous areas a river or a lake is topographically located in a valley bottom or basin bottom, and, therefore, in many cases fog is not distinct from the radiation fog formed as a result of a cold air lake discussed in Section 5.4.3. Because radiation fog does not develop in the daytime, any fog in the daytime is steam fog. River fog most frequent occurs in fine, calm weather. According to an observation of the river fog on the Saigawa, a tributary of the Shinano river, the river fog occurred on 27 clear days or on 22 calm days out of a total of 39 days (Asakawa, 1950).

3.4.4.   Amount of Precipitation

The amount of precipitation is smaller on the coast than inland, but one reason for this is thought to be that the efficiency of catch with the rain gauge is low along the seacoast, where winds are stronger than inland. A more obvious phenomenon is the distribution of the amount of snow. Because winds are strong and air temperature is comparatively high near the seacoast, there is little or no snow cover from several hundred meters to 1 km away from the coast, even in the snowy districts.

Snow depth y (cm) on the morning of February 22, 1953, in the Tokyo metropolitan area within 25 km from the seacoast is given by the equation below (Yoshino, 1953):

$$y = a \log x + b \tag{3.93}$$

where x is the distance (hm) from the sea coast, and the values of a and b are calculated by the least squares method. In this case $a=+11.49$ and $b=+1.09$. This is applicable only in the range of 1 hm$<x<$250 hm and when no influence of a stationary front is recognizable. The value of a is chiefly connected with the duration, t hr, of snowfall. The empirical formula is:

$$a \doteqdot 0.36t \tag{3.94}$$

From Eqs. (3.93) and (3.94):

$$y=(0.36t) \log x+b \tag{3.95}$$

In meso-scale, an increasing effect of the coast on precipitation can be observed. Concerning this problem, the reader is referred to Arakawa's theoretical investigation and Bergeron's detailed description from the dynamic point of view (Arakawa, 1938; Bergeron, 1949). Mano et al. (1954) reported on the coastal maximum of precipitation in early summer and winter in the Tohoku district, N-Japan. The coastal maximum may be attributed to the increase of friction on the coast, the lifting in the air current, and the increase of vertical mixing on the coast.

Lakes also effect precipitation. The diurnal variation and horizontal distribution of the wind and rainfall around Lake Victoria, for instance, are very apparent (Flohn et al., 1966). The maximum of the annual precipitation, 2,300 mm, appears near the center of the lake, and on the lake shore it is 1,000–1,500 mm. This may be attributed to the convergence of the regular nocturnal land breezes above the lake. The easterly prevailing winds shift the area of maximum convergence westwards. The nocturnal thunderstorm maximum appears also during the night and early in the morning. In the daytime, subsidence in association with the lake breeze occurs above the lake, and hence there are no clouds or precipitation on the lake. The contrast in the diurnal variation of thunderstorm activity on the opposite shore of Lake Victoria is also detectable. The effect was considered to be that of local topography (Lumb, 1970).

Lake-effect storms occur frequently on the leeward shores of the Great Lakes, U.S.A., during the cold arctic air outbreaks in early winter. Muller (1966) describes the snow belts of the Great Lakes, i.e., areas where the mean annual snowfall is more than four times as much as that of neighboring regions a few tens of kilometers from the lakes. Peace and Sykes (1966) analyzed a snow storm of Lake Ontario characterized by a single, intense snow band. A pronounced confluence-convergence line, generally 2 km wide, in the low-level wind field is said to coincide with the cloud-band position. Recently, Lavoie (1972) has studied this phenomenon at Lake Erie; he proposes a single-layer numerical model, derived to simulate the meso-scale disturbance induced in cold air moving over an unfrozen lake of irregular shape. Among the conclusions, the followings should be noted: (a) The vertical structure of the planetary boundary layer is relatively unimportant; (b) a depiction of horizontal asymmetries is necessary in understanding the morphology of lake-effect storms; (c) roughness differences between the water and land surfaces, topography, latent heat release, and surface heating over the lake play important roles; and (d) surface temperature excess of several degrees (°C) is a crucial and dominant effect.

It is regrettable that there have as yet been very few studies on the effect of rivers on precipitation. The only paper that deals with this problem is the study on the effect of the Volga River by Bowa (1950).

Evaporation from the lake surface is of importance to study the water balance of the lake and the effect of the lake on the surrounding areas. Yamamoto et al. (1964) estimated the amount of evaporation from Lake Towada (59.8 km²) by means of the turbulent transfer theory and obtained a value of 902 mm per year.

3.4.5. Salty Wind

The amount of salt carried by the wind rapidly diminishes as the distance from the sea increases. In general, the decrease is proportional to the logarithm of the distance, and depends upon the wind velocity on the coast and the terminal velocity of salt particles. The wind direction and whether the wind comes from over ocean or the inland sea are also important. This means that the distance along the (blowing) path of the wind affects the amount of salt particles on the coast. For instance, the amount of salt adhesion (S_{OC}) on the coast facing the ocean is expressed by

$$S_{OC} = (0.008 - 0.024) \times u^3 \tag{3.96}$$

but, S_{IN} on the coast facing an inland sea is expressed by

$$S_{IN} = (0.008 - 0.017) \times u^3 \tag{3.97}$$

where u is the wind velocity (m/sec) on the coast and S_{OC} and S_{IN} are the amounts of salt in units of particles·cm^{-1}·sec^{-2} (Funatsu, 1964).

The salty wind does damage to the coastal region, when it blows with a mean wind velocity greater than 20 m/sec. The amount of salt adhesion on insulators of electric-light poles or of towers for power-transmission lines are expressed in the case of typhoons from September 16–30, 1961, as shown in Fig. 3.67. The horizontal distributions of the salty wind damage in the Kanto plain at the time of the three typhoons are shown in Fig. 3.68. On these occasions the salty wind damage was extensive. In Fig. 3.68 (a), serious damage was observable 40 km inland; in Fig. 3.68 (b), 55 km inland, and (c) 80 km inland. The total damage on the power-transmission lines by the adhesion of salt on insulators amounted to 504, 348, and 780 for the respective cases.

Basic studies on salt particle transport toward the land have been carried on both theoretically and experimentally by Toba and Tanaka (1967–1972). What makes their studies of special significance is that, using the impaction factor, λ, which determines the sedimentation-impaction ratio and has an order of $(1-3) \times 10^{-2}$, they estimated theoretically the number of salt particles and the particle sedimentation rate on the coast and at a point 200 km inland (Tanaka,

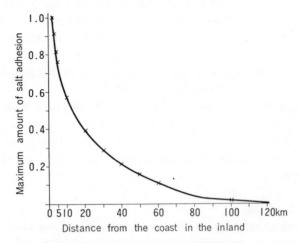

Fig. 3.67 Relation between the maximum amount of salt adhesion and the distance from the coast inland in the Kanto plain, Japan, in the case of typhoons on September 16–30, 1961 (Funatsu, 1964)

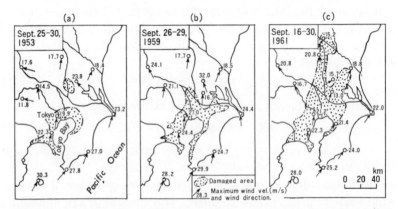

Fig. 3.68 Area with salty wind damage, maximum wind velocity (m/sec), and wind direction in three cases of typhoons in the Kanto plain, Japan (Funatsu, 1964).

1972). In the regions where the winds from the sea are relatively weak, the strong effect of the salty wind is limited only in the coastal area. For instance, Lomas and others (1967) point out that the concentration of windborne salt is considerable up to approximately 2 km from the coast of Natanya (32°20′N, 34°51′E, 35 m) in Israel. In the western areas of Wales, England, the windborne salt of marine origin deposits are significantly greater at stations 1.6 km inland from the coast than at stations farther inland (Edwards et al., 1964).

Chapter 4

TOPOGRAPHY AND CLIMATE IN A SMALL AREA

4.1. Mountains

Mountains greatly affect local climates. One of the most influential factors is the height above sea level of a mountain. In addition, the distance from a mountain and the direction of its ridge or valley play an important role, since they are correlated with the development of the circulation system between mountains and valleys. A mountain also exerts an influence on the neighbouring plain.

This section discusses the cases dependent most heavily on the height above sea level, and Section 4.2 shows cases much affected by the direction of the slope, by its inclination, and also by the mountains nearby.

4.1.1. Air Pressure

Air pressure decreases as the height above sea level increases. In small areas, local differences in pressure are not as great as those in temperature and precipitation. In this sense, it is not necessary to pay much attention to the pressure difference of a local climate, but it should be remembered that pressure is a factor related with the mountain-valley circulation from a purely climatological point of view as well as with mountain sickness from the point of view of applied climatology.

Table 4.1 Vertical change of atmospheric pressure, temperature, and density, together with the change of pressure for each 1°C change of air temperature.

Height above sea level	Standard atmosphere			Press. change by 1°C change
	Press.	Temp.	Density	
0m	1013mb	15.0°C	1.2249kg/m³	0.0mb
100	1000	14.4	1.2083	0.1
500	955	11.8	1.1626	0.2
1,000	899	8.5	1.1071	0.4
1,500	846	5.3	1.0538	0.6
2,000	795	2.0	1.0024	0.7
2,500	747	1.3	0.9531	0.9
3,000	701	−4.5	0.9054	1.0
3,500	658	−7.8	0.8598	1.1
4,000	616	−11.0	0.8159	1.2
4,500	577	−14.3	0.7737	1.3
5,000	540	−17.5	0.7331	1.4
5,500	510	−20.8	0.6943	1.4
6,000	472	−24.0	0.6571	1.5

Table 4.1 shows the change of pressure according to the change of height above sea level in the standard atmosphere. The standard atmosphere, by which is meant a hypothetical vertical distribution of air temperature, air pressure, and air density, is established by international agreement.

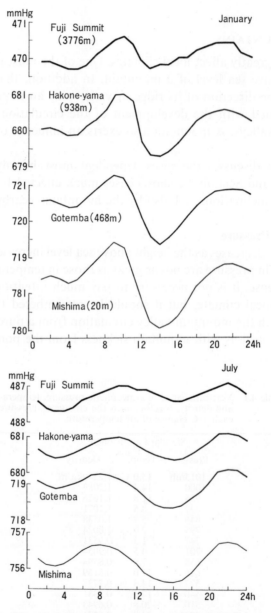

Fig. 4.1 Diurnal change of air pressure on Mt. Fuji (Fujimura, 1971).

The diurnal change of air pressure on high mountains is different from that at lowlands. As shown in Fig. 4.1, two maxima and two minima appear, and they take approximately the same values (Fujimura, 1971). The maxima occur at 9 and 21h and the minima at 3 and 15h, but they occur slightly earlier at lower levels. At stations below 1,000 m a.s.l., the diurnal change is quite similar and the maximum in the morning is greater and the minimum in the afternoon is smaller. The diurnal range is greater in winter than in summer. According to temporary observation on the southeastern slope of Mt. Fuji during January 16–31, 1951, the boundary of the diurnal change of air pressure between the upper-level type and the lower level type lies at 1,300–3,000 m.

4.1.2. Insolation and Sunshine

Insolation is affected by the height and slope inclination of a mountain. The higher a mountain is, the shorter is the distance solar rays travel through the atmosphere. They become stronger at higher altituded, because the amount of water vapor or dust in the atmosphere decreases. In general, the sun rises earlier but sets later on high mountains, so that the annual variation becomes smaller than at lowlands.

Table 4.2 Mean observed intensity of direct solar radiation
(cal·cm^{-2}·min^{-1}) at a surface at right angles
to the radiation (Alissow et al., 1956).

Station	Jan.	Feb.	Mar.	Apr.	May	June	July	Aug.	Sept.	Oct.	Nov.	Dec.	Year
Alma-Ata	1.27	1.31	1.26	1.28	1.31	1.32	1.31	1.25	1.26	1.23	1.21	1.12	1.26
Zugspitze	1.54	1.59	1.63	1.62	1.58	1.55	1.58	1.57	1.59	1.57	1.57	1.49	1.57

Alma-Ata (43°16′N, 850m a.s.l.),   Zugspitze (47°25′N, 2,962m a.s.l.)

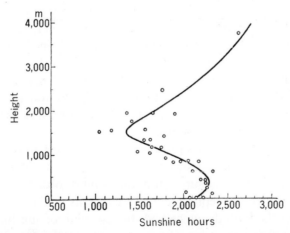

Fig. 4.2 Relation between annual total sunshine (hr) and height in the Japanese mountains (Yoshino, 1961a).

Table 4.2 shows a comparison between Alma-Ata and Zugspitze. Despite the fact that Zugspitze is situated latitudinally north of Alma-Ata, Zugspitze represents a larger annual mean value than Alma-Ata. The annual range is 0.20 cal·cm^{-2}·min^{-1} at Alma-Ata, and that of Zugspitze is 0.14 cal·cm^{-2}·min^{-1} (Alissow et al., 1956).

The inclination angle of slopes has a more significant effect on radiation in higher latitudes. In accordance with the height above sea level, the forest zone or vegetation zone can be seen in places higher than is expected from the temperature only. One reason for this is the influence of radiation on the slopes.

As a rule, it is expected that hours of daylight become greater in accordance with the increase of height, but they are distinctly fewer in places where fog occurs frequently. Figure 4.2 illustrates the relationship between the annual total hours of sunshine and the height above sea level in the Japanese mountains (Yoshino, 1961a). Owing to differences in periods of observation at observation points, differences in latitudes (ranging from 32°N to 43°N), and various conditions of small-scale topography, some points deviate considerably from the tendency curve. It can be said, however, that the curve given in Fig. 4.2 shows a striking change. That is, the minimum is found at about 1,500 m above sea level, and the maximum at 400–500 m. The fact that the minimum occurs at 1,500 m, is related to the maximum frequency of fog or clouds at 1,500 m, as shown below.

The distribution of sunshine hours in Austria (see Fig. 4.33) has a close connection with the altitude of the inversion layer in summer and winter. From this it follows that the height of the minimum hours of sunshine differs according to season, i.e., summer or winter.

Turner (1958) has obtained the following results at the observatory in the Ötz Valley (46°53′N, 11°02′E, 1,940 m) in Austria: When it is cloudless, the elevations of the horizon, reaching on the average 22°, cause a loss of 28% in the annual sum of radiation. The maximum of the daily radiation sum reaches 755 cal·cm^{-2}. Midday intensities are always higher by 11–12% than those in the lowland (200 m). Daily sums of sky radiation are, in the months without snow cover, somewhat reduced by the natural horizon, but fresh fallen snow may raise the sum of sky radiation by 40%, because of multiple reflexion and slope influence. On the average during a year, the station gets less radiation with cloudless skies (3%) and more radiation with overcast skies (30–35%) than an open place in the lowland (200 m).

To obtain the duration of sunshine in areas among mountains, methods such as Turner's are useful. First, the angles of elevation of a mountain or a forest in all directions are measured from the point of observation, and then all the data thus obtained are represented as a circle, as in Fig. 4.3. By drawing a solar path line at the latitude concerned, the duration of sunshine is obtained. Figure 4.3 is an instance at an observation site in the Ötz Valley, Austria (Turner, 1958). The solar paths are given in this figure for the winter solstice,

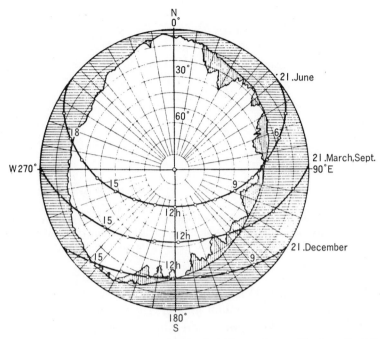

Fig. 4.3 A presentation of sunshine duration in accordance with the path of the sun under the influences of natural forests and mountain horizons (Turner, 1958).

vernal and autumnal equinox, and summer solstice. It can be seen from this figure, for instance, that in the southeastern direction a mountain stands in the way of sunshine up to the angle of elevation of 30°, or a forest does so up to 37°.

Turner (1966) presents a very interesting map (1: 500) in color on the global irradiation on the slope of Davos Dischma, Switzerland. The study area is situated between 2,000 and 2,230 m a.s.l. and comprises an area of 6.86 ha. In this area, the values are divided into eight grades between 10 and 90 kcal· cm^{-2}. Because of the topographical differences, the possibility of sunshine varies between 550 and 1,600 hr in the growing season, May 15–September 30, corresponding to 27–79 % of the astronomically possible sunshine duration. Turner compared the distribution of vegetation types and growth of planted young trees, occurrence of *mycorrhizal fungi*, etc., and the distribution of global irradiation that the effects of local differences in global irradiation are reflected clearly.

Recently, Fliri calculated the energy budget in the central Alps crossing Bavaria, Tyrol, and Venetia and showed the cross-sections of relative sunshine duration, global radiation, albedo, direct global radiation, atmospheric backward radiation, heat used for evaporation, and long-wave outgoing radiation (Fliri, 1971). Figure 4.4 shows some of the interesting cases, such as (a) Annual mean relative sunshine duration, expressed as percentage of the maximum

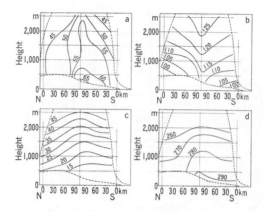

Fig. 4.4 Cross sections of (a) relative sunshine duration (%), (b) global radiation (kcal·cm⁻²·year⁻¹), (c) annual mean albedo of global radiation (%), and (d) long-wave outgoing radiation (kcal·cm⁻²·year⁻¹) along the Alps in Tyrol (Fliri, 1971).

possible duration at the respective points, is 65% on the valley floor in the central part, as opposed to 45% at the outer and higher parts of the mountain mass, where the cloudiness is much greater. (b) Global radiation (kcal·cm⁻²·year⁻¹) increases in accordance with increasing altitude and, due to the limited cloudiness in the central part of the mountain mass, the values are approximately 10% higher in the central part than in the outer parts when the same levels are compared. (c) Annual mean albedo of global radiation on the cross-section shows a striking contrast between the northern and southern sides of the mountain mass. (d) Cross-section of long-wave outgoing radiation also shows a contrast, which depends strongly on the ground surface temperature.

Radiation and heat balance observed in the high mountains in Russian Central Asia, Karakoram, Nepal-Himalaya, and Ecuador are given in Table 4.3 (Flohn, 1971). In this table, R is the radiation balance of the atmosphere;

Table 4.3 Radiation and heat economy in high mountains (Flohn, 1971).

	Height	Albedo	R	P	$L·E$	A	Month
Sary-Tasch, Alai-Valley	3,150m	0.17	321	178	89	54ly/d	September
Karakul-Lake, Pamir	3,990	0.22	324	243	29	52	August
Koshagyl, Aksu-Valley, Pamir	3,710	0.21	294	244	0	50	August
Fedchenko-Glacier, Firn	4,900	0.66	142	−37	116	7	July–August
Fedchenko-Glacier, near tongue	2,900	0.16	374	202	95	65	July–August
Turkestan Mountains, Kum-Bel-Pass	3,150	0.14	348	248	50	50	September
Turkestan Mountains, N-slope 33°	3,150	0.20	145	85	19	41	September
Turkestan Mountains, S-slope 31°	3,150	0.15	426	304	47	75	September
Chogo Lungma Gl. (Karakoram)	4,300	0.23	391	−41	10	434*	July
Paramo de Cotopaxi (Ecuador)	3,570	0.22	201	47	154	0	July
Chukhung (Nepal-Himalaya)	4,750	0.16	332	199	121	12	April

  * Ablation as the rest term.

P is the heat flux caused by the turbulence between the air and ground surface; $L \cdot E$ is the exchange of latent heat by evaporation, and A is the heat flux between the ground surface and the deeper layer of the earth. The radiation balance equation has been explained in Eq. (3.1). It was indicated that the heat balance in high mountains depends on the cloud development and rainfall occurrence in the afternoon hours caused by the secondary circulation of valley winds.

The vertical distribution of turbidity factor is important. Linke's turbidity factor Θ is given by the following equation:

$$\Theta = \frac{\ln I_o - \ln I}{\ln I_o - \ln I_{m,w}} \qquad (4.1)$$

where I_o is the flux density of the solar beam just outside the atmosphere; I is the flux density measured at the earth's surface with the sun at the zenith distance, which implies an optical air mass m; and $I_{m,w}$ is the intensity observed at the earth's surface for a pure atmosphere containing 1 cm of precipitable water viewed through the given optical air mass. In summer, the turbidity factor over 3,000 m a.s.l. is 2.0, and the solar radiation becomes 5 times weaker between 3,000 and 1,500 m and 10 times weaker between 1,500 and 500 m than the pure air mass. In winter, the turbidity factor over 3,000 m a.s.l. is 1.5, and the solar radiation becomes 4 times weaker between 3,000 and 1,500 m (Lauscher et al., 1932).

In addition, the local turbidity, such as is caused by summit clouds, must be taken into consideration. Mobile observations with cars or motorcycles were made during the years 1931 and 1932, and the intensity of solar radiation and turbidity factor were analyzed in relation to altitude and cloudiness (Lauscher, 1934). The turbidity factor at 1,600 m a.s.l. was 2.06 when it was cloudless but 3.05 when there was a cumulus cloud. It is pointed out by Lanscher that the difference is greater at higher places than at lower places.

4.1.3. Air Temperature

(A) Temperature Lapse Rate

Air temperature decreases with height. The dry adiabatic lapse rate is 0.976°C/100 m, that is, about 1°C per 100 m. The actual value in a mountainous region differs somewhat from this value because of the influences of the topography, vegetation, and the complex air movement due to small-scale circulation. Moreover, water vapor in the air also produces a lower lapse rate: First, the air temperature falls at the dry adiabatic lapse rate as the air goes up, until the water vapour attains the state of saturation (this height is called the *condensation level*), and above this level the vapour condenses. In the course of this condensation, the vapour releases a latent heat. In this case, the fall of the air temperature is less than at the dry adiabatic lapse rate. This lapse rate is termed *wet adiabatic lapse rate*. The value of this rate varies according to temperature and pressure. The values are given in Table 4.4.

Table 4.4 Wet adiabatic lapse rate (°C/100m).

Air pressure	−30	−20	−10	0	10	20	30°C
1000mb	0.93	0.86	0.76	0.63	0.54	0.44	0.38
900	0.93	0.85	0.74	0.61	0.52	0.43	0.37
800	0.92	0.83	0.71	0.58	0.50	0.41	
700	0.91	0.81	0.69	0.56	0.47	0.38	
600	0.90	0.79	0.66	0.52	0.44		
500	0.89	0.76	0.62	0.48	0.41		

The lapse rate deduced from the observed values of a mountainous region varies with areas, seasons, or height; a diurnal variation can occur even at the same point (Peattie, 1936).

The temperature lapse rates in Japanese mountains are shown in Table 4.5. This table shows that the maximum often appears both in February and August. The annual average is 0.56–0.62°C/100 m except for Mt. Tsukuba, as mentioned below.

Table 4.5 Observed temperature lapse rate (°C/100m) (Yoshino, 1961a).

Summit and mountain foot	Difference of height	Feb.	May	Aug.	Nov.	Annual mean
Mt. Iwate-Morioka	1,615 m	0.70	0.60	0.57	0.60	0.62
Mt. Tsukuba-Mito	839	0.39	0.33	0.44	0.30	0.36
Mt. Fuji-Kôfu	3,498	0.61	0.60	0.57	0.54	0.58
Mt. Hakone-Mishima	915	0.63	0.57	0.51	0.40	0.57
Mt. Ibuki-Shunjo	1,213	0.60	0.53	0.58	0.49	0.56
Mt. Aso-Kumamoto	1,104	0.61	0.54	0.62	0.51	0.57

Mt. Iwate (39°51′N, 141°01′E, 1,771 m), Mt. Tsukuba (36°13′N, 140°06′E, 869 m), Mt. Fuji (35°21′N, 138°44′E, 3,772 m), Mt. Hakone (35°11′N, 139°01′E, 940 m), Mt. Ibuki (35°25′N, 136° 24′E, 1,376 m), Mt. Aso (32°54′N, 131°04′E, 1,142 m).

Table 4.6 Temperature lapse rate (°C/100 m) of Mt. Zugspitze on the northern side of the Alps (Hendl, 1966).

	Jan.	Feb.	Mar.	Apr.	May	June	July	Aug.	Sept.	Oct.	Nov.	Dec.	Year
Zugspitze-Bad Tölz	0.39	0.45	0.52	0.61	0.61	0.63	0.63	0.59	0.55	0.47	0.44	0.39	0.52

Bad Tölz (47°46′N, 11°34′E, 654 m), Zugspitze (47°25′N, 10°59′E, 2,962 m).

Table 4.6 shows the monthly mean temperature lapse rate of Mt. Zugspitze on the northern side of the Alps (Hendl, 1966). The maximum appears in June and July, and the minimum appears in December and January. Such an annual march is normal in Middle Europe.

On Mt. Maljen, Yugoslavia (about 44°N, 20°E, 960 m), the lapse rate of mean daily temperature was calculated under the conditions of N-, NW-, SW-, and S-winds (Milosavljević, 1961). The results of the observation show that the lapse rates are greater in the case of N- and NW-winds than in the case of SW- and S-winds throughout the year.

This contrast between northerly and southerly winds is most evident in winter. The lapse rate on Mt. Kirishima, Japan (31°55′N, 130°51′E, 1,324 m), is 0.86°C/100 m on the slope 650 m a.s.l. and 0.45°C/100 m below that point in February (Yoshino et al., 1963). On the other hand, the rate on both the upper and lower parts of the slope is 0.68°C/100 m in August.

According to Alissow et al. (1956), the lapse rate on the slopes near Alm-Ata (Myn-Dshilki, 3,036 m) is greater (0.5–0.8°C/100 m) in summer, but it shows a strong inversion (−0.1−−0.7°C/100 m) below the Kamenskoje high-land (1,350 m) and 0.4–0.6°C/100 m above that place in winter. From May through October, 1955, the lapse rate of air temperature was also studied in detail by Baumgartner (1961) on Mt. Gr. Falkenstein, West Germany (49°05′N, 13°17′E, 1,307 m). As the inversion develops up to 150–200 m above the valley floor, a thermal belt or a warm zone appears on the slope about 750–850 m a.s.l. However, the inversion is so weak in summer that the thermal belt descends to the bottom of the valley. Recently, Stoenescu (1963) studied the temperature in the Bucegi mountain areas of the Carpathian ranges, Rumania. The mean lapse rate calculated between Virful Omul (2,507 m) and Predeal (1,082 m) was 0.50–0.80°C/100 m for the daily maximum and 0.26–0.43°C/100 m for the daily minimum. The lapse rate was smaller in the winter half-year because of the high frequency of inversion.

On the lower parts of the slopes of Mt. Tsukuba, Japan, there a marked inversion develops very frequently and the lapse rate at Mt. Tsukuba has smaller values. The large values for other mountains in February (see Table 4.5) are thought to be the result of frequent appearance of Pc-air masses from Siberia, which has a relatively large gradient of air temperature in the lower troposphere layer. According to observations at Mt. Fuji and Funatsu (35°30′N, 138°46′E, 860 m), the temperature lapse rate is 0.63°C/100 m in the case of a well-developed NW-winter monsoon (fresh Pc-air mass), 0.54 in the case of a migratory anticyclone, 0.45 in the case of a trough, and 0.49 in the case of a cyclone in winter (Sugawara et al., 1939).

Bleibaum (1953) has shown that in southern Rhön, West Germany, the lapse rate is 0.8–0.9°C/100 m in the case of a NW-airstream, 0.5–0.6 for a SE-airstream, 0.6–0.8 for an E-airstream, and 0.7–0.8 for a W- or WSW-airstream.

The relationship between the lapse rate and the cloudiness as an indicator of synoptic situations was studied on Mt. Kirishima (Yoshino, 1961b). The lapse rate of the minimum and maximum air temperature is always smaller in the case of few clouds on the upper part of the slope (between the mountain top and Hayashida on the mountain slope). On the other hand, a reverse relationship is found on the lower part of the slope (between Hayashida and Makisono). When the minimum and maximum air temperatures are compared, it turns out that the former is smaller than the latter. On the upper part of the slope, however, the lapse rate of the minimum temperature, when the cloudiness is 10, is slightly greater than that of the maximum air temperature, when the cloudiness is 0–1, both in winter and summer. Thus, on the lower part of the slope, the lapse rate of the minimum temperature is very small in winter. Tanner (1963) showed, based on a study in the Great Smoky Mountains, U.S.A., that air temperature lapse rate is generally a function of temperature as well as of the difference in elevation, and that the function changes with the season.

Summarizing the results observed in Japan and other countries, general features in the mountains in the temperate zone of the northern hemisphere are as follows: (a) the lapse rate is greater in the case of northerly than southerly winds, (b) the lapse rate is smaller on the lower part of the slopes or in low mountains, because of the high frequency of the low-level inversion, and (c) in most cases these tendencies are stronger in winter than in summer.

(B) *Diurnal Change*

The diurnal temperature range is small in mountains. The annual variation in the diurnal ranges on the summit of Mt. Fuji (3,776 m) and in Tokyo (6 m) are compared in Fig. 4.5 (Yoshino, 1961a). In this figure, the month is taken for the abscissa, the time of day for the ordinate. The smaller the diurnal temperature range is, the more vertical the isopleths become. In tropical regions where the diurnal range is large and the annual range small, the isopleths lie horizontally (Troll, 1943). From this figure it may be clear that the maximum temperature comes earlier in mountains than in lowlands; i.e., about 12 to 13h on Mt. Fuji and 13–14h in Tokyo. This is because the clouds formed by the updraft spread over the mountains or a fog envelops them, preventing the rise of temperature in the afternoon.

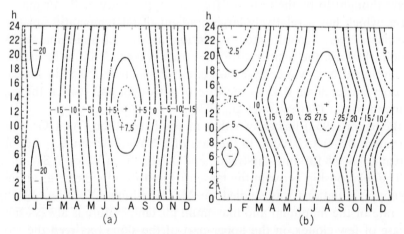

Fig. 4.5   Annual change of diurnal variation of air temperature (°C)
(a) on the top of Mt. Fuji and (b) in Tokyo (Yoshino, 1961a).

The diurnal range of air temperature becomes small as with height increases. This is clearer in winter than in summer on Mt. Fuji, as shown in Fig. 4.6 (Fujimura, 1971). In summer the type of diurnal change on the summit is quite similar to that at the mountain foot.

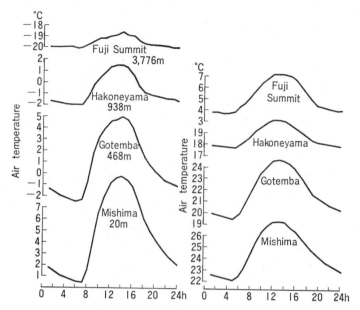

Fig. 4.6 Diurnal change of air temperature on Mt. Fuji (Fujimura, 1971).

The eastern slopes of the Andes in Central Peru, lying between 8° and 12°S, present an ideal situation for studying the lapse rate from tropical lowlands to high mountains (Drewes et al., 1957). The temperature lapse rate here is 0.46°C/100 m on the average, but it differs according to the time of day and altitude. The altitudinal variation in the lapse rate of the maxima is much greater than in the lapse rate of the minima. At the lower elevations, 650 to 1,800 m, the mean monthly maximum temperature decreases on an average of 0.88°C/100 m; from 1,800 to 3,000 m the average decrease is 0.20°C/100 m; and from 3,000 to 4,200 m it is 0.49°C/100 m.

On the other hand, we have as yet very little knowledge of the temperature lapse rate in the polar region. On the Greenland Ice Sheet, the lapse rate of the annual mean air temperature is 0.74°C/100 m, from the values at the stations on the coast and at heights of 2,000–3,000 m on similar latitudes (Diamond, 1960).

(C) *Annual Range and Monthly Range*

In mountain regions, the annual range of air temperature, i.e., the difference between the mean monthly air temperature of the warmest month and that of the coldest, presents very complicated features. As summarized by Hann-Knoch (1932) and Conrad (1936), in general, the higher the altitude is, the smaller the annual range becomes in the alpine regions of Europe and the U.S.A.

In Japan, the annual range of air temperature does not show a simple change with height, but an interesting curve like the one in Fig. 4.7 (Yoshino,

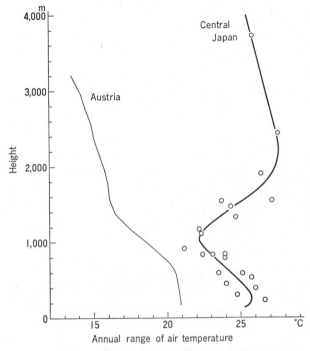

Fig. 4.7 Annual range of air temperature (°C) and height a.s.l. on the mountains in central Japan and in Austria (Yoshino, 1965).

1965). This figure is drawn by plotting the data observed at the mountain-top stations in central Japan (135–140°E, 35–37°N) during 1939–1948. The curve for central Japan in this figure provides an apparent maximum around 2,200 m and a sharp minimum around 1,000 m a.s.l. Geiger (1961), summarizing Lauscher's reports, gives a curve for the mountains in Austria, which may be compared with the one for Japan. The comparison will confirm the peculiarity of the tendency in central Japan, because the maximum and the minimum in the curve for Austria are somewhat vague. However, to explain the maximum and minimum is rather difficult, because the vertical changes of the monthly mean air temperature on the mountain have not yet been thoroughly studied climatologically. One possible explanation may be that the effect of low-level inversion in winter is strong, as discussed above, the air temperature in February around 1,000 m a.s.l. is warmer than that assumed by the lapse rate of 0.6°C/ 100 m, so that the annual range of the mountains around 1,000 m is smaller. In August, the inversion layer does not develop so strongly, but the radiation of the ground surface of the mountain has a great influence in the daytime on the air temperature. The characteristics of air masses in August must also be taken into consideration. On this basis, the air temperature up to 2,200 m, as far as the data presented here are concerned, is warmer than that assumed by the lapse rate 0.6°C/100 m. Perhaps this is the reason that the annual range is greater at this altitude. In Germany, the annual range is somewhat greater than the values given by the curve for Austria below 2,000 m a.s.l. (Maisel, 1931).

At any rate, the relationship between annual range of air temperature and the height of the mountain must be studied in detail not only in the temperate zone, but also in humid tropics. Even in Japan, there are great deviations: On the mountains in southern Japan (32–34°N), the annual ranges are 3–4°C smaller and, on the other hand, in northern Japan (38–43°N) they are 2–7°C greater than that given in the curve in Fig. 4.7.

The monthly range of air temperature is defined here as "the long year mean of the monthly highest air temperature minus the long year mean of the monthly lowest air temperature." The monthly range of air temperature is always greater than the monthly mean diurnal range of air temperature, but the former may also represent characteristics of the latter.

The relations between the monthly range of air temperature and height a.s.l. in central Japan in February and August can be summarized as follows. (a) A marked minimum appears at about 1,500 m in February and about 2,000 m in August. (b) Excluding the part around the minimum of the curve, the range of air temperature increases with increasing height in February, but no change is observable in August. (c) The ranges in February are 5–10°C greater than in August. (d) The ranges on the mountains in northern Japan are different from those in southern Japan.

On the "thermoisopleth-diagram" given by Troll (1943), which shows the diurnal and seasonal changes of air temperature, even on the mountains about 3,000 m a.s.l. there are strong characteristics of the thermoisopleth patterns of the climatic region where the mountain is located (Troll and Paffen, 1964). However, it is still doubtful whether the evidence concerning the annual range

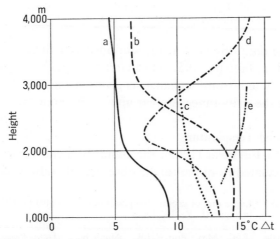

Fig. 4.8 Daily range of temperature (°C) and the height a.s.l. a: Alps, b: U.S.A., c: Equatorial Africa, d: Himalaya, and e: Ethiopia (Lauscher, 1962/64).

and height is only applicable to the Japanese mountains or to the mountains from humid tropics to the polar region.

(D) *Daily Range*

The daily range of air temperature varies in accordance with altitudinal change. The relationship between them is, however, different from region to region. Figure 4.8 shows the relations for the various station groups in the European Alps, U.S.A., equatorial Africa, Himalaya, and Ethiopia (Lauscher, 1962/64). There is as yet no plausible explanation for these different curves, but it seems that aridity of climate plays an important role in these cases.

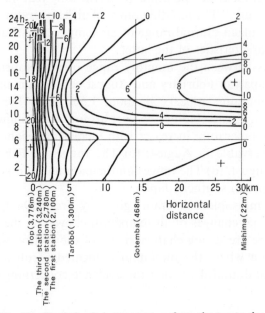

Fig. 4.9 Isopleth of air temperature from the top to the foot of Mt. Fuji in January, 1951 (Owada et al., 1972).

By using the data observed at the stations from the mountain top to the foot of Mt. Fuji in January, 1951, Owada et al. (1972) showed an interesting feature of the daily temperature change as given in Fig. 4.9. From this figure it will be seen that the daily range is great up to 1,300 m a.s.l.

4.1.4. Ground Temperature

According to Fukuoka (1966), the monthly mean lapse rate of ground temperature, calculated from the results observed at stations up to 936 m a.s.l.

Table 4.7 Lapse rate of ground temperature (1 m depth) in °C/100 m in Japan (Fukuoka, 1966).

Jan.	Feb.	Mar.	Apr.	May	Jun.	Jul.	Aug.	Sep.	Oct.	Nov.	Dec.	Year
0.29	0.28	0.33	0.40	0.30	0.42	0.37	0.36	0.39	0.40	0.37	0.36	0.36

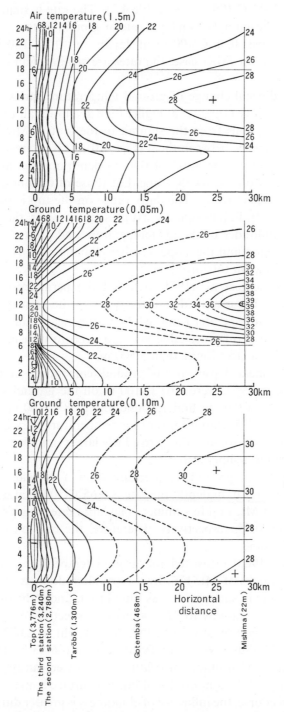

Fig. 4.10   Isopheths of air temperature and ground temperature (0.05 m and 0.10 m depth) from the top to the foot of Mt. Fuji in August, 1952 (Owada et al., 1972).

in Japan, is 0.28–0.42°C/100 m (see Table 4.7). The rate is small in winter and large in summer. The annual mean lapse rate of ground temperature (1 m depth) is, as this table shows, 0.36°C/100 m in Japan, but 0.22–0.30 in Angola and 0.33–0.37 in Turkey. In either case, the rates are smaller than the lapse rate of air temperature.

The type of diurnal change of ground temperature varies with increasing altitude. The daily range is greater at higher than at a lower altitudes. The results of observation at stations from the summit to the foot of Mt. Fuji show that there is a clear difference between them (Fig. 4.10). It is interesting to note that the maximum ground temperature of 0.05 m depth appears at noon and the minimum about 3 h on the whole slope, but those of 0.10 m depth appear at 16 h and at 6–7 h respectively. The daily range on the top is twice as great as that on the foot of Mt. Fuji in summer. At present, however, there is no detailed comparison of amplitudes of the air temperature and ground temperature on the high mountain-top and the slope in other climatic regions. It is hoped that research will be done on this problem.

Recently, Higuchi et al. (1971) detected the permafrost on the top of Mt. Fuji (3,776 m). The mean value of the depth of the permafrost table at ten observation points at the crater rim is 64.3 cm; the mean value of their altitude being 3,732 m. The permafrost table was too deep to measure by the simple sounding method at places lower than 3,500 m.

The monthly mean lapse rates of ground temperature on the SE slope of Mt. Fuji in August, 1952, were 0.67°C/100 m on the ground surface, 0.66°C/ 100 m at the depth of 0.05 m underground and 0.58°C/100 m at the depth of 0.10 m underground. The lapse rates of air temperature vary according to the height as mentioned above, but they are a little larger in most parts of the slope than those of ground temperature.

The lapse rates of ground temperature show a diurnal change: For instance, in the case of 0.10 m depth, the rate is 0.65°C/100 m at 6 in the morning and 0.44 at 15 h on Mt. Fuji in summer. The lapse rates of air temperature are 0.55°C/100 m and 0.75, respectively. Therefore, the relation is inverse in the nighttime and daytime.

Fujita et al. (1968) measured the ground temperatures on the slope of volcanic ash and rocks of Mt. Fuji by a portable radiometer from an aircraft in the early morning in summer. It was found that, under the morning sun, the E-slope heats up very rapidly to 32°C almost irrespective of the elevation, but the other slopes remain mostly 14°–20°C. At a height of 2,072 m a.s.l. on the WNW-slope of the Gurgl Valley in Tyrol, Austria (46°53′N, 11°02′E), a continuous observation of ground temperature was carried out by Aulitzky (1961, 1962 a,b). He showed the amplitudes of ground surface temperature in relation to cloudiness. Of course, the influence of cloudiness is greater during the summer and early autumn, but it is greatest in July and second greatest in September. In August, the influence of vegetation reduces the difference of the amplitudes between a clear day and an overcast day. Olecki (1969) also studied the effect of cloudiness on the diurnal variation of ground temperature in summer at Gaik Brzezowa in the Carpathian submontane zone and came to a similar conclusion. But it must be noted that the correlation between the amplitude of temperature and cloudiness is the highest at 5 cm depth.

In most cases the relation between the ground temperature and air temperature in middle latitudes in warmer seasons is expressed by a linear regression equation. For instance, Owada et al. (1972) have obtained the following equation from an observation on Mt. Fuji in August, 1952:

$$T_s = 5.6 + 0.91 T_a \tag{4.2}$$

where T_s is the monthly mean ground temperature of 0.10 m depth and T_a is that of the air temperature 1.50 m above the ground. However, the relation between the ground temperature and air temperature in the case of a cold climate is complex. Aulitzky (1962b) presented a relation between the ground temperature (0–1 cm and 10 cm depth) and the air temperature (200 cm) at a station 2,072 m a.s.l. from June, 1954, to June, 1955. The results are shown in Fig. 4.11. The values observed during the snow-free season are plotted on this figure, but the values in spring and late autumn make the left parts of the lines complex. This is because of the following facts: First, in spring the air temperature becomes higher, but the ground temperature under snow accumulation stays relatively lower, and second, in late autumn, the ground temperature becomes lower by stronger radiation cooling caused by the lower sun height.

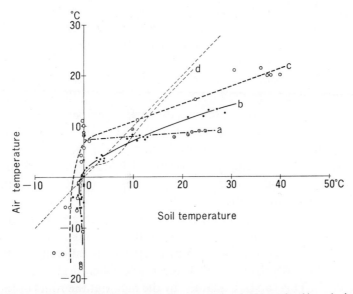

Fig. 4.11 Relation between ground temperature (0–1 cm and 10 cm depth) and air temperature (200 m) at a station, Obergurgl/Poschach-Forest limit, 2,072m (Aulitzky, 1962b). a: mean amplitudes (0–1 cm), b: mean extremes (0–1 cm), c: absolute extremes (0–1 cm), and d: mean extremes (19 cm).

Table 4.8 Difference between the air (150 cm) and the ground surface temperature at 13h during various weather conditions in summer at W-Carpathian Mountain, Poland (Hess, 1971).

Vertical climatic zone	Height a.s.l.	Difference, Δt_{150-0}		
		Anticyclonic SW-airstream (Aug. 22, 1961)	Anticyclonic NW-airstream (Aug. 24, 1961)	Average (Aug. 20–27, 1961)
Cold	2,400m	+10.3	+0.2	−3.5°C
Moderately cold	2,000	+7.5	−1.5	−3.7
Very cool	1,700	+5.4	−2.8	−3.8
Cool	1,350	+3.0	−4.3	−4.0
Moderately cool	900	−0.2	−6.3	−4.0
Moderately warm	500	−2.9	−8.0	−4.3

In the higher places on the mountain, the difference between the air and ground temperatures depends largely upon the weather. An example of this phenomenon may be seen in Table 4.8, which shows the temperature difference at 13 h in the West Carpathian region of Poland (Hess, 1971). The deviation according to the weather, that is, the variation from August 22 to 24, is the greatest at 2,400 m and the smallest at 500 m, but of special interest is the average value of eight days from August 20 to 27, which provides the smallest value at 2,400 m. The difference depends upon the vertical climatic zone (Hess, 1965) of the W-Carpathian Mountains.

Aulitzky (1961) also gives the duration of ground temperature in percentage of hours, based on observations from June 1954 to December 1958, as in Fig. 4.12. The period from November to April is a winter season with snow cover, and the period from May to October a vegetation period. During the summer months, the ground surface temperature becomes higher than 20°C in about 20% of the hours. The results also show that, during the autumn months from September to November, the surface temperature falls to −5 to −10°C. This means frost in these months, which might influence the development of vegetation, structural soil, etc., in the high mountains. During the winter months, on the other hand, the snow cover prevents the surface temperature from lowering; it stays at −5 to 0°C. Thus, the ground temperature of −5 to 0°C appears as deep as 20 cm during 10–40% of the hours in a year if there is snow cover, but if there is no snow cover, it appears as deep as 1 m during 35–50% of the hours in a year, and the ground temperature of −10 to −5°C occurs near the surface for 10% of the hours in a year.

The local difference of ground temperature is much greater on high mountains than that of air temperature. According to results of observations in Haggen, Tyrol, Austria (47°13′N, 11°6′E, 1,830 m), the difference of the ground temperature (10 cm depth) between Haggen and Obergurgl was most striking (Kronfuss, 1972). This can be explained by the fact that the roots of the younger plants spread over this soil layer.

The surface condition of the ground greatly affects the air temperature in

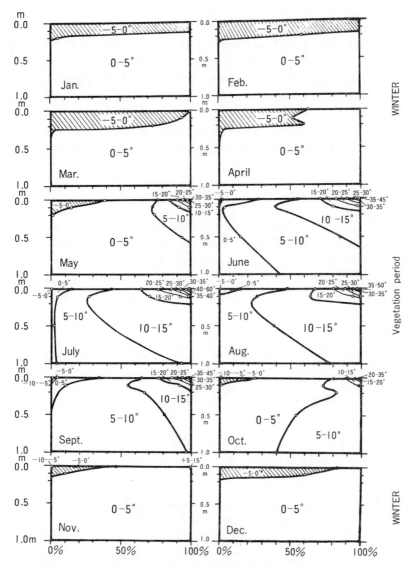

Fig. 4.12 Monthly mean duration of ground temperature in % of hours at a station, Obergurgl, 2,072 m a.s.l. (Aulitzky, 1961).

higher places. For instance, a glacier plays a more decisive role on thermal conditions in high mountains in the glaciated area than in the nonglaciated area (Hess, 1962). Hess has made clear that the cooling effect of the Fedchenko Glacier in Pamir increases with altitude: The glacier lowers the annual mean air temperature by 0.4°C at a height of 3,000 m, but by 2.8°C at 4,000 m and by 5.2°C at 5,000 m (Hess, 1967).

4.1.5. Humidity

Humidity is determined by temperature and the amount of water vapour, and the water vapour content decreases rapidly with altitude due to the rapid decrease in temperature. The value of the pressure at 5–5.5 km a.s.l. is half that at the sea level, but the amount of water vapour in the free atmosphere decreases by half at about 1.5 km a.s.l. The water vapour content may more properly be represented by the absolute humidity or water vapour pressure, and the relationship between water vapour pressure and height is given in the following formula:

$$l_h = l_o \times 10^{-\frac{h}{6}\left(1+\frac{h}{20}\right)} \qquad\qquad (4.3)$$

where l_h and l_o are the water vapour pressure at a height of h km and on the surface level, respectively. Relative humidity is in inverse proportion to temperature, and it is low during the daytime and high at night, except in the alpine zone. Actually, however, it is high where clouds or fog arise, and suffers a seasonal change with the arrival of various air masses, as discussed below.

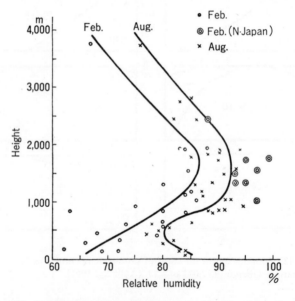

Fig. 4.13   Relative humidity and height on the Japanese mountains in February and August (Yoshino, 1961a).

Figure 4.13 shows the relationship between the monthly average relative humidity and the height above sea level in February and August in Japan (Yoshino, 1961a). Generally, the humidity is low in winter and high in summer. On mountains of 1,000–2,500 m in the Kanto, Tohoku, and Hokkaido districts, Japan, the relative humidity is high in winter. The most striking features in

Fig. 4.13 are the maximum at 1,500–2,000 m in February and August, and the minimum at 400–500 m in August. As are the cases with the sunshine duration and the annual temperature range stated above (4.1.2 and 4.1.3 (c)), this state corresponds approximately with the vertical distribution of the number of fog days shown in Fig. 4.15. Below 400 m, however, humidity increases as the altitude decreases, though the change in the number of foggy days is slight. This phenomenon can be explained by the fact that the distance from the coast plays a greater role than the height above sea level in Japan. It may be worth noticing that this phenomenon occurs only in summer when temperature and therefore evaporation from the sea are at their highest, but not in winter.

The winter maximum in N-Japan can be explained by the effects of the monsoon. The winter monsoon comes from the cold continent of Eurasia, but it takes a great amount of water vapour from the warm Japan Sea; thus, it is humid when it arrives in Japan. The average monthly humidity in the mountains in N-Japan is approximately 15% higher than those in S-Japan.

The relative humidity in mountains also varies greatly in accordance with the prevailing wind direction, which is caused by the different synoptic weather patterns. Milosavljević (1961) reports on relative humidity at different wind directions at Divcibare, 960 m a.s.l., near the summit of Mt. Maljen (1,103 m

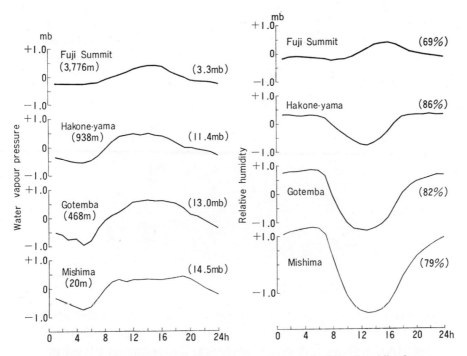

Fig. 4.14 Diurnal change of water vapour pressure and relative humidity from the summit to the foot of Mt. Fuji, expressed by departure from the daily mean value given in brackets (Fujimura, 1971).

a.s.l.) in Yugoslavia: In the case of N-winds it is 85–96%; in the case of NW, 89–91%; in the case of SW, 53–78%; and in the case of S, 47%.

The water vapour pressure and relative humidity in high mountains show interesting diurnal changes. Generally, the change is of a mountain type at altitudes higher than 2,000 m a.s.l. Mt. Fuji is shown as an example in Fig. 4.14. The water vapour pressure and relative humidity have similar type of change with maxima in the afternoon. But, on the slope below 1,000 m, the water vapour pressure has its maximum in the daytime, but the relative humidity has a maximum in the early morning. The minimum of the water vapour pressure occurs in the early morning, but that of the relative humidity in the afternoon (Fujimura, 1971). Thus, the diurnal change of humidity in an alpine zone has specific characteristics.

4.1.6. Fog and Clouds

Fog as a weather phenomenon is a visible aggregate of minute water drops floating in the air. Fog reduces horizontal visibility to below 1 km. "Mist" in English and "damp haze" in American English are the terms for a state with a visibility between 1 and 2 km. The relative humidity in mist is lower than in fog. When a low-altitude cloud covers part of a mountain, an observer in the cloud regards it as fog. In a mountainous district, air current cools through adiabatic change as it flows up a slope, producing water droplets called cloud particles or fog drops. Clouds are likely to form in some layers of a certain altitude, but not in other layers of other altitudes even when there are no mountains nearby, so that the occurrence of fog and clouds on a mountain is affected by its height, which corresponds to the average condensation level.

Photo 4.1 Low stratus layer formed by the ground inversion observed from Mt. Kobushi (2,483 m) in central Japan (September 29, 1968, 5h, by M. M. Yoshino).

The layers with the maximum amount of clouds are as follows (Ishimaru, 1952): The first is a layer of a fog or stratus, appearing below 400 m near the inversion layer, with a depth of 100 to 200 m. The second is a layer of clouds, cumulus or cyclone nimbostratus, having maximum at about 1,500 m. The height of these cloud bases H is represented approximately as follows:

$$H = 125(t_o - t_d) \qquad\qquad (4.4)$$

where t_o is the temperature on the ground (°C), and t_d is the dew-point temperature of the air. Several other layers of clouds lie over these layers. The layers of minimum cloud quantity are at heights of 300–1,500 m and 2,500–3,500 m. The height of these layers varies according to season: in general, they are high in summer and low in winter.

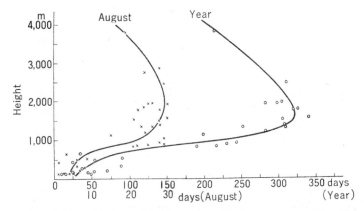

Fig. 4.15 Number of days with fog and height a.s.l. on the Japanese mountains (Yoshino, 1961a).

The vertical distribution of the number of foggy days in Japan shows a distinct maximum at 1,500 m, as illustrated in Fig. 4.15. The annual total number of foggy days exceeds 300 days; especially in August, fog occurs almost every day at a height of 1,500–2,000 m. The height at which the maximum and the minimum appear differs from one season to another and from one district to another.

Photo 4.2 Cap cloud over P. Bagna (3,129 m) observed from Monti della Luna (2,204 m) in W-Alps (September 5, 1962, by M. M. Yoshino).

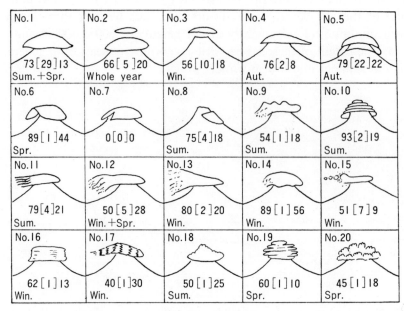

(a) Cap Cloud

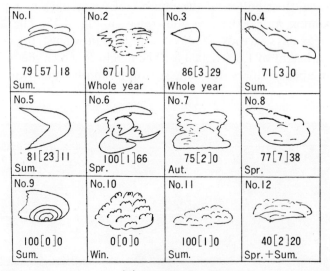

(b) Hanging Cloud

Fig. 4.16 The varieties of cap-form clouds on Mt. Fuji and hanging clouds on the lee of Mt. Fuji. Square brackets show occurrence frequency (%), values on the left of square brackets are precipitation probability (%) and values on the right of square brackets gust probability (Yuyama, 1972).

A close investigation carried out in the Alps (Ekhart, 1950) reveals the following relationship between the average annual amount of clouds and the height above sea level: The minimum appears on the ground surface, the second minimum at about 1,500 m, the maximum at 3,000 m, and the second maximum below 500 m. The condensation level in the E-Alps was studied in relation to precipitation distribution by Fliri (1967c).

The occurrence of fog in a mountain, (i.e., clouds covering mountains) at certain heights affects every climatological phenomenon in the mountainous regions, including insolation, sunshine, air temperature, ground temperature, humidity, etc. This is shown by a comparison of the curve of variation in fog days with height shown in Fig. 4.15 and the curves in Figs. 4.2 and 4.13.

Furthermore, mountains have their own forms of a cap-form cloud near the summit, a flag-form cloud on the one side of the ridge, and a hanging cloud on the leeward of the ridge, among others. The hanging cloud on Mt. Fuji, Japan, is well known for its beauty. In Italy, this cloud is known by the name *Contessa del Vento* (Countess of the Wind). Specific measurement of and experiments on the cloud on Mt. Fuji were done by Abe (1937), who classified the cap clouds into 20 varieties.

Yuyama (1972) observed the cap-form clouds on Mt. Fuji and the hanging clouds on the leeward side of Mt. Fuji and compiled statistics (1944–1949) of occurrence, precipitation probability, and gust probability for each cloud form, as shown in Fig. 4.16. The most frequent cloud forms are Nos. 1 and 5. When these cap-form clouds appear, the precipitation probability is very high, 75–90%. These clouds persist 3–4 hours in most cases.

Today, the cloud quantity over high mountains can be studied by satellite pictures. One of the most interesting studies is the one in the central Sahara mountains from the satellite pictures (Winiger, 1972). On the Tibesti mountains, the annual average cloud amount is 10%. The cloud amount of each mountain in Tibesti is 8% at Emi Koussi (3,415 m), 15% at Mouskorbé (3,376 m), and 23% at Toussidé (3,265 m). From these values, the annual precipitation estimated is 80–120 mm at Emi Koussi, 100–150 mm at Mouskorbé, and 150–250 mm at Toussidé. Meridional breaks of the equatorial air are of great importance for the formation of a cloud and rainfall. They reach the Hoggar mountains, where the annual average cloud amount is 23%, about ten times more often than the Tibesti. This is called a *cloud-bridge* from the Niger to the Sirte.

Ives (1941) described the crest clouds over the Colorado Front Range, which occur in association with chinook. The top of the crest cloud is approximately 5,000 m a.s.l. and the lower limits of the crest cloud are 3,000 m on the windward side and 3,500 m on the leeward side of the range. The base of the cloud is 30 m above the land surface. In the crest clouds above this height, there appears a chaos of swirling snowflakes, pine needles, and gravel particles, as Ives wrote.

The cloud formation along the local front, the convergence or the conflict zone between the prevailing wind and local circulation, such as the sea breezes, is discussed in Section 5.1.2 (E). Probably, the most striking are the clouds on the Hawaii Islands (Henning et al., 1967). The amount of fog precipitation was measured by a Hohenpeissenberg fog collector (Grunow, 1960) on Gr. Falkenstein, W-Germany (Baumgartner, 1958). According to his measurement, the amount is 176% of the normal precipitation on the summit, 1,312 m a.s.l. and the fog maximum in the zone from the summit to a height of 1,150 m. In the zone at 800–700 m the minimum appears, whereas in the zone from 700 m to the valley bottom valley fog (ground fog) appears. Using the same type of fog collector, an observation was carried out on Mt. Sonnblick (3,106 m) from November 1959 to May 1965 (Grunow et al., 1969). Excessive fog deposits amounted to from 88% to 353% of the monthly precipitation observed by normal rain gauge. The highest amount of fog deposits is 6.9 mm/hr.

Nagel (1956) is concerned with the effect of wind velocity on the fog precipitation. He made following experimental equation:

$$N_w = [(R+N)-R] - \frac{v}{15} \cdot R \qquad (4.5)$$

where N_w is an addition of fog precipitation under the influence of wind velocity; $(R+N)$ is the amount of precipitation measured with a fog collector; R is the amount of precipitation measured by normal rain gauge, and v is wind velocity (m/sec). Worldwide measurements using the Hohenpeissenberg fog collector are now under way (Grunow, 1963).

An observation on the frequency distribution of fog particles has been made on Mt. Nisekoannupuri in Hokkaido (Kinoshita, 1952). The results on the summit (1,300 m) were compared with those on the mid-slope (1,100 m), where the base of a cap cloud often lies. It is shown that the fog particles grow larger with the ascent of mass air and that in particular particles with a larger diameter increase in number. The experimental formula of frequency distribution of fog particles is expressed by Best (1951) as follows:

$$1 - F = \exp\left[-\left(\frac{x}{a}\right)^n \right] \qquad (4.6)$$

where x is the diameter of fog particles; F is a ratio of fog-water quantity with smaller particles than x to the total amount of fog-water; n is a constant having a value of about 3.3, and a is a constant that increases with the amount of fog-water. According to the investigation on Nisekoannupuri, a in Eq. (4.6) is 11.6 on the mid-slope and 16.7 or 20.8 on the summit.

4.1.7. Precipitation

(A) *Rainfall Zone*

Observations made it clear already in the 19th century that the amount of precipitation is larger in mountains than in lowlands. A theoretical elucidation of the phenomenon, however, was attempted for the first time by Pockels (1901).

Pockels asserts that the amount of precipitation increases with height and that the maximum comes at the steepest incline or at a point slightly removed toward the summit. Pockels's theory, however, was only a first approximation, and later Ono (1925) made a more detailed theoretical study, arriving at a result that is also applicable to the leeward areas of mountains. The values were ascertained by observation of snow in Yamagata Prefecture, Japan.

The relationship between monthly or annual precipitation and the height above sea level was investigated in Europe chiefly at the beginning of the 20th century. The conclusion was that the maximum amount of precipitation comes on the middle part of the slope of a high mountain. The height of the maximum precipitation zone is below 1,000 m in the tropics and is 1,400–1,500 m in the temperate zone. Later investigations have proved that there are rather great regional differences in height. The place on a slope where the maximum appears is a "precipitation zone" or "rainfall zone". The reason why the maximum appears on the middle part of the slope of a high mountain is that as the air ascends along the slope it forms water droplets through adiabatic change, which at a certain height fall as raindrops. From this it should be clear that the height of a rainfall zone is related both to the temperature and the amount of water vapour. Above 2,000 m, the decrease of water vapour in the air reduces the amount of rainfall, although the zone is higher in summer than in winter, since the air temperature is higher in summer.

Recently, Weischet (1969) investigated the vertical profiles of precipitation on tropical mountains. He summarizes his results from the Columbian Andes as well as the obervations by Hastenrath (1967) on Central America, by Domrös (1968) on the Central Highland of Ceylon, by Flohn (1968) on Venezuela and on the Peru-Bolivian Andes by Ortolani (1965). Further Flohn (1970) adds an example from the SE-slope of Mt. Kilimanjaro (5,895 m). Weishet (1969) concluded that: (a) In the humid tropics the maximum precipitation zone is located between 900 and 1,400 m; toward the higher levels the annual rainfall decreases at a rate of 100 mm/100 m; and (b) Where a high plateau is extended in the mountainous region, a secondary precipitation maximum occurs at the marginal ranges of the plateau.

One of the "wettest spot of the earth" is found at a station on Mt. Waialeale on the island of Kauai, Hawaii (Henning, 1967). This station (22°04′N, 159° 30′W) is situated at a height of 1,547 m and here the average annual rainfall from 1920 to 1958 was 12,344 mm. Other famous points are found at Cherrapunji (25°15′N, 91°44′E, 1,313 m) and its neighborhood, Mawsynram (25° 18′N, 91°35′E, height unknown) in Assam, India. Annual rainfall at Cherrapunji is 10,869 mm (49 years' average) and at Mawsynram 12,666 mm (5 years' average). It must be noted here that these stations are situated on the maximum rainfall zone on the respective mountain slopes under the macroclimatological conditions of continuous tropical airstreams with abundant water vapour. Its relation to the local front is discussed in Section 5.2.3 (G).

As has been discussed by Henry (1919) and Wagner (1937), the rainfall zone on the slope is obscured at the middle and high latitudes. Weischet (1965) proposes that an advectional type of precipitation is normal in extratropical regions, in contrast to the convectional type of precipitation in the tropics. He presents cross sections for the extratropics: (a) E-Bavaria-Drau-Valley, (b) Swiss Mittelland-Sesia-Valley, (c) Canada along 49°30′N, (d) U.S.A. along 39°N and (e) U.S.A. along 34–35°N. As far as these cross sections are concerned, the precipitation is high at higher places on the mountains. That is, the shapes of the cross section of precipitation and the topography are quite similar. For the tropics, the cross sections of Hawaii along 19°40′N, Java along 113°E, Colombia along 5°N, Ecuador along 0°50′S, and Chile-Bolivia along 20°S are presented, and the maximum zones on the slopes are clear.

Table 4.9   Rainfall on the mountains in Japan (Hirata and Tamate, 1922; Kato, 1922).

Height a.s.l.	Station	Rainfall (mm)
	Mt. Mitsumine, June–Oct., 1916	
1,080m	Summit	1449
900	NE-slope	1524
700	″	1497
500	″	1406
300	NE-foot	
	Mt. Haruna, June–Oct., 1915–1918	
1,391m	Summit	1957
1,179	E-slope	1789
961	″	1525
200	Foot	1023
	Mt. Shiragami, May–Nov., 1917–1919	
1,470m	Summit	1658
960	E-slope	1793
450	E-foot	1575
	Mt. Nantai, June–Oct., 1914–1918	
2,480m	Summit (W)	1253
2,484	Summit (Center)	1141
2,480	Summit (E)	1349
2,453	Summit (Shrine)	1455
2,273	Slope	1644
2,133	″	1538
1,798	″	1692
1,495	″	1706
1,296	Mountain foot	1835
1,270	″	1628
1,270	″	1434
	Mt. Tsukuba, June–Oct., 1902–1910	
870m	Summit	897
240	Slope	905
30	Foot	775

Observations of mountain climate have become extensive in Japan since the end of the 19th century, especially after 1910, when the bitter experience of serious country-wide damage from floods showed the necessity for observations of the precipitation distribution in mountains. Table 4.9 shows the results of some of those observations (Hirata and Tamate, 1922; Kato, 1922). According to these observations, the height of the maximum amount of rainfall during the warm period is 900 m on Mt. Mitsumine, 1,200 m on Mt. Haruna, 1,000 m on Mt. Shiragami, 240 m on Mt. Tsukuba (where height plays no remarkable role), and 1,300 m on Mt. Nantai. Thus, the rainfall zone generally lies at 1,000–1,500 m, except for lower mountains (lower than about 1,000 m) like Mt. Mitsumine or Mt. Tsukuba. These observations were helpful to some extent, though they were not completely accurate; the observation points were not carefully selected, and the influence of wind turbulence, which caused raindrops to get into rain gauge, was overlooked. The problem of rainfall on a slope incline is discussed below.

The total annual amount of rainfall on the summit of Mt. Fuji is unknown due to lack of observations during the winter. From the extrapolated value based on the measurement on serveral points from the mountain-foot to 1,300 m a.s.l., together with the measured value of 5,600 m on the summit at the time of Typhoon Cathleen (September 12–15, 1946), the annual amount of rainfall on the summit is estimated at 26,000–28,000 mm (Fujimura, 1950, 1952). The latter value was determined using a Fujimura-type rain gauge, which is so designed as to admit raindrops even from the sides of the cylinder with collecting funnel. A rain gauge with a horizontal collecting funnel is of little use on a mountain top, because the funnel catches few raindrops where the wind is very strong, as it always is on a mountain top. This is a very perplexing obstacle to accurate measurement of the amount of rainfall, the somewhat precarious foundation on which the definition of the amount of rainfall in mountainous districts is based.

Annual precipitation on the eastern slopes of the Colorado Front Range surprisingly increases with height to the highest station, 3,750 m, where it is at least 1,020 mm (Barry, 1972). The vertical distribution of precipitation shows not only major seasonal contrasts, but it also varies markedly in the same month in different years according to synoptic situations such as upper cold lows or troughs. Williams et al. (1962) reported that the precipitation profile across the mountain range of Utah, U.S.A., differed with synoptic situations, such as cold lows and frontal activities.

Bergeron (1949) made a thorough review of the coastal orographic maxima of precipitation on the Scandinavian and Dutch coasts in autumn and winter.

(B) *Precipitation Distribution*

An investigation in and around Mt. Nantai (Ebara and Momoi, 1915, 1917) has already made clear that the relation between the height above sea level and the amount of rainfall changes with various causes of precipitation. This problem has also been the objective of observations on Mt. Fuji (Fujimura, 1948), on Mt. Iwate (Yoshida, 1951), and on Mt. Ibuki (Kodama, 1954). On the southern slope of Mt. Iwate, the maximum rainfall zone is on the mountain slope at the time of a cyclone, a *bai-u* front, or a typhoon, but the maximum appears either on the summit or on the foot, depending on the wind speed and

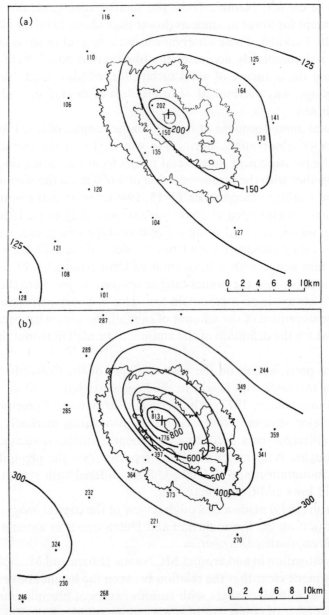

Fig. 4.17 Average rainfall distribution in the Mt. Kirishima region, Kyushu, Japan, in (a) February and (b) August (Yoshino, 1961b).

the height of the cloud base. Low wind speed, among other factors, plays a pre-eminent part in cases where the amount of rainfall varies very little with height; the maximum rainfall is on the summit, and there is little rainfall at the foot when the wind is northwesterly after the passing of a cyclone or a cold front. A thunderstorm or a shower can be the cause of irregular rainfall distribution with height.

The distribution of rainfall in the Kirishima area, Kyushu, Japan, is as follows: The annual precipitation reaches 4,500 mm at the NW part of the region (Yoshino et al., 1963), and the distribution pattern differs according to the causes, as shown later in Fig. 5.6 (Yoshino, 1961b). In the case of extratropical cyclone, the maximum appears on the windward part of the mountain region, but in the case of a typhoon, the maximum appears slightly leeward of the center of the mountain region because of the stronger wind, which brings raindrops further leeward. Therefore, the monthly isohyets do not coincide with the

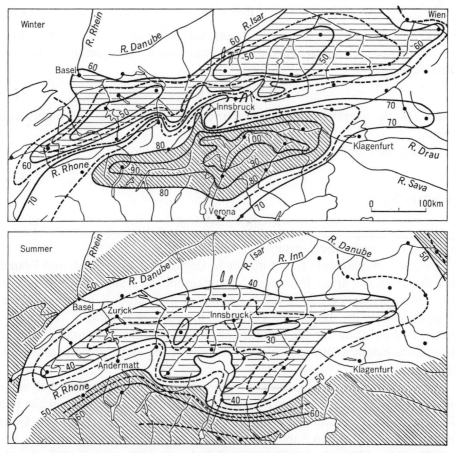

Fig. 4.18 Precipitation variability in the Alps for 1931–1960, (a) in winter (Dec. Jan. and Feb.) and (b) in summer (June, July, and Aug.) (Fliri, 1967a).

topographical contour lines. The monthly rainfall distributions for February and August are shown in Fig. 4.17.

At a number of stations in the Alpine area of Austria, Bavaria, Italy, and Switzerland, the monthly precipitation variability was calculated (Fliri, 1967a, b). In Fig. 4.18, the distributions of the variability in winter and summer are given in percentages. On the southern side of the Alps, the variability reaches 90–100% in winter, but the whole Alps region is covered by low variability in summer, due to the synoptic situation contrasts.

(C)   *Empirical and Theoretical Representation*

Many experimental equations showing the relationship between the monthly or annual amount of rainfall and the height above sea level on mountains have been proposed. From them some equations can be deduced, like the following, although there may be many regional variations. To take one of the simplest examples, the equation below indicates a linear relation between the amount of rainfall and the height:

$$R = a + bh \tag{4.7}$$

where R is the amount of rainfall (mm), and h is the height (m). However, when the slope inclination is taken into consideration, the equation is often like the following:

$$R = a + bh + c \tan z \tag{4.8}$$

where z is the average slope angle of the areas surrounding the observation point.

Sometimes an equation takes the form of a parabola:

$$R = a + bh + ch^2 \tag{4.9}$$

or a hyperbola:

$$R = a + \sqrt{b + ch + dh^2} \tag{4.10}$$

where a, b, c, and d are constants.

These formulas were found mainly in Germany, Switzerland, and France, and the difference arises from complex topographical features as well as from different directions of slopes, various causes of precipitation, or from different synoptic conditions during the rainfall. Therefore, it is difficult to obtain a generalized equation for the monthly or annual amount of rainfall as a function of the height above sea level alone, even if the slope angle is taken into consideration.

There are many theoretical studies on the orographic rainfall. A dynamic model of orographic rainfall is posited by Sarker (1966). He assumes the two-dimensional wet neutral atmosphere which is applicable to the West Gahts Mountains of the Indian Subcontinent in the case of summer monsoon.

Gocho and Nakajima (1971) studied the orographic rainfall on the Suzuka Mountains by assuming the mountain shape to be:

$$\zeta_s(x) = \frac{a^2 b}{a^2 + x^2} \tag{4.11}$$

where x is the horizontal distance from the ridge (leeward is positive), b ridge height $= 0.8$ km, and $a = 4$ km. Vertical velocity of the airstream crossing over the ridge is expressed by

$$w(x, z) = \mathrm{R}_e \left(\frac{\rho_0}{\rho_z} \right)^{\frac{1}{2}} W(z) \cdot e^{ikx} \tag{4.12}$$

Here, w is obtained by solving approximately the following differential equation:

$$\frac{d^2 w}{dz^2} + (l^2 - k^2)\, w = 0 \tag{4.13}$$

where z is height, and l^2 is Scorer's parameter. Figure 4.19 shows one of the results in the case of the wind maximum of 20 m/sec at a height of 1 km and the wind velocity 10 m/sec at $z = 0$, and $l^2 = 1$ km^{-2}. Assuming that the raindrops condensed in the air layer between 0.5 and 5 km fall down with the falling velocity of 4.5 m/sec flowing along the basic airstream, the precipitation amount (mm/hr) can be estimated (see Fig. 4.19 (b)). This figure shows that the maximum rainfall reaches about 30 mm/hr just behind the ridge.

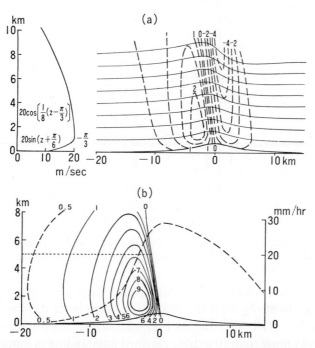

Fig. 4.19 Vertical distribution of wind (a, left), streamline and distribution of vertical velocity (m/sec) over a mountain (a, right), distribution of amount of condensed water (as depth in mm/hr) in a layer 500-m thick (b) and precipitation mm/hr (b, broken line) (Gocho et al., 1971).

(D) *Rainfall Intensity*

Rainfall intensity in mountains is one of the most important factors in erosion problems in geomorphology. In high latitudes with rather low temperatures, rainfall is generally not intense, but when the mountain-top is compared with the foot, it does not necessarily follow that rainfall is weaker in the areas with lower temperatures.

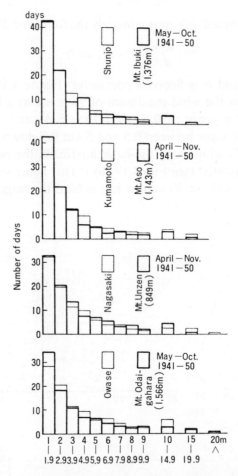

Fig. 4.20 Occurrence frequency (number of days) of daily maximum 10-min rainfall on the mountains and their bases in Japan (Yoshino, 1961a).

It will be seen from Fig. 4.20, which shows daily maximum 10 min rainfall according to its intensity, that the mountains have more days with rainfall intensity below 2 mm/10 min. But, as Fig. 4.21 shows, the number of days with rainfall intensity more than 10 mm/day is most outstanding in mountains. Therefore, the measurement duration should be taken into consideration, as well as the rainfall intensity for a short or long period, in order to determine the dis-

Table 4.10 Constants, b, k, and n of Eq. (4.14) on the mountains and their feet in Japan (Yoshino, 1960).

	Height a.s.l.	b	k	n		Height a.s.l.	b	k	n
Mt. Tsukuba	869 m	10	11.0	0.64	Mt. Unzen	849 m	20	8.9	0.52
Mito	29	30	42.4	0.83	Nagasaki	27	30	25.0	0.70
Difference	840	−20	−31.4	−0.19	Difference	822	−10	−16.1	−0.18
Mt. Odaigahara	1,566	30	8.1	0.38	Mt. Aso	1,143	20	9.8	0.54
Owase	14	60	40.0	0.69	Kumamoto	38	20	14.9	0.65
Difference	1,552	−30	−31.9	−0.31	Difference	1,105	0	−5.1	−0.11
Chūgushi	1,335	10	7.3	0.50	Mt. Ibuki	1,376	10	8.4	0.62
Utsunomiya	120	10	18.5	0.69	Shunjo	161	20	13.4	0.66
Difference	1,215	0	−11.2	−0.19	Difference	1,215	−10	−5.0	−0.04

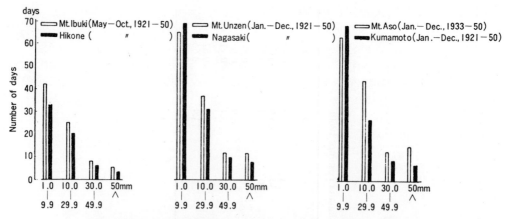

Fig. 4.21 Occurrence frequency (number of days) of daily rainfall on the mountains and their bases in Japan (Yoshino, 1961a).

tinctive characteristics of rainfall intensity in mountains. The following equation indicates the relationships between the rainfall intensity i and the measurement duration t (Yoshino, 1960):

$$i = \frac{k}{(b+t)^n} \qquad (4.14)$$

where b, k, and n are constants that take different values in different places. The values for b, k, and n obtained both on the mountains and their foots nearby are shown in Table 4.10. The values on mountains are smaller than, or the same as, those of the foots. The rainfall intensity for any time between 10 min to 36 hr can be obtained by putting the values in Table 4.10 into Eq. (4.14).

4.1.8. Snow
(A) Snowfall and Snow Accumulation

Air temperature decreases with height above sea level, so that snow rather than rain is the usual form of precipitation on mountains in winter. The rate (%) of snowfall of the total amount of precipitation during the winter or throughout a year is expressed by a linear function with the height above sea level in the Austrian and Swiss Alps, Black Forest, Harz, and Erz mountains (Conrad, 1935; Roux, 1951).

The number of days with snowfall increases linearly with height in Silesia (Kosiba, 1954). The same is true of the number of days with snow cover in Slovakia (Konček et al., 1964). These relationship are applicable to other parts of the world.

In Japan, the number of days with snowfall increases on the mountains of the Hokkaido and Tohoku districts. For example, 139 days on Mt. Akan (1,343 m), compared to 67 days at Obihiro (39 m), Hokkaido, and 162 days on Mt. Iwate (1,771 m), but 101 days at Morioka (155 m), Tohoku district. Snowfalls even in summer on Mt. Fuji (3,776 m) but it has only 100 snow days a year, because of its location (35°21′N).

The depth of snow cover also increases with height, but there is a distinct local variation. In mountain areas in Austria, where inversion layers often develop in winter, the snow depth is considerably influenced by the height of the inversion layer. The average maximum depth of snow cover increases up to 1,000–1,100 m at the rate of 10 cm/100 m. However, the depth of snow cover decreases abruptly at an altitude of 1,000–1,100 m, and increases again above this altitude. This phenomenon is closely related with the height of the inversion layer (Steinhauser, 1948).

Photo 4.3   A heavy snow landscape at Tôkamachi in Niigata Prefecture, Japan (February 22, 1974, by M. M. Yoshino).

The depth of snow cover in the mountain area in Bulgaria is expressed by an experimental formula as linear functions of precipitation and air temperature in the winter months (Stanev et al., 1970). It has been confirmed that the depth of snow cover varies greatly from place to place, but the density is relatively constant.

The local variation of snow depth is especially striking in high mountains. It is caused mainly by wind and radiation differences, which are the results of relief orientation, micro-topography, as discussed in detail by Friedel (1961). An example of snow accumulation and micro-topography in the Ötz valley, Austria, is shown in Fig. 4.22. The relation between snow depth and micro-topography is described below.

The duration of snow accumulation is expressed also as a linear function of height above sea level. This was reported from the N-French Alps by Poggi (1959).

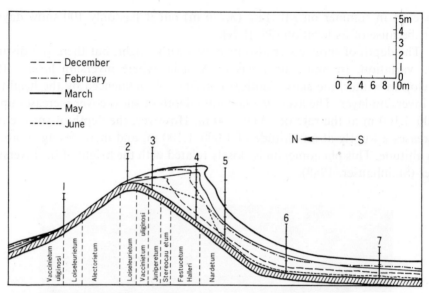

Fig. 4.22 N–S profile of snow accumulation crossing the Daunmoräne (2,230 m) in Ötz valley, Tyrol (Turner, 1961).

The world record depth of snow cover is 11 m 82 cm, which was observed at Mt. Ibuki, Japan, on February 14, 1927 (Arakawa, 1955).

(B)   *Distribution of Snow Accumulation*

In Japan, studies on snow accumulation in mountain areas have been made intensively. The first comprehensive study was made in the Mt. Daisetsu region in Central Hokkaido, under the guidance of Nakaya (1949), and the results of this study were published by Sugaya et al. (1949). The first research on water content of snow accumulation with the aid of air photographs in Japan was done in the basin of the Chûbetsu River at the foot of Mt. Daisetsu. The procedure for this research was as follows (Sugaya, 1949b): (a) The area under examination was classified into several groups, such as an alpine region, alpine wooded region, middle altitude wooded region, etc., as in Table 4.11. (b) Each region was further subdivided into smaller segments by means of air photographs, and then the area of each segment was calculated. (c) The mean water content of snow was obtained through a field survey in each segment, and the total amount of snow cover was obtained by multiplying the mean value by the area.

The conclusion was that the water content of snow over the area of 256 km^2 amounts to 19.85×10^7 m^3, in other words, approximately 2×10^8 m$^3 = 200$ million t, on April 1, 1948. This value is extremely large, and if it is converted into the mean depth of water (rainfall) on the whole basin of the Chûbetsu River, it attains a height of 780 mm. In view of the fact that the total amount of precipitation at a point in a cultivated region was 410 mm for 5 months from November, 1947, through March, 1948, it will be clearly seen how enormous is the

Table 4.11 Water content of snow accumulation in the basin of the Chûbetsu River, W-slope of Mt. Daisetsu in Hokkaido on April 1, 1948. (Summarized by Yoshino based on the data by Sugaya, 1949b).

Region	Subregion		Area (km²)	Mean water content of snow accumulation (cm)	Total water content of snow accumulation (m³)	Total of subregion (m³)
Alpine Region	Shallow snow cover	Rock-Grass	5.3	15	0.08×10^7	0.08×10^7
	Snow drift	Gravel-Grass	4.0	30	0.12	
		Creeping pine	10.7	43	0.12	0.58
	Deep snow cover	Lee of ridge and valley	25.6	135	3.50	3.50
	Standard snow cover	2,000–1,800m	3.6	67	0.24	
		1,800–1,600	17.6	87	0.24	
		1,600–1,400	13.5	107	1.45	3.22
Alpine wooded region	1,800–1,600 m		0.3	125	0.04	
	1,600–1,400		12.4	125	1.55	
	1,400–1,200		9.3	113	1.05	
	1,200–1,000		0.7	94	0.66	3.30
Middle altitude wooded region	1,400–1,200 m		15.5	113	1.75	
	1,200–1,000		31.4	94	2.95	
	1,000– 800		32.1	76	2.44	
	800– 600		16.3	63	1.03	8.17
Low altitude wooded region	1,200–1,000 m		0.7	85	0.06	
	1,000– 800		7.0	68	0.48	
	800– 600		12.6	57	0.72	
	600– 400		22 4	38	0.85	
	400		0.7	25	0.02	2.13
Cultivated region	Hill	430 m	4.2	31	0.13	
	Lowland	360 m	10.1	24	0.24	
Total			256.0			19.85

amount of snowfall in the whole mountainous region. Furthermore, the fact that it is reserved in the form of snow cover is of profound importance in the problem of utilization of water as a natural resource.

As seen in Table 4.11, the distribution of water content of snow cover in the basin of the Chûbetsu River by altitude shows a linear relationship up to a height of 1,400 m a.s.l. and increases at the rate of 9.4 cm/100 m, but above that altitude, it begins to decrease at the rate of 10.6 cm/100 m.

Onuma (1956) used a snow recorder in Japan in his research on the snow accumulation in mountain areas. He calculated the amount of melted snow, the surface water supply, the runoff of melted snow, and the rate of melted snow runoff in a basin of the mountain area.

Snow survey by means of aerial photographs has developed extensively since the heavy snowfall in January, 1963. About 8,600 aerial photographs, covering an area of 20,000 km², on a scale of 1: 20,000 were prepared and

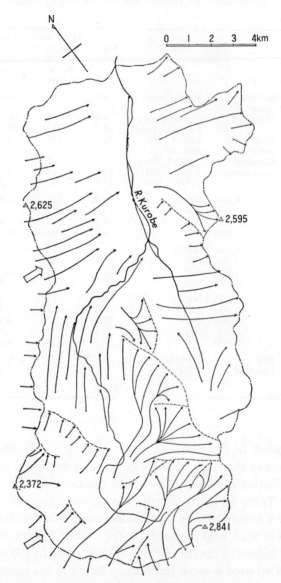

Fig. 4.23 Streamline of the Kurobe River Basin in winter, interpreted through
aerial photographs on March 30, 1964 (Maruyasu et al., 1968).

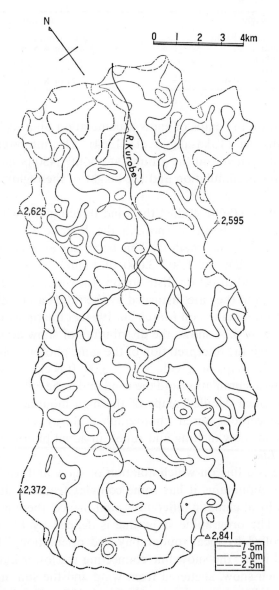

Fig. 4.24 Distribution of snow accumulation of the Kurobe River Basin, interpreted through aerial photographs on March 30, 1964 (Maruyasu et al., 1968).

Table 4.12 Snow accumulation and the height of the region in the Hokuriku district
on the Japan Sea side of Honshu, through aerial photograph interpretation. Photographs were taken during Feb. 14–March 19, 1963 (Takasaki
et al., 1964).

Terrain height	Snow accumulation
Higher than 1,200 m	Varies from place to place
800–1,200 m	3–4 m or 4–5 m
300– 800 m	2–3 m or 3–4 m
100– 300 m	1–2 m or 2–3 m
Lower than 100 m	1–2 m or less than 1 m

interpreted based on the maps of snow accumulation on a scale of 1:20,000 of
the Hokuriku district (Takasaki et al., 1964). The interpreted values were
corrected by measurement on the ground at about 1,450 points. The corrected
values of the snow accumulation and the height of the region are presented in
Table 4.12.

Maruyasu et al. (1968) studied the wind direction by wind marks on the
snow surface and the depth of snow in the Kurobe River drainage basin,
Central Japan, by the same method. This basin has an area of approximately
9×23 km with a height of 1,300–3,000 m. Figure 4.23 shows the streamline
derived from the wind direction distribution. In winter westerly winds prevail
in the upper layer, but they are disturbed by the ridges and valleys. The photographs (1:15,000) were taken on March 30, 1964, at the time of maximum snow
accumulation. Figure 4.24 shows the distribution of snow accumulation in the
same region through the interpretation of the same photographs for each $250 \times$
250 m square on the maps on a scale of 1: 50,000. It is clearly seen that the relation between snow accumulation and the wind under the influence of topography
is most striking. Such detailed maps of the high mountain area could never
have been made without aerial photographs. Maruyasu et al. (1969) made
further research on snow melting and runoff in aerial photographs.

(C) *Snow Line*

A snow line is a line on which ablation of snow and ice is balanced with
snowfall. On the mountains it has some complex features, due to the micro-
topography and local climatic differences, and, hence, there is some confusion
in terminology. The orographic snow line is defined by Flint (1957) as "the
snow line controlled by local topography and orientation," but it would be
better to defined it as "the snow line connecting the lower limits of the actual
snow line at which snow, sheltered from wind and the sun, may survive even
at lower altitudes than exposed snow (Nogami, 1970). On the other hand, the
regional or climatic snow-line is a hypothetical line excluding local irregularities
caused by topography. A schematic illustration is given in Fig. 4.25. The actual
snow line varies from year to year, and the long year average of this line is
called the firn limit or firn line.

The climatic snow line and the regional snow line are problems of macro-

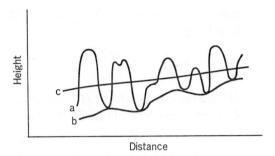

Fig. 4.25 Schematic section of three types of snow lines (Nogami, 1970).
a: Actual snow line, b: Orographic snow line, c: Regional snow line.

or meso-scale climatology, but the actual snow line might be a subject of local- or micro-scale climatology.

Nogami (1970) has obtained the following equation for calculating the change of snow line from the fluctuation of ice-tongue:

$$\Delta h = \frac{\Delta s\,(h_o - n)}{s + \Delta s} \tag{4.15}$$

where Δh is the change of h_o; h_o is the height of the snow line; s is the total area of the glacier, Δs is the area of advanced or retreated ice-tongue, and n is the mean altitude of Δs.

The actual snow line is, as defined above, the line at which the accumulation of snow is in equilibrium with the ablation of snow. Hoshiai et al. (1957) used this criteria in studying the snow line in the Japanese Alps. Arai et al. (1973) proposed the equilibrium line at which the annual snowfall is in equilibrium with annual ablation of snow with reference to the mass balance.

Mass balance at the equilibrium line is expressed by the following equation, in which the amounts are reduced to the water equivalent:

$$A = F_o \tag{4.16}$$

where A is annual ablation, and F_o is the annual snowfall. In the lower half of a glacier,

$$A > F_o \tag{4.17}$$

In this part of a glacier or at the lower limit of a perennial snow patch, the mass balance equation is

$$A = F_o + F_l \tag{4.18}$$

where F_l is the lateral supply of snow or ice, such as glacial flow, avalanche, and wind drift. On the other hand, the amount of ablation is simply obtained by

$$A = \Sigma f \cdot \theta_a = n \cdot \bar{f} \cdot \bar{\theta}_a \tag{4.19}$$

where f is the degree-day factor; $\bar{f}$ is the mean degree-day factor for the ablation period; θ_a is mean daily air temperature (°C); $\bar{\theta}_a$ is the mean air temperature

(°C) for the ablation period, and n is the number of days in the ablation period. f is calculated by the following equation,

$$f=\frac{1}{80}\cdot\frac{R_n+H+LE}{\theta_a} \qquad (4.20)$$

where R_n is the net radiation on snow surface; H is sensible heat exchange, and LE is latent heat exchange. Arai has obtained the following experimental equations:

For Honshu: $F_o+F_l=30\bar{f}(0.59\bar{\theta}_a{}^2+1.6\bar{\theta}_a{}^2)$
For Hokkaido: $F_o+F_l=30\bar{f}(0.52\bar{\theta}_a{}^2+1.6\bar{\theta}_a{}^2)$ $\qquad (4.21)$

Rewriting Eq. (4.21) using altitude,

for Honshu: $F_o+F_l=30\bar{f}(4.96z^2-54.6z+149.1)$
for Hokkaido: $F_o+F_l=30\bar{f}(3.80z^2-38.6z+96.9)$ $\qquad (4.22)$

where z is altitude in km. From these results, it is assumed that the perennial snow patch in Honshu is maintained by the accumulation of more than 1,000 cm of the water equivalent of snow, and the value slightly decreases in Hokkaido.

The snow line in the Alps, especially on Mt. Sonnblick, Austria, was dealt with thoroughly in relation to glacier ablation, temperature, and solar radiation by Morawetz (1961). He found a close relationship, referring to the solar radiation and particularly to the values of albedo. There are many studies on the heat budget, radiation balance, and ablation of glaciers in the Alps (for instance, Sauberer et al., 1950; Hoinkes et al., 1952; Hoinkes, 1953, 1955; Sauberer, 1953), but these glaciological problems will not be discussed here.

(D)   *Avalanche and Remaining Snow*

The mechanism of the occurrence of avalanches is not dealt with in this book, but the local difference of occurrence is dealt with here. Using aerial photographs, distributions of avalanches in the mountainous regions have been clarified (Takasaki et al., 1964). A map showing the relationship between avalanches and the slope gradient was compiled for the Hokuriku district at the time of the heavy snowfall in winter 1963. The slope gradient was measured by contour lines with 20 m intervals on topographical maps and classified into 10° intervals. The conclusion was that 36% of avalanches occur on the slopes facing south and 27% on slopes facing southeast; 85% of slopes with 30–40 gradients are the sites of avalanches; 43% of all avalanches occur at heights of 400–500 m and 32% at 500–600 m; 76% occur on concave slopes; 35% occur in a broad-leaved forest region and 33% on shrub forest area. It must be noted that in this region, on slopes facing south or southeast are the lee side slopes of the ridges for the prevailing winter winds.

An interesting problem is the asymmetric distribution of snow accumulation or snow remaining on isolated volcanic cones. This is of great importance, when considering the erosion of cones from the standpoint of geomorphology,

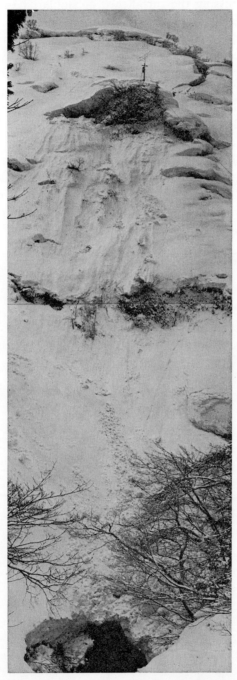

Photo 4.4 A small avalanche on the valley slope, observed at Tochiomata, Niigata Prefecture, Japan. Depth of snow accumulation was 4–5 m, partly over 6 m. (February 24, 1974, by M. M. Yoshino).

and also when discussing the turbulence or similarity to model experiments from a meteorological standpoint.

The aerial photographs in Japan have been taken mainly to interpret land form, forest resources, etc., so that they have been generally taken in all seasons except winter. On the aerial photographs (1:40,000) published by the Geographical Survey of Japan, the areas covered with remaining snow were plotted on contour maps (1 : 50,000) as accurately as possible (Suzuki, 1969). The results, presented in Table 4.13 can be summarized as follows: (a) If the volcanic cone as a whole is taken as a geomorphic unit, the remaining snow seems to be distributed unequally in different directions, in descending order, on the eastern, northern, southern and western sides of the cones, except Mt. Iwate. The altitude of the lowest limit of the remaining snow distribution is different in the four directions: In the west, it is highest, in the south, middle, in the east and north,

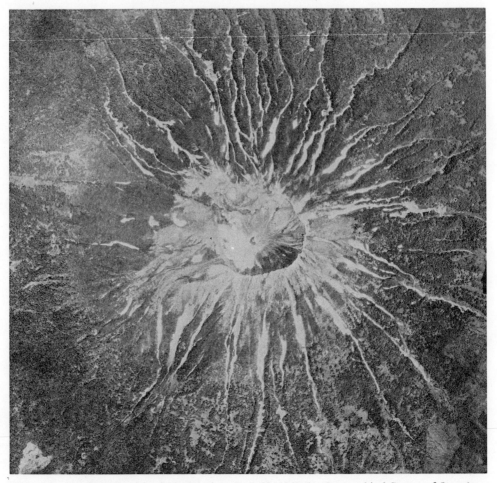

Photo 4.5   Aerial photograph of Mt. Yôtei on June 29, 1948 (by Geographical Survey of Japan).

it is lowest. (b) The watershed of each radial valley is then taken as a geomorphic unit. The remaining snow along the radial valleys running roughly in the E–W direction is distributed much more on the valley slopes facing north than on those facing south. Along the radial valleys running roughly in the N–S direction, much more snow remains on the valley slopes facing east than on those facing west. (c) In every direction, the remaining snow is found in a lower part altitude and much more snow is apparent on the valley floors than on the ridges. (d) Regarding Mr. Fuji, the areas covered with fresh snow seem to show the same unequal distribution as described above. A sample photograph and a map interpreted are given in Photo 4.5 and Fig. 4.26.

This unequal distribution seems to be due to the differences in the depth of snowfall which is heavier on the eastern than on the western side of the cones, and to the difference in the rate of thawing which is greater on the southern than on the northern side of the cones.

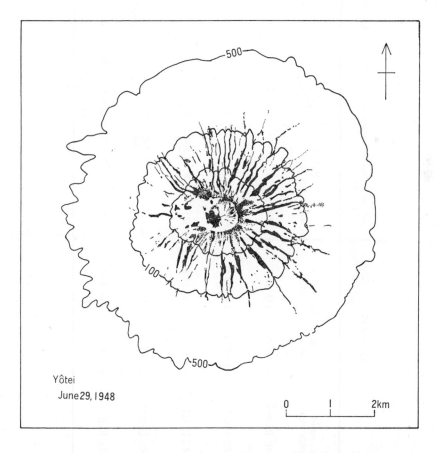

Fig. 4.26 Distribution of the remaining snow on Yôtei volcano (1,893 m), SW-Hokkaido, on June 29, 1948 (Suzuki, 1969). Black areas and spots are those covered with remaining snow.

Table 4.13   The distribution of the remaining snow on four strato-volcanoes and that of the fresh snow on Mt. Fuji (Suzuki, 1969).

Name of volcanoes	Location Lat.	Long.	Altitude	Relative height	Date of air-photo used here	Distribution of the remaining snow			
						as a whole	Altitude of the lowest limit		
							Direction	on ridges	along valleys
Mt. Rishiri	45°10'N	141°15'E	1,719 m	1,719 m	May 25, 1956	east>west	east	500 m	350 m
							south	800	400
							west	900	450
							north	700	500
Mt. Yōtei	42°50'N	140°49'E	1,893	1,700	June 29, 1948	east>west	east		750
							south		1,000
							west		1,200
							north		950
Mt. Iwaki	40°50'N	140°18'E	1,625	E. 1,500 W. 1,200	May 15, 1948	east>west	east	700	550
							south	800	600
							west	800	650
							north	750	550
Mt. Iwate	39°51'N	141°0'E	2,400	E. 1,700 W. 500	May 22, 1948	north->south-east>west	east	650	400
							south	700	550
							west		
							north	—	—
Mt. Fuji	35°21'N	138°44'E	3,776	3,700	Oct. 31, 1951 & Oct. 2, 1962	east>west	east		
							south		
							west	—	
							north	850	800

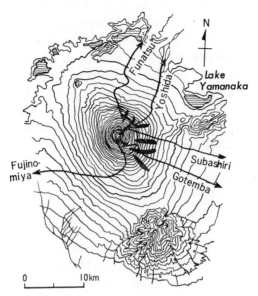

Fig. 4.27   Distribution of avalanches on Mt. Fuji (Ishida et al., 1960).

Mt. Fuji had experienced seventeen large-scale avalanches in total from the spring of 1947 until the spring of 1959. Twelve of them were avalanches of a whole layer of snow accumulation, which occurred in late winter and spring, and five were of fresh snow, usually between late autumn and early winter. The avalanches on Mt. Fuji often occur when a cyclone is approaching, the wind is southerly, and temperature rises above 0°C, while a snowslide takes place when the amount of snowfall is exceedingly large. The areas where avalanches occur frequently are illustrated in Fig. 4.27, which shows that avalanches are restricted to the NNE–E–SE slopes (Ishida et al., 1960). It is hard to study without the help of a snow cover chart of the slopes around the summit, or of a minute distribution of the wind direction and velocity, but it can be assumed from the interpretation of aerial photographs by Suzuki (1969) that the snow falls much on the leeward side of the summit for the prevailing westerlies or northwesterlies, where the wind is not so strong.

In a high mountain region, a peculiar form of snow surface called *snow of penitents*, *Büsserschnee*, or *Nieve de los penitentes* can be observed. This is one of the ablation forms of snow accumulation on high mountains (Troll, 1942). It develops most strikingly around 3,000–5,000 m a.s.l. between 32–35°S in the Cordillera de los Andes in the region at Argentina and Chile.

4.1.9.   Thunderstorms

Strictly speaking, what causes thunderstorms (excepting frontal and squall-line thunderstorms) in mountainous areas in the temperate zone is not a heated slope alone. According to Koch (1955), a thunderstorm that hits a slope of the

Thüringen mountainous area occurs in the area where both surface temperature is highest and air currents converge. In northern Italy, thunderstorms are most frequent in the mountain foot district of the Alps. This fact is correlated with the thermal condition of the slope and the state of the air current there.

Some characteristics of the thunderstorms in Central Europe in winter, which are very rare (2 % of the total), are different from those of warmer seasons. They occur exclusively in NW-situations with strong N-winds and the top rarely surpasses the 600 mb level. It is thought that the long and narrow cloud systems are seemingly caused by the mountains of N-Bohemia and Moravia (Förchtgott, 1969).

Henz (1972) studied the formation of thunderstorms along the lee slopes of the Rocky Mountains from Montana to New Mexico. The process of formation is as follows: (a) The mountain slopes, especially those facing the sun, strongly absorb incident radiation. (b) Air immediately adjacent to the surface is heated by conduction from below, and a valley breeze circulation results. (c) Once the valley breeze circulation is established, "bubbles" of heated air rise from the slopes. (d) As the lifted condensation level of these bubbles is reached, they become visible as cumulus clouds. (e) If considerable amounts of water vapour and sensible heat are brought to the vicinity of the pre-existing cumulus cloud by thermals originating on the plains, it grows quite rapidly into a thunderstorm.

The frequent occurrence of thunderstorms in Gumma and Tochigi prefectures in the Kanto plain, Japan, may be accounted for by the thermal condition on the mountain slopes, the convergence of the southerly current and the invasion of cold air in the upper layer.

The moving velocity of a thunderstorm in the Kanto plain is mostly 6–15 km/hr and the moving distance from generation to disappearance is generally 21–40 km in the case of a thermal thunderstorm (Takeuchi, 1964).

On high mountains, lightning sometimes causes accidents for mountain-climbers. A group of 46 members of a senior high school met an accident caused by lightning at the Peak of Doppyo, 2,660 m a.s.l., in central Japan, on August 1, 1967. According to a report by the Investigation Committee of Matsumoto Fukashi Senior High School, nearly all the members were struck by electricity, 11 persons were killed instantly and 13 persons were seriously injured. However, it occurred on the northern slope quite near the summit and not on the top or on the southern slope with a chain area. The causality extended over the 26 m length of the whole north face. The reason why only the northern slope was struck by lightning is as yet unknown.

4.1.10. Wind

In general, wind velocity is great on mountains. In Japan, the maximum instantaneous wind velocity of 71.1 m/sec was recorded on the summit of Mt. Tsukuba (884.3 m) during the passing of a typhoon on September 28,

1902, and a 72.6 m/sec (the record wind velocity in Japan) SSE wind velocity was observed at the time of Typhoon Kitty on August 31, 1949. On Mt. Washington in the eastern part of the United States a 231 mile/hr (103.3 m/sec) wind velocity which is the world record, was recorded on April 12, 1934.

When the wind velocity on a mountain in Japan is expressed by the number of days of individual classes of wind velocity, it becomes evident that the wind is strong every day in winter, but it is weak in summer. This may be a general characteristic of high mountains in the middle and high latitudes in the Northern Hemisphere. On the other hand, it should be noted that in August, September and October the wind is relatively weak every day, with an occasional, extremely strong wind due to passing typhoons. Table 4.14 gives examples of such a tendency found in January and August. The number of days with wind over 50 m/sec at Mt. Fuji, 25 m/sec at Mt. Ibuki, and 20 m/sec at Mt. Aso, Mt. Tsukuba, and Mt. Unzen is larger in August than in January.

Table 4.14 Number of days* of various wind velocities on the mountains in Japan (Yoshino, 1961a).

Mountain	Month	Wind velocity more than											
		10m/s	15	20	25	30	35	40	45	50	55	60	
Mt. Fuji	Jan.	248	247	239	208	135	60	18	6	0	0	0	days
(3,772 m)	Aug.	182	98	64	39	14	10	5	2	2	1	1	
Mt. Ibuki**	Jan.	242	188	80	22	5	1	0	0	0	0	0	
(1,376 m)	Aug.	189	94	44	23	10	6	1	0	0	0	0	
Mt. Aso	Jan.	119	20	0	0	0	0	0	0	0	0	0	
(1,143 m)	Aug.	69	16	2	1	0	0	0	0	0	0	0	
Mt. Tsukuba	Jan.	220	87	7	0	0	0	0	0	0	0	0	
(869 m)	Aug.	143	34	8	3	1	0	0	0	0	0	0	
Mt. Unzen	Jan.	179	56	8	0	0	0	0	0	0	0	0	
(849 m)	Aug.	91	35	16	9	2	0	0	0	0	0	0	

* Total of eight years, 1950–1957.
** Total of seven years, 1951–1957 for the cases more than 35 m/sec.

Wind velocity increases with height above sea level, but there are wide local variations according to the micro-topographical features, as shown below. The equation below applies to Japanese mountains:

$$\frac{V_2}{V_1}=\left(\frac{H_2}{H_1}\right)^{0.69} \tag{4.23}$$

where V_1 and V_2 are the monthly mean maximum wind velocity (m/sec) at a height of H_1 and H_2 (m) in January. The observed values in some mountains are considerably smaller than those calculated by Eq. (4.23). As the wind velocity in August differs according to the influence of typhoons, no generalization like Eq. (4.23) can be made concerning the relation between wind velocity and height above sea level. The equation for the annual mean wind velocity is:

$$\frac{V_2}{V_1}=\left(\frac{H_2}{H_1}\right)^{0.54} \tag{4.24}$$

where V_1 and V_2 are the annual mean wind velocities (m/sec) at an altitude of H_1 and H_2 (m).

The diurnal change of wind velocity on high mountains differs from that at low land. In general, the maximum appears at night and the minimum in the daytime on high mountains. This pattern is clearly seen on the top of Mt. Fuji in July, as shown in Fig. 4.28. In winter, this is obscured because of strong winter monsoons, but still this tendency is detectable. On the mountain foot, as seen at Gotemba and Mishima, a striking daytime maximum is observed both in January and July.

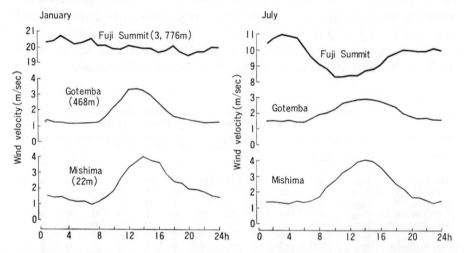

Fig. 4.28 Diurnal change of wind velocity (m/sec) on the summit of Mt. Fuji (Fujimura, 1971).

On the summit, the wind velocity strengthens (Steinhauser, 1950). The height of the effect on the upper airstream above the mountain varies according to the temperature gradient (Georgii, 1923). Georgii has obtained the following experimental equation:

$$H_e = 0.28 H_b + 650 \, (\gamma - \gamma_m) \tag{4.25}$$

where H_e is the height of the effect (m); H_b is the height of the mountain; γ is lapse rate in the atmospheric layer up to a height equal to the mountain height, and γ_m is the mean value; for instance, γ_m is 0.45 for 0–1,000 m, 0.46 for 1,000–2,000 m, 0.50 for 2,000–3,000 m and 0.56 for 3,000–4,000 m. It might be said that the constant 0.28 is too small, although there is no detailed information.

Steadiness of wind is defined as:

$$\text{Steadliness} = \frac{\text{Vector mean wind velocity}}{\text{Scalar mean wind velocity}} \times 100\% \tag{4.26}$$

Pleiss (1951) calculated the annual steadiness at the stations in Saxony and showed that it was 43% at Fichtelberg (1,214 m), which was greater than at low land. Steinhauser (1938), Schüepp et al. (1963), and Flemming (1964), calculating seasonal change of wind steadiness on high mountains, have shown that it is greater in summer than in winter at Davos, Switzerland, and the other stations, and that the maximum appears in October and the minimum in August at Sonnblick (3,106 m).

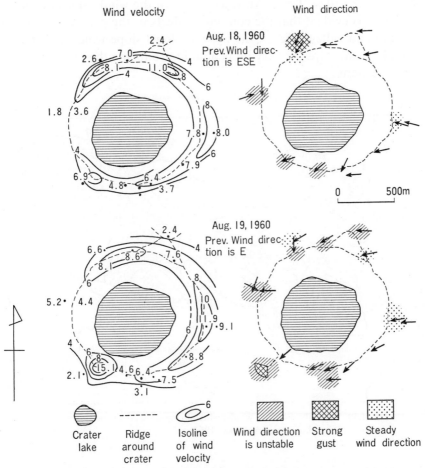

Fig. 4.29 Distribution of wind velocity and wind direction on the top of Mt. Ohnaminoike, S-Kyushu, Japan (Yoshino, 1961b).

There is a great local difference of wind velocity around a mountain-top. An example from Mt. Ohnaminoike, a volcano with a crater lake, 1,300–1,400 m, is illustrated in Fig. 4.29 (Yoshino, 1961b). The observation by means of a radiosonde, made on August 18, 1960, shows that the upper wind is about 12 m/sec, ESE, and on August 19, 20 m/sec, E. Therefore, the highest velocity on the windward side of the mountain-top was 2/3 to 3/5 of that in the free atmosphere, and a semicircular isoline runs around the crater. On the other hand, on the leeward side of the crater, the wind velocity was 1/4–1/5 of that in the free atmosphere. Where the crater is cut down in a saddle-shape, an extremely strong wind blows. For example, in the southwestern part of the crater, 15 m/ sec is marked, but it is only 2 m/sec at a point shielded from the wind 10 m below the saddle point.

Aulitzky (1955) reported the results of wind measurement at stations (1,940 m) in Ötz valley, Tyrol, Austria. Although the wind velocity is not very strong here, it is evident that the combined effects of the periodic wind system of mountain and valley breezes, the up- and down-slope winds and the cross-valley winds of the Alps are very influential in producing the local peculiarities of wind conditions.

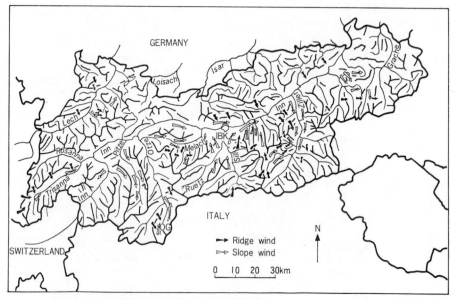

Fig. 4.30 Prevailing surface winds, which bring snow, over the ridges and slopes in N-Tyrol, Austria (Aulitzky, 1963).

In recent years it has become possible to study wind distributions over high mountain ridges by interpreting aerial photographs in the snow season. An example has already been given in Fig. 4.23. Another example for N-Tyrol is presented in Fig. 4.30. In this map, the prevailing surface winds over the ridge and the slopes are illustrated by the shape of the snow surface in the late winter (Aulitzky, 1963). The prevailing surface wind direction is very complicated, and the winds on the slope sometimes blow opposite to the winds over the ridge.

Whereas wind direction prevails at right angles to the running direction of a ridge, the reverse is the case on the leeward side of a narrow ridge. Such a situation can be seen in Photo 4.6 and Fig. 4.31, in which there are deformed

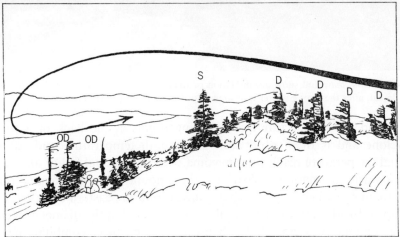

Photo 4.6, Fig. 4.31 Wind-shaped trees on both sides of a ridge of Mt. Azuma, central Japan. Deformed trees (D) on the windward side of the prevailing wind, symmetrical form trees (S) just behind the ridge, and deformed trees, which show the opposite direction (OD) to the prevailing wind. It is thought that the eddy formed behind the ridge is the cause for this situation. The eddy is not formed exactly on X-Z plane, but has some twisted character under the influence of microtopography. (August 1959, by M. M. Yoshino).

trees on both sides of the ridge on Mt. Azuma located on the borders of Nagano and Gumma Prefectures, both in opposite directions. The wind-shaped trees in high mountains are discussed below in Section 6.1.3.

4.1.11. Icing and Hoarfrost

Hoarfrost in mountainous regions has an important bearing on means of transportation and communication. It is also ecologically interesting as a cause of deformed coniferous trees (Yoshino, 1973).

A detailed investigation of hoarfrost on Mt. Fuji, Mt. Ibuki, and Mt. Iwate, Japan, has been made. The results are: An observation of the fog and ice on Mt. Fuji in connection with the passing of nineteen cyclones during the

Photo 4.7 *Silver thaw* and snow accumulation of the slope, about 2,200 m a.s.l., of Mt. Azuma, central Japan, after the strong winter monsoon (April 2, 1960, by M. M. Yoshino).

icing period has made clear that (a) Icing begins at the back of a traveling anticyclone and is often observable when both temperature and equivalent potential temperature reach their maxima with the gradual fall of air pressure; (b) in a warm air mass with S-winds, the amount of icing increases; (c) icing comes to an end when a cyclone has gone beyond the eastern coast of N-Honshu, Japan. In addition to these results, it is known that the diameter of each of the fog is about 30 μ, and that, when these fog particles combine, they form larger particles, which increase the amount of icing. An observation on Mt. Ibuki, on the other hand, has shown that a fog particle increases in size with the rise of temperature. In this case, the specific gravity of icing also increases. The adhesion power of icing is at its peak when the size of fog particles is 10–15 μ in diameter and becomes small with larger sizes. There are three basic types of icing on Mt. Ibuki, i.e., soft rime, hard rime, and glaze. When a fog particle is small in diameter, it becomes soft rime, but as it grows larger, it turns into hard rime, and finally into glaze. Silver thaw is a colloquial expression for a deposit of glaze built up on trees, shrubs, and other exposed objects. The developing direction of silver thaw is a good indicator of prevailing winds on the mountains.

In Czechoslovakia, automatic measurement of the hoarfrost in mountains is being made by means of a geligraph (Konček, 1960). He measured the amount of hoarfrost at the observatory of Lomnitzer Spitze (49°12′N, 20°13′E, 2,635 m) and proved the clear diurnal change with a maximum during 16–18h and a minimum at 8h to noon.

In Japan, hoarfrost, formed apparently toward the direction of the wind is called *hanaboro*, and that produced by the sublimation of water vapour is termed *kibana*. The former looks a little dark against the snow cover, presenting the appearance of an aggregate icicle.

4.2. Hills, Basins, Valleys, and Bases of Mountain

A hilly or low mountain area where less variation of elevation is seen than in a mountain area, a comparatively flat area within the mountains, a transitional region between a mountain and a plain—the local climate in these places exhibits the most complicated features. In Japan, it is such places, rather than plains, that pose serious problems with regard to utilization of land.

Unlike mountain climate, the climate in these places does not depend so much on their height above sea level as on variation in relative height within the area, the direction of the slope, its gradient, the prevailing winds, or condition of the earth's surface. Of course, the relative importance of these factors varies with different local climatic phenomena, and even in the same place, it may vary according to the weather.

4.2.1. Insolation and Sunshine

(A) *Insolation*

If the influence of clouds is not considered, the amount of radiation on a slope is determined by the direction and gradient of the slope. The following equation was obtained by Okanoue (1957) and Okanoue et al. (1958):

$$R = \int_{t_1}^{t_2} I_s dt = I_0 \left[\frac{1}{\omega} \{1 - (\sin k \cos h \cos \varphi + \cos k \sin \varphi)^2\}^{\frac{1}{2}} \right.$$
$$\times \sin(\omega t + \alpha) \cos \delta + (\sin k \cos h \cos \varphi + \cos k \sin \varphi)$$
$$\left. \times \sin \delta \cdot t \right]_{t_1}^{t_2} \tag{4.27}$$

where, R is the daily solar radiation, I_s is the intensity of solar radiation on a slope, I_0 is the intensity of solar radiation on the plane perpendicular to the insolation, ω is angular velocity around the earth's axis, k is the angle of gradient of the slope, and h is the direction of the slope, t_1 and t_2 are the time of sunrise and sunset, respectively, noon being zero and the morning and the afternoon being negative and positive, respectively, φ, δ, and t are latitude, declination and hour angle, respectively. I_0 varies according to place and season, but for simplicity's sake, let it be equal to 1 cal·cm^{-2}·min^{-1}.

Lee and Baumgartner (1966) outlined the topography and insolation climate and proposed a *radiation index*, which is a percentage of I_s in Eq. (4.28) to normal radiation, the quantity theoretically available to a surface oriented normally to the sun's rays during the exposure period or simply $I_0/2e^2$.

$$I_s = \frac{I_0}{e^2} (\sin \phi' \cdot \sin \delta + \cos \phi' \cdot \cos \delta \cdot \cos \omega t') \tag{4.28}$$

Table 4.15 Radiation indexes at 50°N (Lee
and Baumgartner, 1966).

Slope inclination (%)	Aspect				
	N	NE–NW	E–W	SE–SW	S
0	42	42	42	42	42
10	38	39	42	44	46
20	34	36	42	47	49
30	30	33	42	49	52
40	26	31	42	51	54
50	23	29	42	52	56
60	21	27	42	53	57
70	19	26	42	54	58
80	17	24	41	54	59
90	15	23	41	54	59
100	14	22	41	54	59

where I_s is the maximum insolation extending on a sloping surface for a day neglecting atmospheric effects, I_0 is the solar constant, 2.00 ly/min, e^2 is the radius vector of the earth, ϕ' and $\omega t'$ are the latitude and hour angle of an equivalent horizontal surface (Lee, 1962), δ is the solar declination. Table 4.15 shows an example of the radiation index at 50°N.

Rouse et al. (1969) determined the direct beam radiation on a slope from graphs and combined it with diffuse radiation to give the global solar radiation for all slope facets of Lake Hill, Mt. St. Hilaire, E of Montreal. Global radiation I is composed of direct beam solar radiation Q and diffuse or sky radiation q:

$$I = Q + q \qquad (4.29)$$

Measurements of Q and q from horizontally mounted sensors allow the calculation of I for any slope. The direct radiation on a slope Q_s can be obtained from various formulae.

Recently, a device has been developed for estimating the short-wave radiation income on slopes (Garnier, 1968; Garnier et al., 1968). This method uses a knowledge of surface geometry, latitude, and the sun's declination. This proposal seems noteworthy in two respects: (a) It presents integration over any period of time, whereas most others are for instantaneous values only, and (b) in developing the integration, atmospheric transmissivity has been allowed for. An example of a table for use at latitude 45°N on a clear equinoctical day with an atmospheric transmissivity of 0.75 is given in Table 4.16. Kondratyev (1965) presents a formulae q_s in terms of q_h as follow;

$$q_s = q_h \cos^2 \frac{k}{2} \qquad (4.30)$$

where q_s is the diffuse solar radiation falling on a slope, q_h is the diffuse radiation measured on the horizontal, and k is the angle of the slope. He concludes that (a) the greatest slope differences in I occur in autumn and winter, whereas in summer, variations due to exposure are minimum; (b) on the N-slopes, low

Table 4.16 Daily totals of direct solar radiation (ly·day⁻¹) at the equinoxes for 45°N for an atmospheric transmissivity of 0.75 (Garnier and Ohmura, 1968).

Angle of slope	Azimuth of slope									
	0	10	20	30	40	50	60	70	80	90
0.0	369.6	369.6	369.6	369.6	369.6	369.6	369.6	369.6	369.6	369.6
10.0	299.8	300.7	303.6	308.4	314.8	322.8	332.1	342.3	353.1	364.2
20.0	220.8	222.8	228.7	238.6	252.3	269.0	287.9	308.3	329.6	351.2
30.0	135.2	138.3	149.1	166.9	189.7	215.7	244.0	273.7	304.0	334.0
40.0	45.4	54.4	77.1	105.1	136.3	170.0	205.2	241.5	277.9	314.0
50.0	0.0	9.3	33.6	63.8	97.5	134.0	172.4	212.0	251.9	291.4
60.0	0.0	2.4	16.4	40.7	71.6	106.8	145.1	184.8	225.6	266.2
70.0	0.0	1.0	9.3	27.6	53.7	85.5	121.4	159.5	198.9	238.2
80.0	0.0	0.5	6.0	19.5	40.9	68.4	100.2	135.3	171.3	208.1
90.0	0.0	0.3	3.9	13.9	30.8	53.8	81.4	111.7	144.0	176.1

Angle of slope	Azimuth of slope									
	90	100	110	120	130	140	150	160	170	180
0.0	369.6	369.6	369.6	369.6	369.6	369.6	369.6	369.6	369.6	369.6
10.0	364.2	375.4	386.1	396.2	405.3	413.2	419.6	424.3	472.2	428.2
20.0	351.2	372.4	393.0	412.4	429.5	444.5	457.0	466.2	471.9	473.8
30.0	334.0	363.6	391.8	418.2	442.1	463.2	480.7	493.9	502.2	505.0
40.0	314.0	349.3	383.0	414.6	443.4	468.8	490.3	506.8	517.3	520.8
50.0	291.4	330.0	366.9	401.5	433.2	461.4	485.5	504.3	516.6	520.8
60.0	266.2	305.7	343.5	379.2	412.0	441.2	466.5	486.8	500.2	505.0
70.0	238.2	276.8	313.8	348.2	380.4	409.1	434.0	454.5	468.7	473.8
80.0	208.1	243.9	278.2	310.2	339.7	366.2	389.4	409.0	423.0	428.2
90.0	176.1	208.2	238.2	266.3	291.5	314.1	334.2	351.4	364.5	369.6

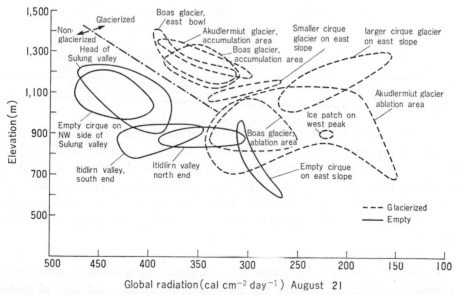

Fig. 4.32 Comparison of glacierized and nonglacierized valleys and cirque basins in an area between the heads of Quajon and Narpaing fiords, E-Baffin Island, N.W.T., Canada, with respect to elevation and computed global radiation for August 21. Each curve encloses the spread of values of elevation and global radiation for the individual basins (Williams et al., 1972).

radiative thermal energy in spring retards snow melting and this results in wetter N-slope soils throughout the year; and (c) the greater consumption of available heat for evapotranspiration from N-facing points results in lower air and ground temperatures than on their S-facing counterparts, notably in spring and autumn.

Following Garnier and Ohmura (1968), mentioned above, computations of the global radiation on a slope for an area on Barbados (Garnier and Ohmura, 1970) and in E-Baffin Island, Canada (Williams et al., 1972) were made. One of the results is shown in Fig. 4.32, which presents a relationship between the effect of elevation on glacierization in a mountain region. This figure illustrates how the presence or absence of glaciers in existing cirques and valleys can be discriminated by the combined influence of global radiation and elevation. It is of interest that the upper part of the larger glaciers corresponds to the accumulation area where over the budget year net gain of snow and ice exceeds net loss and, on the other hand, the lower part of the larger glaciers roughly corresponds to the ablation area where loss of snow and ice exceeds gain except for addition of ice by glacier movement. Similar problems have been discussed from other viewpoints as mentioned in Section 4.1.8 (D)

Table 4.17 Radiation budget (ly/year) on Gr. Falkenstein, according to the results of Baumgartner (Miller, 1965).

	Short-wave radiation		Long-wave radiation	Whole-spectrum radiation
	Direct	Diffuse		
Downward	53,000 *30,000*	42,000 *5,000*	224,000 *3,000*	319,000
Upward		−20,000 *2,000*	−264,000 *5,000*	−284,000
Net		75,000	−40,000	35,000 *15,000*

(Italic numbers represent the influence of topography on each flux of radiation.)

Miller (1965), reviewing the studies on the heat budget of the earth's surface on slopes, summarized the magnitude of the influence of terrain on Gr. Falkenstein, Mittelgebirge, Germany, which was studied by Baumgartner (1960), as shown in Table 4.17. The effect of terrain is the greatest on the direct short-wave radiation, as seen on this table. Differences in terrain can be expressed in terms of heat supplied by radiation and turbulent fluxes to melt snow (Miller, 1956a).

(B) *Sunshine*

There is a great local difference in the duration of sunshine in areas among mountains. One reason is that the astronomically determined time of sunrise or sunset is influenced by topographical features, and the duration of sunshine is reduced, and another reason is the frequent occurrence of fog or clouds. To what extent the topographical features are responsible for the delay in sunrise

(a)

(b)

Photo 4.8 (a)   Contrast of remaining snow between the sunny side (right) and shadow side (left) slopes in winter (February 6, 1972, by M. Ishii).
    (b)   Conditions at same place in summer. In east-west running valley in Mitagawa, Saitama Prefecture, Japan (July 29, 1971, by M. Ishii).

or advance in sunset varies from place to place in mountainous regions. Even a small saddle or peak of ridges in the way of sunshine has a strong influence on the time. As for fog and clouds, they are subject to the influence of the development of inversion at night and an up-current on a slope and valley wind in the daytime. The altitude for each of these phenomena varies according to whether it is summer or winter.

Steinhauser (1956) presented a distribution of the duration of sunshine in

Austria, in which the rate of sunshine (%) is divided into six grades in summer and seven grades in winter, as shown in Fig. 4.33. During the summer, eastern lowlands enjoy a longer duration of sunshine than the mountainous area in the west, which is covered more often with clouds. In winter, however, the western mountainous areas are exposed to the sun longer than the eastern lowlands since the former stands higher than the altitude of the inversion layer. The percentage of possible sunshine in the lowlands along the Danube River is especially low during the winter, decreasing to 25% or below. A detailed distribution chart for the sunshine is also useful for environmental problems.

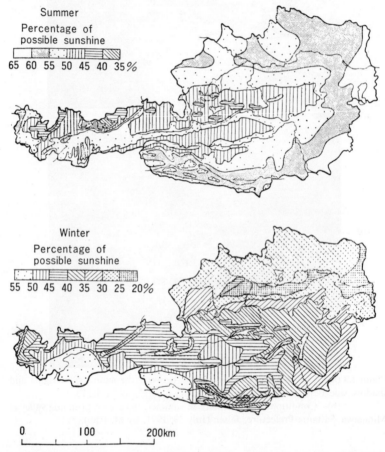

Fig. 4.33 Distribution of percentage of possible sunshine in Austria, 1928–1950 (Steinhauser, 1956).

Huttenlocher (1923) and Garnett (1935, 1937) made authoritative classic studies on the problems of shadow and sunny slopes in high mountain regions. These conditions are important for the situations of settlements and land use in mountain regions. Böhm (1966) applied Garnett's calculation method in making a shadow map of Paznaun valley (800–2,400 m), Tyrol, on a scale of 1 : 50,000.

He points out that the degree of contrast between shadow and sunny slopes increases with altitude. Such a contrast is more pronounced in the difference between air temperature and soil surface temperature.

4.2.2. Air Temperature

(A) *Minimum Temperature or Nighttime Temperature*

As for hills, basins, valleys, and bases of mountains, the minimum temperature is lowest where the altitude is lowest within the area. As discussed in detail, the air just above the ground surface is cooled by radiation at night, and this cooled air gradually goes down toward the lower part and accumulates in the bottom. When it does not stay there, on the other hand, the cooled air gathers the surrounding cold air as it goes down, being cooled further through a continuous cooling from the ground. As a result, the temperature becomes lowest at the lowest altitude in the valley or the basin.

The lowest minimum temperature occurs in most cases under micro-topographical situations like the basins. For example, the lowest temperature, −38.5°C, was observed in the Tzara Bârsei Basin, Bod, Rumania, in the E-Carpathian Mountains, on January 25, 1942 (Gugiuman et al., 1970). Also, the lowest temperature, −41.0°C, was observed at Asahikawa in the basin in central Hokkaido, Japan, on January 25, 1902.

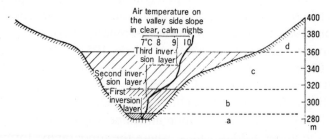

Fig. 4.34 Frost danger zone along the cross section of Finkenbach, Odenwald, Germany. a: Frequent frost, b: Frost once or twice a year, c: Frost experienced once or twice by old men, d: Frost is rare (Schnelle, 1950a).

The effect of micro-topography is more striking in small valleys or basins. An investigation in Finkenbach Valley, Odenwald, Germany, shows that there is an inversion of 3°C between the valley bottom and at a height of about 100 m above the bottom as an average of nine clear, calm nights (Schnelle, 1950a,b). As shown in Fig. 4.34, the first inversion, above the valley bottom up to a height of 30 m, exhibits a difference of 0.5°C, but, in the second inversion layer, with a thickness of 40 m, the difference becomes as great as 2°C.

The temperature distribution at an altitude of 70 cm above the ground in the vicinity of Lake Constance on the borders of the southern part of West Germany and Switzerland is given in Fig. 4.35. There is a clear relation between temperature and topography—low temperature at low altitude. The variation

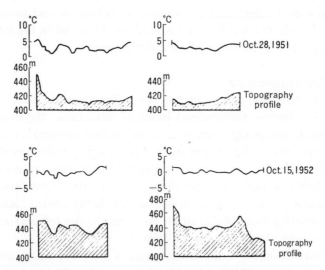

Fig. 4.35 Air temperature profiles and the micro-topography profiles near Lake Constance on a clear, calm night (Aichele, 1953b).

with height becomes 5°C for 30 m, 2–3°C for 20 m, and 3°C for 10 m in extreme cases. The maximum height of the cold air lake is 2–3 m in a small valley, and 7–8 m in a larger one. The depth of a cold air lake depends on such micro-topographical factors as the shapes of longitudinal section or cross section of a valley (Aichele, 1953 b). These circumstances reflect, of course, the phenology such as the flowering date of apple trees (Aichele, 1953 c). In General, in larger valleys 400–500 m wide at the upper parts of valley slopes, the depth of a cold air lake is 30–75 m with occasional differences varying from ±5 to ±10 m due to the prevailing meteorological conditions (Schnelle, 1956).

Lomas (1968) reports on the minimum temperature distribution in the Beit Shean basin, the upper Jordan valley, where the topography, forming a series of river terraces extending from north to south, is relatively closed. He found −1.5°C on the basin floor (−275 m), +1.0°C on the river terrace (−175 to −200 m) and +2.0°C on the lower part of valley slopes (0 to −25 m). Such topographically closed situations always suffer from low minimum temperature. Shitara (1965) studied the minimum temperature in the Kakuda Basin, Tohoku district, Japan, where the Watari Hills (150–300 m) cause a barrier effect, damming up the cold air. Based on the observation of the frost damage distributions in the basin (see Fig. 4.41) on mulberry leaves and Japanese persimmon leaves, he classifies the degree of cold into seven grades.

Inversions in a valley are generally most intense in the deepest parts of the valleys as Goldreich (1971) reports from Johannesburg. The relationship between the shapes of the valley slope and the nighttime temperature distribution is discussed in some detail in Section 5.5.5.

Differences in temperature within a relatively small valley are caused by the length as well as the depth of the valley. The results of measurements on the minimum temperature at 50 cm above the ground at thirty-seven points over 38 km² near Heidelberg, Germany, are as follows (Kreutz and Schubach, 1952): The temperature gradient from valley to plain on a clear night was 7°, 8°, 4°, 5°, and 3°C in five small valleys slightly differing in size. Although it is difficult to describe the topography numerically, it can be inferred from a topographical map that the first two valleys are about twice as large as the latter three.

In a valley, the lowest temperature is naturally found in the basin. An observation made for two months in a valley in Herefordshire, England, with the aid of eight screens placed at various topographical situations has shown that there is a difference of 5.6°C/100 m between the two points with relative heights of about 9 m; the mean value of eleven measurements on clear nights reached 8.5°C/100 m (Lawrence, 1956). This is thought to be a very large increase rate.

Analysis of a detailed observation of the Fukushima Basin, Japan, shows that, concerning the relationship between the minimum temperature and the height above sea level, the areas at 100–200 m above the floor of the basin are warmer. Above these areas, the temperature gradually decreases at the rate of 0.6°C/100 m (Umeda, 1958). The difference (ΔT) between the temperature at 18h on the previous day and the minimum temperature of the day attains a maximum at a height of 40–80 m above the floor, and below this height it decreases. ΔT is closely related to the inclination of slopes; it becomes smaller as the gradient becomes steeper.

According to the results of the observation in the Baar Basin, S-Germany, which is in the shape of a triangle with each side 8 km, an inversion of air temperature is observable at the rate of 6°C/100 m on clear nights, and on cloudy days little difference is recognized. The maximum altitude of frost during the growing period is 6–8 m above the floor of the basin. In general, the minimum temperature on the basin-floor is low, e.g., the ground surface temperature was—4.5°C on June 26, 1949 (Aichele, 1951).

Obrebska-Starkel (1970) studied the height of the upper limit of the inversion layer in a valley or a basin. She classifies the thermal zonation of the slope as follows: (a) the zone with the inversion layer in the valley bottom, (b) the warm slope zone, and (c) the summit zone. Table 4.18 is based on the table in her paper which is limited to slopes of valleys and basins in central Europe. This table shows that the upper limits of the inversion layer in the valley or its basin appear at heights between 120 m and 370 m. As will be discussed below, the thermal belt on a mountain slope is generally located at levels between 200 and 400 m (see Section 5.5.6). It is obvious from the results of the measurement at Quickborn, Holstein, Germany, that the minimum temperature distribution at the surface is largely determined by the nature and state of the soil, but at 70–80 cm above the ground surface, it is influenced not so much by the nature and state of the soil as by the micro-topography or the height above sea level

Table 4.18 Height of inversion layer on the slope of valleys or basins in central Europe (Obrebska-Starkel, 1970).

Region	Type of relief (Relative height)	Average height of upper limit of inversion layer	Researcher
Nowy Targ, W-Carpathian Mts.	Basin, 400–700 m	370 m	Michalczewski (1962)
Gr. Arber, Bavarian Forest	Mittelgebirge, 900 m	300 m	Geiger et al. (1933/34)
Gr. Falkenstein, Bavarian Forest	Mittelgebirge, 900 m	200–300 m	Baumgartner (1960, 1961, 1962)
N-Slope, Kirgiz and Transilenian Ala-Tau	Front mountain of high mountains 650–850 m	200–250 m	Gelmgoltz (1963)
Central Balkan Mts.	Hilly terrain, 200 m	200 m	Tiskov (1963)
Ötztal, Tyrol	High mountain, 800–1,300 m	180–220 m, Winter 120–140 m, Summer	Aulitzky (1967, 1968)
Raba Valley W-Carpathian Mts.	Hilly terrain, 120 m on the foot of mountains, 900 m	150 m	Niedźwiedź (1970)
Beskiden, Jaszcze-Jamme and Valley, W-Carpathian Mts.	Middle mountain 400–600 m	120–140 m	Obrebska-Starkel (1969a, b, 1970)

(Eimern, 1951). However, the effect of the ground surface is observable, if the micro-topographical conditions are equal. An observation of the minimum air temperature at the eighteen points in the moor, peat land along the Danube River and its surrounding region, has made clear that the lowest minimum temperature always occurs in the center of moor land (Kern, 1951a,b).

Local differences in the minimum temperature are more apparent in the early morning after clear and calm nights than after cloudy and windy nights. As stated above, an inversion of about 5°–6°C/100 m was recognized on the morning after a clear, calm night between the floor of a basin and the adjacent slopes with a relative height of approximately 50 m. When the sky had been completely overcast with low clouds throughout the night, little difference was perceived on the following morning. The difference also becomes smaller under a high wind, which prevents the development of inversion. An example is that the difference of 1.0–2.0°C/100 m on calm nights is reduced to 0.4°C/100 m when the nocturnal wind velocity is over 5 m/sec (Eimern et al., 1954). According to an observation in Frankop near Hamburg, Germany, the cold air lake begins to be formed when the wind velocity becomes 2 m/sec or weaker at a level of 10 m above the ground (Franken, 1959). Particularly, under the influence of a N-wind, the air temperature falls markedly in Frankop, because the topographical conditions of the prevailing wind become favorable for the formation of a cold air lake there.

The minimum temperature distribution in comparatively restricted areas is related more or less to the cultivation of crops. An example is the damage from late frost in spring or first frost in autumn. It is important, therefore, to observe in this connection how the spring or autumn distribution of minimum temperature differs from the distribution in the other seasons of the year. According to an observation of the distribution of the minimum temperature at 1.5–2.0 m above the ground, it was confirmed that no particular difference in distribution patterns is recognizable between the temperature in spring and that in other seasons (Franken, 1955/56). However, the absolute values of temperature difference change in accordance with the height above the ground, the topographical conditions, and seasons. In summer months, the minimum temperature differences between two nearby stations on clear, calm nights are less than in autumn and those are less than in spring (Eimern, 1964).

Dykes or small hills also affect the minimum temperature. A dyke of railroads of 12–15 m height made differences with the range of 0.1 to 0.5°C in the minimum temperature 50 cm above the ground at ten points on a sloping area behind the dyke from April 15 to July 7, 1955, and those ranging from −1.8° to +0.3°C on clear, calm nights (Scultetus, 1964). A study on a hill in Kaleto, Bulgaria, revealed that a hilltop station (200 m) was on an 8-day average and warmer than the surrounding plain (0 m) in June and July, and on a 17-day average 5.2°C higher in October, 1960, on clear, calm nights (Bluskowa, 1965). For detailed information on frost distribution, cold air drainage, cold air lakes, and inversions, all of which have bearing on the distribution of the minimum temperature, see the respective Sections.

(B)   *Maximum Temperature or Daytime Temperature*

The distribution of the maximum temperature, usually having smaller local differences than the minimum temperature, is under the control of the temperature lapse rate, the direction and gradient of slopes—in other words, the extent of insolation and hours of daylight. A place exposed to the prevailing wind has comparatively low temperatures.

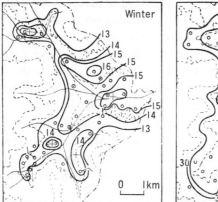 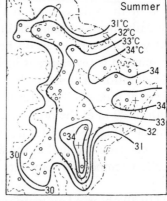

Fig. 4.36   Distribution of maximum air temperature (°C) in the Aoki region in winter and summer. Left: Dec. 20, 1951. Right: Average of 4 days, July 25, and Aug. 3, 19 and 25, 1952. (Ozawa and Tsuboi, 1952, 1953).

Under similar conditions of surface cover, the temperature is high at lower altitudes except for V-shaped valleys with steep slopes on both sides. An observation at the village of Aoki, Nagano Prefecture, where the distribution of the maximum temperature was measured at forty points, shows that the temperature differs by 2°C on calm, clear days in winter and by 3°C in summer, on an average, with the height difference of 200 m, as shown in Fig. 4.36 (Ozawa and Tsuboi, 1952, 1953). According to a study in Bath and the surrounding areas, England, the maximum air temperature is 1.7–2.2°C higher in the valley in winter and 2.8–3.9°C higher in summer than on the hill of 250 m (Balchin and Pye, 1947). The higher maximum air temperature in valleys and basins in relation to calculating the daily mean is discussed by Böer (1952). He presents the values for Kaltennordheim (valley) and Kalteneber (plateau) in Thüringen, Germany. The feature of the type of this distribution become more distinct in fine weather. A monthly average has a smaller value of difference. The result of a nine-month observation on Harburger mountain and the nearby Elbe lowlands in that the difference in the monthly mean is only 0.6°C, although on some days the difference is about 2°C (Eimern and Kaps, 1954).

The lapse rate of the air temperature on the slopes in the hilly terrain in the daytime is larger than the normal lapse rate, 0.6°C/100 m. Oliver (1964) states that the lapse rate in the daytime between the upper station (671 m a.s.l.) and the lower station (408 m a.s.l.) on an upland in South Walse, England, is on an average 0.77°C/100 m from November 1958 to April 1960.

To summarize, the facts mentioned above, it can be said that the temperature decrease apparently with the height in the lower part of the slopes of a valley or basin in the daytime on a fine, clear day. Its decreasing rate, lapse rate, is greater than the normal lapse rate.

As Mäde (1956) has shown on a cross section along the cities of Wernigerode, Aschersleben and Gatersleben, Germany, the frequency distribution of deviation from the daily mean air temperature is greater at a station located in the bottom of the valleys than at the stations on the surrounding slopes, although the daily mean air temperature is almost the same.

In a valley with steep slopes on both sides, the maximum temperature as well as the minimum temperature is high in the middle part on the slopes. This is because the duration of sunshine becomes shorter and soil moisture becomes more abundant, as it descends nearer to the bottom. An observation in a valley, with the width of 20 m at the bottom and the gradient of slopes being 15–25°, in W-Steiermark, 20 km SW of Graz, Austria, by Morawetz (1952) shows that the lowest temperature is found at the bottom with the relative height of 65 m, while the highest temperature appears at 40 m above the bottom except the early morning.

Elomaa (1970) observed air temperature distribution on an esker at Lammi, S-Finland. His conculusion is that the maximum air temperatures are fairly uniform as compared with minimum air temperatures, and the time of the maximum temperature is more closely related to the exposure than to the height of the observation point. The SW-slope of the esker is 0.1–0.2°C warmer than the N-slope.

The temperature difference according to the direction of a slope is brought out by the difference in ground temperature, which in turn is brought about by the slope direction. The maximum temperature is sometimes recorded on the S-facing or SSE-facing slopes from May to July, but in the other seasons on the SW-facing slopes in the northern hemisphere. For further details, see Sections 4.1.4 and 4.2.3 dealing with ground temperature.

Microclimatological observations of the various climatic elements on slopes have been made in many parts of the world, for instance, Kagawa Prefecture, Shikoku, Japan (Uehara, 1949, 1950a, b); on the Cushetunk Mountain, New Jersey, U.S.A. (Cantlon, 1953); on the Central and SW-Harz Mountains, Germany (Hartmann et al., 1959); in Montgomery Ridge and Green Ridge in Ontario (MacHattie et al., 1961); and in Fanshan, China (Fuh, 1962). The investigation on Fanshan, a small hill at 32°08'N, 118°48'E, 210 m a.s.l. with the relative height of 190 m, near Nanking, China, has made it clear that the maximum temperature appears at 15h 25m on the W-slope, but at 13h 50m on the E-slope on clear winter days.

In the southern hemisphere, the N-facing slopes naturally have higher temperatures. In the Dionisia district (Buenos Aires) the maximum temperature at 30 m above the ground is recorded on the N-facing slopes, and both the soil moisture and humidity are higher on the S-facing slopes (Cagliolo, 1951). The observations were made at five sites, on the peak and slopes in four directions and on a hill in the southeastern part of the city from 1939 to 1943. The results of these observations must be examined from the standpoint of heat budget, the relationships between the climatic elements being taken into consideration.

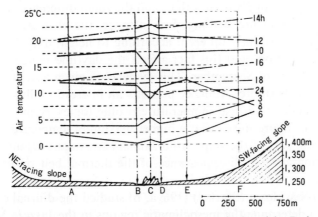

Fig. 4.37 Diurnal change of air temperature at the Sugadaira Basin on clear days in autumn, 1950 (Yoshino, 1961a).

An example of research along this line is a study by Wendler (1971), which will be mentioned below.

(C)   *Diurnal Variation of Air Temperature Distribution*

The diurnal variation of the air temperature distribution at the Sugadaira Basin, Nagano Prefecture, Japan, from September 23 to October 2, 1950, is given in Fig. 4.37 (Yoshino, 1961a). The inversion had already taken place in the morning and the air temperature was 5°C at Point F on the outskirts of the basin with a relative height of 50 m, and at the foot of the basin it was 0.5°C, the difference between the two points being 4.5°C. Frost fell at Points B and D. At 8h there was a considerable rise in temperature, but within a forest where a small increase rate was expected it was lower by 2°C than outside the forest. Point F on the SW-facing slope, due to the delay of sunrise, had a lower temperature than the other points. A further rise in air temperature was observed at 10h, when the temperature showed small difference except within the forest. At 12h, the temperature within the forest was higher than that outside the forest, because it maintained a great increase rate. This was the same at 14h. On the other hand, at Point F the temperature was highest at 14h since it was on the SW-facing slope. A similar tendency was still noticeable at 16h, though the temperature was lower by 7–8°C than at 14h. At 18h, the whole basin became isothermal, and at midnight, 24h, the temperature everywhere fell by about 2°C. Though the inversion had occurred before 3h, the air temperature remained about 1°C higher inside the forest than outside because of the small decrease rate of air temperature. The air temperature continued to fall until 6h. The distributions of air temperature in the early morning (Yoshino, 1961a) and in the afternoon (Sekiguti, 1951c) are given in Fig. 4.38.

Aulitzky (1968) reports the air temperature conditions on a slope in Obergurgl, Tyrol, in the Central Alps in detail. The isopleths of monthly change of the diurnal variation of air temperature at four points on the WNW-facing slopes are shown in Fig. 4.39. Even though the differences in the patterns are not great, the change of absolute values and the small differences in accordance with change of height are striking. For instance, the changes of sunrise and sunset hours are apparently different according to height and the surrounding topographical situation, and this fact affects the daytime temperature change, especially from June to September. This evidence is given as a closed isotherm in the figure. Furthermore, Aulitzky presents an interesting illustration on the diurnal change of the vertical temperature profile for every month, as shown in Fig. 4.40. This figure shows that a higher maximum temperature appears on the lower middle part of the slope from January to August around 13h; it also shows the formation and disappearance of the thermal belt on the slope, discussed below.

Obrebska-Starklowa (Starkel) (1969a, b) studied the diurnal change of air temperature to determine the mesoclimatic regions in the Jaszcze Valley in the Gorce Mountains, Polish Beskids. She uses various features for criteria: (a) the

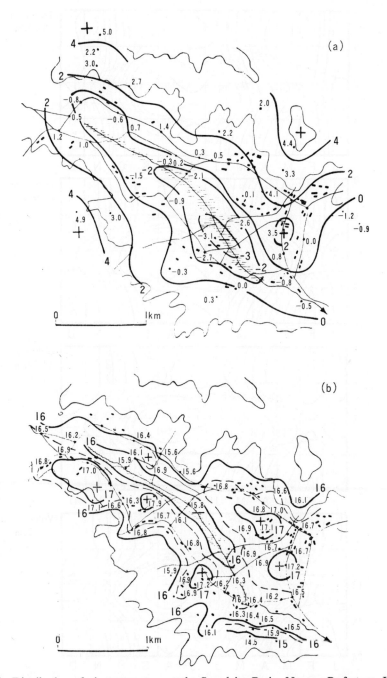

Fig. 4.38 Distribution of air temperature at the Sugadaira Basin, Nagano Prefecture, Japan, in the nearly morning (a) and afternoon (b) on a clear day (Yoshino, 1961a; Sekiguti, 1951c).

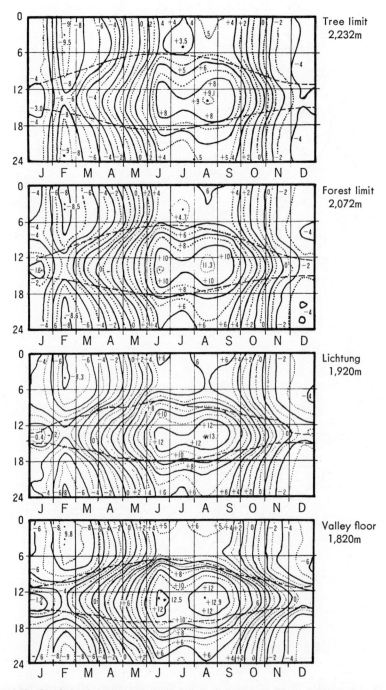

Fig. 4.39 Isopleth of monthly change of diurnal variation of air temperature (°C) at four stations on the slope in Obergurgl, Tyrol, Austria (Aulitzky, 1968).

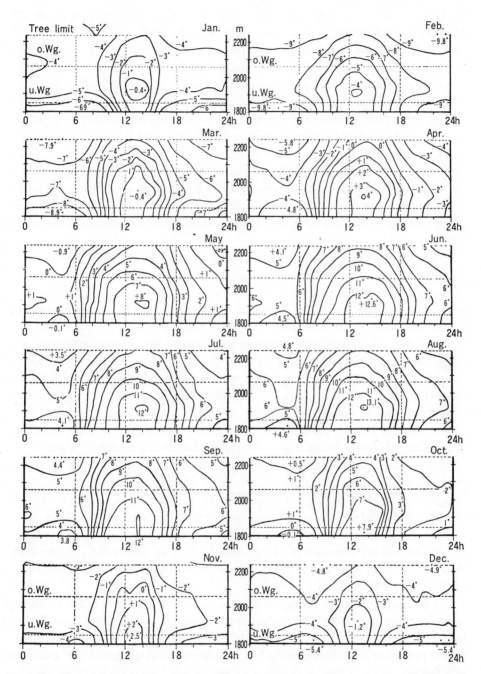

Fig. 4.40 Diurnal change of the vertical profile of air temperature (°C) on the slope in Obergurgl, Tyrol, Austria. o.Wg.=upper (actual) forest limit   u.Wg.=lower forest limit (Aulitzky, 1968).

spatial distribution of daily mean temperature and of mean temperatures in the daytime as well as at night, (b) the shaping and range of inversion of air temperature in the evening, after sunset, and early in the morning during the period of minimum temperature, as well as the intensity of heating by radiation of various parts of the valley at noon, (c) the values of mean and absolute amplitudes in a given series, and (d) the percentage frequency of deviations of hourly values from the mean daily temperature. On the basis of these thermal indices, valleys of middle-high mountains have three mesoclimatic regions: I. cool summit ridges, II. warm overinversive slopes, and III. an inversion layer parts, averaging 120–140 m above the valley bottom. This has been discussed above in relation to Table 4.18. The region of summit ridges remaining under the influence of frequent advections of air masses shows a variation of climatic elements with the height above sea level. On the other hand, the climatic conditions in the other two regions relate to the existing configuration of the terrain.

A comparative study of diurnal temperature variation at various points in a limited area was carried on by the author in the following manner. Variations in temperature at each point were read on thermograms at regular intervals (6 min in this case), and the standard deviation for temperature fluctuations during each interval was obtained. Large deviations indicate great fluctuations in the course of temperature variation in a small area. The observations were made at several regions in Japan with the following conclusions: (a) The standard deviation usually has values by day, especially around noon, two or three times larger than at night. (b) Temporal maxima are to be observed just after dawn, before sunset, and during the night when a cold air current flows. The maximum after dawn is most conspicuous when there is a considerable lag in the time of sunrise because of topographical features, especially the direction of slopes. (c) In a similar topography the value becomes smaller in inverse proportion to the wind velocity. (d) When the wind is not strong, topographical complexity, rather than dimensions, of the area under study is influential. This is due to the local difference of wind turbulence affected by topography.

(D) *Distribution of Frost*

When the temperature of the ground falls to or below 0°C, frost occurs. Strictly speaking, there are various forms of frost, but here only the distribution of frost is considered.

The cold air current flows just like the stream of a river in accordance with micro-topography, as will be discussed in Section 5.4. The place most favored by a cold air current is termed the *frost path*, and the reservoir of cold air is a a *frost pocket* or *frost hollow*, which is a topographical depression or a clearing in a forest.

The frost distribution appears to be roughly same to the minimum temperature distribution. The vertical temperature distribution near the ground surface during the night shows an inversion condition, but it varies with the micro-topographical situation of a spot, the state of the ground surface, or the time of year. Therefore, the frost distribution does not accord exactly with the horizontal distribution of minimum temperature, e.g., at 150 cm above the surface.

The result of a comparison between measured minimum temperatures in a screen and minimum temperatures at a height of 120 cm at 27 points over an area 10 km×6 km in Owada, Saitama Prefecture, Japan, shows that frost occurs when the minimum temperatures are from −1°C to +8°C in the screen. Although the conditions vary from place to place, frost is expected at any point when the minimum temperature in the screen is +3°C on an average (Nose et al., 1953).

Frost probability on the cross sections of a valley has been studied in Finkenbach, Odenwald, as mentioned above (Schnelle, 1950a, b, 1956) and in the upper Modau Valley, Odenwald (Schubach, 1960). There is a great difference in the frost probability with height differences of 40–50 m on the lower part of the valley side slopes.

The minimum temperature was observed at a hundred points in a small valley, Neotoma in Ohio (250 m in width at the top, 6 km in length), from 1941 to 1943 (Wolfe et al., 1943, 1949). The observation of the frostless periods in frost pockets, various points in the valley, caves, and joints in rocks, showed that there is a remarkable variation with the position of respective points (Table 4.19). For instance, the frostless period was 100 days shorter in frost pockets. The difference in period between the points may vary as much as 30 to 50 days. The year by year differences in the vegetation and growing season, especially the defoliation period, as well as in meteorological conditions, are added further to this variation.

Table 4.19   Difference of frostless period in a small area
in the Neotoma Valley, Ohio, U.S.A.
in 1941 (Wolfe et al., 1943).

	Last frost (spring)	First frost (Autumn)	Number of days of frostless period
Frost pocket	May 25	Sept. 26	124 days
Lower slopes	May 14	Oct. 11	150
Upper slopes	April 22	Oct. 29	190
Some ridge tops	April 3	Oct. 29	209
Cave	April 3	Nov. 11	221
Crevice	April 3	Nov. 25	235
Grotto	April 3	Dec. 13	254
Lancaster W. B.	May 5	Sept. 26	144

The most outstanding example of a frost pocket appears in a doline in limestone regions. Aigner (1952) reported results from a doline in Gstettner subalpine meadow (approximately 47°30′N, 15°E) near Lunz, Austria. The plateau has a height of 1,450 m a.s.l. and the relative height from the valley bottom is 1,270 m. The doline has a depth of 150 m. In the severe winter from

1929 to 1930, the thermometer registered −51°C, and after that, −52.6°C, a record in Europe, was measured. Measurements along a cross section of this doline on January 21, 1930, obtained the following results: −28.8°C at the bottom and −1 to −2°C on the upper part of the slope. The difference attained almost 30°C (Schmidt, 1930a). Sauberer and Dirmhirn (1953, 1956) observed air temperature change, wind velocity, and heat balance in this doline and showed the complex formation process of a cold air lake in the frost pocket. Details of this frost pocket are described by Geiger (1961, 1965).

Observations concerning frost hollows have been carried out in England; the frost hollow in a narrow, dry valley near Rickmansworth is the most conspicuous (Hawke, 1944; Manley, 1944). At Rothamsted odds C against unity of a frost-free period of a given length or less are given by Smith (1954) as the following Equations.

$$\text{length (days)} = 178.6 - 24.2 \log C \qquad (4.31)$$

But, in a frost pocket at Rickmansworth it is

$$\text{length (days)} = 97.3 - 31.7 \log C \qquad (4.32)$$

In the Bükk mountains, NE-Hungary, Wagner studied the formation of cold air layers in the dolines for several years (Wagner, 1963, 1965, 1970). In a doline of Középbérc (48°N), he observed −2°C at 4h in the morning on August 5, 1960. Formation processes of cold air lakes (or ponds) are discussed in Section 5.4. According to a survey on frost in the highlands of Ceylon, 6–8°N, up to 2,526 m a.s.l., by Domrös (1970), frost occurs at Nuwara Eliya (1,896 m), Pedro (1,900 m), and Park (2,000 m) during the period from December to April. Statistically, the mean monthly number of frost days on grass recorded at the meteorological stations, Nuwara Eliya, are 2.5 days in January, 3.6 in February, 1.5 in March, 0.2 in April, 0.1 in November, and 0.8 in December, 1897–1967. But the year by year fluctuation is very great. These frosts in the highlands of Ceylon are local phenomena and occur in the frost hollows.

The length of the frost-free or frostless period decreases with increasing altitude. Its relation is in most cases expressed by a linear regression equation. For instance, the following relations were obtained from the results at 46 stations located from the coast to 1,300 m a.s.l. in Slovenia, Yugoslavia (Hočevar, 1961):

$$\text{Last occurrence of frost} = 107 + 0.025H \qquad (4.33)$$

$$\text{First occurrence of frost} = 301 - 0.047H \qquad (4.34)$$

$$\text{Frostless period} = 194 - 0.072H \qquad (4.35)$$

For instance, Eq. (4.33) means that in spring the last frost comes later at the rate of 25 days/1,000 m a.s.l. Equation (4.35) shows that the frostless period is about 90 days at a height of 1,500 m, where the main forest limit is situated in this region.

It must be pointed out that the relation between the phenomena concerning

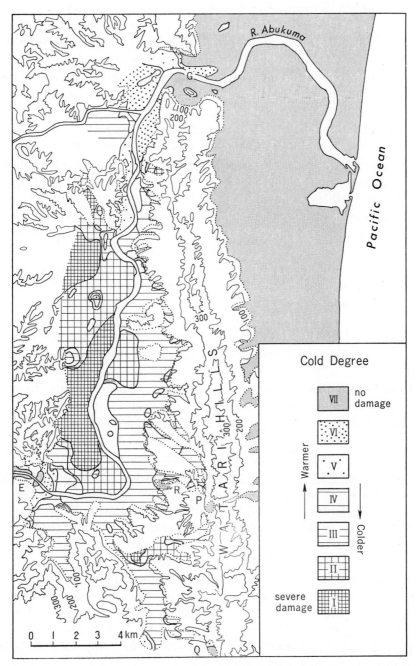

Fig. 4.41 Distribution of degree of cold at the Kakuda Basin, Fukushima Prefecture, Japan (Shitara, 1965).

air temperature and height is expressed by a linear function in a range of relatively small height difference like Eqs. 4.33, 4.34 and 4.35, but by complicated curves in a range over 2,000 m, as discussed in the earlier part of this book.

The height (above the ground) of frost injury on trees depends upon the micro-topography, because cold air accumulates deeper in a depression, valley and basin. According to an observation in the forests in Hokkaido, the height of the accumulation was 3.1 m in the flat valley bottom, 1.8 m on the lower part of slope, and 3.6 m in the concave valley bottom (Sasaki, 1962).

Shitara (1965) studied the frost distribution in the Kakuda Basin, Fukushima Prefecture, Japan, by observing the frost damage of mulberry leaves and Japanese persimmon leaves. He classified the frost damage into the following four classes according to the degree of damage: For mulberry damage: (a) Severe damage. All leaves fell and injured leaves could not grow for two weeks after the damage, because many stalks themselves withered due to freezing. (b) Heavy to moderate damage. Complete leaf fall but compound leaves soon began to regrow because there was no stalk damage. (c) Partial damage. Some leaves fell, while other leaves remain though partly damaged. (d) No damage. All the leaves remained perfect. No leaf damage was observed. For Japanese persimmon damage: (a') Severe damage. All leaves fell due to the frost. (b') Heavy to moderate damage. Many leaves fell and those remaining became partly discolored. (c') Partial damage. Some leaves fell or became partly discolored. (d') No damage. Leaves suffered no change. No leaf damage was observed.

According to these classifications, the damage degree in the basin was surveyed. Combining the two patterns, the distribution of the degree of cold was ascertained as given in Fig. 4.41. From this figure, the following characteristics are note worthy: (a) The severely cold Kakuda Basin is separated by the Watari Hills from a comparatively warmer coastal region. (b) The basin, however, has two warm areas: one around the SW-corner near the entrance of the Abukuma River (E in Fig. 4.41) and the other around the N-corner area near the gap (G in Fig. 4.41). (c) Excepting the former area, the climate is on the whole warmer to the north. (d) It is warmer on the E-side of the Abukuma River than on the W-side in the region where the river runs from S to N. (e) Warm islands are found in places (Shitara, 1965).

In a comprehensive work on frost from the local and microclimatological standpoints edited by Schnelle (1963), Schneider (1963) discusses the concept and classification of frosts, Burckhardt (1963a, b) the metorological presupposition of night frost and the research method for frost damage distribution, and Baumgartner (1963a, b) the heat balance of soil and plants and the cold air drainage in relation to frost formation.

Discussions on cold air drainage and cold air lakes appear in Sections 5.4.1–5.4.3.

4.2.3. Ground Temperature

The amount of insolation received by the ground surface differs according to the gradient and direction of a slope. In the Northern Hemishere, a S-facing slope is supposed to catch the maximum insolation, but in fact the maximum ground temperature tends to appear on a SW-facing slope, since the rise in the ground temperature depends on the soil moisture. The greatest amount of solar energy received during the morning is spent for the desiccation of the soil. It is after the evaporation of the soil moisture that the solar energy absorbed is able to raise the ground temperature. Consequently, the maximum ground temperature appears on a SW-facing slope, and not on a S- or SE-facing slope. In May, June, and July with high solar altitudes and early sunrises, however, this tendency becomes obscure. In a recent investigation on the ground temperature on E- and W-facing slopes, Boros (1971) suggested that there are still many problems to be solved.

Provided that the soil moisture is constant, the ground temperature on a slope is closely connected with the amount of solar radiation the slope receives. If the amount of solar radiation is represented by the insolation coefficient (cf. Eq. 4.27), the ground temperature at a depth of 5 cm demonstrates a clear relation to the insolation coefficient. On the basis of an observation on the slopes at Nakayama, Ibaraki Prefecture, Japan, on October 30, 1959, a regression line is drawn that exhibits the ground temperature of 17.2°C with the insolation coefficient of 150 cal·cm^{-2}·min^{-1}, 17.5°C with 250 cal·cm^{-2}·min^{-1}, and 18.4°C with 350 cal·cm^{-2}·min^{-1}.

Wendler (1971) estimated the heat balance and compared the climatic elements at two stations within 5 km and 200 m altitude in the Fairbanks region in Alaska, U.S.A. The heat source at the valley station (130 m a.s.l.) is mainly radiation at 61 cal/cm^2 for an average day of the year from August, 1966, to July, 1967, and to a smaller extent, the sensible heat flux, at 17 cal/cm^2. Most of this energy is needed for evaporation (-74 cal/cm^2). The heat flux in the soil is about zero over the year, and -4 cal/cm^2 is used mainly for the melting of snow cover. On the other hand, the heat fluxes are generally smaller at the hill station (332 m a.s.l.). The major source for an average day is the radiation balance (24 cal/cm^2), but the sensible heat flux is slightly negative (-2 cal/cm^2), and the latent heat flux is much smaller (-19 cal/cm^2) than at the valley station. The surface temperature at the hill station is higher, and the air is warmed slightly. This higher surface temperature also raises the outgoing long-wave radiation. From this, the less positive radiation balance at the hill station can be explained partially. The heat flux in the soil is almost zero, and the value of -3 cal/cm^2 represents the energy needed to melt the snow cover. The measured ground temperature is as given in Fig. 4.42 and the estimated heat balance in Fig. 4.43. It was proved that the radiation balance with long-term means cal-

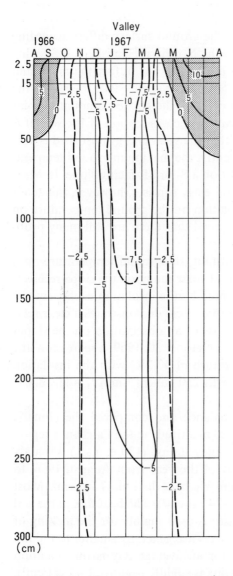

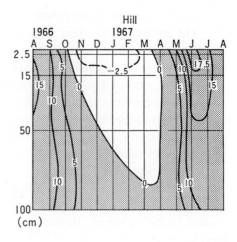

Fig. 4.42 Ground temperatures for the year from August 1966 to July 1967 at the valley station and at the hill station in the Fairbanks area, Alaska, U.S.A. Heat balance at the same stations is shown in Fig. 4.43 (Wendler, 1971).

culated by Gavrilova (1966) was found to be satisfactory: For example, she estimated a long term mean of 43 cal·cm⁻²·day⁻¹ for Fairbanks.

4.2.4. Wind

On the ridge of a mountain, the wind direction at right angles to the running direction of the ridge is predominant, and the wind velocity in that direction is stronger. In a deep valley, the wind blows exclusively along the valley. The area just outside a valley or a comparatively wide and flat bottom in a valley is the most complicated.

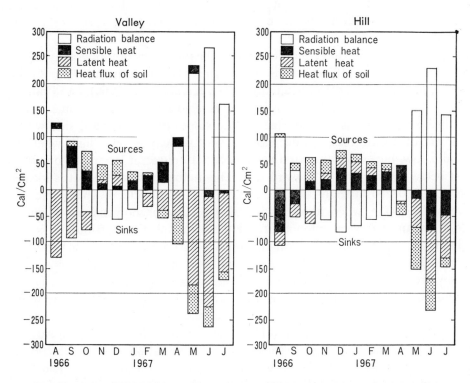

Fig. 4.43 Monthly mean of the daily sums of the heat balance for the year August 1966 to July 1967 at the valley station and hill station. Compare with Fig. 4.42. (Wendler, 1971).

Here the velocity and the direction distributions of the wind under the influences of topography—valley, basin, ridge, slope etc. with relatively small height differences—are dealt with first. And second, the wind systems with diurnal variation, namely the mountain and valley breezes are discussed.

In some parts of the world, there are very strong winds under definite synoptic conditions. These strong local winds and the special winds systems such as föhn, chinook, and bora, together with the air flow over mountains and the lee waves are discussed in Section 5.3.

The wind conditions under consideration have also been described by Aubert de la Rüe (1955), Alissow et al. (1956), Geiger (1961, 1965), and Flohn (1969). Their descriptions, will afford the reader further information.

(A) *Wind Distribution in Mountains and Hilly Regions*

The wind distribution in mountains is an exclusively local phenomenon. Not only does it affect considerably the distribution of rain or snow, it also acts on such meteorological factors as cloud, fog, air temperature, air humidity, and the distribution of frost. Since the topography affecting the state of the wind at a certain point varies according to the direction of the prevailing wind, the wind direction is an important factor. Even if the wind direction is constant, the topography affecting the state of the wind at the point varies with wind velocity.

First of all, the wind conditions under the influences of topography with a relative height below 100 m are discussed. Kaiser (1954), Eimern (1955), Kreutz (1955), Kreutz et al. (1955), Scultetus (1959), Kreutz et al. (1960), and Franken (1962) studied the wind conditions in various regions in Germany. Some aspects of their results are as follows: (a) On the slopes on the leeward side with a gradient of 2–4 degrees, the decrease in wind velocity is very slight when the wind is low, but, when the wind velocity is 5–10 m/sec, it lessens by one-third just below a ridge on the leeward side. The minimum velocity spot is located between 200–350 m lee of a ridge, and the maximum velocity spot comes between 500–600 m, where a down wind blows. These phenomena can be understood by an analogical inference from the wind velocity distribution behind wind breaks or fences. The maximum appears in a convex part on the windward slopes (Kaiser, 1954). (b) The wind conditions on the slopes and in a small valley are schematically illustrated by the results observed at Hausen, Untertaunus, Germany, as shown in Fig. 4.44 (Kreutz et al., 1955). This figure shows where the wind is strong or weak. At the place where the wind is strongest, severe wind erosion is expected. (c) Another result on both sides of a ridge in Langenseifen and Untertaunus shows a striking deviation of wind direction as well as wind velocity, as shown in Fig. 4.45 (Kreutz et al., 1955). The wind direction on the windward slope deviates clockwise to the prevailing wind, which flows at right angles to the running direction of the ridge; whereas the wind direction on the leeward slope deviates counterclockwise. In general, this tendency seems to apply in the cases in the northern hemisphere.

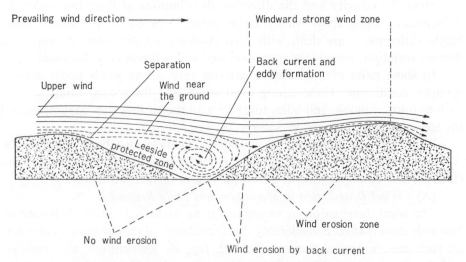

Fig. 4.44 A schematic illustration of wind and eddy in the valley in the case of crossing prevailing wind (Kreutz et al., 1955).

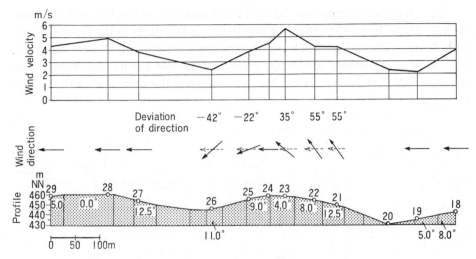

Fig. 4.45 Wind velocity and direction affected by the micro-topography in Langenseifen, W-Germany (Kreutz et al., 1955).

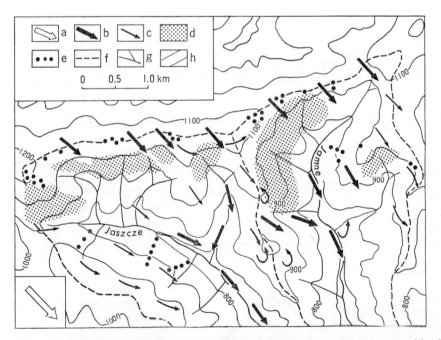

Fig. 4.46 Distribution of wind directions and velocity at a height of 1 m above the ground level in the upper sections of the Jaszcze and Jamne drainage areas, Gorce Mts., Poland, during the NW-advection (on the basis of observations carried out of June 27 and 28, 1962). Explanations of signs: a—general direction of advection above the Gorce Mts., b—wind directions with greater velocity, c—wind directions with smaller velocity, d—places situated on the lee side, e—areas of wind breaks and "flagging tree forms," f—divides, g—creeks, h—contour lines (Obrebska-Starklowa, 1969c).

Eimern (1968) summarizes the method of research on winds in mountainous and hilly areas. In a comprehensive study on local climates in the Gorce Mountains, Poland (Obrebska-Starklowa, 1969c), the wind conditions were observed. As a whole, the summer conditions given in Fig. 4.46 are like the winter conditions in that they both indicate NE-N winds over the ridge and in the valleys

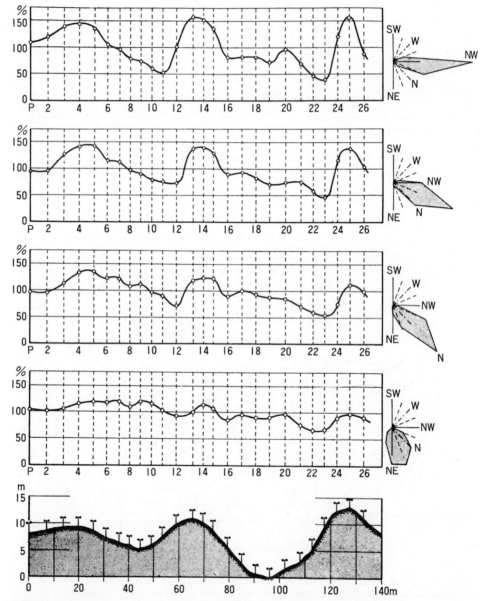

Fig. 4.47 Distribution of wind velocity (%) affected by ridges and gullies in Dischma valley, Switzerland, in relation to prevailing wind direction. 100% means the average of all observed values (Nägeli, 1971).

of Jamne and Jaszcze. On the S-facing slopes in the center, there are opposite directions to the prevailing wind directions. Of interest is the distribution of wind-shaped trees on the ridge, where the wind is strong.

One of the most outstanding works on wind in relation to micro-topography was made by Nägeli (1971) in the subalpine zone of Stillbergalp in the Dischma valley, Switzerland. Observing the wind velocity at 127 points located at every corner of meshes 20×20 m, he presented color maps (scale 1:750) of wind velocity distributions for the SE-wind and NW-wind. Probably these maps are the finest ones ever-published concerning wind conditions in relation to micro-topography. An example of the results of his measurements is given in Fig. 4.47, which shows the local variation in the wind velocity according to the prevailing wind direction. When the prevailing wind blows at right angles to the ridge direction, the difference ranges from 40% to 160%, but, when it blows in parallel, the difference ranges from 70% to 120%, where 100% is the average for the area. The wind maps indicate that the plant communities in the area have a close relationship with the wind velocity. For example, on the ridge where the wind is strong, 127% of the average wind velocity, *Cetrario-Loiseleurietum* develops, but, on the other hand, in the gullies where the wind is weak, 84% of the average wind velocity, *Calamagrostietum villosae*, develops.

Similar observations made in the Ötz valley, Tyrol, Austria (Aulitzky, 1955), were mentioned above. In a study by Reiter (1966) on the air pollution in connection with airstreams in the valley around Garmisch-Partenkirchen, S-Germany, the airstreams in the main valley as well as side valleys are investigated in detail. According to this study, marked differences in the wind conditions in the side valleys were observed between the mountain breezes and SW-föhn or SW–W gradient winds, even though they are similar in the main valley.

Kaps (1955) expresses the degree of ventilation in a valley, D, by the following equation:

$$D = \frac{d}{d+b} \cdot \frac{d}{t} \qquad (4.36)$$

where d is the width of a valley at its top, b the width at its bottom, and t the depth. In a small V-shaped valley in the upper courses, the wind velocity diminishes in proportion to the decrease in the width of the valley floor. According to the result of an investigation at the Azusa River in Nagano Prefecture, Japan, the wind velocity at a point with an altitude h up to 10 m above the floor is roughly proportional to the width of the valley at an altitude of $10h$.

The relation is evident between the topography in a limited area and the wind direction. The occurrence frequency of the wind direction at a station in the mountainous areas in the Kanto district, Japan, is expressed by the following equation:

$$F = -208.25 K_2 + 19.81 \qquad (4.37)$$

where F is the maximum frequency of the wind directions for the total of a year in %, and K_2 is the deviation of the angle of inclination of the same or opposite directions on an area of 2 km radius (Yoshino, 1952). In this statistical study, K_2 was obtained in the following way: First, using a topographic map (1:50,000), lines along 16 directions from the observation point on the map were drawn. Secondly, the altitude of the highest point along the 2 km line, or the lowest if the observation point was on the summit, was read and the values of the angle of inclination [{(the altitude of the highest point)—(the altitude of the observed point)}/2 km] along every line were calculated. Thirdly, the deviations of the values of the angle of inclination of the 8 directions were calculated.

Fuh (1963) studied the wind velocity in valleys, based on the results observed in the upper courses of the Yangtze and Unhang Rivers. His conclusions were: (a) The relative speed of the wind in the valley is dependent on the angle β between the wind direction and the valley as well as the ratio of the width L of the valley to the height H of the mountain on both sides of the valley. The wind velocity in the valley is greater than that in open country when the angle β is smaller than 15°–35°, but less when β is larger than 15°–35°. In the former situation, the smaller the angle β and the ratio L/H, the stronger the wind speed in the valley becomes, but in the latter situation, the larger the β and the smaller the L/H, the weaker the wind becomes. (b) The variation of the wind speed with local elevation can be expressed by

$$\frac{u-u_1}{u_2-u_1}=\frac{\log (10+z)-\log (10+z_1)}{\log (10+z_2)-\log (10+z_1)} \tag{4.38}$$

where u_1 and u_2 are the known wind velocity (m/sec) at heights of z_1 and z_2 (m) and u is the wind velocity (m/sec) at a height of z (m). (c) The increase of wind velocity with height above the ground can be expressed by the equation

$$\frac{u-u_1}{u_2-u_1}=\frac{\ln z-\ln z_1}{\ln z_2-\ln z_1} \tag{4.39}$$

(B) *Influence of Topography on Wind Structure near the Ground*

Here some problems in meso-scale are dealt with first. Dammann (1960) reports the wind conditions in the Rhein-Main region in relation to the temperature profile in the lower atmosphere, air pressure fields, and wind field determined by the synoptic pattern and the topography under consideration.

According to the results of studies on air pollution in the neighborhood of a nuclear power factory, air trajectory in the boundary layer is characterized primarily not by the topography, ridges, and valleys of hills, but by the airstreams in the lower atmosphere when there is turbulence in the daytime (Gifford, 1953). Wind rose shows different patterns in the cases when the Richardson number Ri $\geq$ 1 and Ri $\leq$ 1 (Holland, 1952). As shown in Fig. 4.48, the wind rose shows the influence of local topography or exposure in the case of a stable air layer, Ri $\geq$ 1, but it shows stronger effects of gradient wind direction in the case of an unstable air layer, Ri $\leq$ 1. The same tendency is found in the wind rose at the stations in Saxony (Flemming, 1966b). Frenkiel (1962)

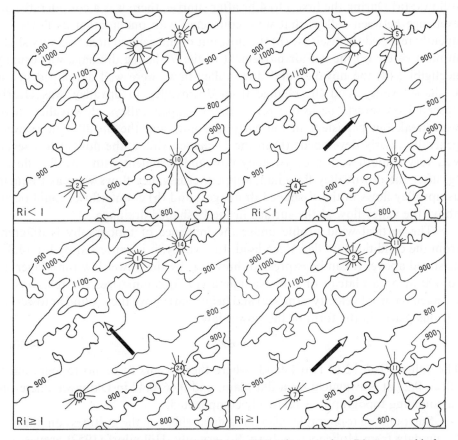

Fig. 4.48 Patterns of wind rose in relation to Richardson number, Ri, topographical situation, and gradient wind direction. Thick arrow shows the gradient wind direction at the 1,670 m level (Holland, 1952).

studied the wind profiles over hills and came to the following conclusions: (a) The wind profiles are largely determined by the hill profile in the immediate neighborhood of the measurement site. (b) The effect of thermal stratification is important for all wind speeds.

A study on diffusion over coastal mountains of S-California (Hinds, 1970) came to a similar conclusion: The daytime (unstable) conditions minimized the importance of terrain, whereas the nighttime (relatively stable) conditions led to apparently significant interactions between terrain and synoptic scale weather events.

The aerial dispersion of aerosolized insecticides for forest insect control in a mountainous region is also a problem related with diffusion under the influence of topography. This problem was studied in the Sawtooth National Forest just north of Fairfield, Idaho, U.S.A. (DeMarrais et al., 1968). The thermal stratifications were more striking, but the wind conditions were as follows: (a) During the first 3 to 4 hr after sunrise, there was a considerable air exchange between valleys that were connected with one another as flows frequently reversed themselves; (b) the flow out of the general area was very slight in most cases; (c) the net air movement in the V-shaped drainage was usually negligible; (d) the upvalley and downvalley flows were well developed in the U-shaped valley, and the change from downvalley to upvalley was clearly observed on several occasion; (e) in the V-shaped valleys, downvalley flows were weak but persistent in the early morning, and the upvalley flows were generally poorly developed; (f) in the U-shaped valley, the flow at low levels was very smooth with a downvalley flow and turbulent with upvalley flows; (g) after the downvalley flow had stopped, crossvalley flows were as frequent as upvalley flows in the V-shaped drainages; and (h) counterflows could occur above the valley flow at the surface in the U-shaped drainage.

The vertical wind profile under the influence of topography is different from the conditions on flat, open land. In a valley, it is expressed by Eq. (4.39) as stated above. It can be expressed, however, by the power law for a layer up to 400 m on a plateau or hill. Nakajima (1973) summarizes the relationship between the wind velocity u (m/sec) and height (m) above the ground at towers in various places in the form of the power law:

$$u \propto Z^k \qquad\qquad (4.40)$$

The value k is 0.12 at Iwo Island, where there seems to be no topographical effect, 0.23 at Kawaguchi on a flat plain, and 0.46–0.47 in Okinawa where the surrounding topography is relatively complicated.

Surrounding topography also has a great effect on the gustiness on hilltops. Using data from Hohenpeissenberg, S-Germany, Höhndorf (1952) argues that the fluctuation of wind direction has a close relation to the topography in different directions from the observation point.

Flemming (1966a, 1967b) reports observations of wind velocity, direction, and lateral gustiness in the Weisseritz valley at Tharandt and on the surrounding plateau, Germany. A comparison of the lateral gustiness in the valley bottom and on the plateau shows that the lateral gustiness is generally reduced by about 1.8° in the valley, particularly in the case of the wind blowing parallel to the valley, but it increases by about 0.7° when the wind blows at right angles to the valley.

Yoshino (1957a, 1959) made a micrometeorological observation on the wind velocity, direction, intensity of turbulence, and fluctuation angle of wind direction (lateral gustiness) in a small valley in Nishiura, Numazu City, Japan. As shown in Fig. 4.49, observations were made simultaneously once a minute; the sampling time was 10 min for each set. The results obtained are summarized here: At the ridge stations, the value of intensity of turbulence, $g = \sqrt{\overline{u'^2}}$, higher than 1.00 in some cases and shows great deviations. Mean wind velocity does not show any striking tendency arising from the difference of prevailing

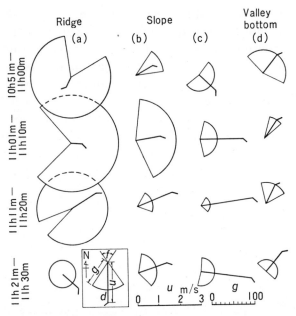

Fig. 4.49 An example of the results observed on March 21, 1952, at the stations on the ridge (a), upper part (b), and lower part (c) of the valley slope, and in the valley bottom (d) at the middle coures of a river in Nishiura, Numazu City, Japan. u: mean wind velocity, d: mean wind direction, g: intensity of turbulence, and f: fluctuation angle of wind direction (Yoshino, 1959).

wind direction or topographical situation from the ridge to the adjacent valley bottom. Its absolute value, however, differs with the prevailing wind direction. The point exposed topographically has a negative relationship between the angle of fluctuation of the wind direction and the mean wind velocity.

Tyson et al. (1970) investigated the spectra of velocity fluctuations in mountain winds observed in two Transvaal valleys of very different geometry. The measurements were made in the shallow Jukskei River 24 km NE of Johannesberg and in the deep Blyde valley 16 km N of Pilgrims' Rest, South Africa. It has been confirmed that surges with periods of the order of 1 hr contribute significantly to total turbulent energy. At the same time it is obvious that a second peak with periods varying between 8 and 20 min may occur in spectra for mountain winds in clear weather. Total turbulent energy of both mountain and valley winds depends on the configuration of the parent valley, and the form of the spectrum is sensitive to the occurrence and amount of cloud cover. Horizontal velocity spectra of the mountain wind show that the maximum turbulent energy is generated by waves of the order of 10 km in length and a period of 1 hr in valleys around Pietermaritzburg (Tyson, 1968).

To ascertain the influence of micro-topography on the wind structure near the ground, I observed the variation of wind velocity up to a height of 5 m above the ground at various stations on flat open land and on the side slopes or in the bottom of a small valley in Sugadaira, Japan (Yoshino, 1958). Differences of the profiles from the so-called logarithmic law are evident. The profiles can be roughly grouped into three types according to their micro-topographic situation. When the prevailing wind flows at right angles to the running direction of a small valley with steep side slopes, the maximum wind velocity at 2–3 m above the ground appeared on the lower part of the valley slope or in the valley bottom, owing to the counterflow of the lower part of the eddies formed in the valley. The counterflow had some twisted form: in other words, the eddies are not formed on an exact x–z plane. The ratio of its speed to that of the prevailing wind showed a relation to the prevailing wind speed and the time of day. The results are given in Fig. 4.50. This figure shows the eddy formed in a small valley clearer than Fig. 4.44.

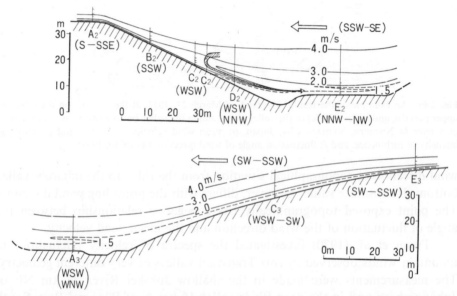

Fig. 4.50 An average state of the distribution of wind velocity in a small valley in Sugadaira, Japan. Mean wind direction (1 m above the ground) at each station is shown in brackets. The arrow indicates the prevailing wind direction. The valley runs from E-NE to W-SW (Yoshino, 1958).

Of special interest is an analogical consideration of the topography, vegetation, and building effects (Flemming, 1967a). Flemming named the layer modified by these effects the *morphogenous layer*. It can be said that the concept of the morphogenous layer permits quite a large number of phenomena to be satisfactorily interpreted, without neglecting physical relations. But it is hoped that a more complete treatise will be done to confirm this consideration.

Observations of the vertical variation of wind speed on the top of the plateau of the ice cap in Devon Island in the Arctic (Holmgren, 1971) show an increase of the roughness parameter z_0 from $5 \cdot 10^{-5}$ to $1 \cdot 10^{-3}$ m in near-neutral stratifications when the wind speed at the 5 m level decreases from 6–7 m/sec.

The increase of z_0 suggests that it is related to the flow conditions at the rear of the snow dunes. With increasing stability the deviations from a log-linear relationship become increasingly apparent.

Doumani (1967) classifies the surface morphology of snow accumulation into two types, erosional and depositional. The former is subclassified into sastrugi, irregular pattern, pit, footprint, etc., and the latter into drift form, current ripple mark, barchanoid, etc.

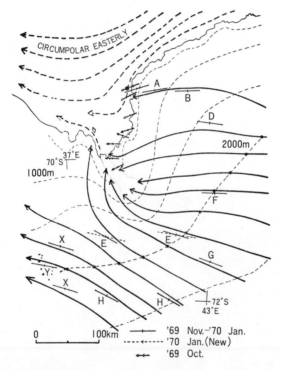

Fig. 4.51 Orientation of sastrugis and pitted patterns, and streamlines of prevailing winds near Showa Station in Antarctica (Ageta, 1971).

Ageta (1972) observed the net ablation of sastrugi and drifts in an area of 100–200 km inland from the Antarctic coast near Showa Station, where strong katabatic winds blow. For instance, the net ablation of hard sastrugi during the period from November, 1969, to January, 1970, was 2.8 cm/month. Using these sastrugi and pitted patterns shown in Photo 4.9, Ageta (1971) estimated the streamline of the prevailing winds from the orientation of sastrugis and pitted patterns as given in Fig. 4.51. Though the area mentioned is broad as compared with those in middle latitudes, this map shows local topographical effects on the katabatic wind directions.

Photo 4.9 (a) Drift in Antarctica (November 9, 1969, by Y. Ageta).
(b) Sastrugi in Antarctica (September 20, 1969, by Y. Ageta).

Another interesting phenomenon in Antarctica is the *McMurdo Oasis*, the most broad region without ice and snow. The McMurdo Oasis is located roughly in the area 77°10'S–77°45'S and 160°20'E–163°00'E and is 2,500 km². In this area, there are several dry valleys. Yoshida et al. (1972) made a statistical investigation of the climate of the Wright Valley (Fig. 4.52), one of these dry valleys. The fact that the Wright Valley with a length of 50 km from 160°45'E to 162°45'E has no glacier, ice, or snow in the bottom can be attributed to the low cloudiness in the valley, which may be caused by dry, warmer westerly

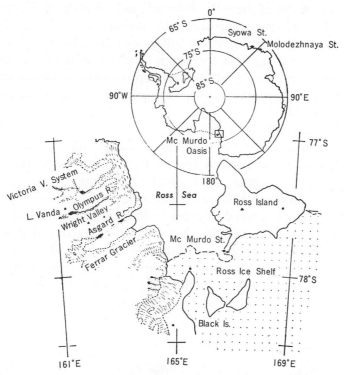

Fig. 4.52 The Wright Valley and the adjacent area (Yoshida et al., 1972).

winds descending from the high continental ice plateau, 2,000–4,000 m a.s.l., under the influence of disturbance on a synoptic scale. The mechanism must be studied in more detail from the standpoint of local climatology as well as low layer aerology. Loewe (1974) described the dry valleys of southern Victoria Land in Antarctica, as well as the interior of Peary Land in NE Greenland as *Polar dry deserts*. He writes that they are more likely to occur where land topography favors descending air currents of high speed and low humidity or blocks the free access of neighboring ice masses.

In Japan, beginning with the study by Katsuei Misawa, detailed studies on the distribution of wind have been carried out using plant indicators. For example, an investigation was made of the winds over the Akaho alluvial fan where the bending directions of top twigs of persimmon trees served as an indicator (Sekiguti, 1951a). The measured bending direction, which has a confidence interval of about +15°, corresponds to the direction of the wind that prevails during the growing season of persimmon trees from April to June. Sekiguti measured the direction at 230 places, and obtained the following results: (a) In early summer, the southerly wind prevails predominantly all over the area of the broad fan surface. (b) Strictly speaking, however, its direction is affected by the features of the Tenryû valley in this region. Its direction diverges where the width of the valley is broad, and it converges where the valley is narrow. (c) When it encounters some obstruction, the wind generally changes direction toward the right. (d) In the valleys that cut into the fan surface, the direction of the prevailing wind is different from that of the fan surface above. That is, in shallow and small valleys, the wind blows up along the stream, and in deep and large valleys, the wind blows opposite to that of the fan surface, as a result of an eddy.

(C) *Mountain and Valley Breezes*

Since the end of the last century, many theories on mountain and valley breezes have been proposed, and these were culminating in the studies by Wagner (1938). He designates the level of equalization as the *effective ridge altitude*, which is an altitude above the valley, usually at the average height of the surrounding ridges, where the otherwise different diurnal pressure variations are completely equalized. Wagner also made clear further that the diurnal temperature variation in valleys up to the effective ridge altitude is two times larger than the variation in a similar layer over a plain. Thus, there must be a pressure gradient from the plain to the valley during the daytime, causing the valley breeze, and a reverse gradient at night, causing a mountain breeze.

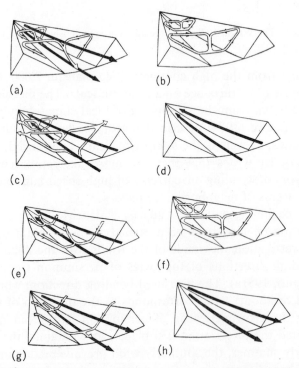

Fig. 4.53 Schematic illustration of the normal diurnal variations of the air currents in a valley. The explanation is given in the text (Defant, 1949).

Defant (1949) stresses on the thermal slope winds on the side slopes of a valley and presents a famous schematic illustration, Fig. 4.53, on the normal diurnal variations of the circulation systems in the valley. He also reviews the theories in connection with the problems of the other local wind circulations (Defant, 1951). The explanation of the diurnal change of wind circulation given in Fig. 4.53 is as follows: (a) about the time of sunrise, (b) 9 in the morning, (c) about noon, (d) afternoon, (e) in the evening, (f) at the beginning of night, (g) midnight, and (h) dawn. The black arrow shows a mountain or valley breeze, and a white arrow up- or down-slope wind and its associated circulation.

The actual state of the mountain and valley breezes varies according to factors such as topographical conditions, seasonal change of the height of the sun, duration of daytime and nighttime hours, and vegetation and surface conditions. These factors are discussed separately below.

The topographical conditions for the valley breeze may include the running direction of the valley, width of the valley floor, length of the valley, height of the surrounding ridge, flatness of the valley, and distance from the plain. The topographical conditions for the mountain breeze are, in addition to the conditions stated above, the area of the drainage region where the cold air concentrates, and probably the inclination of the valley bottom along the river.

When these factors are favorable for the development of the mountain and valley breezes at a point, the mountain and valley breezes occur very frequently and the wind velocity becomes very high. For instance, the SW–SE mountain breeze develops with a frequency of 70–80% at night almost throughout the year and the NW–NE valley breeze occurs 60–80% of the daytime (about 10h–16h) from March to October at Bucheben, Rauris Valley, north of Sonnblick, Austria (Steinhauser, 1968). Another good example of valley breeze is seen in the records of the Rhône valley in Switzerland (Billwiller, 1914; Bouët, 1961; Yoshino, 1964). The change of the wind velocity on clear days at Sirre, where the valley bottom has a width of 1.5–3.0 km and the valley runs straight for almost 50 km, shows a striking maximum in the afternoon of over 4 m/sec, as a monthly mean, from March to September, and even stronger winds of over 5 m/sec prevail in summer months. The peak gust observed in upvalley winds was 10–11 m/sec from March to September.

By observing wind-shaped trees (Photo 4.10) as indicators of wind conditions, I investigated the extent and the local characteristics of the valley breeze in the Rhône valley, Switzerland (Yoshino, 1964). A valley breeze is observed from the shoreline of Lake Léman (380 m a.s.l.) to the upper course of the river near Mörel and Deisch (about 1,000 m a.s.l.). The strongest winds are to be found especially north of Martigny (480 m a.s.l.). Their horizontal distribution along the river course shows a close relation to the topographical conditions of the valley bottom: the width and topographical irregularity of the valley bottom, such as fans and moraines. In this investigation a region of calm was found between Deisch and Reckingen (1,330 m a.s.l.) and a northerly wind region in

Photo 4.10 Wind-shaped trees in the Rhône valley in Switzerland. Left: larch (*Larix decidua*) east of Eisfluh (1,510 m). Right: poplar at Martigny (June 9, 1963, by M. M. Yoshino).

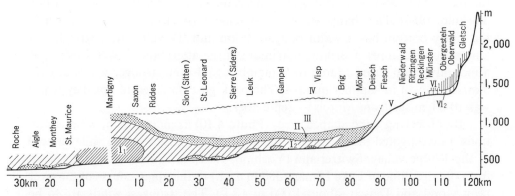

Fig. 4.54 The wind conditions in the valley bottom and on the slopes along the longitudinal profile of the Rhône valley, Switzerland. Region I_1: the component of the upvalley winds is strongest. Region I_2: the component of the upvalley winds is strong. Region II: the component of the upvalley winds is obvious, but weaker than I_2. Region III: the components of the upvalley winds and upslope winds are comparable. Both winds are relatively weak. Region IV: the component of the upslope winds is more obvious. Region V: the winds are weakest. Region VI_1: the northerly winds are strong. And Region VI_2: the component of the northerly winds is traceable (Yoshino, 1964).

the uppermost part of the valley from Reckingen under the influence of upper winds. The vertical distribution on the valley slopes along the river is shown on the illustration map in Fig. 4.54, dividing the area into the following six regions by the wind conditions: Region I_1, where the upvalley wind is strongest; Region I_2, where the upvalley wind is strong; Region II, where the upvalley wind is obvious; Region III, where the upvalley wind and the upslope wind are comparable, both being weak; Region IV, where the upslope wind is more obvious; Region V, where the winds are weakest; Region VI_1, where the N-wind influenced by the upper winds is strong; and Region VI_2, where the N-wind is traceable. From Fig. 4.54, several interesting phenomena can be seen: (a) The upper limit of the valley breeze region, the boundary between Regions III and IV, is situated at a height of 1,100–1,300 m, which is much lower than that in the free atmosphere over the valley bottom; (b) The height of the upper boundary of Region II might be correlated closely to the height of the maximum wind velocity, which causes the driest part on the slopes along the middle course of the Rhône river at an elevation 80–120 m above the valley bottom, as has also been revealed by botanical investigation; (c) The heights of the lower limit of Regions VI_1 and VI_2 and the upper limit of Region IV are quite similar to the state in the Azau valley, which will be described later.

The phenomenon (b) mentioned above has been reported in other valleys of the world. The vegetation studies have pointed out that the dry condition in the valley bottom, where the valley wind develops, prevails in the valleys of the La Paz River, Bolivia (Troll, 1952) and of the Himalayas (Schweinfurth, 1956). This valley wind effect is called *Troll-effect*. Schweinfurth (1972) gives a detailed description of this effect in the region of E-Himalaya.

The models shown in Fig. 4.53 was derived from the assumption that there is no prevailing wind in the upper layer, not being taken into account the direction or gradient of a slope. Even if the valley slope is similar geometrically, the absolute size plays a decisive role in developing the valley breeze. The following example is the case of a large valley: in the Azau valley in the U.S.S.R. the mountain and valley breezes were closely observed from August 2 to September 10, 1954 (Vorontsov and Schelkovnikov, 1956). The Azau valley in the Caucasus Mountains is surrounded by the mountain ridge with a height of 3,500 m and the width of valley bottom 1.5 km. The results observed up to 1,000 m above the valley bottom are shown in Fig. 4.55. According to this figure, the up-slope wind velocity on the slope at 500 m above the valley bottom is several times stronger than the wind velocity at the same altitude in the valley. Upper winds flowing over the ridges also prevailed in the uppermost part of the valley in the daytime. At a height of 3,500 m above the floor, for example, the valley wind has the velocity of 0.5 m/sec in the valley, whereas the up-slope wind is 2.7 m/sec on the slope at 500 m above the floor. When a mountain wind is blowing the wind velocity is 1.0 m/sec in the valley and 4.0 m/sec on the slope, although both are at the same height. Accordingly, a boundary line between the

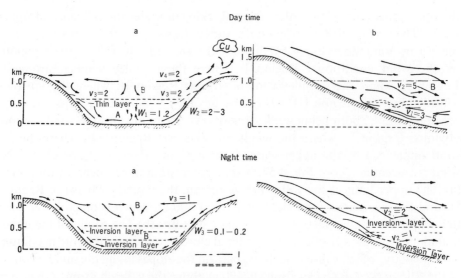

Fig. 4.55   Circulation system in the Azau valley, August 24–25, 1954.
a: Cross section, b: Longitudinal section, A: Air current in the valley, B: Upper winds
from the other side of the ridge, 1: Horizon of the side slope of the valley, 2: Boundary
layer, v: Horizontal velocity (m/sec), w: Vertical velocity (m/sec) (Vorontsov et al., 1956).

prevailing wind in the upper layer and the air current within the valley lies low
in the central part of a cross section of the valley and high near the valley slopes.
During the day, especially around noon, it is as low as 200–600 m, and at night
it rises to 500–900 m. Furthermore, under conditions favorable for the development a circulation system in the valley, the above-stated daytime conditions
continue to exist from 5 h, that is, immediately after daybreak, until 23 h, that
is, 4 or 5 h after sunset. In Fig. 4.55, A stands for the circulation in the valley,
and v_1 its velocity; B stands for the upper winds flowing from the other side of
the mountain, and v_2 is its velocity, and v_3 is wind velocity from the valley
slopes (wind component at right angle to the running direction of the valley).
The borderline between A and B is too complicated to distinguish, unlike the
distinct model in Fig. 4.53.

Thyer et al. (1962) and Buettner et al. (1966) observed the valley winds in
the Mt. Rainier area and confirmed that the maximum velocity above the valley
floor occurs at an altitude of 1/4–1/3 of the relative height between the ridge
and the valley floor. The maximum occurs later in a larger valley.

In a very shallow valley, the valley winds are subdued by gradient winds,
whereas the mountain winds may be quite well developed. Summarizing the
results mentioned above, it can be pointed out that the following points are
different from the model described in Fig. 4.53: (a) There were no periods either
in the morning or evening when only the slope winds or the valley winds are
flowing; (b) there is no vertical component in the midvalley either in the daytime or nighttime, and (c) the slope winds in the daytime ending in strong up

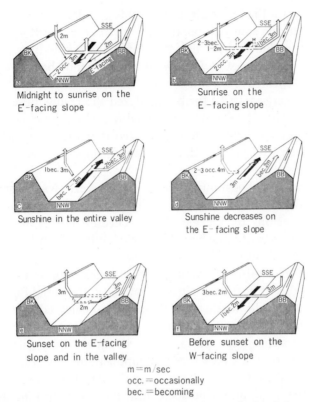

Midnight to sunrise on the
E'-facing slope

Sunrise on the
E – facing slope

Sunshine in the entire valley

Sunshine decreases on
the E– facing slope

Sunset on the E–facing
slope and in the valley

Before sunset on the
W–facing slope

m = m/sec
occ. = occasionally
bec. = becoming

Fig. 4.56 Slope winds and mountain and velley breezes in the air layer near the
ground in the V-shaped Dischma valley in Davos, running in a SSENNW
direction (Urfer-Henneberger, 1964).

winds at the ridges feed the antiwinds. The air thus lost is replaced by the valley
wind. There is practically no delay between the systems.

   In the Dischma valley near Davos in Switzerland, a V-shaped valley oriented in a SSE–NNW direction, Urfer-Henneberger (1964) studied the mountain
and valley breezes and slope winds by observing wind and air temperatures at
5 stations. The ridge stations on both sides are located about 2,540–2,560 m
a.s.l. and the valley bottom at 1,700 m a.s.l. The distance from each station to
the ridges is approximately 35 km. The model presented is somewhat different
from the one given in Fig. 4.53. As shown in Fig. 4.56, the daily course of
circulation is subdivided into the following seven phases. (a) *Midnight until
sunrise on the E-facing slope*: downvalley winds in the valley bottom, down-
slope winds above the upper limit of the temperature inversion. (b) *Sunrise on
the E-facing slope*: no change on the valley bottom and on the W-facing slope.
(c) *The entire valley in sunshine*: winds turn upvalley in the bottom and up
winds begin also on the W-facing slope. (d) *Decreasing radiation on the E-facing
slope*: on the E-facing slope winds turn parallel to the valley, first at the upper,
then at progressively lower parts of the slope. (e) *Sunset on the E-facing slope*

and in the bottom: in the lower air layers winds circulate across the valley from the shady slope to the sunny slope. (f) *Before sunset on the W-facing slope*: the cross circulation includes about half the slopes and winds turn downvalley in the bottom. (g) *After sunset until midnight*: down winds begin on both slopes, and the cycle is repeated from (a).

Urfer-Henneberger (1970) further analyzed the results observed at the 13 observation points and presented a modified model of the circulation in eight stages in the Dischma valley. She emphasized that the valley wind system is strongly influenced by factors such as radiation, slope exposure, and direction of the valley.

A theoretical model of mountain and valley breezes was presented in three dimensions by Thyer (1966), who was trying to develop the previous models in one or two dimensions by Fleagle (1950), Gleeson (1953), and Rao (1960). Thyer's model consists of a V-shaped valley of finite length leading into a flat plain, and a given rate of heat flow from ground to air is assumed. Of the seven equations considered, three are components of the equation of motion and the others are equations of heat conduction, continuity, state, and Poisson. Coriolis terms were omitted. His mathematical model has succeeded in producing a system of simultaneous slope winds, valley wind, and anti-valley wind by simulating heating at appropriate parts of the boundary of a three-dimensional region. The results on the valley breeze component are given in Fig. 4.57. The model shows several characteristics such as a thin layer of slope winds, updrafts over the ridge, the maximum wind speed near the valley bottom, an anti-valley wind layer of about the same thickness, etc. The similarities between the model and reality are obviously convincing, and this indicates the validity of the model assumptions and the usefulness of such numerical experiments (Flohn, 1969).

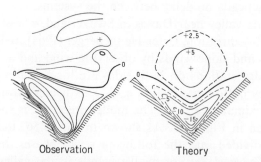

Fig. 4.57 Comparison of observation with theory of valley breeze on a cross section of valley with isopleth of longitudinal wind component (m/sec). Observation was made in the Carbon River valley, the Mt. Rainier area, at 13h, July 9, 1959 (Thyer, 1966).

Knudsen (1961) dealt with the mountain and valley winds in S-Norway from a synoptic point of view, and Sterten (1961) reported the results of extensive local meteorological observations on mountain and valley winds during field programs in S-Norway. The results indicate that there is a close relationship between the mountain and valley wind system and the interdiurnal variation in atmospheric pressure. It seems likely that the same conditions that result in a mountain wind also cause a rise in pressure in the valley atmosphere, which in turn is of primary importance for the initiation of the valley wind. For the valley wind to develop further, a temperature rise in the bottom layer of the valley atmosphere is of great significance. On the whole the observations in this area are similar in space and time to the former investigations (Sterten, 1965). These conclusions were derived from observations in areas with comparatively small relative height of 300–400 m. It is thought, therefore, that effects similar to local cyclones and anticyclones, which are formed thermally during the daytime and nights respectively, must play a significant role in producing the mountain and valley wind systems.

MacHattie (1968) analyzed the record at three stations in the Kananaskis valley, Canada. He pointed out that the effects of subvalleys on surface winds are less susceptible to being overridden by synoptic scale influences than the flow along the main valley and that daytime upvalley winds suffer more gradiend wind interference than nocturnal downvalley winds.

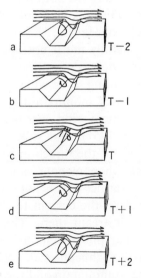

Fig. 4.58 Interaction between the eddies formed by the crossing upper winds and the upvalley winds in a small valley (Yoshino, 1957b).

In a small valley, there is an obvious interaction between the prevailing upper winds and the mountain and valley breeze systems in the valley. The small V-shaped valley in Sugadaira, Japan, is used as an example here (Yoshino, 1957b). Precisely speaking, the interaction occurs between the upvalley winds and the eddies formed by the prevailing upper winds, which arise at right angles to the running direction of the valley. The eddies in the valley are not constant, but are deformed at certain intervals, and the special wind conditions accompanying upvalley winds appear in the valley when the eddies are deformed. The periodic deformation processes of the eddies are illustrated in Fig. 4.58. From this figure the following conclusions can be drawn: (a) Surface winds in the lowest layer descend at almost the central part of the valley, forming eddies at the shadow side slope of the valley, and these eddies have some twisted shapes; in Fig. 4.58 is seen a backward current deflecting (hitherward in the illustration) on the valley slope. On the other hand, at the exposed side of the valley slope, the winds flow up in a meandering way. (b) The eddies at the shadow side of the valley weaken the above-mentioned tendency, and the meandering upflows also weaken their fluctuation. (c) Upvalley winds develop especially at the shadow side of the valley, although the prevailing cross winds are still flowing over. Upflows at the exposed side of the valley slope meander only a little. (d) The characteristics of the eddies and upflows become more prominent. This is almost the same as the conditions in (b) described above. (e) Fully developed eddies and upflows show the same states as those described in (a). From this state to that in (a), it takes about 8 min.

Under relatively simple conditions, both topographically and meteorologically, the local wind systems develop markedly. For instance, on the broad slopes of a volcano, Mauna Loa, Hawaii, U.S.A., which faces the trade wind, the warm, turbulent, and generally moist upslope wind extends up to 600 m, and the nocturnal, cool, drier downslope wind extends to only about 55 m under the influence of the prevailing trade wind (Mendonca, 1969).

There are many variations of mountain and valley winds. Among them, the most conspicuous one is Maloja wind over the area between Engadine and Bergell in Switzerland (Defant, 1951). This valley breeze blows during days of clear undisturbed weather, not in the direction toward the upper part of the valley, but downstream. This phenomenon, called the *Maloja wind paradox*, is thought to occur when the warmer part is not on the vally side slopes, but in another deeper valley. Therefore, the current to the deeper valley takes place over the pass and blows downstream there. The problem has not yet been satisfactorily solved, but similar phenomena are reported from several other stations, for instance, at Davos (Schüepp et al., 1963).

Various names are given to mountain and the valley breezes; for instance, *la Breva* for the valley breeze and *Tivano* for the mountain breeze near Lake Como, *Ora* for the valley breeze near Lake Garda, which occasionally blows very strongly. The nocturnal wind in the vicinity of Lake Garda is called *Sover, Sopero, Torbole,* or *Paesano*. This wind blows irregularly and is weaker than *Ora*. Around certain lakes in Austria, the daytime wind is termed *Unterwind*, 7nd the nocturnal one *Oberwind*. In the Himalayan valleys leading to Tibet on the Indian side, strong winds blow in quite opposite directions in the day and at night. A native tribe calls the valley with such strong winds in the Tista valley *the den of wind*.

In Japan, a mountain wind that blows down along the River Shô in Toyama Prefecture from midnight till early in the morning is called *arashi*, and the usual air current from the south sweeping over the mountain is called *dashi*.

Where a valley leads to a plain from a mountain range, a strong local wind sometimes blows. This phenomenon occurs in restricted areas. Strong local winds will be discussed in Section 5.3.

4.2.5. Precipitation

(A) *Distribution of Rainfall in Mountainous Regions*

The effect of a mountain on the precipitation distribution is obvious, but very complicated. For the effects of high mountains, that is, expressing the elements of mountains by altitude only, the relations are relatively simple, as discussed in Section 4.1.7. However, for low mountains, for example those whose relative height is smaller than several hundred meters, and of hills or valleyside slopes, the relations are too complicated to provide any generalizations, because many other elements of mountains play crucial roles.

In a study at Bath, England, 27 temporary observation points were set up within an area of 50 km², and 250 rainfalls were observed for 15 months (Balchin and Pye, 1948). The amount of precipitation increases at the rate of 50–60 mm/100 m up to a height of 130 m on slopes facing prevailing winds. As far as each precipitation is concerned, greater the absolute amount of rainfall, greater the increase rate. The distribution of rainfall on a small area in a mountain depends greatly on the cause of the rain, not to mention the direction of the prevailing wind at that time.

Based on the results of observations in restricted drainage areas in Japan, the distribution of rainfall can be classified into three types (Sugaya, 1949a, 1950): (a) Type A topographical distribution, (b) Type B even distribution and (c) Type C scattered distribution. As shown in Fig. 4.59, Type A shows a linear relation with height, which takes place when a typhoon or a cyclone is approaching, and hence the air currents come almost parallel to the major direction of a slope. Type B is of such rainfall as bears no distinct relation to height under the influence of a warm front or a cold front. Type C, quite independent of height, has a wide range of variation as compared with Type B. Type C is often found in case of rain due to a stationary front. In both Types B and C the surface of discontinuity with a slight gradient and slow moving velocity reaches a relatively higher level than the mountain top, so that the distribution becomes rather uniform or dispersed over a wide area independently of topography. The rate of occurrence of these three types is given in Table 4.20. This is the result of observations at 30 points within an area of 520 km². The predominance of Type A over Types B and C in the basin of the Iwai River is accounted for by the fact that the topography of this area is simple, and that the major direction

Table 4.20 Occurrence frequency of rainfall of Types
A, B, and C shown in Fig. 4.59, during
the period from August 13–November 3,
1949 (Sugaya, 1949a, 1950).

Basin	Type			
	A	B	C	AC
Isawa River	4	2	6	2
Iwai River	9	1	1	0

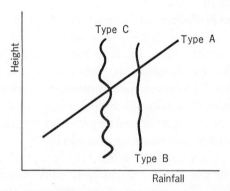

Fig. 4.59 Three types of relations between the rainfall and height in the
small area in the mountainous region (Sugaya, 1949a, 1950).

of the slopes is in accord with the prevailing wind direction.

Kauf (1950) has done observation of precipitation distribution in the Jena
region, Germany. He reports that the difference in distribution patterns of
W-, NW-, NE-, and SW-airstreams and thundershowers make clear the difference in the windward and leeward slopes even in a small area. He also
presents the precipitation distribution caused by passing cold and warm fronts.
In most cases, a greater amount is received at higher places, but in the case of
the NE-stream, the amount increases in the valley.

In a study on the rainfall of Hong Kong, Peterson (1964) pointed out that
much of the rain falls in the form of showers or thunderstorms due to local
convection in summer, and the most rainfall occurs near the tops of the mountains. The rainfall is greater in deep valleys or at mountain feet on the leeward-
side than on the ridges at the same level in the neighboring region. Though there
is a slight difference in degree from one case to another, a rough outline is as
follows: Let h be the difference in altitude between a point and the mountain
ridge on the windward side and d the horizontal distance between the point
and the ridge. The rainfall is slightly affected by a mountain on the windward
side if h/d is over 0.05, but much affected if it is over 0.1. The strongest influence
is expected in the case of 0.3 and over.

This is because raindrops produced on the windward side are carried over the top down to the leeward side by horizontal transport of an ascending air current along the slope. This phenomenon, called *spillover*, plays a very significant role in the amount of rainfall in a limited area in mountainous regions. An average of 10% of the rainfall within an area of 200–300 km^2 is attributable to this phenomenon (Fletcher, 1951).

There is a close relationship between the amount of rainfall and the distribution of wind velocity. As stated in Section 4.1.7, the amount of rainfall has much to do with the method of rainfall measurement. For instance, it is thought that the trees in a mountain receive more raindrops than those coming into a rain gauge with a level receiver.

In order to correct for this defect, a new model of a rain gauge, with its receiver parallel to the gradient of a slope, was tested for the measurement of rainfall distribution. Usually, the gauge is buried vertically under the ground so that the receiver forms the same gradient as the ground surface. The values obtained thereby indicate *hydrologic rainfall*, whereas *meteorological rainfall* is denoted by the values obtained through a normal rain gauge with a level receiver. Needless to say, the absolute values of hydrologic rainfall are greater in mountain ranges. Furthermore, the pattern of regional distribution itself varies in accordance with the method of measurement. The measured values of hydrologic rainfall become greater nearer the summit, but the amount of meteorological rainfall is reduced at the summit under the influence of high wind velocity there, and the middle part of the slope receives a larger amount of rainfall instead. For example, the results of measurements at Baye de mon-

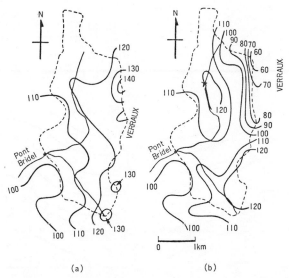

(a)   (b)

Fig. 4.60 Distributions of hydrologic rainfall (a) and meteorological rainfall (b) in the Baye de montreux presented at the IUGG Conf. in 1948 by Hoeck (Schirmer, 1951).

treux from May 1 to October 31, 1946, as shown in Fig. 4.60, have revealed the above-mentioned phenomenon near the summit in the northeastern part (Schirmer, 1951).

Godske (1953) made a detailed study on the distribution of the amount of precipitation in the Hardanger fiord, Norway, (70×70 km) from May to September in 1945 and 1946. According to his results, mountain ranges with a height of 1,000 to 1,700 m between fiords exert such a strong influence that the leeward side of the mountains was kept from rain as a *rain shadow*. This holds true both in individual cases of rainfall and in the total amount of rainfall of each season. An examination of representativeness shows that, concerning the regional representativeness, it is small at points such as the recess of a narrow fiord or at points where the local rainfall amount is large and concerning the seasonal representativeness, it is small in early summer when the amount of local precipitation is large. The representativeness at an observation site becomes more important in measuring the amount of area rainfall.

Sauberer (1948) distinguishes six zones of rainfall, based on observations at 31 stations distributed on slopes at heights between 610 and 1,850 m in the Lunz region Austria. He discusses the effects of slopes on these rainfall zones.

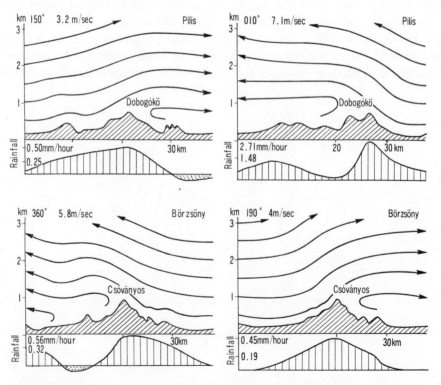

Fig. 4.61 Distribution of orographic rainfall along the topographical cross section of the Pilis and Börzsöny mountainous regions in Hungary (Szepesi, 1960).

Dimitrov and Vekilska (1966–68) discuss the influence of the Balkan Mountain on the formation and distribution of rainfall in N- and S-Bulgaria, based on the observed results at 18 stations. The annual rainfall decreases in the sub-Balkan valleys, due to rain shadow effects. Szepesi (1960) reports the effect of the mountains of the Carpathian Basin on the orographic rainfall. Analyzing the results observed at about 400 stations according to 43 synoptic situations, he points out the fact that there is a clear relation between the rainfall and the topography in the Bakony, Pilis, Börzsöny, Mátra, and Bükk mountainous regions, Hungary, as shown in Fig. 4.61: The rainfall intensity reaches a maximum on the top of mountains in all cases studied where the mountains are 700–1,000 m high.

Takeda (1960) has obtained the following equation, which reflects a topographical profile:

$$Y = b \sin mx \cdot e^{-my} \qquad (4.41)$$

where b and m are the parameters determined by the boundary condition, and the wind components are expressed by

$$\left. \begin{array}{l} u = V_0 \left(1 + bm \sin mx \cdot e^{-my}\right) \\ v = V_0\, bm \cos mx \cdot e^{-my} \end{array} \right\} \qquad (4.42)$$

The rainfall intensity on the profile can be obtained by the following equation:

$$nW = \frac{n_0 W_0}{\sqrt{u^2 + 1}} \sqrt{(u - \beta \cos \xi \cdot e^{-\eta}) + (1 + \beta \sin \xi \cdot e^{-\eta})^2} \qquad (4.43)$$

where n is the total number of raindrops on the slope, W is the terminal velocity of raindrops on the slope, and n_0 and W_0 are the total number of raindrops and the terminal velocity of raindrops on the plane at infinitive distance respectively. The dimensionless parameters $u = w/V_0$, where w is the falling velocity of raindrops and $\beta = bm$. The dimensionless coordinates $\xi = mx$, and $\eta = my$. The horizontal rainfall (meteorological rainfall) is expressed by

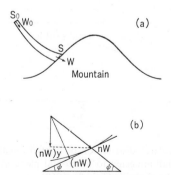

Fig. 4.62 The signs in Eqs. (4.42—4.45) (a): Divergence (or convergence) of rainfall streams affected by the topography. (b): Meteorological rainfall $(nW)_y$ and hydrologic rainfall nW (Takeda, 1960).

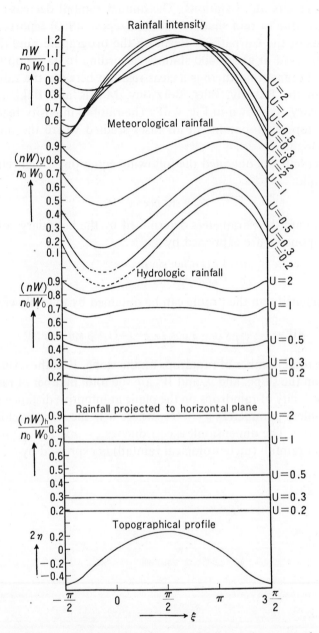

Fig. 4.63 Theoretically calculated distribution of rainfall intensity, meteorological rainfall, hydrologic rainfall, and rainfall projected to horizontal plane in accordance with change of U values along the topographical profile (Takeda, 1960).

$$(nW)_y = \frac{n_o W_o}{\sqrt{u^2+1}} \; (u - \beta \cos \xi \cdot e^{-\eta}) \tag{4.44}$$

and the slope rainfall (hydrologic rainfall) by

$$nW = \frac{n_o W_o}{\sqrt{u^2+1}} \cdot \frac{u(1+\eta)}{\sqrt{1+2 \, \beta \sin \xi \cdot e^{-\eta} + \beta^2 \, e^{-2\eta}}} \tag{4.45}$$

The symbols in the above equations are explained in Fig. 4.62. The calculated results are given in Fig. 4.63 in the cases where $U=2, 1, 0.5, 0.3$, and 0.2, which values correspond to $V_o = 3, 6, 12, 20, 30$ m/sec, respectively, if $w \doteqdot 6$ m/sec and $\beta = \pi/10$.

The maximum rainfall intensity appears on the tops of mountains when the wind is strong. Other interesting characteristics can be read from Fig. 4.63; e.g., the meteorological rainfall is more on the leeward than in the windward side.

Masatsuka (1960, 1962, 1964) observed the meteorological rainfall and hydrologic rainfall on Mt. Yokomine and Mt. Takabotchi, 138°02′E, 36°08′N, 1,680 m and 1,664 m, respectively, in central Japan. His results are given in Section 5.1.2.

An attempt was made to obtain the topographically adjusted normal isohyetal maps for W-Colorado, U.S.A., by applying a coaxial correlation technique (Russler and Spreen, 1947). In this study, topographical parameters considered are: (a) *Elevation*: station elevation mean sea level. (b) *Rise*: difference in elevation between the station and the highest point within a 8-km radius (expressed in thousands of feet). (c) *Exposure*: the sum of those sectors of a 32-km radius circle, centering around the station, not containing a barrier 1,000 feet (about 330 m) or more above the station elevation (expressed in degrees of the azimuth). (d) *Orientation*: direction of the exposure defined in (c), expressed by 8 points of the compass. (e) *Zone of Environment*: those areas having the same general direction of moisture inflow and similar relation to the principal mountain ranges are included. Detailed description of this method is omitted here for the sake of space, but it would probably be useful in obtaining detailed isohyetal maps for other regions.

(B) *Area Rainfall*

Area rainfall or areal rainfall is the amount of total rainfall for a given area, in most cases for a drainage basin or catchment area.

It was discussed above that the amount of rainfall in a mountainous area differs considerably according to the method of observation, i.e., type of rain gauge, and equally great differences arise according to the density and position of observation points, because of the great local variability of rainfall.

An increase in number of observation points generally results in an increase in the total amount of rainfall measured in an area. According to an investigation into the thunderstorm rainfall in Muskingum Basin, Ohio, on August 3, 1939, an isohyetal map based upon data from 449 stations, 46 km² per station, indicates very minute distribution patterns, but an isohyetal map based on data from 22 stations, 960 km² per station, reveals only an obscure center of thunder-storm rainfall (U.S. Weather Bureau and Corps of Eng., 1947). In Japan, the rainfall amount in a still narrower area must be examined, because of the more irregular topography. Table 4.21 shows the results of studies in the vicinity of Isohara with an area of approximately 1,160 km², in Ibaraki Prefecture, Japan (Central Met. Obs., 1952). This table shows the statistics of every rainfall of over 30 mm per one cyclone summarized for September, 1950, to September, 1951. Mention will be made here of the calculating errors in area rainfall.

Table 4.21 Density of rainfall stations and the error
of calculating area rainfall
in the Isohara area
in Ibaraki Prefecture, Japan
(Central Met. Obs., 1952).

	3	6	9	12	15	18	21	24	Stations
Density (Area/1 Station)	280	140	90	70	60	50	40	35km²	
Error	25	18	13	10	8	6	5	3%	

In an area of 16 km × 35 km at a height of 75–330 m a.s.l., located to the south of Wilmington, Ohio, 55 points of observation were set up at regular intervals of approximately 5 km (Linsley et al., 1951). In order to determine the relative reliability of the mean amount of rainfall on the basis of differing density of observation stations, the average amount for each of 68 rainfalls from 1947 to 1948 was calculated, based on the values at 1, 2, 3, 4, 10, 18, and 55 observation stations. The error of the mean rainfall, E, in this area is given by an experimental equation as follows:

$$E = 0.186 \ P^{0.47} \cdot N^{-0.60} \tag{4.46}$$

where P is the amount of rainfall (inches), and N is the number of observation stations.

According to the equation, the error of the calculated mean is 5 mm in the case of 1-station observation, 3 mm in the case of 2-station observation, and 1 mm in the case of 10-station observation as compared with the value of 55-station observation, provided that an area 560 km² has the mean rainfall of 30 mm.

In Japan, similar investigations have been carried out in many catchment areas. For instance, in the upper streams of Tagokura (745 km²) in the Tadami River basin, the error of the calculated area rainfall, Δp, is expressed by

$$\Delta p = 0.819 \ \sigma^{0.99} \cdot N^{-0.64} \tag{4.47}$$

where σ is the standard deviation of the average point rainfall, and N is the number of observation points (Kobayashi, 1960). In general, the error becomes stable if the density of the observation points becomes smaller than 30 km² per station in Japan (Masatsuka, 1965).

At first sight, the error seems insignificant, but this is the regional mean for a rainfall during the passage of one cyclone. When the total amount of rainfall is calculated for a month or year, this error becomes very significant. Again, when the amount of rainfall is hundreds of milimeters at a time, as is often the case in Japan, this error is serious.

The rainfall distribution for each thunderstorm was studied in three small watersheds, 716.8, 241.4, and 13.3 km², in Illinois, U.S.A. (Huff and Stout, 1952). Curves for the area rainfall in relation to size of the area were obtained for each storm. The relation is;

$$y = a + b \sqrt{x} \qquad (4.48)$$

where y is the mean rainfall, and x is the area (km²). Incidentally, this relation cannot be applied to a larger area, such as Muskingum Basin mentioned above, which has an area of approximately 20,000 km².

As the spatial distribution of storm rainfall can be seen as function of the interplay between surface wind and topography, an attempt was made to estimate the average area storm rainfall in a catchment by information from a limited number of rain gauges, together with the surface wind information, in the Tyne catchment, NE-England (Coolinge et al., 1968).

Czelnai (1972) emphasizes that the accuracy of the areal estimates from sample point rainfall data should be considered to be a function of the size of the area concerned, the number of sample points at one's disposal, and the statistical structure of the given precipitation field. He deals with the accuracy and representativity of sample point rainfall observations from the standpoint of stochastic analysis. He asserts that the micro-scale variations and the stochastic errors of the observations are the major components of the noise-level in the data-field.

Areal coverage of precipitation over a subsynoptic-scale and a meso-scale network of stations in NW-Utah, U.S.A., was discussed by Williams and Heck (1972). As would be expected, 100% areal coverage is rare in summer, but, surprisingly, this is also true in autumn, winter, and spring. This can be attributed to the fact that the precipitation in the mountainous west in the U.S.A. has a scattered nature.

(C) *Snow Accumulation*

One characteristic of the snow accumulation in mountainous areas is its extremely local distribution. The distribution of snowfall is found to be more local on the whole than that of rainfall, since the influence of the wind is more evident on the falling snowflakes than on raindrops. Moreover, the snow accumulated on the ground is carried away by the surface wind toward a shelter from wind. Accordingly, the snow distribution in mountainous areas shows a local aspect, and the state of snow accumulation on and around small undulations is transfigured to a smooth surface.

If there is any variation in the height above sea level in the regions under consideration, the period of snow cover, just like the ratio of snow in the total

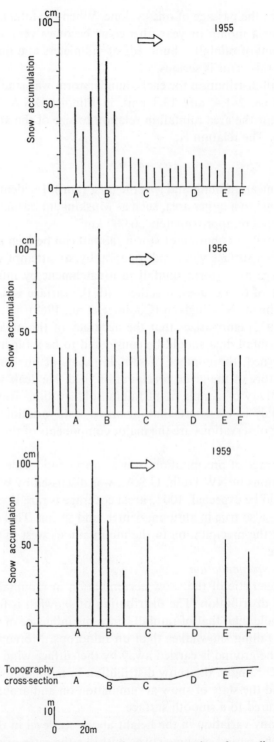

Fig. 4.64 Snow accumulation along a cross section of a small valley
at Sugadaira, 1,300 m a.s.l., central Japan (Tateishi, 1969).

amount of precipitation during the winter in mountainous regions, maintains linear relations to the height above sea level. In the mountains lower than several hundred meters, linear relations are also usually found. In Japan, it is also confirmed that the depth of snow accumulation and the date of last continuous snow cover bears a linear relation to the height above sea level. A similar linear relationship exists between the height above sea level and water content of snow, the latter being in most cases linearly related to the snow depth.

Sugaya (1953) divided the area in the vicinity of Towada Lake, Hakkôda, NE-Japan, into relatively minute parts, and proved that for each part a linear relationship exists between the height above sea level and water content of snow. A liner relationship, however, holds true for precipitation during the summer and winter on the condition that the area is divided into three parts, and for water content of snow on the condition that it is divided into eight parts. This is probably because the distribution of snow cover is more local. The linear relation for water content of snow is a rate of 10–20 cm/100 m.

A closer observation, made possible by dividing an area into more minute parts, shows that in many cases the relationship between the snow depth or water content of snow and the height above sea level does not form a straight line on a figure but rather two lines linked together or a complicated curve.

When the shape of the mountain slope is rectilinear or concave, the snow depth describes a curve showing a small rate of increase in the lower altitudes and a sharp increase at higher altitudes (Onuma, 1955). According to the results of the observation on the NW-slope of Tsuganomori (1,640 m), SE of Yonezawa City, at an altitude of between 500 and 1,300 m, the snow depth describes a quadratic expression below 1,000–1,100 m a.s.l., but almost a straight line above that altitude (Yoshida, 1960).

As stated at the beginning of this Section, the surface snow on the ground is transported by the wind, so that local variation in the depth of snow accumulation depends mainly on the local variation of wind velocity near the ground. An example of this phenomenon can be drawn from the results observed in Sugadaira, central Japan (Tateishi, 1969). Measuring the snow accumulation along three cross sections of small valleys, Tateishi has made clear that there exists a relationship between the snow accumulation and the wind velocity. Figure 4.64 shows some of his results. According to this figure, maximum appears at a point between A and B, just below the separation point on the wind shadow slope, and, at a point between B and C leeward of this maximum point, there is a minimum. At a point between D and E facing the prevailing wind, another minimum point appears. Calculating a deviation at each point from the mean wind velocity, the local variation of snow accumulation is expressed by a linear function of wind velocity like the following:

$$\text{for 1956} \qquad y = -33x - 0.13 \qquad\qquad (4.49)$$

$$\text{for 1959} \qquad y = -33x - 5.17 \qquad\qquad (4.50)$$

where y is the deviation of snow accumulation (cm) from the mean value for all points, and x is the deviation of wind velocity (m/sec) 1 m above the snow surface from the mean values for all points. Equations (4.49) and (4.50) are obtained in a range of wind velocity deviation with ± 1.0 m/sec and snow accumulation deviation with ± 20 cm.

A study on the snowfall on level land and the neighboring S-facing and N-facing slopes with a gradient of 40° at Tôkamachi Experimental Station has made it clear that the water content of snow cover even in a small area differs remarkably with the direction of the slope (Takahashi, 1953). The values obtained on the S- and N-facing slopes from measurements repeated seven times at intervals of 10 days from January 6, 1953, are given in Table 4.22. The values are represented in percentages of the value of flat land, which is 100%.

Table 4.22 Characteristics of snow accumulation on the S-facing and N-facing slopes in Tôkamachi, Niigata Prefecture, Japan (Takahashi, 1953). Percentage compared to a flat open place.

	S-facing slope	N-facing slope
Snow depth	66%	111%
Water content of snow cover	63	100
Average density	96	91
Hardness	89	87

From the standpoint of water balance or water resources, snow accumulation is the storage of water in winter. Roughly speaking, at the end of March or April, the amount of storage in the form of snow in mountainous areas in central Japan is approximately 20% (480–720 mm) of the annual precipitation. At the end of July, it decreases to lower than 10%. However, an accurate estimation awaits further studies, especially on the calculation of the area snowfall and the measurement of the accumulation in mountainous regions.

4.2.6. Evaporation

Evaporation in relation to micro-topography in a mountainous region depends on the wind distribution and insolation. Evaporation increases with the increase of wind velocity, but it decreases sharply in the range of weak wind velocity. The relationship between evaporation and wind is opposite that between precipitation and wind, as shown in Fig. 4.65. This figure was obtained from the results observed in Obergurgl-Poschach, 2,070–2,210 m a.s.l., in the Ötz valley, Austria (Prutzer, 1961). The distribution of evaporation coincides perfectly with the cross section shape of the topography and, therefore, the vegetation distribution corresponds to the state of evaporation. The average total evaporation for 28 days from August 11 to September 7, 1955, was 62.7 mm for *Loiseleurietum*, 56.1 mm for *Vaccinietum*, and 51.6 mm for *Rhododendretum*.

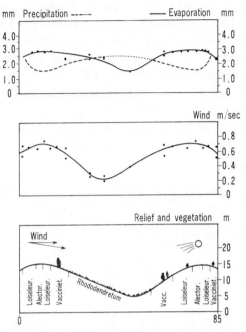

Fig. 4.65 Evaporation, precipitation, and wind velocity in relation to micro-topography in Obergurgl-Poschach, Austria (Prutzer, 1961).

Müller (1964) estimated the evapotranspiration of a drainage basin, 5,000 km², of Drau near Villach in the S-part of the Austrian Alps. He calculated the evaporation for the months with snow cover by Sverdrup's formula (Sverdrup, 1936):

$$E = \frac{0.623 \rho K^2 v \, (e_L - e_S)}{p \ln \dfrac{a}{z_0} \ln \dfrac{b}{z_0}}$$

(4.51)

Table 4.23 Comparison of calculated and measured monthly evaporation (mm) in the drainage basin of Drau in Austria (Müller, 1964).

Month	Klagenfurt (446 m)		Kanzelhöhe (1,500 m)	
	a	b	a	b
March	17.2	—	3.1	—
April	48.5	87.0	27.4	—
May	92.0	115.8	68.0	55.7
June	115.0	129.5	90.2	84.5
July	126.2	141.5	106.7	89.7
August	107.2	119.2	95.0	93.3
September	73.8	77.6	65.0	61.0
October	37.2	38.8	35.8	—
November	8.2	—	6.2	—

a) Evaporation calculated by the method of Thornthwaite, b) Potential evaporation measured with the U.S. Weather Bureau Class A-pan.

where $K=0.24$, ρ is air density (g/cm³), v is wind velocity (cm/sec) at a height of a cm above the ground, e_L is water vapour pressure (mb) at a height of b cm above the ground, e_s is saturated vapour pressure (mb), and $z_0=0.25$ cm. The governing air pressure is p (mb). On the other hand, he calculated the evaporation for the months without snow cover by Albrecht's formula (Albrecht, 1962). Müller has proved that the results calculated coincide fairly closely with the measured values and that the amount of evaporation calculated by Thornthwaite's method is slighly higher than the measured potential evaporation with a U.S. Weather Bureau Class C-pan, especially in mountainous areas in summer, as shown in Table 4.23.

Table 4.24 Evaporation (mm) at a height of 1 m above the ground in Dyjskosvratecky úval in Czechoslovakia (Quitt, 1961).

Date (1960)	Direction and gradient of the slope				Plain	Remarks
	N-20°	S-20°	W-22°	E-18°		
July 31	8.5	13.2	10.3	13.0	12.1	Clear, SE–S winds
Aug. 1	6.3	7.1	7.0	6.8	6.8	Clear–Cloudiness 5, W-wind
Aug. 25	12.1	14.0	13.1	14.1	13.6	Clear, SE-wind
Aug. 26	13.1	14.2	13.6	13.7	13.7	Clear, SE-wind
Aug. 27	13.0	14.6	13.9	14.1	14.2	Clear, Calm
Average	10.6	12.6	11.8	12.3	12.1	

Quitt (1961) investigated the evaporation in connection with the mesoclimatic conditions in the central part of Dyjskosvratecky úval, Czechoslovakia. The results show an apparent difference between different directions of a slope. This difference is great when the weather is fine and calm, as shown in Table 4.24. The evaporation amounts at a height of 10 cm above the ground are 0.2–0.4 mm less than those given in Table 4.24.

In a very small catchment area, 0.18 km² and 312–505 m a.s.l., along a tributary of the Nozu River, Japan, the maximum evapotranspiration rate is 7 mm/day at the time of the cease of a rainfall, but the rate decreases exponentially 1.5–1.3 days after that, and the final rate is constantly 1.2 mm/day after 13 days (Ishihara et al., 1971, 1972). In this area, the total annual amount of rainfall, 1,840 mm, minus runoff, 1,160 mm, is 680 mm, which is considered to be the total loss, mainly caused by evapotranspiration.

In the arid zones, the evaporation problem is of great importance, but there are only a limited number of studies. Topographical effects must be taken into consideration in connection with permeability of soil, as pointed out by Oliver (1969). Steep slopes increase the probability of a higher ratio of runoff to infiltration. In the irrigated area and in the oasis, the evaporation measured by means of a pan is great, eventhough advection has not small effect (e.g., the annual amount in Khartoum, Sudan, 1961–1962, was 3,492 mm).

Gavrilović (1970) reports on two floods that occurred in the Wadi Bardagué in the NW-part of Mt. Tibesti, central Sahara. On this occasion, no evaporation was observed. On the afternoon of June 7, 1968, an unusual rainfall occurred and only some minutes after the first white foam wave at the Oasis Bardai, the river was 60 m broad and 1 m deep. The flood speed was 2.5 m/sec, the discharge 70 m³/sec. On the next day the river was only 20 cm deep and the river water speed was 0.1 m/sec. On June 9 and 16, there was no river water, but only small ponds here and there. By the way, the "climatic value for June, 1957–1968" of the rainfall at Bardai, 1,020 m a.s.l., was 0.8 mm. The problem now is how to estimate the actual areal evaporation for these lands.

4.2.7. Fog

As is described in Section 5.4.3, a cold air lake tends to be formed in basins within mountainous regions, and accordingly fog is apt to occur through ground inversion in the basins. The thickness of a fog layer serves as an indicator of the thickness of the cold air lake, so that a careful observation of the flow of fog will afford a better understanding of cold air drainage.

The outflow of fog from a basin along a valley is a phenomenon common among various districts. For example, at Nagahama at the mouth of the Hiji River in Ehime Prefecture, Japan, the fog that has risen back in the Ohzu basin begins to flow together with a violent wind. This type of fog, called *asagiri*, and the kind of wind, called *arashi* in this district, occur very frequently in the morning from the end of September to the end of December: The average monthly total of the number of days with fog from September to December is 10–16 days (1956–1969). Because of this violent cold air outflow, which reaches 10 m/sec or more frequently, the pine trees at the mouth of the Hiji River are bent toward the sea, whereas the pine trees at the other part are bent toward inland. Fog comes out on the sea with the wind until about 10 h in the morning. A typical example is shown in Photo 4.11. A similar outflow of fog from a basin can be observed in the vicinity of Kawanakajima, where the Sai River drains into the Shinano River, central Japan (Asakawa, 1950). According to Asakawa, in 1947, for instance, fog often occurred in May and September, and naturally, on mornings following clear, calm nights.

Along a valley bottom where fog occurs locally with great frequency, sometimes very localized phenomena can be observed. For example, a big fog-belt called *Malojaschlange* (Maloja-snake) creeps down the valley from the Maloja pass and covers the slopes up to 2,000 m (Holtmeier, 1966). In this area, spruce (*Picea abies*) is distributed instead of the normal larch-pine (*Pinus cembra*) forest that is typical for Oberengadin.

Recently, because of dams across valleys in mountainous districts, conditions for the formation of fog have changed. For example, the annual total number of days with fog at Ohzu, mentioned above, is 143.7 days for the average of 1956–1960, but 93.7 days for 1961–1969. This is thought to be because of the dam constructed at the upper course of the Hiji River in January 1960 (Fuji-

Photo 4.11 Morning fog with *arashi* at Nagahama from the Ohzu basin (upper right hand) in W-Shikoku, Japan, at 7h20m (November 3, 1973, by Shosei Watanabe).

moto et al., 1971). Not only the frequency of occurrence or density of fog, but also the maximum altitude for fog, is subject to variation in relation to the height of the water surface, so that the utilization of land on a slope is influenced more by fog recently than in the past.

In the Ljubljana Basin, fog occurs very frequently due to inversion. The study on the air temperature inversion in this basin by Bernot (1959), based on the records at Ljubljana Observatory (300 m a.s.l.) and at Smarna gora (667 m), has classified the inversion into four types. Among them, the winter inversion type accompanies fog most frequently. Petkovšek (1969) studied the monthly fog frequency for 12 years (1956–1967) at 50 stations situated in the lowlands, valleys, and basins of Slovenia. The radiation fog is most frequent in this area. Although the lowlands are divided by a high ridge, the geographical distribution of average fog frequencies in the lowlands is rather uniform. The highest frequency of fog days is to be found in the central valleys and basins of Slovenia that cover the Alpine foreground. Lower frequencies are found on the boundaries toward the Adriatic coast, in the Alpine highlands, and on the margin of the Pannonia plain. Although the annual differences are generally small and not significant, a trend to a slight increment of fog frequency appears in this 12 years period.

Schindler (1961) observed the fog distribution along the road in a car every morning from middle October to middle April, 1958/59 and 1959/60, except on sundays and holidays. Dence fog occurs most frequently in October and November in the valley, which he called *fog hollow*.

From the tower of the observatory (1,005 m a.s.l.) on the top of Mt. Hohenpeissenberg, Heigel (1967) observed fog and haze in the valley and made climatological statistics. Of interest is the monthly change of height of inversion shown in Table 4.25. This table shows that the thickness of ground fog is 40–70 m in October and November, when the occurrence of fog is most frequent.

Table 4.25 Height of inversion around
Mt. Hohenpeissenberg*
(Heigel, 1967).

	Morning	Afternoon
Jan.	810 m	848 m
Feb.	775	819
March	838	907
April	834	908
May	806	917
June	828	906
July	852	917
Aug.	855	927
Sep.	808	898
Oct.	794	884
Nov.	760	821
Dec.	783	807

* The level around Hohenpeissenberg is 720–725 m a.s.l.

Photo 4.12 Fog in a basin, observed near Friedberg,

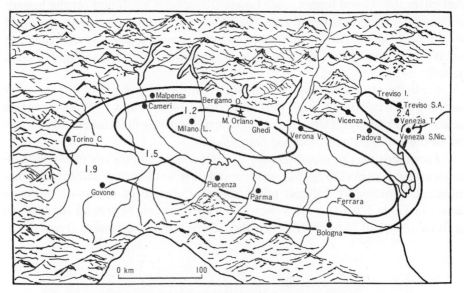

Fig. 4.66 Average visibilities (km) in the Po Valley, 08–18h during the period November 22, 1968, to February 17, 1969 (Montefinale et al., 1970).

Montefinale et al. (1970) reported the results of intercorrelations between daily average visibilities at 18 stations in the Po Valley. As shown in Fig. 4.66, the area including Milano and Ghedi is the center of low visibility. They pointed out that there are several areas concerning the local fog not originated by the city or industrial regions, but by topographical situation in the Po Valley.

4.2.8. Phenology
On S-facing slopes, phenological phenomena in spring, such as the opening of leaves, flowering, and full blooming, come earlier than on N-facing slopes in

Austria (November 29, 1970, by M. M. Yoshino).

the northern hemisphere. Of couse, phenological phenomena in autumn, such as the turning and falling of leaves, come later on S-facing slopes than on N-facing slopes. Such phenomena are well known all over the world qualitatively, but it is hard to say quantitatively how much the difference is as a general rule.

Heigel (1955) observed on the Hohenpeissenberg the effects of exposure, height, wind shelter, kind of soil, and soil moisture on the plant phenology and concluded that the exposure plays the most important role. As shown in Table 4.27, the upward shifting velocity (phenological gradient) of flowering of dandelions is much later on N-facing slopes than on S-facing slopes. This difference in the phenological gradient can be attributed to the thermal conditions of the slopes. Heigel showed a parallel relation of sunshine duration to the phenological gradient. Simonis (1960) observed a difference of 2–3 weeks in the phenological stages of various plants, as shown in Table 4.26.

Table 4.26 Difference of plant phenological stages between the S-facing and N-facing slopes. Plus means that those on the S-facing slope are earlier. Nos. 2–6 were located on slopes with an inclination of 45° (Simonis, 1960).

	Stage	1935	1936	1937	1938
1. Lime-tree (*Tilia petiolaris*)	Half leaves	1	0	2	3 days
	Full leaves	1	0	0	3
2. Clover (*Anthyllis vulneraria*)	Budding	16	8	4	—
	First flowering	16	13	6	—
3. *Hypochoeris radicata*	Budding	21	3	18	16
	First flowering	3	16	4	24
4. *Lolium perenne*	Budding	19	11	2	22
	First flowering	14	12	2	11
5. *Ervum hirsutum*	First flowering	−7	9	9	10
6. *Lolium multiflorum*	First flowering	16	—	0	7

The relation between the altitude and the delay of phenological stages is expressed generally by a linear regression equation as long as the vertical relation height is not great, say within several hundred meters. The rate of delay is

Table 4.27   Phenological gradient (days/100 m) on the slopes (Yoshino).

Name of plant	Phenological stage	Phenological gradient (days/100 m)	Topographical situation	Region	Liter- ature
Dandelion	Flowering	22 / 7	N-facing slope / S-facing slope	Hohenpeis- senberg,	(1)
Sweet cherry	Flowering	2	Slopes	S-Germany	(1)
Winter rye	Harvest	7	Slopes		(1)
Apple (*Schöner von Boskop*)	First flowering	4	W of Lake Constance	S-Germany	(2)
Apple (*Schöner von Boskop*)	First flowering	2	Höri Peninsula in Lake Constance	S-Germany	(2)
Cherry (*Prunus yedoensis*)	Flowering	2.1 / 3.5	Slopes from the foot up to the middle part / Slopes from the mid- dle part up to the summit	Mt. Tsuku- ba, Japan	(3)
Cherry (*Prunus yedoensis*)	Flowering	3.8	From plain to moun- tainous area	Miyagi Pref., Japan	(4)
Cherry (*Prunus yedoensis*)	Flowering	1.6	From plain to moun- tains	All Japan	(5)
Cherry (*Prunus serrulata*)	Flowering	2.7	From plain to moun- tain	All Japan	(5)
Cherry (*Prunus subhirtella*)	Flowering	4.8	Alluvial fan	Akaho, Nagano Pref., Cen- tral Japan	(6)

(1) Heigel (1955), (2) Aichele (1953c), (3) Ueno (1938), (4) Ueno (1941), (5) Nakahara (1944), (6) Sekiguti (1950).

called phenological gradient and is expressed in days /100 m. In Table 4.27, some values of phenological gradients are compiled from the results in Germany and Japan.

The phenological gradient, like the temperature lapse rate, varies in a com- plicated way from place to place according to the topographical situation. As Schnelle (1955) has pointed out, the first flowering of apple trees was 2 days later in the bottom of a cold air lake formed in a 150–200 m deep valley than on the thermal belt of the slope. The delay of phenological stages in a valley bottom in Austria is described by Rosenkranz (1951). Balchin and Pye (1950) also reported the delay of plant growing states of 10–20 days until the end of March in the inversion layer as compared with the thermal belt over the inversion layer with a relative height of 200 m.

Table 4.28   Change of beginning of phenological season in days/100 m in Rheinland-Pfalz, W-Germany (Aichele, 1964).

Above sea level	Before	First Spring	Full	Early	High Summer	Late	Early	Full Autumn	Late
350–450 m	+2	+3	+3	+3	+8	+1	−4	−2	−6
250–350	+3	+3	+4	+3	+8	+1	+3	+2	0
150–250	+4	+3	+5	+5	+3	+1	+1	+3	+3

Aichele (1964) summarized the phenological events from spring to autumn in the Rheinland-Pfalz, 1950–1963, in a time-height diagram and presented the change of beginning of phenological season per 100 m as shown in Table 4.28.

In spring and early summer it changes at a rate of 3–5 days/100 m but in mid summer it changes at a great rate of 8 days/100 m. In autumn, the reverse condition occurs above 350 m; that is, autumn begins at a height of 450 m and comes down. On the other hand, however, autumn begins slightly earlier at lower altitudes, due to the inversion. Therefore, autumn comes latest at a height between 300 and 350 m.

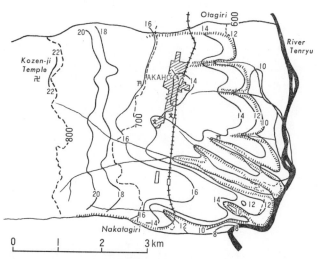

Fig. 4.67 Distribution of flowering dates of cherry blossoms (*Prunus subhirtella*) at the Akaho Fan, central Japan (Sekiguti, 1950).

In Japan, there are many phenological studies on the distribution of cherry blossoms. Misawa et al. (1940) observed it in Nagano Prefecture and derived the phenological gradient 2.2 or 3.8 days /100 m for the 80% flowering date of cherry blossoms (*Prunus subhirtella*). Sekiguti (1950) carried out a detailed observation of the same phenological event at the Akaho alluvial fan, 620–840 m a.s.l., Nagano Prefecture, central Japan. The reason for using the 80% flowering date is that the fluctuations of the date are smaller than those of the first flowering date or full blooming date. Moreover, as to the degree of flowering, 80% flowering is easy to judge. The distribution based on the results observed at 142 points in an area 5 km × 5 km, is shown in Fig. 4.67. The effect of altitude is predominant with a rate of 4.77 days/100 m. Furthermore, micro-topographical effects are made clear; namely earlier flowering is seen (a) on the wind-shadow sides of valleys, (b) at places immediately below cliffs, (c) at the wind shadows of the hills, and so on.

Shitara (1954) studied the seeding season of rice-plant in the Shinjo basin, NE-Japan [Sci. Rep., Tohoku Univ., 7th Ser. (3) 29–40]. He made clear that the local temperature differences are reflected sharply in the local differences of the seeding season and the drying method after the harvest of rice in the basin.

The thawing times were also compared with the seeding times. It was found, however, that, in the western part of the basin, the correlation of distribution is positive, but negative in the eastern part. Its cause is attributed to the difference in agricultural techniques between two parts.

Chapter 5

LOCAL AIRSTREAMS AND WEATHER

5.1. LOCAL AIRSTREAMS AND DISTRIBUTION OF CLIMATOLOGICAL ELEMENTS

This chapter deals with the local airstream and its effects on the distribution of climatological elements in a small area. Such treatment on a relatively broad scale has been developed under the name of synoptic climatology since the time of World War II (Jacobs, 1946, 1947). The steps in the development of the study for a given region are as follows: (a) To divide, if necessary, the geographic region into subareas of such a size that each can be described by a prevailing local airstream of a single direction. (b) To examine long series of weather maps and classify each of them for each area. (c) To assemble all available weather observations at the same date and hour with each weather map. (d) To summarize these data obtained at each station both singly and combined, with respect to the flow pattern, wind speed, character of flow and air-mass type.

Such a method is applicable to studies on phenomena in a small area, and thus it is important to conduct our study along this line.

5.1.1. Local Airstreams and Distribution of Wind Direction

The distribution of surface winds over Hokkaido in winter was studied from the standpoint of synoptic climatology (Kawamura, 1961). Surface data of 163 climatological stations were classified into ten flow patterns in accordance with the gradient wind direction on the sea level weather map. Maps of streamlines based on prevailing winds at each station are given in Fig. 5.1.

Solid arrows show the direction of gradient winds. Broken arrows are drawn for streamlines that are difficult to ascertain due to the scarcity of stations and their locations at the valley bottoms. The characteristic pattern changes with the direction of gradient winds but is influenced by major topographical features. Low hills and small valleys are unimportant in determining the distribution of surface winds. The frequency of calm and the variation of wind directions increase in the interior basins, such as in the Nayoro and Kamikawa Basins, where air flow is insulated by the surrounding mountain ranges. Moreover, wind directions at each station in a basin are strongly influenced by the trend of the valley direction.

The same approach was used for a study on the surface wind distribution over central Japan (Kawamura, 1966). A schematic illustration of Types I, II, III, and IV among the six types is given in Fig. 5.2. A brief additional explana-

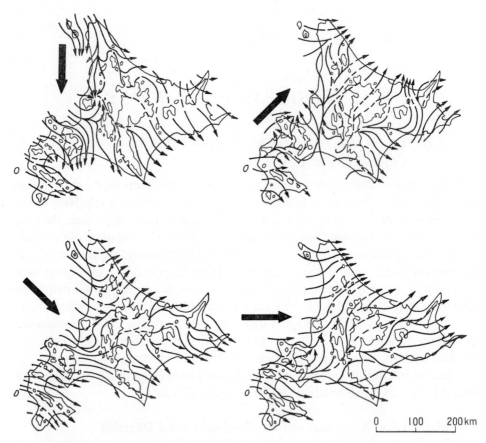

Fig. 5.1 Surface winds according to the direction of the gradient wind in Hokkaido in winter (Kawamura, 1961).

tion is as follows. Type I: The gradient wind over this area exceeds 10 m/sec and comes from the direction of 250° to 330°. (The wind direction represents angles measured clockwise from north.). Type II: The gradient wind speed is the same as Type I. The gradient wind direction is included in the fanshaped domain from 330° to 20°. The curvature of isobars is cyclonic. Type III: The same as Type II, but the curvature of isobars is anticyclonic. And Type IV: The gradient wind speed is the same as above, but its direction is between 30° and 60°. The pattern of local wind systems and the distribution of wind speeds change with the direction of the gradient wind. Areas with strong winds chiefly tend to lie along the rivers in plains and valleys in a mountainous region. On the other hand, areas with calm or light winds are distributed in the regions affected by a mountain range. The locations of wind systems and the situation of convergence or divergence lines resulting from them differ in detail in each case, even with the same flow pattern.

Kawamura has recently applied the same study method to the other parts

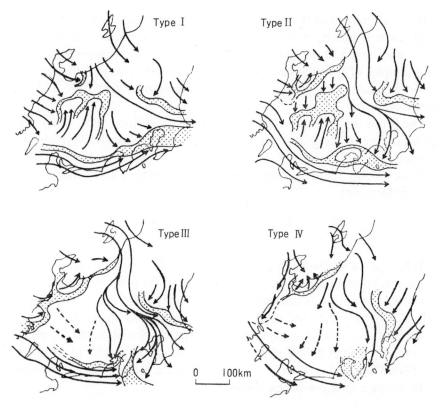

Fig. 5.2 A schematic representation of local surface winds in central Japan in winter. Arrow: surface winds. Dotted area: shifting zone of daily discontinuous line between wind systems (Kawamura, 1966).

of Japan (not yet published). It was pointed out, further, that the distribution of surface winds on a local scale is affected not only by the flow pattern of the synoptic scale, but by the local circulation, such as land and sea breezes, as a result of diurnal variations in the local pressure pattern. Generally speaking, since the local circulation is superimposed on synoptic scale circulation, the wind distribution is more complicated on a local scale.

The difference in the direction of the gradient wind or the prevailing upper wind often indicates a difference of thermal stability in some regions. An example of studies on the local wind conditions in Saxony, Germany, showed that the stability is small in the case of NW-streams and large in the case of SE-streams (Flemming, 1966b). The local wind systems were much more apparent in the latter case than in the former case. Where pronounced thermal stability occurred, two or three maxima were found on the frequency distribution of the wind direction. In such cases, the problems can be changed to those on the surface wind distribution and the stability of the lower air layer, which is discussed in Section 5.5.5.

It has already been mentioned that in fine weather when the general pressure gradient is gentle, mountain and valley breezes develop in mountainous regions, and land and sea breezes on coastal zones. These phenomena present good examples of how the conditions of local air currents and the distribution of their directions change along with the change of the weather and general pressure gradient.

The following is a statistical study on the areal fluctuation of wind directions according to the prevailing wind directions. The results were obtained at 200 points in the city area of Tokyo on days when either S- or N-winds prevailed (Yoshino, 1955b). Then, the fluctuation (y) at a fixed time was obtained from the following formula:

$$y = \left\{ \frac{1}{n} \sum_{i=1}^{n} (D_i - \bar{D})^2 \right\}^{\frac{1}{2}} \qquad (5.1)$$

Where D_i is the wind direction at each observation point, and $\bar{D}$ is the average wind direction at a fixed time calculated from the data obtained at all the observation points. If the general wind velocity (here this means the average wind velocity of 7 climatological stations) at a certain observation time is expressed by x, the following experimental formulae are obtained for each wind direction. The values collected by observations carried out 60 times during the period from September 1943 to August 1944 were used for the formulation.

In the case of S prevailing winds:

$$y = 4.998 \; x^{-0.57} \qquad (5.2)$$

In the case of N prevailing winds:

$$y = 2.669 \; x^{-0.36} \qquad (5.3)$$

The unit of y above is $1 = 22.5°$ and that of x is m/sec. It is clear from these formulae that when the general wind velocity becomes lower, the range of dispersion grows abruptly, and if the wind velocities of the S- and N-winds are equal, the S prevailing winds are apt to have a wider fluctuation range than N-winds. The values obtained from these formulae are listed in Table 5.1. The table shows that the difference in dispersion between N- and S-winds becomes exceedingly great, especially as the wind velocity nears 0.

This relationship between the dispersion and the wind velocity is clearly shown in the maps showing streamlines (Fig. 5.3). A greater disturbance of streamlines is seen in the case of weak winds, and relatively more paralleled streamlines are observed in the case of strong winds. As mentioned above, the wind velocity being equal, the local disturbance of wind direction becomes greater in S- than in N-winds. It is thought that this is due to the fact that the atmospheric layers near the ground are less stable and the disturbance due to thermal effect is greater, when S-winds rather than N-winds prevail in this area. The reason for greater fluctuation when the wind velocity is low is believed to

Table 5.1 Areal fluctuation of wind direction in accordance with prevailing wind velocity for S- and N-winds in the city area of Tokyo (Yoshino, 1955b).

Wind velocity	Mean wind direction	
	Southerly	Northerly
10 m/sec	± 61°	± 52°
8	± 69	± 57
6	± 81	± 63
3	±120	± 81
1	±225	±117

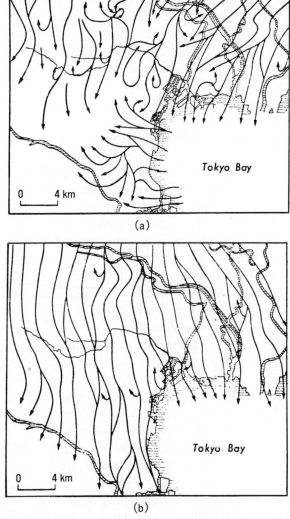

(a)

(b)

Fig. 5.3 Streamlines based on observations at about 200 stations in the city area of Tokyo. (Upper) 14h, October 12, 1943. (Lower) 14h, February 1, 1944. (Yoshino, 1955b).

be because, when the wind velocity decreases, the local difference in the occurrence of mechanically caused eddies comes great, and, as a result, various thermal eddies, small and large, grow quite locally. It is also assumed from other observation results (Yamashita, 1954) that eddies of various extents occur as a result of increasing thermal effects when the wind velocity decreases below 6 m/sec.

Chopra and Hubert (1965) and Chopra (1972) explained that the vortex systems leeward of the Gran Canaria, Tenerife and Maderia Island observed by weather satellites were atmospheric analogs of the vortex street phenomenon well known in fluid mechanics. The core radius of the various vortices in wakes associated with the islands ranges from about 15 to 40 km. The corresponding values of the peak peripheral wind speed vary from 2.6 to 6 m/sec.

Findlater (1971) described a strange wind of Ras Asir, formerly Cape Guardafui, Somalia, East Africa. The pronounced horizontal and vertical variation of wind near Ras Asir is consistent with a persistent and meso-scale eddy linked to the topography of the environs of Ras Asir and embedded in the edge of the strong flow of the SW-monsoon. The eddy has a diameter at the surface of about 200 km, though it is associated with distortion of the wind field over an area of about 500 km diameter. The eddy extends upward to between 500 and 1,000 m, and the axis slopes toward the NE with increasing altitude.

5.1.2. Local Airstream and Distribution of Precipitation

The distribution pattern of climatological factors is more or less affected by local air currents in a small area, but among them precipitation is the one that reflects the influence of air currents most precisely. Thus, an explanation follows here on the relationship between precipitation and local air currents, both on a micro and meso-scale. The problems will be discussed separately under the following themes: (a) synoptic climatological treatment, (b) relation to cyclones, (c) band structure, (d) relation to local convergences, fronts, or instability lines, (e) topographic effects, and (f) weather divides.

(A) *Synoptic Climatological Treatment*

Kawamura (1961) studied the distributions of daily precipitation amounts in Hokkaido from the viewpoint of synoptic climatology, which was described above (Section 5.1.1). The results show a very clear localization of precipitation of more than 5 mm, which corresponds roughly to new snow accumulation of 10 cm/day (Fig. 5.4). The winter monsoon situation with NW gradient wind direction occurred in 16% of the 500 sheets of distribution maps for days with precipitation, 1951–1955. In the same way, WNW was 19%, W 21%, WSW 8%, and SW 5%. It is unquestionable that the results can be used in quantitative forecasting of local precipitation.

The occurrence frequency of heavy rainfalls in the Pacific Ocean side of the Tohoku district, Japan, has been studied from the standpoint of synoptic climatology (Yamashita, 1970, 1972). In this region, 80–90% of the heavy rainfalls of

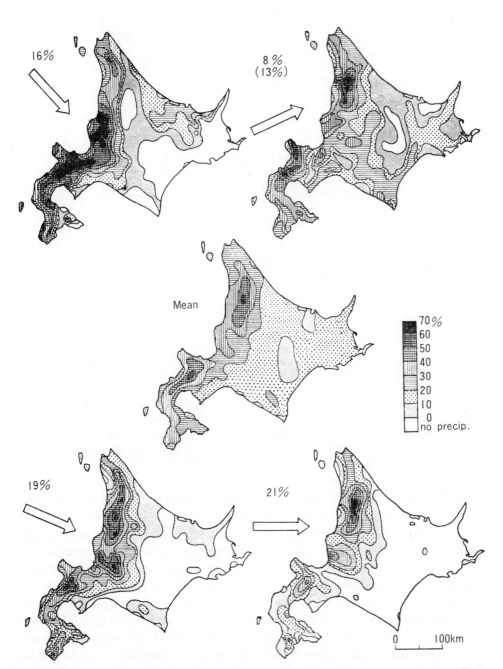

Fig. 5.4 Frequency of precipitation of more than 5 mm/day in Hokkaido according to the gradient wind direction at the 850mb level in the case of a winter monsoon (Kawamura, 1961).

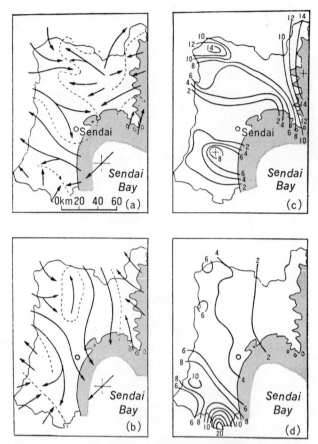

Fig. 5.5 Streamlines in the case of prevailing northeasterly winds (a) in spring and early summer and (b) in summer and early autumn. The corresponding rainfall distribution in mm during (c) 6h~ 9h on July 6, 1958, and (d) 9h~12h on September 26, 1958 (Kusano, 1960).

more than 100 mm/day are brought by typhoons and extratropical cyclones. The rest cases are: The points where a great amount of rainfall is observed frequently do not correspond to the topographical situation, that is, a slope angle of the mountain range, but appear on the coast when the northeasterly maritime airstream prevails. After comparing air temperature with sea temperature on the coast, the following mechanism was considered. The NE-airstream, which is originally cold, is heated from the sea near the coast of the Tohoku district, where the warm *Kuroshio* current prevails, and hence convective clouds develop in this air current. As the heat supply is cut off on land, convective clouds are dissipated in the coastal region. Therefore, the heavy rainfall is concentrated on the coast.

A distribution map of streamlines at the ground level was prepared for Miyagi Prefecture, Japan, where general winds were classified into 8 groups according to direction (Kusano, 1960). Figure 5.5 (a) is the stream map for NE general winds in spring and the rainy season in early summer, and Fig. 5.5 (b) is for NE general winds in summer and early autumn. Figure 5.5 (a) shows that NW-currents flow around Sendai and in the northwestern part of the prefecture, and S-currents prevail on Oga Peninsula and in the southern part of the prefecture. Broken lines are the borders of these winds with different directions. Figure 5.5 (b) shows that SW-currents exist partially around the margin of the SW-border of the prefecture, SE-currents north of it and also along the NE-coast, and N- and NW-currents in other areas. Although all of the broken lines, the borders of streamlines with different directions, do not necessarily coincide with the convergence lines, the following two examples show that they have something to do with precipitation distribution.

Figure 5.5 (c) shows the precipitation distribution during 3 hr from 6 h to 9 h on July 6, 1958. The locations of the wind convergence line and the heavy rain zone coincide rather well in the NE of the prefecture and to the S of Sendai. Figure 5.5 (d) shows the precipitation distribution within 3 hr from 9 h to 12 h on September 26, 1958. A comparison between this figure and Fig. 5.5 (b) reveals that the heavy rain zone and the NW-SE convergence line, which extends from NW to SE in the southern part of the prefecture, coincide extremely well. Although no other coincidence can be seen, particularly around the weak convergence line, the cause of the heavy rain zone, which could never be explained by topographical uprising currents alone, has been explained by taking into consideration the convergence line clearly revealed in the current distribution map. Of course, the relationship between the convergence line and the heavy rain zone can be considered only when no fronts in synoptic scale exist around the area.

(B) *Relation to Cyclones*

The Mt. Kirishima area on the Island of Kyushu, slightly smaller in area than the above-mentioned Miyagi Prefecture, is an example that shows the relationship between air currents and precipitation distribution. This area has an elliptical shape with a long axis of about 30 km extending from NW to SE and a short axis of 20 km. Total precipitation in this area in 1944 was classified by cause, into cyclonic, typhoon, frontal, and topographic rains, and daily precipitations during the year were totaled for each type of rain at each observation point to obtain the average precipitation (Yoshino, 1961b). Figure 5.6 shows the average precipitation distributions caused by extratropical cyclones that passed along the Pacific coast and by typhoons. In this area the mountainous region received the heaviest rain because of the current being raised by the mountain. The locations of the rain area, however, were different depending on whether the rain was caused by a cyclone or typhoon. In the case of a cyclone the heavy rain area appeared in the E of the mountainous region, and in the case of a typhoon, in the NW of the region. In both cases, cyclones and typhoons, the prevailing direction of the winds that caused rainfall in the area was E through SE. Therefore, the difference in the distribution of precipitation may

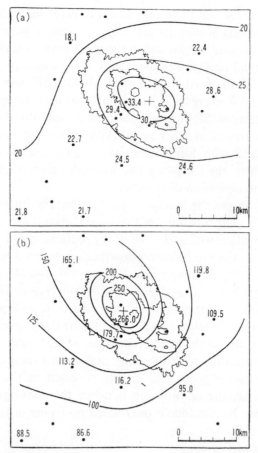

Fig. 5.6 Distribution of average rainfall by an extratropical cyclone (a) and a typhoon (b) at Mt. Kirishima (Yoshino, 1961b).

arise from the difference in the wind velocity. In other words, it is thought that strong E or SE winds caused by a typhoon push the heavy rain area to the NW lee side slope of the mountains, and a cyclone produces the heavy rain area on the windward side of mountains as in the case of ordinary topographical rainfall.

The relation between the rainfall amount and the passage of extratropical cyclones moving along the Pacific coast was studied also from the synoptic climatological standpoint (Yoshino, 1955a). First, the distribution of rainfall caused by the passing of the cyclone was expressed by a linear function of the distance from the track of the cyclone center. From the results calculated for fourteen cases, it can be said that the greater the amount of rainfall, the greater the coefficient of the linear function. Second, an anomaly, R_d, at a station in the mountains region, surrounding the Kanto plain, is defined as

$$R_d = R_m - (R_{tp} + k) \tag{5.4}$$

where R_m is the observed rainfall at the station, R_{tp} the theoretically calculated amount assuming that the level of the station is the same as that of the plain, and k the interval of error belt, namely σ, as calculated from the deviation of the observed value from the theoretical Kanto plain level value. In conclusion, it can be stated that the increasing rate of the amount of anomaly in mountainous regions changes from 1 to 3 mm/100 m, due to the location of the regions to the prevailing winds bringing rain during the passing of the cyclone.

In the region where the low level jet stream tends to converge topographically and to be intensified a very heavy rainfall concentrated in space and time is frequently observed. Such a case was reported in a study of the Kinki districts, central Japan (Nakajima and Gocho, 1968). After analyzing several cases, it was pointed out that these heavy rainfalls occurred in the warm sector of cyclones in relation to the low level jet stream. For instance, the heavy rainfall with 200 mm/day as its maximum in the Kinki district observed on June 25, 1969, occurred over the area where the surface warm front and the low level jet stream crossed each other (Edagawa, 1971).

Matsumoto and Tsuneoka (1969) reported that relatively small scale disturbances with a wavelength of 100–200 km and a cycle of 3 hr were observed in the case of heavy rainfalls over W-Japan on July 9, 1967, superposing over the meso-scale or synoptic scale disturbances.

An extremely heavy rainfall (the maximum hourly rainfall was 122 mm) was observed in a small area around Shimada City in Shizuoka Prefecture, Japan, on August 26, 1959. Analysis of it produced the following conclusions: (a) The rainfall was caused by the small-scale disturbances that developed along the front on a synoptic scale; (b) the heavy rainfall was associated with a moving system with strong convergence (3×10^{-3} to 8×10^{-4} sec^{-1}), and (c) the heavy rainfall concentrated in a zonal area with a width of 2–3 km north of the front (Nakayama, 1960).

An extraordinary world record of daily rainfall, 1,109.2 mm, was observed at Saigo, Nagasaki Prefecture, Japan, during 24 hr on July 25–26, 1957. The distribution of the daily rainfall is shown in Fig. 5.8 (b), and the hourly rainfall in Fig. 5.7. Syono et al. (1959) pointed out that the heavy rainfall occurred where strong wind convergence was observed on a synoptic scale, which moved eastward.

According to a study by Ozaki (1973), a stationary convergence area on a small scale was located over Chijiwa Bay when it was in the warm sector of cyclones. He drew streamlines at the ground level and 100 and 300 m height for the SW-wind and calculated the amount of convergence as shown in Fig. 5.8 (d) for the case with 10 m/sec wind speed. An example of the streamlines at the ground level is shown in Fig. 5.8 (c). The distributions of hourly rainfall amount shown in Fig. 5.7 correspond roughly to the strong convergence areas shown in Fig. 5.8 (d). The convergence area might be caused in the topographical situation of this region. Radar echoes, which tend to be generated in a topographi-

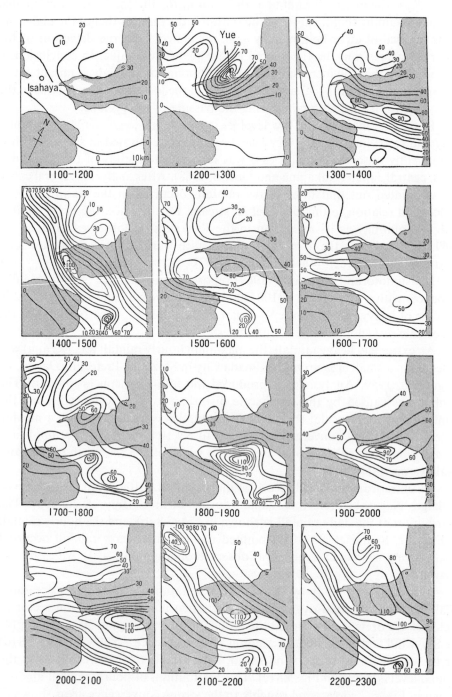

Fig. 5.7 Hourly amount of an extraordinarily heavy rainfall (in mm) in the Isahaya region on July 25, 1957 (Ozaki, 1973).

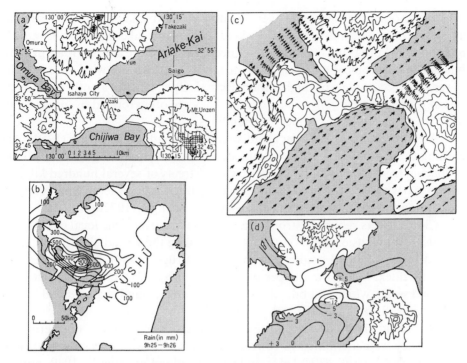

Fig. 5.8   (a) Geographical sketch of the Isahaya region. (b) Total rainfall in mm for 24 hr from 9h on July 25 to 9h on July 26 (Syôno et al., 1959). (c) The estimated streamlines at the ground level for the SW-wind (Ozaki, 1973). (d) The distribution of convergence ($\times 10^{-3}$ sec-4 km^{-2}) (Ozaki, 1973).

cally fixed location, develope fully during the time traveling about 30 km and further 30 km leeward, heavy rainfall occurs. This means the heavy rainfall areas appear normally in a fixed location.

There was a heavy rainfall in the Uetsu region on the coast of the Japan Sea during August 26–29, 1967. The region with daily rainfall more than 300 mm reached an area of 60 km $\times$ 50 km, and its center received as much as 700 mm rainfall during 24 hr from August 28 to 29. During these hours peaks of heavy rainfall were found. It was found that these peaks were brought about by a rainfall cell, which had a diameter of 20–40 km and a lifetime of 5–8 hr (Shinohara, 1970). Local cyclones are discussed also in Section 5.2.2. The band structures and the local convergences have close relations to them as discussed later.

(C) *Band Structure*

When cumulus clouds are arranged in files oriented approximately parallel to the prevailing upper winds, they are called *cloud streets*. The cloud streets sometimes extend in parallel bands across the whole visible sky. In some cases, like a weak lee wave of a mountain range, a bank of cumulus clouds develops in parallel with the running direction of the mountain range but at right angles to the prevailing upper winds. This is not a case of cloud streets. Individual cumulus streets sometimes originate over and downwind from a heat source, such as an industrial plant, a fire, a heated island, or a warm lake, but they have little relation to the orography. Cloud streets occur in temperate and arctic air masses, but not in subtropical air masses. With rare exceptions the wind speed in the lower layer is considerably higher than normal. Open lanes in an otherwise over-cast sky are termed *negative cloud streets*. They are often formed by air descending over a cold river in warmer surroundings.

A comprehensive study on cloud streets was made by Küettner (1959). According to his summary, typical cloud streets are 50 km in length and 5–10 km in spacing from axis to axis. However, cloud streets of several hundred kilometers have been reported. The lifetime of cloud streets is an order of 1 hr. A review concerning the cloud streets was also given by Stringer (1972b).

Cloud streets require a strong gradient wind to overcome local circulation, a vertical temperature profile equal to or exceeding the wet adiabatic, and inter-calated layers of small gradient or inversion (Prügel, 1949). Cloud streets are vertically thick. When their upper part is limited by an inversion, it becomes a flat cloud series, which is subdivided into cloud series, cloud chains, cloud bands, and cloud rolls or waves (Prügel, 1950).

Rodewald (1950) described the cloud streets over Holstein, northern Germany. They were attributed to convergence of sea breezes over a warm penin-sula. The cloud streets extended from Hary to Jutland on June 21, 1949, causing rainfall of over 5 mm. In the second case on July 23, 1949, a 20 mm rainfall with thunder and squall was observed in the areas northwest of Hamburg. As a showery type rainfall occurs sometimes along cloud streets, the distribution pattern of rainfall tends to show a stripe, a form called *rainfall stripes*. In other words, this is a trace of cloud streets. A case observed in southwest Ger-many on September 18–19, 1946, had a belt of rainfall with 5–15 km wide and 50–130 km apart (Schirmer, 1951). A map of the daily rainfall amount for the summer months in lower and middle Franken made clear a band-like pattern of the maximum and minimum zones (Schirmer, 1952). This is because of the showery type of rainfall in this area in summer. The width of the band was 5–15 km, and the running direction of the band has no definite relation to the topo-graphy. An investigation on the rainfall belts in the Tôkai district, central Japan, showed that the axes of the belts run on the coastal plains in parallel with a mountain range in 70% of the total cases (Suzuki, 1956).

The cloud streets in the winter monsoon region in N-Japan, named *snow-fall band* by Higuchi (1963), are observed frequently in the northern part of the Japan Sea and in the lower Ishikari plain, Hokkaido. The formation of such cloud streets could be explained as due to the convection in heated air flows with restoring force by the wind shear. The temperature difference between the air and sea surface is 10–25°C, which is the normal condition in the northern Japan Sea in winter. Magono and Yamazaki (1965) studied these conditions experi-mentally, and Magono (1971) made clear the localization of the cloud streets in the lower Ishikari plain, Hokkaido. This will be mentioned below in a discussion of the *Ishikari front* in Section 5.2.3 (F).

Okabayashi (1972) reviewed thoroughly the cloud patterns in East Asia through an interpretation of meteorological satellite pictures. He defined two major cloud patterns in the outbreaks of the winter monsoon around Japan. (a) One is cloud streaks, which might correspond to the cloud streets discussed above. The streaks have an order of 10–60 km in width and the space between streaks is about 25 km on the average. At least several streaks are parallel. (b) Second is a thick cloud zone called a convergence band cloud, which has a width of 50–200 km and a length of 500–1,000 km, in some cases 1,500 km. This cloud coexists mostly with the streak clouds, but it is not discussed here because of its scale.

The cloud streak extends gradually leeward and increases in height and width. If the height becomes more than 1,000 m and the width several to 10 km, snow begins to fall. The intensity of snow fall increases with an increase of distance from the place where the streak formed initially. Cloud streaks develop densely—the number of streaks per unit area is greater—over seas where the temperature difference between the air and the sea is small. The streak is thicker and the space between the streaks is wider over seas where the temperature difference is greater.

Of interest is the band structure of rainfall occurring in the case of the

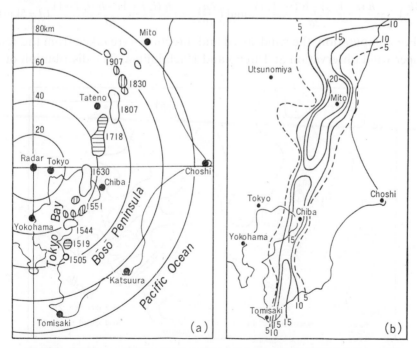

Fig. 5.9 (a) Time variation of radar echo on August 5, 1957. (b) Total amount of rainfall in mm from 9h to 24h on August 5, 1957 (Yanagisawa, 1961).

moist, warm air inflow in association with a tropical cyclone. Probably the first description of spiral bands from radar data on hurricane and typhoon circulation was given by Wexler (1947). The spiral bands are also called radar rain bands or radar bands. An analysis was made of the formation and the features of stationary rain bands that had appeared in the Kanto district (Yanagisawa, 1961). An example is shown in Fig. 5.9. Small elementary cells with a height of 4 km were generated at 15 h 05 m (JST), and their time variation is given in Fig. 5.9 (a). These cells developed slightly on reaching the western coastline near Chiba and at 17 h 18 m apparently developed thunder and lightening. The maximum height of echoes observed by RHI was about 10 km. The total amount of rainfall, as given in Fig. 5.9 (b), shows a striking band structure. A precipitation rate of 10–20 mm/hr was recorded between 11 h–14 h at Mito.

The formation of small elementary cells requires some trigger action, in most cases, by the topography. In order to examine the effect of topography, the distribution of updraft by topography was calculated. Five km mesh was adopted to compute the topographical updraft for the formation of small elementary cells. Updraft (ω) is described approximately by the following equation:

$$\omega = -\rho g \left(u \frac{\partial h}{\partial x} + v \frac{\partial h}{\partial y} \right) \tag{5.5}$$

where ρ is the air density, and g is the gravitational acceleration,

$$\left(\frac{\partial h}{\partial x} \right)_{i,j} = \frac{h(i+1,j) - h(i-1,j)}{2d}, \text{ and } \left(\frac{\partial h}{\partial y} \right)_{i,j} = \frac{h(i,j+1) - h(i,j-1)}{2d} \quad (d=5\text{km})$$

For computation, the wind speed and direction were taken from the results of upper wind observation at Tateno and Hachijôjima. The distribution of ω is

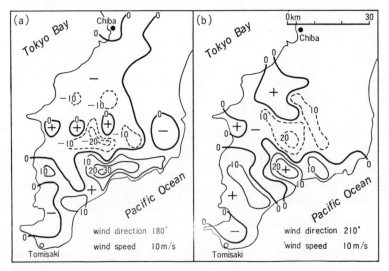

Fig. 5.10 Distribution on vertical velocity (ω) in cm/sec in the Bôsô Peninsula. Prevailing wind direction is (a) 180° and (b) 210° (Yanagisawa, 1961).

shown in Fig. 5.10. A comparison of the ω distribution and echo-generating points suggested that these points were distributed on the lee of the stronger updraft area in the ω distribution. A precise comparison of the two maps indicates that the generating points were distributed from 20 to 50 km on the lee of the updraft area, and the time lag was estimated to be about 20 min with $V=$ 10 m/sec. Because the minimum detectable precipitation rate of the radar was about 1 mm/hr at the radar range of 100 km, it is estimated that the radar reflectivity (Z) of small elementary cells was about 200 mm$^6 \cdot$m^{-3}. Furthermore, the radar reflectivity of cumulus clouds was found to vary between 10^{-3} mm$^6 \cdot$m^{-3} and 10^{-1} mm$^6 \cdot$m^{-3}, and it is assumed that these time lags were due to the growth rate of the cumulus clouds. Namely, after small nuclei entered due to topographic updraft, cumulus clouds were pushed to the lee by the upper wind before radar detection.

A significant radar rainband of a typhoon was analyzed with radar and surface observations (Tatehira, 1961, 1962). The general features are as follows: (a) The rain band constituting the eye wall is not always identical, but also is replaced by a new one about twice an hour. (b) The rain band of a typhoon is accompanied by a meso-scale zone of low pressure along its frontal edge, and surface winds seem to be confluent to this low pressure zone. The rain band is also accompanied by a strong convergence zone (-60×10^{-5} sec^{-1}), as calculated from surface winds, and by a marked increase of precipitation intensity.

Sekiguti (1965) pointed out that the orographic rain bands appear in several parts of central Japan under the influence of typhoon situations. Yaji (1969) found 6 types of rain band that extend from the SE slopes of the Kii mountains toward the NE in the case of heavy rain days.

(D) *Relation to Local Convergences, Fronts, or Instability Lines*

The problems of the local airstream and the distribution of precipitation are related more or less to the local convergence or local front. Heavy rainfall concentrated in a zonal, small area, cloud streets, snow bands, and radar rain bands are examples of phenomena closely related to the local convergence as discussed in the preceeding pages. The local front or discontinuous lines in meso- and micro-scale will be discussed later.

The instability line, the *sumatras*, associated with the southerly winds brings heavy rainfalls along the Kuala Langat coast and also frequently along the S-facing slopes in and around Kuala Lumpur and Tanjong Malim in the north, the Malaysian State of Selangor on the west coast of the Malay Peninsula, in July (Sien, 1968). In the early morning, lines of cumulonimbus clouds along the *sumatras* over the Straits of Malacca develop during the southwest monsoon season, from about May to September. The cloud system moves across the west coast of the peninsula producing squally conditions and often intense rainfall (Sien, 1970). These heavy coastal showers in the early morning hours are distributed in the belts lying east and west. Spells of the *sumatras* may affect the Island of Singapore 3 or 4 days consecutively.

In Singapore, a study of rainfall distribution for 30 min was carried out in 1953, using 32 rain gauges over an area of 435 km² (Watts, 1955). A similar observation was made over Hong Kong, distributing 21 rain gauges over an area of 435 km² with the following conclusion (Watts, 1959): Conditions of general rain over a small area occur in some tropical regions, but in others the rain is invariably scattered and occurs in individual rainstorms of short life and little movement. Watts suggested that the difference could depend on latitude or on the proximity of large land masses, but in the cases studied, Hong Kong and Singapore, topography within the small area appears to be unimportant.

It was found in Singapore that many storms were complex with several centers of maximum intensity and merging isohyetal patterns. The diameter of storms is 10 km or less, though a few were as much as 18 km across. The duration of storms is shorter than 1 hr in most cases.

Linear isohyetal patterns, i.e., rain bands, were not seen, even though a continuous line of convergence aloft is frequently accompanied by separate small intense rainstorms. *Sumatras*, known as a local line-squall, do not play any role in forming the band structure of rainfall distribution.

Thunderstorms in the Natal interior differ both in frequency and time of occurrence from those at the coast in the E part of South Africa. Preston-Whyte (1971) presented a schematic diagram of thunderstorm formation associated with the instability line in Natal. It is suggested that these storms may reach the coast if regenerated by the in-phase occurrence of a passing coastal low. Tyson (1968, 1969a, b) described the wind systems that are induced by the topography of Natal. The *Umzansi*, the local name of a plain-mountain wind, is one of the wind components that contribute to the formation of thunderstorms along the instability line in the Natal interior.

A statistical survey of the effect of the Scandinavian mountains on the precipitation by fronts was made by Rossi (1948). For instance, dependence of precipitation on the moving direction of the front was described as follows: in the case of WSW, the precipitation is the greatest on the east side of the mountains, because the air masses ahead of the front arrive in Finland not coming over the mountains but turning on the southern edge of the mountains. In the case of W, the lower parts of the mountain influence the distribution of precipitation; that is, on the eastern side of the mountains in Sweden, considerably greater precipitation is to be found in the area where the elevation of the mountain is least. In the case of WNW and NW, the air mass flowing after the front has a greater influence on the frontal precipitation than the air mass ahead of the front. Summarizing for W, WNW, and NW, the precipitation distributions are similar on the windward slopes but different on the leeward slopes. It was concluded that the more the direction of the arrival of the front turns to the north, the weaker the precipitation is in Sweden and Finland.

Tang (1963) analyzed the durnal variation of weather in the Nan Shan region (96°E–104°E, 36°N–41°N, WNW of Lanchow), China, in association with the diurnal change of the position of the local convergence. Of interest is the precipitation distributions in the daytime and at nighttime; the former concentrate on the ridge where the valley breezes and the up-slope winds converge and the latter in the valley where the mountain breezes and the down-slope winds converge.

(E) *Topographic Effects*

It can be said that a world-famous example of the topographical effect on the contrast of rainfall distributions between two different airstreams, monsoons, is from the island of Ceylon. As the east-west distance of the island of Ceylon is about 220 km, the total area 65,610 km², and the highest peak 2,526 m a.s.l., the phenomena might be considered to be mesoscale. The SW-monsoon prevails from mid-May to the end of September, and the NE-monsoon from December to February. During each monsoon period rainfall is greatest on the windward slopes of the Central Highlands and least on the lee side, where föhn effects dominate (Domrös, 1971). As shown in Fig. 5.11, the concentration of the highest rainfall and the contrast between the windward and the leeward sides are more conspicuous in the SW-monsoon period. On the leeward slopes, the *kachchan*, föhn wind, develops in this period.

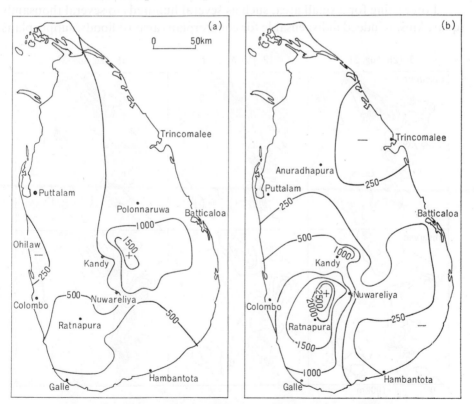

Fig. 5.11 Rainfall distribution in mm over Ceylon, (a) for the period of the NE-monsoon and (b) the SW-monsoon (Domrös, 1971).

Another good example of the relation between topography, clouds, and precipitation under relatively simple airstream conditions is the phenomena on the Hawaii Islands. In this region, the trade winds and the winds originating in the daily secondary circulations blow regularly. The interaction between these two wind systems are described elsewhere. Here it is pointed out that the annual rainfall amount reaches 5,000 mm or more at heights of 900–1,200 m a.s.l. on the northeast slope of Mt. Kohala, but it is only 250 mm on the summit of the high volcanos and 125 mm on the coast WSW of Mt. Kohala (Leopold, 1951). These distributions are brought about by the topographical situation to the trade wind direction.

Above the saddle between Mouna Kea and Mouna Loa, both having a height of about 4,200 m a.s.l., a sea breeze front from the west develops quasi-stationarily for 5 to 8 hr. Beneath the sea breeze front, calm is observed on the ground level, but the wind force is 6–7 at a distance of less than 4.8 km within the trade wind region, and the wind force is 4–5 within the sea breeze region (Henning and Henning, 1967). Undoubtedly, these circumstances reflect on the distribution of rainfall, although some seasonal change in the trade wind might exist.

Forecasting for a small area, such as several hundreds to several thousands square kms, is indeed indispensable for counterplans against floods, supply plans

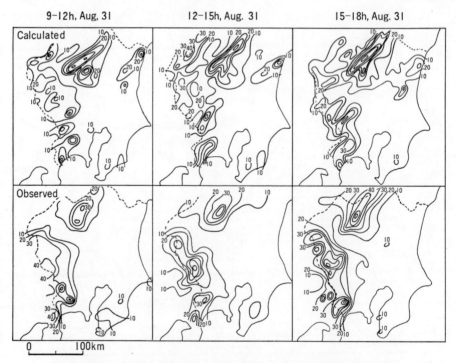

Fig. 5.12 Distributions of calculated and observed 3-hour rainfall on August 31, 1949 (Ishihara et al., 1957).

for hydroelectricity, and plans for agricultural irrigation. As the first step in establishing a method of small-scale quantitative rainfall forecasting, the effect of topography on rainfall was investigated (Ishihara et al., 1957). The atmosphere was divided into two layers—one above 850 mb, the other below it. Vertical velocities for each layer were computed assuming the dynamical vertical motion model of synoptic pattern for the upper layer with a mesh size of 300 km and the topographical vertical motion model for the lower layer with a mesh size of 10 km. By multiplying the water vapour condensation function by these vertical velocities, the amounts of rainfall for each layer were obtained, and by adding them, the total rainfall amount observed at the surface was obtained. The method was applied to Typhoon Kitty which passed the Kanto district from August 31 to September 1, 1949. The results, as shown in Fig. 5.12, were satisfactory in the distribution, but as the correlation coefficient of calculated and observed values of 24-hr rainfall amount was 0.63, it is not satisfactory quantitatively. In order to get better results, the effects of friction and isallobaric wind must be considered.

The contrast between the windward and leeward slopes is not sharp, but is still apparent on the mountains in middle latitudes. An example on the NW-SE cross section of the Rhön region in Germany shows greater precipitation amount on the upper part of the windward slope under the condition of a NW air current (Bleibaum, 1953).

It can be concluded that precipitation maxima is found on the windward slope as mentioned above if considered mesoclimatologically. It is, however, not always true microclimatologically. Kreutz (1952) reported the microclimatological conditions of airstream and precipitation in the upper Vogelsberg, Germany. The relationships between air currents and precipitation distribution were studied from June to November, 1959, at five points—Point A on a gently sloped mountain top, Points B and C 40 m and 80 m, respectively, below the top on the windward slope of the mountain, Point D 25 m below the top on the lee side slope, and Point E 40 m below the top on the slope parallel to the prevailing air

Table 5.2 Ratio (%) between the rainfall at the top of the mountain and the points on the slopes (Masatsuka, 1960).

Point	Topography	Rainfall ratio	
		a) Meteorological rainfall	b) Hydrological rainfall
A	Mountain-top	100 %	100 %
B	40 m below the top, windward slope	76	106
C	80 m below the top, windward slope	78	102
D	40 m below the top, leeside slope	112	103
E	40 m below the top, on slope parallel to the prevailing currents	103	99

currents—on Mt. Hachibuse-Takabotchi in Nagano Prefecture (Masatsuka, 1960). The ratios between the rainfall at the top of the mountain and those of the other points are listed in Table 5.2. According to the table, the rainfall measured using a normal rain gauge (called the *meteorological rainfall*) shows that the windward slope receives about 20% less rainfall and the leeward slope about 10% more rainfall than the top of the mountain. On the other hand, the rainfall measured using a special rain gauge for slopes (called the *hydrological rainfall*) shows that both the windward and leeward slopes receive more rainfall than the top of the mountain. No great difference in rainfall was seen between on the slope parallel to the currents and at the top of the mountain.

Observations of this kind were also carried out at Mt. Waita (1,500 m a.s.l.), a part of the Kujû mountain region in the middle of Kyushu. The detailed observation made around the top of this conical mountain using the special rain gauge showed that the leeward slope received more rain (Takeda et al., 1960; Sakanoue, 1969).

Although various observation methods can be used for the survey of precipitation distribution in a small area, whatever method is used, the distribution shows a close relationship with the velocity and direction of prevailing winds, particularly with the latter. In other words, the amount of precipitation at a certain place on the slopes is decided by its position to the prevailing winds. The same can be said about the relationship between the distribution of snowfall and topography.

In the case of a showery rainfall, the relation between the air current and the precipitation distribution on a small scale is very difficult to understand either quantitatively or qualitatively. As was shown in a study on the detailed distribution of rainfall in the island of Singapore (Watts, 1955), half-hourly isohyets showed a localized, short life, and a high-intensity rainfall, and there were observed no definite rules. Airstream precipitation was defined as showery rainfall resulting from convection within quasi-geostrophic winds and an analysis showed that in the W- and SW-Scottish Highlands, the role of the mountains is considered to be one of intensification and prolongation of precipitation (Smithson, 1970). The distance from the coast in relation to the prevailing winds is an important factor in showers.

A localized heavy rainfall amounting to nearly 500 mm/day was recorded in the region north of Osaka Bay on August 30, 1960 (Nakajima and Yoshioka, 1971). The hourly precipitation was 70 mm in the center of the rainfall region. The axis of the rainfall zone, 100 km long and 20–30 km wide, run from Osaka Bay to Lake Biwa, caused topographically by a moist current invasion through the Kii Channel from the SW or SSW.

(F) *Weather Divides*

The boundary line dividing two climatic provinces has been drawn in most cases on the watersheds of the main mountain chains on a macro- or meso-scale. However, this is not always correct from the viewpoint of local climatology in

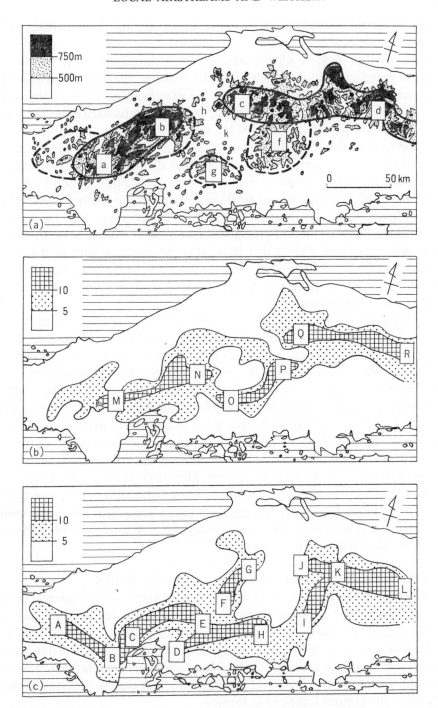

Fig. 5.13 (a) Topography of Chugoku district, W-Japan. (b) Density (frequency per unit area) of the weather divide, the southern limit of the precipitation area in the case of the NW-monsoon in winter. (c) Density of weather divide, the southern limit of the cloudy area under the same conditions (Shitara, 1958).

regions where the prevailing winds in one season are apparently stronger than in the other seasons.

The weather divide is clearly observed in such regions in the season with stronger prevailing winds, which shift the weather divide leeward from the mountain ridge. It was pointed out that the precipitation region of the winter monsoon invades along the valleys to the point 50 km leeward from the water-shed of the main mountain chain of Honshu, Japan (Fukui, 1966).

The weather divide in W-Japan during the NW-monsoon in winter was studied in detail by Shitara (1958). The weather divide frequency per unit area was calculated, and distribution maps were presented. The topography and the distributions of the frequency of weather divide and the southern limit of the precipitation area and the cloudy area are shown in Figs. 5.13 (a)–(c). It is clear that the southern limit of precipitation area M–N is situated on the leeside of the mountains a–b, and further leeward is the southern limit of the cloudy area A–B–C–E. The difference of the position between the mountains c–d and the weather divide Q–R as well as J–L is small, probably due to the running direction of the mountains.

During the *Bai-u* season, a cold and humid easterly wind called *Yamase* prevails over the Pacific side of N-Japan. The distributions of occurrence frequency of cloudy and rainy weather in the Tohoku district under the synoptic conditions with *Yamase* showed a striking influence of the mountains higher than 750 m a.s.l. in the northern part of the Tohoku district, but in the southern part the influence was shown only by mountains higher than 1,000 m a.s.l. (Shitara, 1967, 1969). A similar study was made of the winter weather distribution (Shitara, 1966). In a smaller region, for example, the Sanbongi Plain, in the NE-part of Tohoku district, Shitara (1963) examined the discontinuity of weather produced by two different local airstreams. This was mentioned in Section 3.4.2.

In summary, the influence of topography on the weather divide varies from region to region in accordance mainly with the running direction of the mountain ranges or valleys to the prevailing winds, the average height of the mountains and the intensity of prevailing winds.

5.1.3. Local Air Currents and Temperature Distribution

In a small area where the overall ground condition is looked upon as homogeneous, the greater the wind velocity is, the more the temperature near the ground decreases in the daytime and increases at night. If the wind velocity near the ground is influenced by micro-topography and shows local differences, the temperature also has a tendency to show local differences in relation to the local difference in the wind velocity. This phenomenon is clearly observed particularly when the ground inversion is formed.

One of the most interesting local climatological phenomena in the Antarctica is the *dry valleys*. The *McMurdo Oasis*, located, in the southwestern edge of the Ross Sea (77°10'S–77°45'S, 160°20'E–163°00'E), has an area of 2,500 km².

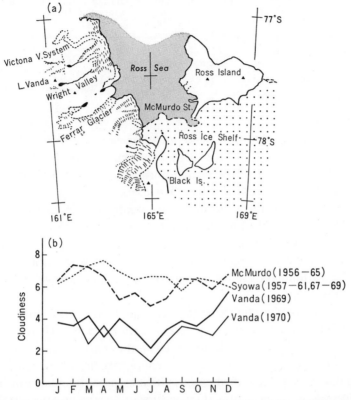

Fig. 5.14 (a) Geographical location of Wright Valley and (b) monthly mean cloudiness in tenth (Yoshida and Moriwaki, 1972).

The dry valleys with no glaciers are found in this region. Wright Valley is one of these dry valleys as shown in Fig. 5.14. In the bottom (2–3 km wide) of this U-shaped valley, there is no glacier from the Wright Lower Gracier to the 50 km upper part of the valley. The characteristic climatic condition of Wright Valley is a low cloudiness and a small number of days with snowfall. The prevailing winds are westerly or easterly. The westerly winds are generally stronger, drier, and warmer than the easterly in summer, and this tendency is also detectable in winter. This westerly wind is probably not a simple katabatic wind; it may relate to a large extent to the disturbances on a synoptic scale (Yoshida and Moriwaki, 1972). The air temperature in Wright Valley is higher than that of McMurdo Station in summer, but lower in winter. Therefore, it is my opinion that these dry conditions might be caused by a local circulation that brings up-slope winds on the side slopes and down currents in the center of the valley in summer, but further study is needed.

As mentioned above, the cold NE-winds, the *Yamase*, invade on the Pacific coast of N-Japan in early summer. These winds blow under the synoptic condi-

Table 5.3 Daily mean air temperature (°C) during *Yamase* days and
frequency (days) of *Yamase* in northeastern Honshu,
Japan, 1954 (Asai and Nishizawa, 1959).

	May		June		July		August	
	Temp.	Freq.	Temp.	Freq.	Temp.	Freq.	Temp.	Freq.
Yamase	12.2°C	16days	11.2	28	15.8	28	21.4	24
Without= *Yamase*	12.9	15	16.7	2	21.8	3	23.3	7

tion with the polar maritime air mass, an anticyclone, over the Sea of Okhotsk.
The air temperature on the coast of the Tohoku district, N-Japan, during the
Yamase days in July is 6°C lower than normal as shown in Table 5.3. The interdiurnal air temperature during the *Yamase* is 0.18°C, in contrast to 0.60°C
during the normal weather period. If the weather changes from the normal
weather to the *Yamase*, the air temperature becomes an average of 2.82°C lower
(Asai and Nishizawa, 1959).

Flemming (1966a) stressed the different effects of wind structure and turbulence on the air temperature on a valley floor for the cross-wind and parallel-
wind conditions.

The statistics of climatic elements according to synoptic patterns or prevailing wind conditions are recorded at many stations today. These are especially
important at stations in middle latitudes or on the coasts, because the conditions
at these stations depend strongly on the characteristics of air masses, whether
they are polar or tropical and whether maritime or continental. For instance, at
the station in Brno, Czechoslovakia, the mean air temperature in the case of
N-winds is 4.9°C, but that in the case of S-winds is 18.9°C in spring. It is
4.9°C and 17.0°C, respectively, in autumn (Nosek, 1964). Such a great difference
can not be disregarded in considering the climate at certain stations.

Lowry (1963) studied the atmospheric structure in the 11 km wide Valsetz
Basin in the Oregon Coast Range, U.S.A., for the three types of days in summer.
The days of Type A are warm, clear, and without a sea breeze; Type B days are
cool, cloudy, and without a sea breeze; and Type C days are comparatively
warm and cloudless with fully developed sea breeze. He presented the diurnal
variation of the vertical distributions of potential temperature and mixing ratio
in the basin and made clear the observable differences between the types.

The change of air temperature distribution caused by passing fronts in a
small area is also of interest. After a cold front has passed, the air temperature
decreases slowly on the coast (Yoshino, 1953) and in the city center (Eriksen,
1966), which resembles the night conditions. After a warm front has passed, the
air temperature increases rapidly at inland and suburban stations.

Thompson (1973) classified the distribution patterns of air temperature in
the Armidale district (30°31′S, 151°36′E), New South Wales, according to the
synoptic pressure patterns and air flow directions. He showed that the temperature variations are influenced by surface heat accumulation and cold air drainage

and the difference of slope directions under the conditions of maritime or continental anticyclonic synoptic patterns.

5.2. LOCAL ANTICYCLONES, CYCLONES, AND DISCONTINUOUS LINES

5.2.1. Local Anticyclones

It is very difficult to define the size of a local ancityclone. It can be said, however, that the local anticyclone has a diameter of more than 100 km and is apparently smaller than a migratory anticyclone on a synoptic scale. Its lifetime may range from a half day to 2 or 3 days.

The local anticyclone is formed mechanically, i.e., by orographical influences, and thermally. In actual cases, it is not easy to separate the causes of formation, because the local anticyclone develops strikingly in a cold mountainous region surrounded by warm sea regions. A good example of a local anticyclone is the one that appears in the mountainous region of central Japan

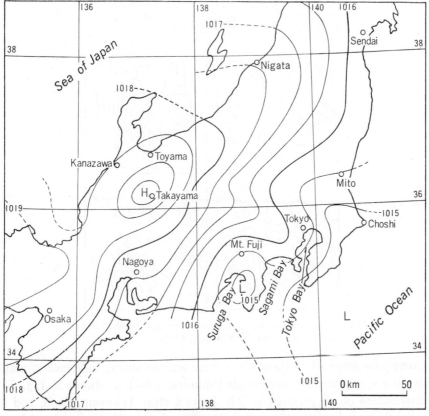

Fig. 5.15 Distribution of monthly mean air pressure in central Japan reduced to the ground level in February, 1931–1960 (Yoshino).

in winter. As its center is located near Takayama City, it is named the *Taka-yama high*. The occurrence of a Takayama high is so frequent (12.3 days/ month as an average from December to February) that it appears also in a monthly mean map, as shown in Fig. 5.15.

The first case study of the formation of Takayama high was made by Ooi (1951) on January 20, 1941. He concluded that the air layer near the surface over the mountainous region in central Japan is cooled by radiation on clear nights, and the Takayama high is formed. The cold air flows down along the large valleys and produces a local discontinuous line on the coast, which will be discussed below.

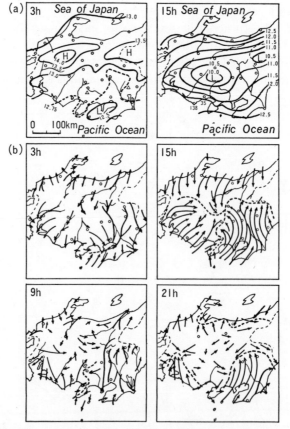

Fig. 5.16   Composite maps of (a) air pressure at 3 and 15 and
(b) winds at 3, 9, 15 and 21h (Shimizu, 1964).

Composite maps of 19 days when there was no strong influence of action centers on a synoptic scale were made (Shimizu, 1964), as shown in Fig. 5.16. The distribution of air pressure at 3 h shows a clear Takayama high, and the winds are flowing down in the valley. On the other hand, a clear local low is seen at 15 h in the afternoon, and the sea breezes and valley winds are blowing.

Utagawa (1964) pointed out that the dynamic effect of the mountains in central Japan plays an important role in forming the Takayama high.

Čadež (1963) showed that the thermodynamic cold anticyclone is a consequence of the transport of internal energy in the cooling region from the surrounding atmosphere. Assuming that this energy transport is performed with the sound velocity, the following equation was derived:

$$\Delta p_o = \frac{\partial p_o}{\partial t} = -\frac{g\rho Z_f}{\bar{T}}\frac{d\bar{T}}{dt} \qquad (5.6)$$

where p_o is air pressure on the ground (height$=0$), ρ is air density, T is air temperature, Z_f is height of the air layer near the ground to be cooled. If the values $p_o=1000$ mb, $Z_f=100$ m, $T=275°$K and $\overline{\Delta T}=-10°$C are used as an example, the result will be $\Delta p_o=4.6$ mb. This value means that the direct influence of cooling on the air pressure at the ground is observable.

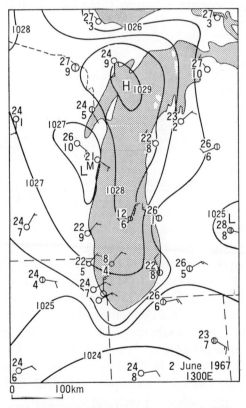

Fig. 5.17 Meso-scale weather in the Lake Michigan area on June 2, 1967, at 13h00m (EST). Isobars are in millibars. Air temperature (upper) and dew point temperature (lower) in °C appear to the left of station (Strong, 1972).

The cold waters of Lake Michigan have an interesting effect on warm air flowing over them. The inversion layer has an average thickness of 1 km. Coupled with this inversion a shallow meso-scale high pressure, the lake anticyclone, often develops (Strong, 1972). The example of the intense lake high shown in Fig. 5.17 is 2 mb, but one intense high was estimated to be as strong as 8 mb on June 25, 1965. A direct consequence of the stable period is the absence of convective cloudiness at the lower level over the lake. If cumulus clouds are advected, the subsidence rapidly dissipates those clouds. With solar radiation as much as 20% higher over the lake than over land, the sun continues to supply thermal energy into the water surface, but low-level wind divergence and mid-lake upwelling prevent any immediate concentration of this heat in the surface waters in the central part of the lake.

5.2.2. Local Cyclones

A local cyclone can be also called a meso-cyclone. The local cyclone has a smaller size than that of a cyclone on a synoptic scale. The components of both of these types of cyclones are shown in Table 5.4, which is based mainly on a study by Saito et al. (1973).

Table 5.4 Comparison of the local cyclone on a meso-scale with the cyclone on a synoptic scale (Saito et al., 1973).

	Local cyclone	Extratropical cyclone
Scale	Meso-scale	Synoptic scale
Scope	About 100 km	Several 1,000 km
Deepness	Smaller than several mb	Several 10 mb
Moving velocity	50–100 km/hr	30–50 km/hr
Vertical velocity	Several 10 cm/sec	1 cm/hr
Intensity of rainfall	Several 10 mm/hr	Several mm/hr
Lifetime	Several hours	About 1 week

A local cyclone is smaller in diameter, but much stronger in vertical velocity, and hence the rainfall intensity is about 10 times stronger than in the case of a cyclone on a synoptic scale in middle latitudes. An exceptional circulation, clockwise circulation, is observed in a meso-cyclone smaller than 75 km in diameter. But, in normal cases, the circulation is cyclonic on a meso-scale, greater than 15 km in diameter. The lifetime of local cyclones ranges from 1 hr to several hours. Due to such characteristics, some of local cyclones cause heavy rain in a restricted area within a short period. *Shûchû-gô-u*, a severe local rain storm, which occurs frequently in SW-Japan during the rainy season, is concentrated both in space and time. Examples of Shûchû-gô-u with daily rainfall more than 200 mm were mentioned in Section 5.1.2 (B).

Sometimes, three or more such local cyclones are born simultaneously and move eastwards one after another. As the local cyclone brings heavy rain, a cycle of rainfall intensity is often observed in such a case.

The local cyclone is caused and maintained by phenomena on a synoptic scale. However, the moving velocity of the local cyclone is much faster than that

Table 5.5 Local cyclones in Japan (Yoshino).

Local name	Region	Synoptic situation	Season
Suruga-wan cyclone	Suruga Bay	Monsoon situation	Winter
Sagami-wan cyclone	Sagami Bay	Monsoon situation	Winter
(No special name)	Kanto Plain	North Pacific anticyclone	Summer
(")	"	Anticyclonic weather	Any season
Haboro-cyclone	West of Hokkaido	Monsoon situation	Winter

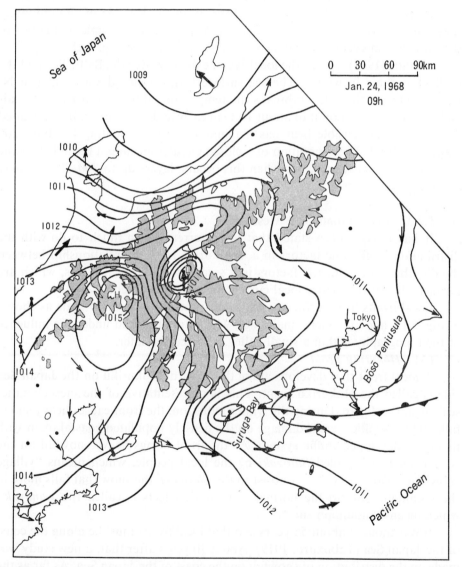

Fig. 5.18 A typical weather map of *Takayama high* and *Suruga-wan cyclone* at 9h on January 24, 1968. Thick arrow shows wind velocity more than 5 m/sec (Nomoto, not yet published).

of the parent system, and the moving behavior seems to be independent in some cases. Further study is needed on such problems.

Some local cyclones that are seen very frequently in various regions in Japan are listed in Table 5.5.

The *Suruga-wan cyclone* on the monthly mean map has been already shown in Fig. 5.15. As was shown in the case of the local anticyclone, the local cyclone is produced both thermally and mechanically. Therefore, the local cyclone is generally seen on the sea in winter and on the plain in summer even where the topography is a predominant in cause of the local cyclone. The local cyclones in winter given in Table 5.5 are always formed on the lee side of the mountain ranges to the prevailing air flows on a synoptic scale. An example of analysis of the Suruga-wan cyclone of January 27, 1967, is shown in Fig. 5.18.

Solantie (1968) studied the local cyclone formed over the Baltic Sea and the Gulf of Bothnia in winter. The sea surface forms a closed warm area in N-Europe. During the cold outbreaks the temperature difference between the air and seawater may amount to about 10°C in early and mid-winter. It is indicated that the flux of sensible heat has maximum values amounting to about 200 cal·cm^{-2}·(6 hr)$^{-1}$. The generation of a closed circulation cannot begin before the upper flow has weakened, but when this occurs, layers up to 800–700 mb may warm up over the sea.

5.2.3. Local Fronts on a Meso-scale

It has already been stated that local currents have much to do with the distribution of climatic factors in a small area. Here the relationship between the distribution of climatic factors in a small area and local fronts, which are formed when two or more local currents meet, is discussed.

Local fronts include various scales of fronts, such as meso-scale as shown in Fig. 5.19, and extremely small ones, which are produced under the influence of the topography within a small valley, as discussed later.

(A) *Hokuriku Front*

A local front that develops along the coast of Hokuriku on the Japan Sea side is called the Hokuriku front. This is representative of the mesoclimatic-scale fronts in Japan. During the winter monsoon, the front causes heavy snow-fall, and thus inflicts great damage on the densely populated coastal plain with its highly developed traffic system. As the front brings heavy snowfall on the coastal plain it is called *Satoyuki* by the local people, which means "village snow" or "plain snow." In contrast to the *Satoyuki*, the snow that falls more in the mountains due to the normal orographic effects is called the *yamayuki*, which means "mountain snow."

It was thought about 55 years ago that local fronts must lie along the coast of the Japan Sea (Tabukuro, 1919). About 10 years after that, a new study was made on the distribution of snowfall on the coast of the Japan Sea. As far as the snowfall due to the westerlies is concerned, *satoyuki* occurs where the coastline

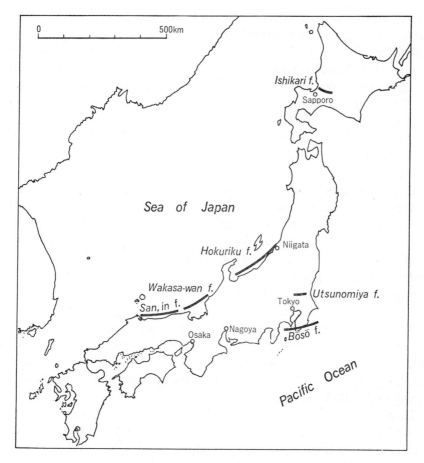

Fig. 5.19 Location of the local fronts in Japan (Yoshino).

runs E–W and *yamayuki* where the coastline runs N–S (Katsuya, 1928). Many studies have been made on this problem. There were several views concerning how the front is produced. One opinion was that the front develops between the cold air current that flows down from the local high located around high mountains in central Japan at night and the damp warm air current over the Sea of Japan. Another opinion was that the front is produced when S air currents, back currents of the NW air currents, which are cooled while blowing down along the slopes of mountain in central Japan meet with the NW air currents which come over the Sea of Japan. A third opinion was that the front is produced by currents from three different directions, such as N or NW, S, and W. This three-current theory seems to be the most reasonable. The three currents have the following characteristics: N or NW air currents are the fresh polar continental air mass, which moves southeastward at a comparatively high speed from the northern part of the Sea of Japan. It has high humidity because enough water vapour supply occurs over the Sea of Japan, and hence it has an absolutely unstable lower layer. S air currents with a thin layer of 200–300 m are characterized by low temperature and high humidity cooled by radiation while remaining inland over central Japan. W air currents are changed in quality while flowing over the southern part of the Sea of Japan. They have characteristics of a rather high temperature and high humidity and show unstable conditions in the middle layer. Both S and W air currents are the modified airmasses when they are blowing over the Sea of Japan and staying over the Japanese Islands. The southerly winds in winter in the Hokuriku district were also studied by Nakamura (1966).

A comprehensive study was made on the mechanism of heavy snowfall concentrated on a narrow area on the Japan Sea coast (Fukuda, 1965) with the following conclusion: A minor disturbance traveling through the area where the upper air is very is indispensable for heavy snowfall concentrated to a small area. A minor disturbance is formed along the boundary between the high-latitude high and middle-latitude high and reaches the coast of the Japan Sea, traveling over the Japan Sea. Simultaneously the jet stream intensifies over the Kyushu and Shikoku districts, and a belt of strong cyclonic vorticity is formed over the southern part of the Japan Sea. The minor disturbances travel in succession along this belt, reach the coast of the Japan Sea, and produce heavy snowfall in association with the instability of monsoon air and the orographic front along the coast. The disturbances form cumulonimbus clouds at a low level and make the local ascending current.

An unprecedentedly heavy *satoyuki* snowfall occurred in the last ten days of January, 1963, globally one of the coldest of recent years. A research project was carried out on these problems by Japanese meteorologists (Japan Meteorological Agency 1968), and the results of this project show:

Heavy snowfall is usually characterized by meso-scale distribution and, as a matter of fact, meso-scale disturbances of various kinds were detected on radar scopes, barograms, wind fields and cloud distributions. The S-winds in the lower layer over the inland areas are undoubtedly an essential component of the Hokuriku front, which is now understood to be a kind of meso-scale disturbance.

According to the results of a synoptic and dynamic analysis of the heavy snowfall on January 16, 1965 (Matsumoto et al., 1967), a cold dome, meso-scale disturbances, and the meso-scale convective system are to be important conditions. The cold dome has been recognized as a necessary condition of heavy snowfall, and a well defined stable layer in the rear part of the cold dome has a steep upslope toward the center with remarkable undulation. A family of meso-scale disturbances bringing heavy snowfall is observed in the rear part of the cold dome, and it precedes the onset of the northwesterly monsoon. The dense rawinsonde network revealed the existence of a meso-scale convective system, with a convergent and positive vorticity field in the lower layers and a divergent and negative vorticity field in the upper layers of the cold dome, in relation to the heavy snowfall. The cold dome is found generally in the upper troposphere.

Heavy snowfalls and the Hokuriku front were studied thoroughly on a meso-scale by Miyazawa (1960, 1962, 1964, 1965, 1966). A further analysis of the *satoyuki*, the heavy snowfall on the coastal plain of the Sea of Japan, was made by drawing, a surface synoptic map on a meso-scale (Miyazawa, 1968). For the period from noon, January 23, to 15 h, January 24, 1963, 3-hourly surface meso-maps are shown in Fig. 5.20 (a)–(f). The cold vortex invaded over the Japan Sea from southern Manchuria on the afternoon of January 22 and passed across the Tsugaru Strait on the morning of January 25. The chief cause of the heavy snowfall was the persistent stagnation and fluctuation of the local front, but from a synoptic point of view the superposition of the following three sources is important:

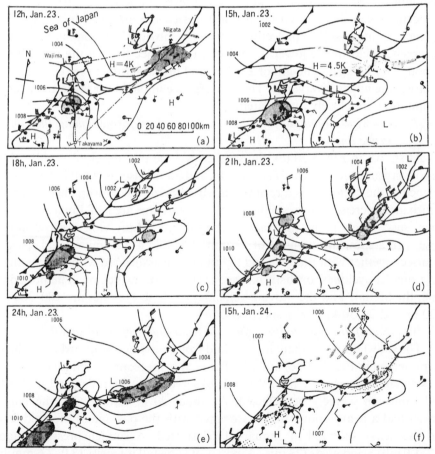

Fig. 5.20 Surface meso-scale map, January 23–24, 1963. The radar echo groups are stippled, and stippled areas enclosed with broken lines indicate 3-hour precipitation exceeding 3 mm. The area of the triangle within chain lines, for which divergence is computed, is 13,600 km² (Miyazawa, 1968).

(a)   Snowfall accompanying the convergence line, "Hokuriku front."

As shown in Fig. 5.20 (a), (b) and (f), three wind systems, southerly from inland, westerly from the San'in district to the coastline of the Hokuriku district, and northwesterly from the sea, converge to a line along the coastline and bring snowfall due to horizontal convergence. Radar echo cells in the line echo diffuse and bring snowfall when they move inland from the sea. The results of mesoanalysis are summarized as follows: Southerly warm air flows into the south side of the convergence line, and a snowfall area is recognized in advance of warm air. The smaller cyclone, whose scale is about 200 km, is accompanied by cold advection in the rear and by warm advection in advance of the front and moves towards the SE with a speed of 40 km/hr. The features of this smaller cyclone are similar to a cyclone system on a synoptic scale. The convergence line, the Hokuriku front, is likely to be formed under an intense lower convergence provided by the cold vertex coming over the warm Japan Sea surface. The maximum value of convergence at the lower level is about -2×10^{-4} sec^{-1}, and this occurs when the cold vortex has just passed the northern Japan Sea. The existence of meso-scale convective systems is obviously recognized within a convergence field in the lower layer and a divergence field in the upper layer, the boundary being at a height of 2–3 km, and the larger the lower convergence, the larger the upper divergence. The convergence in the lower layer and average precipitation are closely interrelated.

(b)   Snowfall accompanying the smaller cyclone formed along the convergence line.

As shown in Fig. 5.20 (e), the smaller cyclone that develops rapidly along the convergence line is located at Takada, and an intense snowfall is observed in advance of the cyclone. As a result of intense horizontal convergence, a smaller disturbance increases to a smaller cyclone, and the area of intense snowfall moves successively along the track of the cyclone. Many violent snow storms, those with a snowfall intensity of 22 mm/ 3 hr and a maximum wind speed of 20 m/sec together with thunder and lightning, are associated with the passing of these smaller cyclones, or disturbances, over the coastal region. The horizontal scale, the lifetime and the moving velocities of these cyclones are 100–300 km, 5–20 hr and 20–50 km/hr, respectively.

(c)   Snowfall accompanying the cold front and prefrontal squall line.

The snowfalls accompanying the cold front moving southward from the sea are presented in Fig. 5.20 (c) and (d). The cold front from the sea takes place on the stagnant convergence line. Without observed data by radar and a ship the cold front can scarcely be detected. At about 20 h, January 23, the Niigata district had a heavy snow storm associated with the passage of the cold front, and the heavy snowfall of January 1963 started. A sudden pressure change (0.7–1.0 mb), wind gusts (23 m/sec), bursts of snowfall intensity (3.0–7.6 mm/hr), and maximum surface wind convergence (-6×10^{-4} sec^{-1}) are recorded on the arrival over the Japan Sea of a few wave-shaped lines denoted as a "prefrontal squall line" preceding a cold front. The cold front and the squall line moving southward over the Japan Sea correspond to the line-shaped radar echo and move with speeds of 20–50 km/hr, but they slow down as they approach the coast of the Hokuriku district and then gain speed after passing over the Hokuriku district.

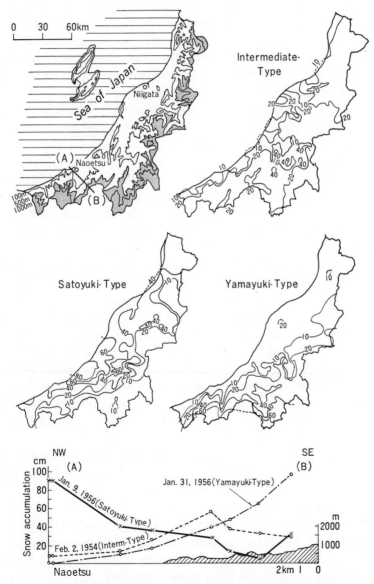

Fig. 5.21 Topography of Niigata Prefecture and examples of *satoyuki*-type (village snow or plain snow), *yamayuki*-type (mountain snow), and intermediate-type distributions of depth of daily new snow accumulation in cm and their cross sections on the line (A)–(B) (Fukaishi, 1961, 1963).

Distribution maps of snowfall were prepared for each heavy snowfall day during the winters in 1952–1953 through 1956–1957 in Niigata Prefecture (Fukaishi, 1961, 1963, 1964). According to the classification, 9% of the days with heavy snowfall were revealed to be of *satoyuki*-type: 49% of *yamayuki*-type: and 42% were intermediate. The relationship between the new snow accumulation and the topography ranging from Naoetsu City southeastward to the prefectural border of Nagano and Niigata is shown in the lower part of Fig. 5.21. The distribution of *satoyuki*-type snowfall due to the presence of the Hokuriku front shows a striking contrast to that of the *yamayuki*-type, as shown in the upper part of Fig. 5.21. This *satoyuki*-type distribution appears more frequently in January and February. *Yamayuki*-type snowfall, on the other hand, appears at the beginning of winter and rarely lasts for days, with little variation of appearance from year to year.

There are more than 300 papers and monographs concerning the Hokuriku front and the heavy snowfall due to it. Here its importance is stressed from the standpoint of local climatology.

(B) *Wakasa-wan Front*

If the NW monsoon in winter prevails over the sea near Wakasa Bay, local fronts associated with small disturbances come down to the coast one after another and bring periodic precipitation. However, the local front is stagnant when the W-wind prevails on the sea and fine weather is found on the south side of the front.

Along this local front, a bank of cumulonimbus develops, and a shower occurs. An observation made from a ship on December 11–13, 1964, recorded the following phenomena (Ino, 1965): On the northern side of the front, the air temperature was 11°C, the relative humidity 60%, and the wind velocity 10 m/sec. An air pressure jump (difference) of 1.4 mb was observed between the northern and southern sides of the front. The air pressure on the southern side, i.e., the sea region near the coast is higher and the air temperature is 2°C lower than on the northern side. The wind speed was 13–14 m/sec in the front. On the northern side of the front, no convectional clouds were observed but altostratus was seen on the southern side. The temperature of the sea surface was almost the same on both sides.

From these data, it can be supposed that the front is sustained by relatively warm air, which is a winter monsoonal flow coming over the Sea of Japan, and the cold air from the south, inland or Honshu.

(C) *San'in Front*

The San'in front develops along the coast of San'in during the period of the winter monsoon, as the Hokuriku front does. Although two types of snowfall in relation to the front were classified many years ago by Katsuya (1925 a, b, c, 1928), no detailed research has been made since then.

The two types of snowfall in this region were called *uranishi*-type and *okinishi*-type. In *uranishi*-type snowfall, the monsoon blows parallel to the coastline, and a S-wind blows on the ground level with the wind force decreasing at night. This type of snowfall brings much snow on the coast and corresponds to the *satoyuki*-type snowfall of the Hokuriku front. This condition often continues from October to January. On the other hand, the *okinishi*-type snowfall, which corresponds to the *yamayuki*-type snowfall of the Hokuriku front, is brought about by the monsoon blowing perpendicularly to the coastline and has a tendency to cause more snowfall in the mountains, with the maximum snowfall intensity around sunset.

In the case of *uranishi*-type snowfall, the intensity of snowfall becomes greatest at 6 in the morning. Therefore, it can be assumed that the San'in front also becomes active from night to dawn.

In the coastal area, when the monsoon changes direction from W, it brings about a shower-threatening sky, soon followed by such weather as thunder, snow, or hailstones called the *serai*. This phenomenon may be attributed to the development of the front. The front lies, according to fishermen, 10 to 15 km off the coast of Karo Port near Tottori City, and a land breeze develops on the coastal side of the front. It is said that the Japanese word *serai* means *seriai* or struggle in force between the land breeze and the W-monsoonal winds. The thunder, called *yukiokoshi* by the local people, may mark the beginning of the winter monsoon accompanied by the front.

(D) *Bôsô Front*

The Bôsô front and the Hokuriku front mentioned above are the two most typical meso-scale local fronts in Japan.

An investigation of the occurrence frequency of fronts on a synoptic scale in the Bôsô region has made clear a winter maximum and a summer minimum using synoptic maps (Sugawara, 1946). The number of days per month when the front remains over the southern part of the Kanto district is more than 20 every month from November to March, if such fronts as extend from the low located around Hokkaido and Chishima are included. Some of the Bôsô fronts develop thermally and others dynamically. The latter case is accompanied by a local cyclone, which occurs over Suruga Bay or Sagami Bay due to dynamic causes. Thermally developed fronts occur when the damp warm SW- or W-currents over the sea meet the cold currents flowing out from the local anticyclone that develops over the inland Chubu and Kanto districts cooled by nocturnal radiation under the conditions of fine, anticyclonic weather with the weaker winter monsoon.

A typical situation is shown in Fig. 5.18, and the average state in relation to the winds is shown in Fig. 5.22. The dry, cold fall wind, *oroshi*, of the NW-monsoon prevails on the northern side of the Bôsô front, and the wet, warm, strong W–WSW-winds prevail on the southern side of the front (Yoshino, 1970). Figure 5.22 was obtained as an average of 15 days with typical monsoon situations.

The Bôsô front may be formed in relation to some other weather pattern.

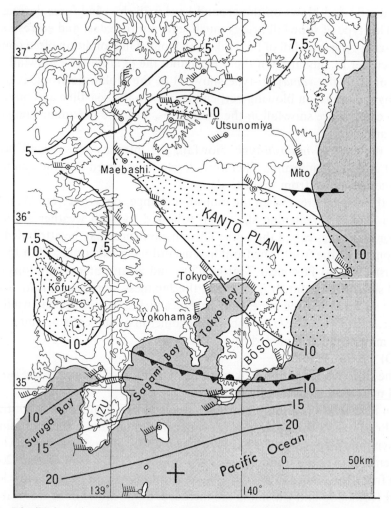

Fig. 5.22 Distribution of wind speed (m/sec) and direction during the developed monsoon in winter. On the Kanto plain, the *oroshi*, a fall wind, and on the Pacific coast, W~WSW winds are prevailing. Betweeen them, the Bôsô front is located. (Yoshino, 1970).

When Honshu is covered by a migratory anticyclone, the southern part of the Kanto district receives local rainfall due to NE-winds. The rainfall on November 14, 1939, was considered to be of this type (Takahashi, 1940). In this case warm S-currents were detected in the upper layer at 100 m above the ground, and the slope of the frontal surface was about 1/130. The gloomy, cloudy weather with strong NE-winds called *kashima* weather by the people at Chôshi City may occur on the northern side of the Bôsô front.

The front on a clear calm night, when the winter monsoon became weak, was analyzed using hourly observation data (Ooi, 1951). According to the results, the Bôsô Front is branched out to the east of a small low with its center over Sagami Bay. Rainfall was observed around the front and SW–W winds prevail on the southern side, and NW-winds on the northern side of the front. A topographical cross section drawn along the line from Naoetsu City and Takada City on the coast of the Sea of Japan down to Hiratsuka City on the coast of Sagami Bay is shown in Fig. 5.23. It illustrates the relationship between the topography and the air currents in the morning following a fine, calm night in the mountainous region in Honshu.

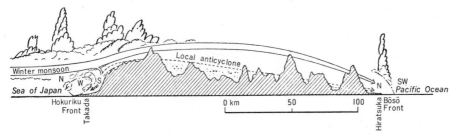

Fig. 5.23   Model of the Hokuriku front and the Bôsô front under the influence of weaker winter monsoon (Yoshino, 1961a).

When the Bôsô front develops well, it extends westward and reaches Ise Bay, south of Nagoya. Strong N-winds, which blow along the River Fuji, also present a very interesting phenomenon (Nakayama, 1953b). Another example is the N-wind *ohkawara*, which blows abruptly in the morning after a night when the W-monsoon has ceased. It is said that fishermen of Heta in Izu Peninsula are quite afraid of the wind. All these phenomena seem to be caused by the westward extension of the Bôsô Front. An investigation made on the distribution of rainfall in the Kanto plain (Kamiko, 1955) shows that there are places along the River Tone and the River Edo where rainfall ceases earlier than at other places, and, on the other hand, places to the west of the eastern part of Chiba Prefecture and the southern part of Ibaraki Prefecture with a prolonged rain period.

A distribution map showing the occurrence frequency of line echoes observed by the radar on the top of Mt. Fuji (Makino and Hitsuma, 1973) reveals clearly the zones corresponding to the Hokuriku front and the Bôsô front, as shown in Fig. 5.24. The zone, where a large line echo longer than 500 km frequently occurs, runs in a direction from WSW to ENE. The position is, however, a little far from the Bôsô Peninsula, and it is most apparent in autumn. Further investigation is needed on the relation between the local discontinuous line on the ground level detected by the wind direction, air humidity, and air temperature, and the line echo, which is closely related to the wind direction at the 600–700 mb level.

(E)   *Utsunomiya Front*

The Utsunomiya front is smaller than the above-mentioned Hokuriku and

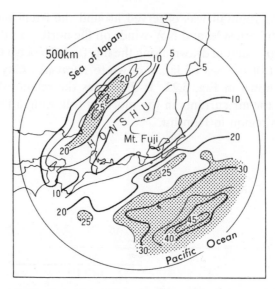

Fig. 5.24 Distribution of large line echo observed by radar on the Mt. Fuji.
Total occurrence frequency 1965–1968. (Makino and Hitsuma, 1973).

Bôsô fronts. A cyclone often develops dynamically around Utsunomiya City under the influence of the winter monsoon. The front that extends eastwards from the center of the cyclone is called the Utsunomiya front. The front sometimes develops on a clear night when the monsoon has weakened, like the Bôsô front.

A detailed analysis of the Utsunomiya front that appeared on February 3, 1946, when a typical winter synoptic pattern was formed, showed three air masses over Tochigi Prefecture (Shinohara, 1949): northern air mass, A; middle air mass, B; and southern air mass, C. Air mass A blows out onto the Nasu alluvial fan from the northern mountains and forms a northerly air current. Air mass B, which comes down across the central mountains of Honshu and the mountains in the western part of the prefecture, forms a calm area to the south of Utsunomiya City in the early morning but gradually takes the form of an air current as the sun rises. In the daytime it grows to become the current with a divergence axis running E–W around Utsunomiya City. Air mass B has a comparatively small amount of clouds and a large diurnal temperature difference and is drier than air mass A. Air mass C is the W- or SW-currents that appear in the south of the prefecture during the daytime and constitute a discontinuity of wind direction between the air masses B and C. Thermally, however, air masses B and C have little difference. Air mass C has the least clouds.

The border between air masses A and B constitutes the Utsunomiya front and characterizes the climate of Tochigi Prefecture. On the northern side of the front where winds blow down, strong winds are often observed, and the southern side undergoes a comparatively high temperature.

(F) *Ishikari Front*

In the Ishikari plain of Hokkaido a local front called the Ishikari front is formed (Kawamura, 1965). After the winter monsoon has blown hard, the wind becomes a little weak in the upper layer (850 mb). A narrow zone with a weather divide develops from Iwamizawa through Tôbetsu to Ishikari Bay. On the southern side of the divide it is cloudy due to the southerly currents, and on the northern side much snow falls due to the westerly currents from the Japan Sea. Sometimes it snows heavily along the front.

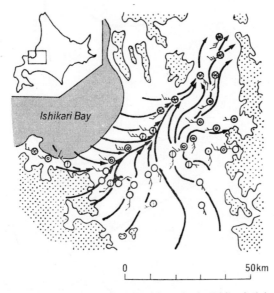

Fig. 5.25   An example of the local Ishikari front in the Ishikari plain, Hokkaido, at 12h, on February 11, 1961 (Kawamura, 1965).

An example of the Ishikari front in the Ishikari plain on February 11, 1961, is given in Fig. 5.25. Other features of the Ishikari front are apparent from the observations of the horizontal distribution of daily snowfall and the surface wind under the northwest monsoon by a meso-scale network in the Ishikari plain (Lee et al., 1972). It was found, as shown in Fig. 5.26, that (a) a few parallel cloud bands were formed over Ishikari Bay, (b) radar echo bands were observed to the lee side of each cloud band when the monsoon was prevailing, (c) southerly winds prevail along the foot of the eastern mountain region, (d) northwesterly winds blew onto the plain, (e) snow crystals generated in the clouds were transported by the prevailing NW-wind, and (f) the snowfall converged to a location near Iwamizawa or Kuriyama. Considering the results described above, it was concluded that the localization of snowfall was caused mainly by the convergence of the wind system, although the convergence was not related to the generation of snow crystals. Comparing the change of the depth of snow

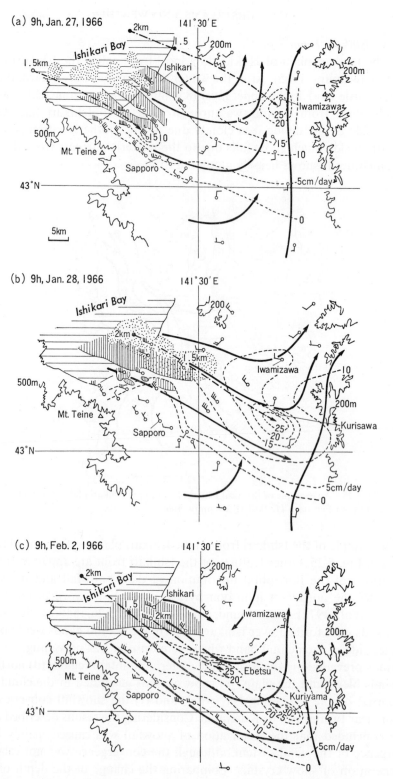

Fig. 5.26 The relation between the origins of snow crystals, the snow cover and streamlines on Ishikari plain. Dotted: Radar echo areas, Shaded: Cloud areas (Lee et al., 1972).

cover on the coast with that on the leeward inland, it was found that the new
fallen snow on the coast was drifted far to the leeward inland, and the amount
of drifted snow was estimated.

Okabayashi (1972) studied the Ishikari front using satellite pictures and
came to the following conclusions: (a) The Ishikari front has a length of more
than 800 km and extends to the north over the Sea of Japan. (b) A small high is
produced in the center of Hokkaido due to the nocturnal radiation cooling. (c)
The pressure difference between this small high and the trough over the Sea of
Japan is 0.5–1.5 mb. (d) A very small cyclone develops along the front and comes
down to the south gradually. (e) The NW-monsoon is stronger in the southern
part, and hence, the front bends to the east and brings heavy snowfall where it
reaches the land. This is the Ishikari plain in most cases. These features are
shown in Fig. 5.27. The phenomena stated above were also confirmed by laboratory experiments and radar echoes (Magono, 1971).

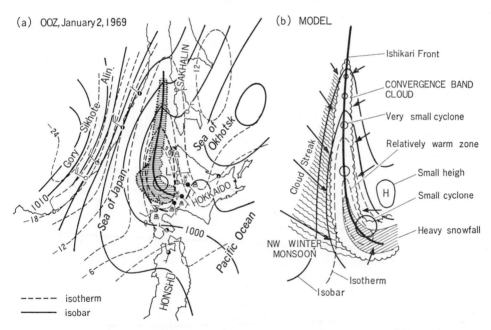

Fig. 5.27 The Ishikari front detected by the satellite pictures at 00Z, January 2, 1969, and a tentative model of the front (Okabayashi, 1972).

(G) *Other Local Fronts*

It was pointed out in the investigations of rainfall distribution in the Osaka
plain, along the River Yodo, and on the coast of Osaka Bay that the torrential
rains were caused by local fronts formed between the topographically disturbed
currents or topographically stagnated cold air masses and the rain-causing warm
currents (Nishimura, 1932). The local frontal surface is produced by a current
about 500 m in height than emerges in the southern part of Lake Biwa and a

current that intrudes from Osaka Bay and is pushed up over the former current. Rainfall of this kind was named *secondary orographic precipitation*, as this rain was attributed to topographically formed fronts. While ordinary orographic rain does not continue for long since it is accompanied by a traveling cyclone, secondary-orographic rain tends to continue for a long time.

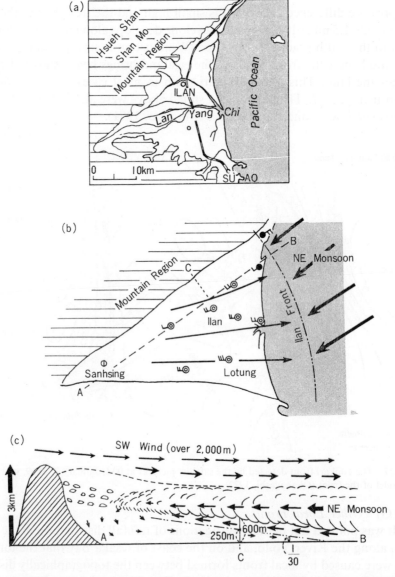

Fig. 5.18 Local front produced by the NE winter monsoon and its backcurrent formed topographically near Ilan, NE-Taiwan. (a) Topographical situation of the Ilan plain. (b) Observations from December, 1942. (c) Cross section along the line A–B on the plain (Modified by Yoshino after Kabasawa, 1950).

In the Kanto plain, a local front of this kind runs from the southwestern to the northeastern part when SE-winds prevail, and a heavy rain zone is formed along the front. The front appears clearly especially when a typhoon brings moist warm SE-winds. It can be called also orographic rain band.

A local front of this kind was reported from the Ilan plain in NE-Taiwan (Kabasawa, 1950). There the NE-winter monsoon with a wind force of 5 to 6 strikes against the mountains and forms a front between the monsoon current and its back current produced topographically, as shown in Fig. 5.28. The lower layer, 200–300 m thick, is a SW-wind with a wind force of 2 to 3. Due to this local front, rainfall during the winter monsoon season is much greater in the NE-part of the plain. For instance, rainfall during the period of one monsoon outburst, December 11–17, 1936, was more than 140 mm on the NE part, but 139 mm at Ilan, slightly north of the center of the plain, and 79 mm at the SW-part of the plain. Such a distribution pattern is also found in the monthly total maps in winter. The inclination of the front was 1:30 on the coast. At Sanhsing in the SW-corner of the plain, calm and partly cloudy weather was observed, unlike the rainy weather on the coast.

Recently, a local front that develops crossing the plain from Shizuoka City to Shimada City, Shizuoka Prefecture, was investigated in detail (Nakayama, 1953a). The front is formed when the cold current that blows down from the inland Chubu district, Honshu, along the River Ohi and other rivers meets with S-currents and brings heavy rainfall.

Two world-famous places with heavy rainfall are Cherrapunji (25°15′N, 91°44′E, 1,313 m) and Mawsynram (25°18′N, 91°35′E) in Assam. These points are located on the southwestern slope of the Khasi Hills with a height of 1,500 m a.s.l. Although there is no detailed analysis on a local scale, the rainfall at these points is probably not caused by a simple orographic effect alone, but is under the influence of a local front that is formed between the moist SW-monsoon and the E–NE mountain winds blowing down from the Valley of Buramaputra. The reason for this speculation is that the rainfall at Cherrapunji in the summer monsoon period has a maximum intensity during the morning hours.

A meso-scale front also occurs along the southeastern New England coast (Bosart et al., 1972). Temperature gradients of 5–10°C over 5–10 km on a length scale of 100 km separating the light N- or NW-flow from the stronger E-flow are common along this front. There is a persistent tendency for coastal fronts to stagnate along a Boston to Providence line. Precipitation appears to be enhanced along and just on the cold side of the frontal zone. The number of occurrences of this front has its maximum in winter: 16–19 from November through February and none from June through August. This is because the land-sea thermal contrast is the greatest in these winter months. The effects of orography, coastal configuration, land-sea temperature contrast, and friction were considered as important factors in their formation, intensification, and dissipation.

5.2.4. Local Fronts on a Micro-scale

When land and sea breezes or mountain and valley breezes begin to blow, the head of the current proceeds just like a front. There is a difference in structure, however, between a warm front and a cold front, and the passing of a cold front can more easily be recognized than that of a warm front. Sea breezes show clearer characteristics in the beginning or their passage than land breezes if observed in the coastal area. Also the beginning or passage of a mountain breeze can be more precisely recognized than a valley breeze. Sea breeze fronts have already been mentioned in Section 3.4.2 (F).

A front is produced at the margin of land and sea breezes or mountain and valley breezes and has a great influence on local climates, as stated above. When the sea breeze front moves inland and stays there, thus causing the stagnation of contaminated air, a great difference in insolation occurs between the inside and outside of the front as explained in Section 3.2.8 with an example observed in the southern part of Tokyo.

A land breeze becomes strong if combined with a valley breeze, and the head of the current is inclined to have the characteristics of a front. An example of this type is observed around Fukuoka City.

In mountainous regions local fronts develop when mountain and valley breezes meet with slope winds or general winds. Following are the results of the observation of a front on a micro-scale (called a *miniature front* in the report) obtained in the southeastern part of Idaho, U.S.A. (Wilkins, 1955).

The studied area has a valley plain about 90 km wide, and the valley flows from NE, and its mouth opens to the SW. A peculiar point of the topography is that at the head of the valley large mountains lie at right angles to the direction of the valley. The valley breeze and the wind that blows down on the slope of these large mountains produce the local front. Here, the local front was observed 177 times in 1953 excluding the fronts accompanied by general depressions. 73% of the 177 came from the north. About the same number of cold fronts and warm fronts appeared in winter, but the annual occurrence of the cold fronts was 2 times as frequent as that of warm fronts. 40% of the fronts from the N and 73% of the fronts from the S stayed around 90–140 km from the heart of the valley and at last receded in the direction from which they came. In winter, January to March, the most conspicuous front was observed at the time of the maximum temperature. This may be because the snow-covered mountain slope causes down-slope winds, and the snowless valley causes the development of a valley breeze. The temperature difference between the upper layer and the lower layer of the frontal surface is generally 5°C or more, and the warm current is always overlaid by the cold current.

The mechanism of occurrence of these two winds in the opposite directions which produce the local front may be explained as follows: The slope of mountains located deep in the valley faces S and therefore has much variation in temperature, thus allowing better development of the slope wind, as shown in Fig. 5.29. As the changing time of the down-slope and up-slope winds is earlier than that of mountain and valley breezes, this variation of time causes the collision of currents. However, the fact that the front continues to exist for 20 to 30 hr suggests that there are some other factors involved.

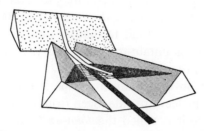

Fig. 5.29 Schematic plan of Snake River valley, showing how miniature front is formed out of conflict between valley wind and katabatic wind down the slope of the continental divide (Wilkins, 1955).

Still, the reason why the down-slope wind is warmer than the southerly valley breeze is not understood. It seems that the application of föhn phenomena to the down-slop ewind is not always practical.

5.3. LOCAL WINDS

This section is concerned with local winds that occur under certain synoptic conditions in particular regions surrounded by specific topography. The daily wind systems, such as land and sea breezes and mountain and valley breezes have already been mentioned above. The topographical effects on the wind velocity and directions are discussed in Sections 4.1.10 and 4.2.4.

5.3.1. Föhn, Bora, and Similar Winds

(A) *Föhn*

Among the local winds, a föhn is the most widely known, and the föhn effect is the most frequent phenomenon of the airstream crossing over mountains. The word *föhn* is derived from *favonius* in Latin, which means "west wind." According to Aubert de la Rüe (1955), this Latin word was imported in Romanche, a language spoken in a part of the Grisons, Switzerland, in the form of *favugn, favuogn, favoign,* or *fogn,* the last being in use in the Livinenthal. On the other hand, the origin of *föhn* can be traced to the old Gothic word *fôn* meaning "fire."

A föhn is a dry, warm wind coming over the mountains and blowing down to the mountain foot. A *föhn* is formed as follows: When an air current goes over a mountain, the air temperature on the windward side becomes lower by adiabatic processes, and the water vapour in the air is condensed into cloud particles and then into precipitation. At this time, the temperature decreases along a curve approximate to the wet adiabat, at the rate of 0.5°C/100 m. The air current that has gone over a mountain gradually increases in pressure on the leeward and becomes warm by adiabatic process. Furthermore, the air is very dry since the water vapour has decreased on the windward side, because most of it has fallen as precipitation. The temperature variation at this time describes a curve equivalent to the dry adiabat at the rate of 1°C/100 m. Such a process in the air current including the cloud formation (and the precipitation) on the windward slope near the mountain-top and the dry, warm wind on the leeward mountain-foot is named the *föhn effect*.

The cloud formed on the windward slope is transported to the top and comes over slightly to the leeward side. Seen from the leeside mountain foot, it looks like a wall; hence it is called the *föhn-wall*. On the Rocky Mountains, it is called a *crest cloud* and on the Dinar Alps along the Adriatic coast, *Kapa*.

Observation at Altdorf, Switzerland, during the föhn period of November 8 to 11, 1934, shows that the air temperature increases from 4–5° to 13–16°C, and the relative humidity decreases from 70–90% to 22–28% (Walter, 1938). Air pressure also decreases. Walter classified the föhn from the viewpoint of the synoptic situation as follows: (a) Cyclonic föhn: The north föhn, which blows in the case of a deep cyclone located south of the Alps, and the south föhn, which blows in the case of a deep cyclone north of the Alps. The south föhn is subdivided into three types according to pressure gradient. (b) Anticyclonic föhn: A high pressure belt is located over the Alps. Hence, the wind is weak and

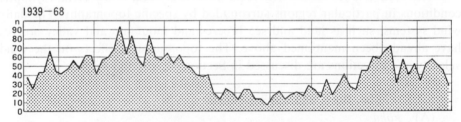

Fig. 5.30 Five-day frequency of föhn in Bad Ragaz in Switzerland, 1939–1968 (Gutermann, 1970).

Table 5.6 Wind velocity of föhn in the Rhein valley, Switzerland, during the period July 1967–June 1969 (Gutermann, 1970).

	Height (m) a.s.l.	Mean wind velocity (m/sec)			Maximum hourly mean	
		Summer	Winter	Year	Wind velocity (m/sec)	Date
(Valley St.)						
Landquart	531	5.6	6.7	6.2	12.9	Mar. 11, 1968, 4 h
Balzers	480	—	—	8.7	23.8	Feb. 18, 1969, 23 h
Buchs	449	5.9	7.0	6.4	14.3	Jan. 7, 1968, 8 h
(Mountain St.)						
Fläscherberg	940	10.0	12.8	11.5	23.8	Feb. 18, 1969, 23 h
Pizalun	1,458	7.0	9.2	8.1	17.1	Feb. 19, 1969, 24 h

cold air lakes develop before the föhn begins to blow. The air from the upper layer descends as the fall wind fills up the valley where the cold air has blown away, and this causes a föhn wind with great gustiness.

Gutermann (1970) studied the föhn frequency statistically and obtained the results of field observation in the Rhein valley in Switzerland. As shown in Fig. 5.30, the five-day mean föhn frequency at Bat Ragaz (47°00′N, 9°30′E, 510 m a.s.l.) becomes maximum at the end of March, and the second maximum comes at the beginning of November. The observed wind velocity, as shown in Table 5.6, is more than 20 m/sec at maximum and has a mean of 6–13 m/sec.

Koch (1960) studied the criteria of the föhn in Thüringen Forest in detail, comparing the air temperature and humidity at the stations on the slopes for the cases of SW-föhn and NE-föhn. He points out that the föhn conditions are more apparent on the slopes than in the bottoms of valleys or basins, where the local small-scale influences are predominant.

Table 5.7 Seasonal and annual mean total number of days with S-föhn at Innsbruck for 1906–1970 (Fliri, 1972).

Period	Winter	Spring	Summer	Autumn	Year
1906–1910	11.7	23.8	11.8	19.4	66.7
1911–1920	17.7	33.3	16.2	16.4	83.6
1921–1930	10.2	28.1	16.0	16.7	71.0
1931–1940	8.0	26.0	11.1	16.1	61.2
1941–1950	8.7	19.8	6.2	11.8	46.5
1951–1960	7.8	17.2	10.2	11.8	47.0
1961–1970	9.4	17.5	9.5	13.8	60.2
Mean	10.6	23.8	11.6	14.7	60.7

A secular change in occurrence frequency of a föhn is observable. For instance, the seasonal and annual total of days with the S-föhn at Innsbruck, Austria, as shown in Table 5.7, was apparently small during the decades between 1941 to 1950 and from 1951 to 1960 (Fliri, 1972). The reason why such a secular change occurred is not clear, but similar tendencies are also found in the statistics for the stations at Bad Ragaz and Altdorf in Switzerland (Gutermann, 1970).

Effects of the föhn are caused by its extreme heat and dryness, although it purifies the atmosphere and produces exceptional visibility. The first effect entails the danger of fires; smoking is forbidden, and people are not even allowed to cook their food. To enforce these regulations, *Föhnwächter* (föhn guards) are appointed by the communes in the Alps, where the föhn is frequent and violent. A heavy rainfall frequently accompanies föhn and causes floods. As the name *Schneefresser* (snow eater) indicates, when the föhn occurs snow cover melts away very quickly, in most cases by sublimation because of the extreme dryness of the föhn. On the mountain slopes, an avalanche sometimes takes place. The melted snow in the mountain region causes high water in the rivers. These effects are especially great in winter and spring. In the valleys called *Föhngasse* (föhn alley way), people can start to cultivate fields earlier than elsewhere, because the snow melts away earlier, and a type of corn that needs much heat, such as maize, is grown in the Ötz valley, Tyrol.

The air temperature of the föhn wind is warmer on the lee side mountain foot than on the windward mountain foot or than before the föhn period, as mentioned above. However, because of its strong velocity and absolute coldness, the föhn itself is not always bioclimatically warm. Lauscher (1956, 1958) pointed out that, among the 129 föhn cases in Innsbruck 1949–1955, 28 cases were warm, 58 cases were indifferent, and 43 cases were cold if the biometeorological cooling effect for the human body is considered.

The föhn in the Alps has been studied intensively since the second half of the 19th century. Further information can be obtained from the catalog of local wind by Schamp (1964), an authoritative description by Blüthgen (1966), and the results of studies that have been made for many years by von Ficker et al. (1948). More recent literature will be found in a review by Brinkman (1971).

A föhn, in one form or another, occurs when an air current crosses over a mountain; examples can be found all over the world. Some of the important ones, e.g., chinook, will be mentioned below. In Japan, when a S-monsoon is usually strong in summer, or when there is a well-developed cyclone in the Japan Sea, into which a strong S-wind blows, this S-wind, going over the central mountainous districts, often causes the föhn effect, like the S-föhn in the Alps. The highest air temperature records in Japan, 40.8°C in Yamagata City (July 25, 1933), 39.1°C in Niigata City (August 6, 1909) or higher than 40°C at many other stations in the mountain regions were caused by the föhn effects.

In Ceylon, the föhn, called *kachchan*, occurs during the SW-monsoon (Thambyaphillay, 1958; Domrös, 1969). Schweinfurth and Domrös (1974) described the effects of the *kachchan* and the "*Uva blowing*," a föhn-like fall wind, on the agriculture in the Central Highlands of Ceylon. During the NE-monsoon, the "*Dimbula blowing*" occurs, but it is far weaker.

(B) *Chinook*

The chinook is the föhn in North America, so that the phenomena associated with the chinook are similar to those of the föhn. Glenn (1961) and Beran (1967) reported that the major chinook region in western North America extends along a strip 300–500 km in width from Alberta, Canada, southward to NE New Mexico and that this chinook region follows the Continental Divide and affects the high plains region along the eastern side of the Rocky Mountains. The region with the strong chinook, however, may be restricted in a zone of 20–80 km. The chinook causes snow cover to melt or sublimate in such a short time that it is well known by the name of *snow eater* and is said to exert great influences on physiological life as will be mentioned below.

The chinook is the old name given by the Indians to a hot, dry, west wind that blows in the valley of the tributaries of the Missouri (Aubert de la Rüe, 1955). It is now used to designate the equally hot and dry wind descending from

the Rockies. Marsch (1968–1969) found that the first real account of the chinook
was given by Mackenzie in January 1793, and later more precisely by the Palliser
Expedition, 1857–1860. In 1881 a geologist, Dawson, pointed out the similarity
between the chinook and the European föhn. In Alberta, the chinook occurs
most frequently in the Lethbridge area with the frequency of 40% of the winter
days; it is 29% at Kananaskis, 27% at Calgary and 10% at Banff (Longley,
1966–1967). Brinkmann (1970) described the chinook in the Calgary region,
Canada. Longley (1967) studied the frequency in winter in detail. The areal pattern shows a maximum frequency along a line just east of the front range of the
mountains with the frequency decreasing slowly eastward, as shown in Fig. 5.31.
At times the warm air sweeps over the mountain and descends on the eastern
slopes without having its full effect on the surface of the mountain passes and
valleys. A synoptic situation in a typical case of the chinook is given in Fig. 5.32.

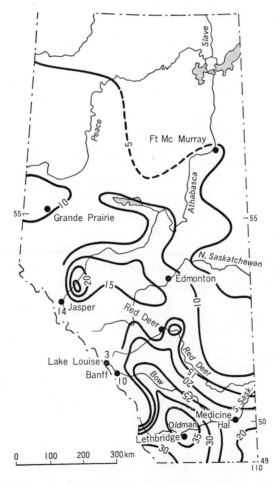

Fig. 5.31 Distribution of mean number of days with chinooks during the winter
(December, January, and February), 1931–1965 (Longley, 1967).

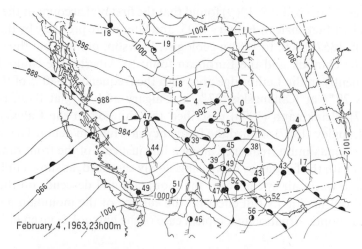

February 4, 1963, 23h00m

Fig. 5.32 Synoptic situation in a typical case of chinook in Alberta. Slightly modified from the map by Drinkwater et al. (1969).

Kendrew and Currie (1955) state that when the chinook occurs, the sky clears abruptly over the W-mountains, and the air temperature rises. In the S of Alberta it rises by as much as 28°C within a few hours, and a rise of 22°C in 15 min has been recorded; farther east the rise may be as large but is less rapid. After one or more days the general warmth gives way again to the normal cold.

It was reported by Glenn (1961) that the chinook was first noted at Spearfish, South Dakota, at 07 h 32 m, January 22, 1943, when the air temperature rose from −20.0°C to +7.2°C, within a period of only 2 min. Such extreme changes in a short period of time are not rare. Perhaps the strongest chinook ever reported is the one with the maximum value of 56 m/sec, observed at Boulder, Colorado (Julian et al., 1969).

In the crest of the waves, the cloud formation known as the *chinook arch* (Thomas, 1963) may form in a linear band of lenticular altocumulus parallel to the mountains. Observation of the arch suggests that its western edge is close to

Table 5.8 Moisture loss and snow loss, lowering of snow surface, during the chinook period January 13–20, 1965, at Calgary, Canada (Ashwell et al., 1967).

Date	Evaporation from surface of 13 cm² (ml/hr)			Max. Temp. (°C)	Lowering of snow surface (mm)	Av. wind velocity 30 cm above surface (m/sec)
	180 cm	60 cm	just above snow			
13	0.30	0.24	0.10	3.9	—	2.6
14	0.13	0.10	0.07	6.8	8.1	1.0
15	0.15	0.10	0.04	4.4	9.4	2.4
16	0.21	0.16	0.07	4.3	8.9	3.6
17	0.18	0.12	0.03	5.0	2.5	2.7
18	0.12	0.10	0.04	4.4	2.0	3.1
20	0.26	0.16	0.10	7.6	1.5	2.1
Average	0.19	0.14	0.06			

the line of the mountains in the morning and that the whole moves eastward during the day (Brinkmann and Ashwell, 1968).

Ashwell et al. (1967) reported the amount of moisture loss to the atmosphere during the chinook period. They measured the evaporation with Piché atmometers at three heights air temperature, lowering of the snow surface, and wind velocity (see Table 5.8). The snow surface is lowered at the rate of 1 cm/day, not a small value. However, the proportion of sublimation and melting amount is still unknown.

Ives (1950b) summarized frequency and physical effects of the chinook in the Colorado high plains region based on his many years' research. Among his conclusions, the followings are worthy of note: (a) all chinooks are westerly, but not all are warm, (b) in some extreme cases, snow cover lowers at the rate of 3 cm/h, (c) crest clouds are over the ridge, and the cumulus zone appears 30 km east of the crest clouds, (d) the drying effect by the chinook occurs in a zone parallel to the foothill with a width of 50 km.

Holmes et al. (1971) observed the chinook using an instrumented aircraft and presented some structure of temperature profiles as well as a map showing location of melting snow. They indicate that a modified Scorer equation is successful to explain the atmosphere and/or topographic inducement of the wave motions observed. Riehl (1974) studied the chinook in the foothills of N-Colorado. About 100 chinook situations occur in 10 years and the highest gust measured during 1964–1971 was 38.9 m/sec. The desiccating effect is estimated at about 0.8 cm equivalent moisture depth on the lee slopes during the winter.

Beran (1967) described the four types of chinook winds. It was shown that the stability of the surface inversion layer on the windward side of the Rocky Mountains is related to the occurrence of chinook to the leeward side. Beran (1967) and Vergeiner (1971b) state that the chinook might be due to large-amplitude lee waves, which will be discussed below in Section 5.3.4.

(C)   *Bora*

The bora is a dry, cold, fall wind on the Adriatic coast and in the inland regions near the coast of Yugoslavia. The term bora is now applied to similar winds in other parts of the world.

The term *bora* is from the Greek βορεασ, which means N-wind. In the Serbo-Croatian language it is called *bura*, but in English, German, Italian, etc., it is called *bora*. The ancient Romans brought this word to the Adriatic and Black Sea coasts. In the island region of the NE-Adriatic Sea, people called it *quarnero*. In Trieste, the strong bora is named *boraccia* and the weak bora in summer, *borino*. The bora is accompanied by a strong gust, which is called *Stösse* or *refoli* (*reffoli*).

The bora is generally classified into two types, anticyclonic and cyclonic. According to the difference in temperature and gradient of atmospheric pressure of air masses located in the east and the west of the Dinaric Alps, it is classified into the following four types (Paradiž, 1959): (a) Dam up bora, (b) dam up—

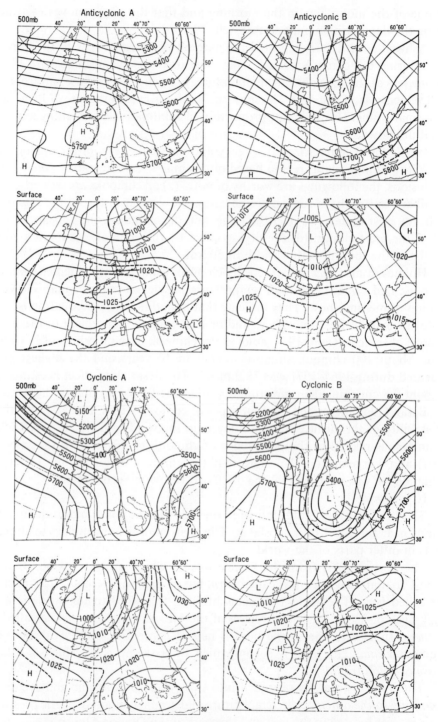

Fig. 5.33 Composite maps of topography at the 500 mb level and the pressure pattern at the surface level during the bora (Yoshino, 1971).

gradient bora, (c) gradient—dam up bora, and (d) gradient bora. In the case of (a), there is a small difference in atmospheric pressures between the air masses on both sides. On the windward side, the bora is anabatic and comes on the cold air masses. (b) is the katabatic type. It becomes stronger in the direction with the pressure gradient. In the case of (c), the bora is anabatic and rises up according to the gradient of atmospheric pressure. In this case, the temperature of the upper air mass on the windward side of the mountains is not very cold compared with the lower air mass. (d) is the bora that blows because of the gradient when the difference of temperature between both sides of the mountains is not great. Where the mountains are relatively low, type (c) bora often becomes the type (d) bora. In certain cases, it results in a föhn-type wind. Therefore, types (c) and (d) are föhns. When the N–E winds are strong in a region where the wind is usually very strong, people call them bora, regardless of their causes. The bora blows differently according to topographical features.

An attempt was made to classify the synoptic patterns of the strong NE-NNE winds with peak gusts exceeding 15 m/sec on the Adriatic coast into four types, cyclonic A and B and anticyclonic A and B, and composite maps of the surface level and 500 mb level were made for each pattern, as shown in Fig. 5.33 (Yoshino, 1971). When the pressure gradient over the region is caused mainly by the high pressure system, these types are called anticyclonic. The location of the anticyclone center determines the subtypes: When the center is situated over Europe, it is named anticyclonic A, and over the Atlantic Ocean, anticyclonic B. On the other hand, when the pressure gradient is caused mainly by a low pressure system over the Adriatic Sea or the Mediterranean Sea, it is called cyclonic type. This type is also divided into two subtypes, which differ in the location of the high pressure zone north of the low. One of them shows the high pressure

Table 5.9 Mean values of the various meteorological elements during the bora in Split, Yugoslavia (Yoshino, 1971).

		Anticyclonic A	Anticyclonic B	Cyclonic A	Cyclonic B
Air pressure	7 h	753.0	745.1	746.4	743.3 mm
	14	752.7	749.5	746.5	753.1
	21	754.1	750.2	747.2	745.6
Interdiurnal Temp. variability	7	−1.3	−1.0	−2.0	−1.6 °C
	14	−1.1	−1.3	−1.9	−2.0
	21	−1.3	−0.6	−1.7	−2.1
Relative humidity	7	45	51	53	68
	14	38	46	50	58
	21	42	49	48	60
Wind velocity	7	NE	NE	NE	NE
	7	8.9	8.0	9.2	6.5 m/sec
Wind velocity	14	NE	NE	NE	NE
	14	8.0	9.0	8.3	6.9 m/sec
Wind velocity	21	NE	NE	NE	NE
	21	9.2	7.1	9.3	7.8 m/sec
Precipitation probability		14	50	56	82 %

zone crossing over central Europe, and the other shows the high pressure zone crossing over the Scandinavian Peninsula. Strictly speaking, a bora day determined by such criteria should be called a "potential" bora day, because the bora must be detected by the meteorological elements observed at the local stations on the Adriatic coast. The mean values of the various meteorological elements at the station in Split on the Adriatic coast for anticyclonic A and B and cyclonic A and B are given in Table 5.9.

Table 5.10 Monthly numbers of bora-days on the Adriatic coast, average of the period 1956–1965 (Tamiya, 1972).

	J	F	M	A	M	J	J	A	S	O	N	D	Year
Anticyclonic A	4.4	4.3	3.1	2.2	3.0	3.2	1.3	2.3	3.7	5.6	2.5	4.4	40.0
Anticyclonic B	0.7	0.8	1.0	1.3	1.5	2.4	4.2	3.6	1.1	1.1	0.5	0.4	18.7
Cyclonic A	2.7	1.5	2.9	2.3	1.3	0.6	0.4	0.7	1.0	3.4	4.2	4.0	25.0
Cyclonic B	3.2	2.7	2.7	1.9	1.5	0.6	0.1	0.4	0.5	0.4	2.3	1.8	18.1

Tamiya (1972) has made a chronology of bora days according to the criteria mentioned above (Table 5.10). He states that they appear frequently from October to March. All types concentrate in winter, except anticyclonic B, which shows a clear peak in summer. He has made clear that the five- and six-day sequences have a tendency to begin with the cyclonic type and end in the anticyclonic type. The longest bora period continued 37 days from November 8, 1958, as the data for 1956–1965 show.

In Trieste, Italy, the bora blows very strongly, as has been reported for many years. Polli (1970) gives the normal values of climate of Trieste, the peak gust and the prevailing wind velocity and directions for every month, 1931–1960. According to him, the observed maximum peak gust of 47.5 m/sec was recorded in Feb., 1929. Yoshimura (1972) surveyed the monthly change of the occurrence frequency of bora days, monthly mean percentage of the bora days to the total observed days, and the peak gust on each bora day at Ajdovščina, Slovenia, and

Table 5.11 Monthly change of the bora-days and peak gusts (Yoshimura, 1972).

		J	F	M	A	M	J	J	A	S	O	N	D
Ajdovščina	Total numbers of bora-days	102	74	64	82	79	52	63	91	70	115	114	104
	Percentage of bora-days to the total observed days	41	37	33	34	35	22	23	33	29	42	47	44
	Mean value of the peak gusts on the bora-days (m/sec)	22	24	19	17	17	16	16	16	16	18	20	20
Trieste	Total number of bora-days	125	120	128	100	97	76	83	102	98	126	135	132
	Percentage of bora-days to the total obseved days	40	42	41	33	31	25	27	33	33	40	45	43
	Mean value of the peak gusts on the bora-days (m/sec)	20	19	17	16	16	16	15	15	16	17	18	18

Trieste. The bora days appear frequently from October to February. The details are given in Table 5.11.

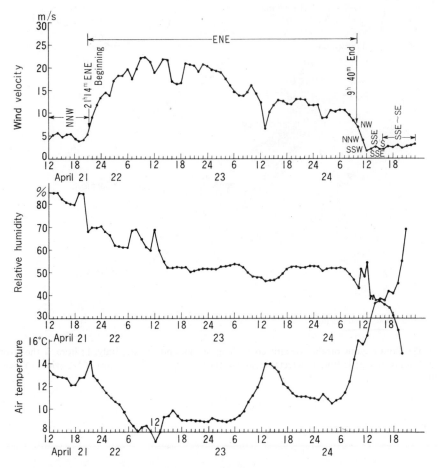

Fig. 5.34 Change of wind velocity, relative humidity and air temperature at Senj during a bora on April 21–24, 1966 (Yoshino, 1969).

An example of the change of meteorological elements during a bora on April 21–24, 1966, is given in Fig. 5.34. It is remarkable that the 1 hr mean wind velocity of more than 10 m/sec continued for so many hours, reaching a maximum of 23 m/see. The bora regions are shown in Fig. 5.35. The cross section along a line approximately at right angles to the running direction of the coast is given in Fig. 5.36. The regions in Croatia where the bora is strongest are the coastal regions from Senj to Karlobag and the small regions near Bakar, Novi Vinodolski, etc. As will be seen from the cross section, a bora is strongest on the coast, and it is very strong up to about 800 m a.s.l., where the Dinar Alps run close to the coast. An example of windshaped trees formed by the bora is

Fig. 5.35 Bora regions on the Adriatic coast, Yugoslavia, and Trieste, Italy. 1: Bora is the strongest; 2: Bora is strong; 3: Bora is detectable; and 4: the boundary of strong gust of NE-winds (Yoshino, 1969).

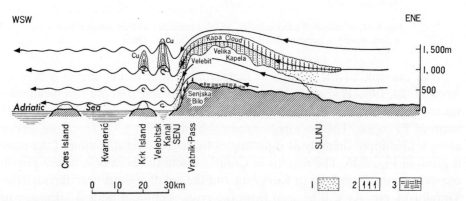

Fig. 5.36 Schematic illustration of conditions during bora along the cross section through Senj on the Adriatic Coast. 1: Precipitation, 2: Wind-shaped trees, and 3: Mosses as indicators of cold air pools (Yoshino, 1969).

Photo 5.1 An example of a wind-shaped tree, *Pinus nigra* (Grade 4), at Karlobag on the Adriatic Coast, Yugoslavia (November, 1970, by M. M. Yoshino).

shown in Photo 5.1. The severe icing that occurs during the strong bora is dangerous for traffic vehicles and ships (Photo 5.2).

The bora is famous at Novorossiysk and Poti on the E-coast of the Black Sea. During the bora conditions a wind force of 11 has been experienced, and the force of 9 is reached during most winter months at Novorossiysk (Met. Office, 1963). Alissow et al. (1956) described the bora at Novorossiysk and on Lake Baykal in detail. The number of storm days was 54 per year at the port of Novorossiysk and, among them, 46 days had bora with a maximum wind speed over 20 m/sec. Bora-like winds are called *norder* at Baku, Azerbaijan. The bora on the west coast of Lake Baykal is named *sarma*, which is a fall wind from NW-NNW. It becomes westerly and weaker in the area south of the Sarma River and is called *charachaicha* there. Other bora-like winds are known in Novaya Zemlya, on the coast of Okhotsuk, and in the Crimea peninsula.

On the southern coast of Central America, the fall winds called *nortes* prevails between October and February or March. They originate from the North American Continent, where they are called *northers* under the synoptic condi-

Photo 5.2   Icing during the strong bora at the Senj harbor in February, 1963 (by Ivan Stella).

tion of cold air outbreaks. Lauer (1973) describes the regions of influence of the *nortes* in the Central Mexican Highlands in detail. When it blows, air temperature becomes lower on the higher and middle part of the mountain slope, but it frequently becomes higher at lower altitudes.

Since the second half of the 19th century, many studies on the bora have been published. An annotated bibliography on the bora contained more than 100 papers, monographs, and books (Yoshino, 1972).

(D)   *Fall Wind; Oroshi*

The *oroshi*, a fall wind, is one of the strongest winds in Japan. It prevails in the Kanto plain in central Japan in the case of the monsoon situation in winter. It appears very frequently in the afternoon as northwesterly winds with a maximum over 10 m/sec. When it blows, relative humidity decreases to lower than 25%, in extreme cases to 5–6%, and the diurnal air temperature becomes 1.5–2.5°C colder than on the preceding day. The oroshi wind develops from the northwestern part of the Kanto plain to the Bôsô front (see Section 5.2.3 (D)) over the plain.

The different vertical structures on both sides of the central mountains are of interest: The wind maximum is found in the layer between the 900 and 800 mb levels on both sides, but there is a humidity maximum of 90–98% on the windward side and the humidity minimum of 30–50% is the lee side in the layer near the wind maximum. The vertical cross sections of wind velocity and relative

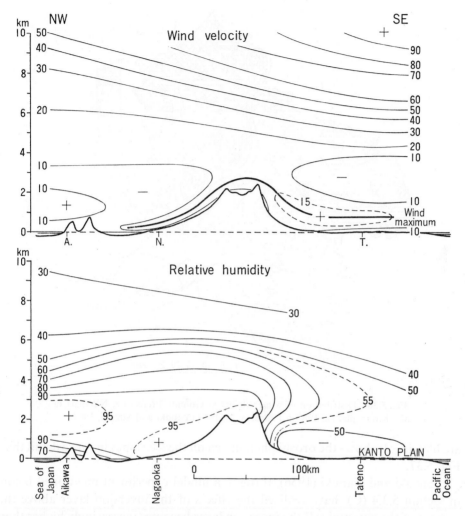

Fig. 5.37 Vertical cross sections of wind velocity (m/sec) and relative humidity (%) along the line from NW to SE connecting Sado Island and Tateno on the Kanto plain on January 14, 1965, 21h (Yoshino, 1970).

humidity from the Japan Sea side to the Pacific Ocean side are given in Fig. 5.37. This figure clearly shows the wind maximum axis and the corresponding high humidity layer on the windward slope and the dry layer on the Kanto plain caused by the fall wind, oroshi.

Yabuki and Suzuki (1967) thought that the oroshi is caused by the föhn effect in association with the *waterfall phenomenon* in the lee side of the mountain. They presented distribution maps of the wind velocity in Gumma Prefecture, in the NW-part of the Kanto plain. Figure 5.38 clearly shows that the wind is strong on the leeward slope and that in the S-parts of Gumma Prefecture, winds with opposite directions, indicating backcurrent caused by eddies,

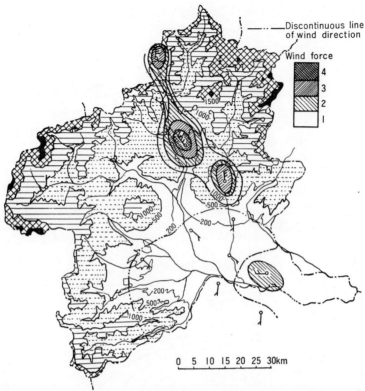

Fig. 5.38 Distribution of wind velocity in Gumma Prefecture, NW-part of the Kanto plain at 19h on February 14, 1949 (Yabuki and Suzuki, 1967).

are blowing. It shows the oroshi phenomena on the lee slopes in more detail than Fig. 5.37.

Suzuki and Yabuki (1956), through a model experiment mentioned below in section 5.3.4 (C), have realized the effects of the inversion layer above the summit of the mountain. If the inversion layer becomes increasingly higher than the summit, the strong fall wind loses its strength and begins to return and go up the mountain slope forming a counterclockwise eddy current. The same phenomenon is observable in the case of the neutral air layer. The wave pattern formed on the boundary of two contiguous atmospheric layers with different densities is characterized by 3 parameters, i.e., density difference, wind velocity and boundary height. Provided that the inherent or subsistent waveform coincides well with the mountain profile, the generated waves come into existence and become maximum both in amplitude and in velocity, but their wavelength is shortest. Under an extremely well developed condition, which is a critical stage, they take a jet form, which induces a strong fall wind at the mountain foot. After passing this critical stage, a counterclockwise eddy current rises along the mountain side, and the upper flow takes a jump over the eddy current and rushed ahead leaving the eddy behind and produces a powerful windy area on

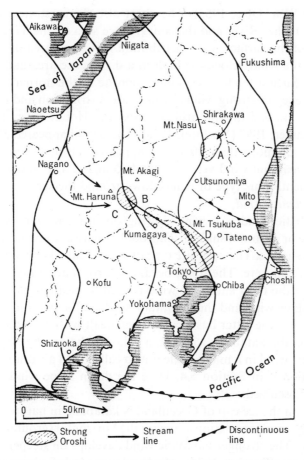

Fig. 5.39 The streamlines and the strong *oroshi* regions in the Kanto plain at 15h, January 12, 1957. A: Nasu-oroshi, B: Akagi-oroshi, C: Haruna-oroshi, and D: Tsukuba-oroshi (Yoshino, 1961a).

the lee side of the mountain. If the atmosphere is stable in the vertical structure, the simple fall winds always develop.

Nasu-oroshi, Akagi-oroshi, Haruna-oroshi and Tsukuba-oroshi are the names given to especially strong N-NW winds that blow over the Kanto plain from November until April. The oroshi is always called by the local people by the name of the mountain nearby. Meteorologically, they do not always come from the direction of the mountains after which they are named, as shown in Fig. 5.39, but they are the dry, strong lee currents of the winter monsoon. The streamlines, the meso-scale lines of discontinuity on the S-coast of the Kanto plain called the Bôsô front and in the NE-part of the plain called the Utsunomiya front were discussed in Section 5.2.3 (D) and (E).

An intensive study of the Nasu-oroshi on the Nasu alluvial fan, N-Kanto plain, shows that the divergence area along the Naka River and the convergence area along the Sabi River are recognizable on the mean streamline chart (Utsunomiya Local Wea. Obs., 1953). The chart also demonstrates the divergence area along the Hohki River and the convergence area on the Yosasa River. It is characteristic of the surface wind distribution in a small area in the case of the fall wind that the wind velocity is high in a divergence area and low in a convergence area. Such a tendency is quite opposite to the normal hydraulic rule.

The frequency of strong winds by pressure distribution and their respective velocities were studied on the bridge across the Hohki River from January, 1956, to September, 1957 (Yoshino, 1958). The oroshi appeared 31 times; those with velocities of more than 20 m/sec appeared 11 times; and the extreme value of the maximum wind velocity was 48.5 m/sec. This shows how strongly and how frequently this type of wind blows around the sea.

(E)   *Katabatic Wind*

The katabatic wind is a wind flowing down an incline, whether on the ground surface or a frontal surface. This book discusses only the katabatic wind on the ground surface. The opposite type of wind is called an anabatic wind, but this is also omitted here. The föhn, bora, and oroshi are all types of katabatic winds, but here the term katabatic wind is used to mean a gravity wind.

A down-slope wind and the mountain wind are the main subjects discussed in this section, but the latter has already been dealt with in Section 4.2.4 (C). A classic study on the katabatic winds in a valley is the one by Heywood (1933). The down-slope wind in a micro-scale is also called cold air drainage or run-off, which is mentioned in Section 5.4. Here the down slope wind in a relatively broad scale is discussed.

The wind on the ice cap of Greenland is katabatic in nature; therefore, the wind velocity has a close relationship with the horizontal (ground level) gradient of temperature. The stronger katabatic wind develops on steeper surfaces at the edge of the ice cap (Putnins, 1970). The katabatic winds frequently develop into gales that exceed 50 m/sec where the wind is channeled into fjords. The katabatic wind in Antarctica has been continuously investigated since the 1950s, because of its dramatic features. The strong intensity of ground inversion over the ice enables the relatively thin layer of air above the ice slopes as katabatic wind to flow down by gravitation. It is very strong with heavy gustiness, reaching a maximum near the foot of the slope.

A typical katabatic wind is characterized as a relatively short duration of a very strong surface wind over a wide area, or by a limited horizontal extent if the wind lasts for a longer period (Schwerdtfeger, 1970). Stations such as Cape Denison, Mawson, or Mirny experience abrupt cessations of the wind, followed by extremely sharp onsets of the katabatic wind (Streten, 1963, Mather et al., 1967). Figure 5.40 shows a schematic illustration of the katabatic wind observed at Vestfold Hills, Princess Elizabeth Land (68°31′S, 78°30′E), on August 12, 1961 (Lied, 1964). The satellite station is situated approximately 29 km inland from Davis. Among the phenomena associated with the katabatic wind, the *standing hydraulic jump* or *Loewe's phenomenon* is the most conspicuous. Nearly

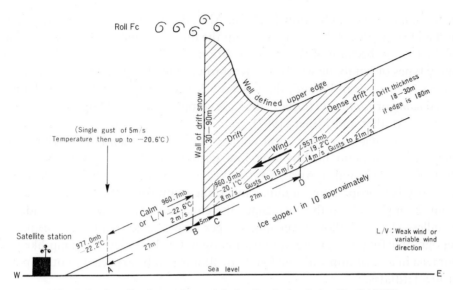

Fig. 5.40 Standing jump observed during katabatic wind at Vestfold Hills, Princess Elizabeth Land on August 12, 1961 (Lied, 1964).

all standing jumps observed from the satellite station on the foot of the slope took the form of a wall of drift snow 30 to 90 m high. On the cooled, thin air layer with strong wind speed, there is a relatively thick air layer with relatively weak wind speed, in which another thin air layer with strong wind speed is seen. The air pressure difference between in and out of the hydraulic jump is 1–3 mb in normal cases, but reaches 20 mb in extreme cases. A wall of drift snow did not always accompany the standing jump. The distinctive roar of the wind up slope, similar to the passing of a distance express train, was always a feature of the standing jump.

The vertical structure of a katabatic wind is characterized also by a relatively smooth snow surface. On the surface, consisting of finegrained frozen snow or large melting ice grains (diameter 1–3 cm), the roughness parameter, z_0 values, is 0.1–1.0 mm (Holmgren, 1971). In the layer of the katabatic wind, the shearing stress decreases with increasing height. The maximum wind speed appears at nearly the same height as the thermocline during the radiation nights with strong inversion.

López and Howell (1967) reported on the katabatic winds in the equational Andes from N-Chile to central Colombia. Even where the maritime air is at its warmest, it overflows the divide and flows down on the east slope of the range as a katabatic wind. The hydraulic jump phenomenon in the valley is often observed. Rainfall distribution is much affected by the wind conditions in this region.

Katabatic wind directions in Antarctica can be estimated by observing the *sastrugi*, mentioned in Section 4.2.4 (B). Also, the prevailing surface wind direction and strength can be estimated from the direction and gradient of the slope of the terrain, because of their close relationship. As has been stated, there are many types of katabatic winds. On Kasprowy Wierch, 2,002 m a.s.l., in the W-Tatras Mountains in Poland, the S-wind has a föhn-type katabatic wind and is called *halny* or *Liptowski* (Orlicz, 1954). A comprehensive bibliography has been made by Mather and Miller (1967).

The *morning glory* is a local name given to a frequently occurring, near-dawn squall, accompanied by long, low, narrow cloud bands, on the S-coast of the Gulf of Carpentaria, Australia (Clarke, 1972). It is concluded that the morning glory is a propagating undular hydraulic jump formed in a katabatic flow from the highlands to the east. Since the conditions (weak pressure gradients, cloudless sky, a shallow inversion, steeper slopes, a low drag coefficient and topographic funnelling), are not uncommon in low latitudes, the jump-like phenomena in association with the katabatic flow are to be expected in other places at low latitudes.

5.3.2. Strong or Characteristic Local Winds

(A) *Catalogue and Distribution of Names*

Schamp (1964) compiled a catalogue of the regular, periodical, and local winds all over the world. Not only the official names of local winds, but also the local names of the winds are contained in this catalogue. Valley breezes are called by various names, such as *Andro* (Garda Lake, Italy), *Bayerischer Wind* (Ötz Valley, Austria), *Breva* (Como Lake, Italy), *Brüscha* (Oberengadin, Switzerland), *Etschwind* (South Tyrol, Austria), *Inferno* or *Inverna* (Lake Maggiore), *Ora* (Garda Lake, Italy, and Etsch Valley, Austria), *Solaures* (French W-Alps), *Unterwind* (Upper Austrian Lakes), *Walliser Talwind* (Upper Rhone Valley), and *Wasatch wind* (Utah, U.S.A.).

Mountain breezes, on the other hand, are called *Albtalwind* (N-Black Forest, Germany), *Chanduy* (Guayaquil, Ecuador), *Elbtalwind* (Saxony, Germany), *Erler Wind* (Bavarian Inn Valley), *Gallego* (N-Spain), *Görlitzer Wind* (Saxony, Germany), *Heidelberger Talwind* (Neckar Valley, Germany), *Höllentäler* or *Höllenwind* (Höllen Valley near Freiburg, S-Germany), *Kynuria* (Plain of Sparta, Peloponnes), *Matinière* (Grenoble, S-France), *Mitternachtswind* (Upper Bavaria), *Oberwind* (Upper Austrian Lakes), *Paesano* (Garda Lake, Italy), *Pontias* (Nyons, S-France), *Rhönwind* (Sinn Valley, Lower Franken, Germany), *Schlernwind* (Etschtal bei Bozen, S-Tyrol), *Sopero* or *Sover* (Garda Lake, Italy), *Tauernwind* (Upper Kärnten, Austria), *Tivano* (Como Lake, Italy), *Viehtanerwind* (Traun Lake, Upper Austria), *Vintschgauer* or *Vintschger* (Ötz Valley, Austria), and *Wisperwind* (Middle Rhein Valley, Rheingau, Germany).

Land breezes are named *Belat* (S-Arabia), *Coromell* (Californian Gulf, U.S.A.), *Karif* (Somalian Coast), *Maskat* (Persian Gulf), *Morget* (Lake Leman, Switzerland), *Nachtwind* (SW-Africa), *Quarajel* (Bulgarian Black Sea Coast),

Terral (Spain and W-Coast of S-America), and *Vaudaire* (N-Coast of Lake Leman, Switzerland).

Sea breezes are called *Balaton-Wind* (Balaton Lake, Hungary), *Boarèn* (Garda Lake, Italy), *Brezza di mare* (Italy), *Brisas* (Uruguay), *Brises solaires* (Provence, S-France), *Emvatis* or *Embatis* (Greece), *Garbé* (Catalonia), *Imbat* (Asia Minor), *Imbatto* (Dalmatia, Yugoslavia), *Kite-* and *junkwinds* (Gulf of Siam), *Marinada* (Catalonia), *Perth doctor* (SW-Australia), *Rebat* (Lake Leman, Switzerland), *Tropaia* (Antique Greece), *Viracao* (Kongo River Mouth), and *Virazon* (W-Coast of S-America).

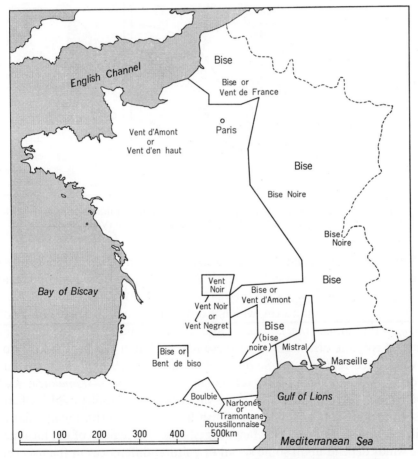

Fig. 5.41 Distribution of local names of N-wind in France (Vialar, 1948).

There are many other catalogues of wind names for countries or regions. Among them, the study by Vialar (1948) is of interest. He showed the distribution maps of local names of winds for eight directions, as well as a catalogue of wind names in France. As an example, the distribution of local names for the

N-wind is shown in Fig. 5.41. The area where the people call the north wind *mistral* is more restricted than expected. However, the area corresponds almost perfectly to the climatological *mistral* area given on the right hand of Fig. 5.42. Incidentally, the synoptic situation is shown on the left hand of Fig. 5.42.

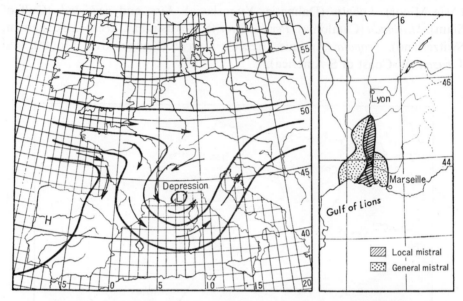

Fig. 5.42 (Left) Schema of synoptic situation of mistral in the lower mistral. (Right) Area of local mistral and general mistral (Vialar, 1948).

Sekiguti (1940) studied the local names of winds in Japan in detail and came to the conclusion that the local names of winds in Japan have developed mainly among the people who work on the sea. *Anaji* (*anaze*), *narai*, and *tamakaze* all mean the NW-winter monsoon, which is disliked by the fishermen. On the other hand, *maji* (*maze*), *hae* (*hai*), *minami*, and *kudari* are names for the monsoon in summer. From the distributions of these names, Japan can be divided into the following three regions: (a) Setouchi Region——*anaji, maji, hae*, (b) Pacific Region——*narai, minami*, and (c) Japan Sea Region——*tamakaze, kudari*. These regions correspond to the three varieties of fish hooks used in these regions. In region (a) the round type of fish hooks used, in (b), the angular type, and in (c) the longish type. This fact suggests that the regions of the local names of the winds depend upon some social or cultural differences in the fishermen's groups.

Ai, distributing along the Japan Sea, stands for the NE-breeze in summer. *hikata*, which has the same distribution as the *ai*, is the land breeze from the Chûgoku Range on summer evenings. The above two names have the longest histories. They are mentioned in the records of the Nara dynasty in the eighth century. Their distributions are very characteristic, that is, they are known only along the Japan Sea coast. *Kochi*, an E-wind, having the widest distribution, is the most popular wind name. It is generally regarded as a spring wind (or cuckoo wind), but fishermen dread it, because it often brings rain or storms. After studying the distributions of the names of winds in SW-Japan, taking into account the frequency of use, boundary lines as in Fig. 5.43 can be drawn.

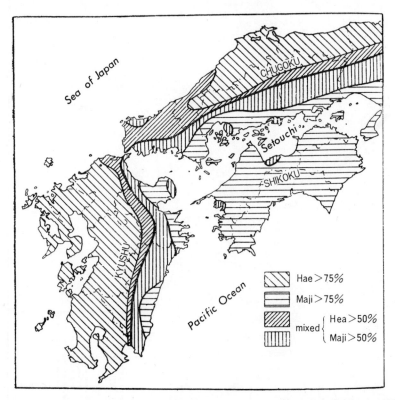

Fig. 5.43 Regions of local names of wind *maji* and *hae* in SW-Japan (Sekiguti, 1940).

Levi (1965) studied the distribution of local winds around the Mediterranean Sea, as shown in Fig. 5.44. As this figure shows, the warm southerly local winds in the hot season prevail south of 37–38°N, with some exceptions. On the other hand, the cold northerly local winds are dominant during the cold season on the N-coast of the Mediterranean Sea, again with some exceptions.

The most frequent and steady winds on the coasts of the E-Mediterranean Sea are the *etesia*, which are caused by the Asian monsoon low in summer. This low pressure trough extends from Persia and Iraq to Turkey and over the E-Mediterranean Sea. *Etesia* means "yearly" or "recurring." In Turkey, this wind is called *meltemi*.

As has been pointed out by Levi (1965), the local winds have considerable influence on the weather and local climate in Israel, in the S-Jordan Rift Valley

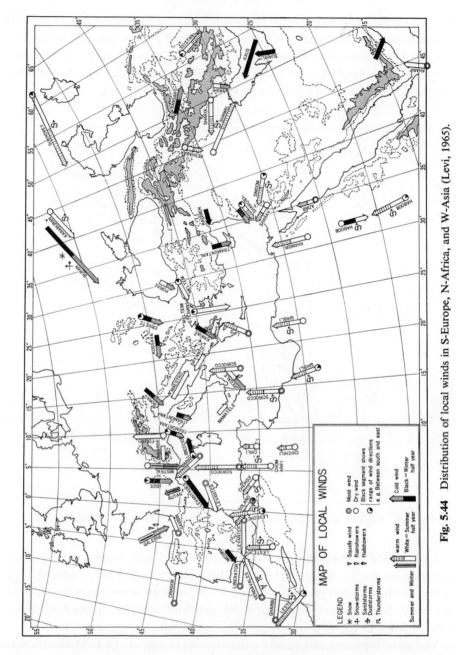

Fig. 5.44 Distribution of local winds in S-Europe, N-Africa, and W-Asia (Levi, 1965).

(the Arava); the etesian summer breezes, N-winds, are stronger there than on the coastal plain. In summer, a cool sea breeze blows from the Gulf of Aqaba, and this cooling effect is particularly noticeable during the approach of a *khamsin depression* when S-winds prevail over Sinai and the Gulf of Aqaba. The valley wind effect is well marked in the Yesre'el Valley in the Plain of Esdraelon with the advent of E-winds, either of the *sharquieh* or the *khamsin* type.

The local names of winds as well as the names of local winds in the other regions of the world have yet to be compiled. These studies are important from the viewpoint not only of folklore or anthropogeography, but also of synoptic, local meteorology and climatology. For example, the dry, hot local winds caused by the föhn effect have special local names in S-Asian countries under the influence of summer monsoon circulation, as will be mentioned below. The reason why the names are given can be attributed to the fact that the damage to agricultural production by dry, hot winds during the summer season is serious in these countries.

(B) *Examples in the World*

Manley (1945) has investigated in great detail the *helm wind* of Crossfell in N-England, a well-known strong and cold NE-wind that blows down on the western slopes of the N-Pennines. The mountain range is 800–900 m a.s.l., and the Eden River valley at the western foot, about 10 km in horizontal distance from the summit of the range, is located at 90 m a.s.l. A strong NE-wind ranges on the surface within the limits of 5 km from the summit, and in the sky at a distance of 6 km from the summit on the lee there appear Rotor clouds, which they call the *helm bar*. A gentle SW-wind, a backcurrent, blows at the ground surface below these clouds.

Meiklejohn (1955) reports from the vicinity of Milfield, England, that a strong W-SW wind blows out from the mouth of a valley into the plain exclusively in a narrow area 3 km wide and 4–6 km long; the vertical cross section of this wind is about 30 m thick on the slope of the mountain. From January to June, 1945, the winds stronger than wind force 6 occurred on 93 days (of 181 days); among them, the wind reached the strength of gales 27 times. This strong wind blows both in the daytime and at night, but is stronger at night probably due to the katabatic effects. It is thought that the chief cause of this strong wind is the local topography. In this region, clouds like the helm bar in the Eden valley mentioned above do not generally appear. The inversion layer is not necessary for the occurrence of this wind.

The strong local ESE-wind at Big Delta, Alaska, is worthy of note for its alarming frequent occurrences in winter and its unusually long duration (Ehrlich, 1953; Mitchell, 1956). There is, for instance, a record of 216-hr duration in November, 1951. Local as it is, the area influenced by the wind almost reaches 300 to 400 km in length, though only 8 km or so in width as shown in Fig. 5.45. It assumes the appearance of a jet stream gushing out of a hose. The thickness of the wind in vertical section is 600–700 m or sometimes 1,000 m. It runs at a speed of 36 m/sec at the ground level and 50 m/sec at a height of 100 m above the ground. The wind makes a meandering like a jet stream, and the wind velocity at Big Delta undergoes a considerable change. This wind occurs under synoptic conditions that induce a southerly air current to blow into the Delta valley from the cols of the Alaska Range, and at the same time when the difference in pressure amounts to 3 mb or more between Big Delta and Northway

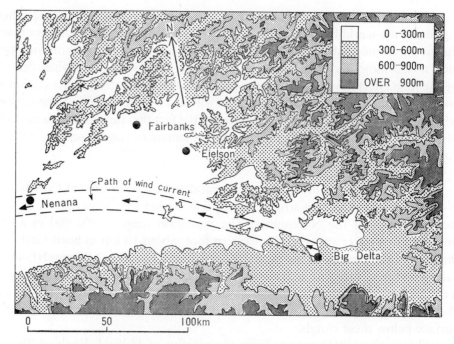

Fig. 5.45 The path of strong ESE-winds near Big Delta, Alaska, and the topographical situation (Ehrlich, 1953).

260 km SE of Big Delta in the upper courses. The path of the strong wind is easily recognized by the blowing snow it carries, sufficient at times to obscure the ground almost completely from aerial observation. In places where the ground could be seen to some extent, the trees were being whipped around, and some broken or uprooted.

The *squamish* is a strong, local wind caused by the cold polar air, which is derived from anticyclones up-country and flusters down the fjords in British Columbia, Canada (Kendrew and Kerr, 1955). It blows strongly within 25–30 km outside the fjords and up to approximately 900 m. The name *squamish* comes from a settlement at the head of Howe Sound, a N–S fjord that joins Burrard Inlet west of Vancouver. According to an investigation from September 1, 1949, through March 31, 1950, in 192 cases (46% of the total occurrences) the winds were NE, and in 73 cases they gained a gale force. Sometimes, the wind continued to blow with a gale force 2–3 days.

A wind called the *koshava* (*kossava*, *koschava*) blows from E–SE on the Pannonian plain, NE-Yugoslavia, under the synoptic situation with an anticyclone situated over Ukraine and Rumania and a cyclone over the Adriatic Sea and the W-mountain region. It continues to blow for several days and in an extreme case blew for 20 days. Milosavljević (1959) reported on the synoptic situation based on the results of observations from March 25 to 29, 1957, and Čadež (1954) investigated theoretically the influence of the Transylvanian Alps.

Table 5.12 Wind velocity of *Koshava*, from March 25 to 29, 1957
(Milosavljević, 1959). V_0: Daily mean, and V_2: Daily
maximum. The wind direction was ESE–SE–SSE.

Station	Height a.s.l.		25	26	27	28	29
Vršac	84 m	V_0	20.0	12.6	10.0	9.8	11.5 m/sec
		V_2	37.3	27.3	24.8	20.8	21.2
Belgrade	244	V_0	21.9	17.6	13.6	11.1	9.1
(Aerol. Obs.)		V_2	33.1	33.8	25.0	17.1	14.3
Novi Sad	134	V_0	15.2	10.2	10.5	7.4	4.0
		V_2	29.6	29.0	21.4	19.0	15.3

The wind velocity during the period is given in Table 5.12. The horizontal pressure gradient is 3.7 mb between Veliko Gradiste and Belgrade.

The *sukhovey* (*suhovei*) is a hot, dry wind that prevails in a broad region covering the desert in central Asia and the north of the Aral Sea and the Caspian Sea, Kazakh, U.S.S.R., and the surroundings. It becomes most severe in the region along the Volga River. The *sukhovey* has been studied intensively by many scholars. For instance, its microclimatic and climatic characteristics in the Caspian plain were summarized by many researchers of the Institutes of Geography and Forestry, Academy of Science (Akademiia Nauk U.S.S.R., 1953). On the problem of the origin and control of the *sukhovey*, 30 papers appeared in the second monograph (Akademiia Nauk U.S.S.R., 1957; Dzerdzervski, 1963). The latter contains a complete bibliography for the period 1917–1955.

The sukhovey develops in the warmer season, especially in June, July, and August, and the annual occurrence amounts to 40–80 days. The strongest sukhovey is found in a region centered around Astrakhan, Volgograd, and Saratov along the Volga River. Synoptically, it develops when this region is occupied by a subtropical anticyclone, which brings prolonged hot weather. In addition, the relatively dry, polar air mass from the Arctic Ocean or the N-Atlantic Ocean invades this region as a modified dry continental tropical airmass. At the ground level, the air temperature is 35–40°C, the relative humidity 15–20%, the wind direction E–SE, and the wind velocity sometimes 20 m/sec.

Evseev (1957) analyzed the atmospheric processes from 1938 to 1952 and selected 21 natural synoptic periods. A result of his study on the trajectories is shown in Fig. 5.46. This figure shows that the air trajectories mentioned above are seen at the 700 mb level. The concentration of the trajectories of the *sukhovey* is striking in the region along the Volga River.

In the region of Kazakhstan, Samokhvalov (1957) studied the time of appearance and the duration of the *sukhovey*. He states that a steady increase of the number of days with the *sukhovey* from the 48th parallel to the south was seen during the two decades from 1930 to 1949. This might be attributed to the warming up observed in the Arctic region during this period.

According to the Russian agricultural record, drought due to the *sukhovey* occurred in 34 years in the 18th century and in 40 years in the 19th century. This

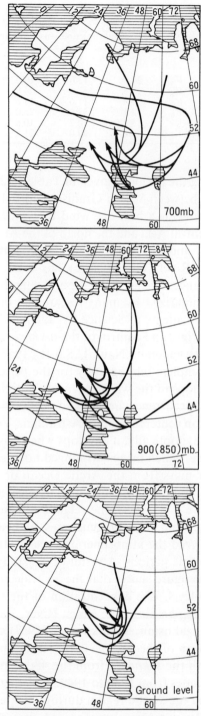

Fig. 5.46 Schematic illustration of air trajectories of the *sukhovey* at the 700 mb, 900 (850) mb and ground level (Evseev, 1957).

means that it occurred on an average once in every 2 or 3 years. The problem of shelterbelts to protect the arable land from this hot, arid wind has arisen.

(C) *Examples in Japan*

In Fig. 5.47, the distribution of the strong local winds in Japan is shown. Generally speaking, the local winds toward the Pacific Ocean on the Pacific side occur during the winter in association with the outflow of the winter monsoon. The local winds toward the Japan Sea blow when a cyclone is traveling over the Japan Sea. Some of the strong local winds in the inland areas are the strong mountain winds or the strong cold air drainage. Some typical winds among them are discussed in detail below.

Fig. 5.47 Distribution of local strong winds in Japan (Yoshino).

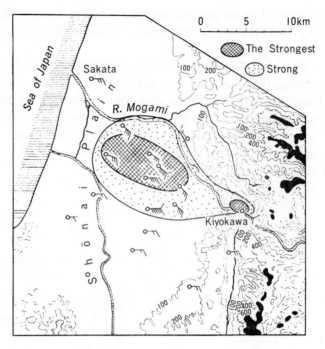

Fig. 5.48 The *kiyokawa-dashi* region, at 10h, January 18, 1950 (Yoshino, 1961a).

Kiyokawa-dashi

The *kiyokawa-dashi* blows where the Mogami River comes out on the Shonai plain, Yamagata Prefecture, from a valley. As shown in Fig. 5.48, it develops in a restricted area of 15×10 km. This wind has often been the subject of theoretical studies and aerological and local observations (Sendai Met. Observ., 1950). According to the results, the *kiyokawa-dashi* occurs most frequently with great persistency in June (See Table 5.13). The synoptic conditions of this strong local wind are: (a) an anticyclone over the Sea of Okhotsk or the Chishima Retto and a cyclone over the Japan Sea. The cyclone functions as a strong suction source. (b) Most of the isobars at the upper level, e.g., the 500 mb level, run in a N–S direction. (c) At the ground level, therefore, the gradient wind direction should be E or SE.

The valley topography of the Mogami River strengthens the wind and makes it a strong SE-wind on the Shonai plain. When the pressure distribution

Table 5.13 Monthly occurrence frequency of *kiyokawa-dashi*, 1940–1949 (Sendai Met. Obs. 1950).

Wind velocity	Apr.	May	June	July	Aug.	Sept.	Oct.	Total
7–10 m/sec	10	24	20	18	27	15	13	127
10 m/sec or more	15	17	37	18	21	17	19	144
Total	25	41	57	36	48	32	32	271

is especially favorable to a violent *kiyokawa-dashi*, similar phenomena appear in ravines of various sizes within the same mountain system, as exemplified by *tachiyazawa-dashi*. As this tendency becomes pronounced, the wind begins to blow not only from the ravines, but also from low cols of mountain ranges, such dashi winds as *chôkai-oroshi* and *gassan-oroshi* (*tsukiyama*).

The mechanism of the growth of the *kiyokawa-dashi* can hardly be said to have been investigated beyond doubt. Furthermore, no full account has been made of the fact that the wind force is not the only property of the dashi, but low temperature and low humidity are also characteristic of it, and that the maximum height of this wind is at 400–600 m, the layer for the strongest wind (15–20 m/sec) coming at 100–400 m. There are some similarities between the *kiyokawa-dashi* and the bora, but the problems mentioned above still remain to be solved.

Hiroto-kaze

The *hiroto-kaze* is a strong wind blowing over the southern foot of Mt. Nagi (1,240 m), on the border of Okayama and Tottori prefectures. It blows in an area of 10 × 10 km, but the most serious damages concentrated in an area, of 5 × 2 km. The *nagi-oroshi*, the *hiroto-kaze*, has other nicknames, like *yokoze-kaze*, *matsubori-kaze*, *kita-kaze*, *hotokoro-wind*, *kazenomiya-no-kaze*, and *yama-shita-kaze*. A shrine with a large hole (cave), named kazenomiya, is located 3 km NE of Hiroto-ichiba. Formerly the local people believed that a strong wind blows out of this hole.

Table 5.14   Monthly frequency of occurrence of *hiroto-kaze*, 1897–1955 (Osaka Met. Obs. 1956).

J	F	M	A	M	J	J	A	S	O	N	D	Year
1	0	0	1	4	5	11	22	42	40	9	0	135

The frequency of occurrence of this wind, as shown in Table 5.14, is highest in August, September, and October, covering 77% of the total frequency. The *hiroto-kaze* is often produced when a typhoon or a small tropical cyclone proceeds to the NE or E along the coast of the Pacific Ocean, so that this type of wind is likely to occur from August to October, the typhoon season.

A comprehensive study on the structure of the *hirota-kaze* has led to the following conclusions (Osaka Met. Obs., 1956): (a) Stratification of the atmosphere is requisite for the birth of the *hiroto-kaze*. At the same time, the wind velocity in the layer just above the summit height must be extremely great in comparison with that in the ground layer, and the wind direction must be northerly. (b) The air current on the lee side first blows down like a waterfall toward the ground and then leaps upward. Consequently, the area affected by the *hiroto-kaze* is shifted to the north or south in accordance with the intensity of the wind velocity in the upper layer. (c) The presence of a valley extending from north to Mt. Nagi, and the topographical condition that the lee side is flat and wide may contribute to the development of the *hiroto-kaze*. (d) It is not always accom-

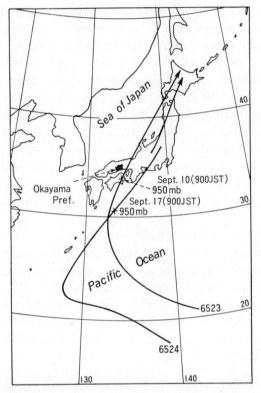

Fig. 5.49   Tracks of typhoons Nos. 6,523 and 6,524 of
September 1965 (Yoshino, 1968b).

panied by the föhn effect. Figure 5.49 shows an instance of the paths of typhoons
in September 1965, which caused a strong *hiroto-kaze*. The first typhoon, with
the lowest pressure of 969.3 mb at Okayama City, passed on September 10. The
maximum gust reached 26.5 m/sec. The second one, 935 mb at the lowest, hit
Japan on September 17. The estimated wind velocity in the area mentioned was
over 10 m/sec between 8 h and 24 h with the maximum of 40 m/sec between 19 h
and 21 h. After these severe local winds, the damage in Nagi-machi amounted
to $880,100: (agricultural products, $592,300; houses, $232,900; the rest
$54,900). This is not a small amount for the economy of this small village.

The distribution of the damage was restricted to an area of 6 km (west-east)
by 3 km (north-south) as shown in Fig. 5.50. The damage of the wind that hit
the area on September 25, 1953, was concentrated slightly to the west, and on
November 23, 1949, the center of the damaged area was found about 4 km north
as compared with the case on September 17, 1965, but the size of the damaged
area was the same. From this it can be said that the size of the area is almost
always 5–6 km (west-east) by 3–4 km (north-south), but the area damaged
changes slightly in each case (Yoshino, 1968b).

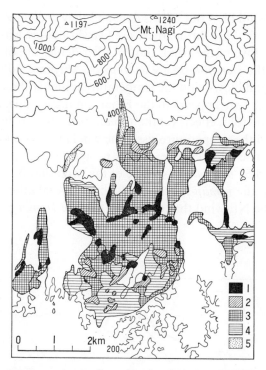

Fig. 5.50   Distribution of damage by the *hiroto-kaze* local wind on September 17, 1965. 1: Serious damage to houses. 2: Light damage to houses. 3: More than 80 percent. 4: 40 to 80 percent. 5: 20 to 40 percent (Yoshino, 1968b).

Yamaji-kaze

The *yamaji-kaze* or simply *yamaji*, though quite contrary to the *hiroto-kaze* in wind direction and topographical conditions, seems to be structurally identical with the latter. If the location of the cyclone in the case of the *hiroto-kaze* is like the N-föhn in the Alps, that of *yamaji-kaze* is like the S-föhn in the Alps.

The *yamaji-kaze* blows on the N-foot of the Ishizuchi Mountain Range, Shikoku—i.e., the area covering the cities of Iyo-mishima, Kawanoe, and Doi—from February through October. A strong wind velocity of over 10 m/sec is expected in April, May, and June, when a cyclone develops short time in the Japan Sea, or in September and October, when a typhoon passes over the W or NW of this region. Wind velocities of over 15 m/sec are observed 2–5 times annually. Crest clouds hanging over a mountain, and the rumbling of a mountain, as is the case with the *hiroto-kaze*, portend the advent of the *yamaji-kaze*. At first a wind of a frontal nature begins to blow, so that the wind direction at the foot varies from time to time, producing an intense gust and a dust storm (Osaka Met. Obs., 1958). The residents of this district, therefore, also call it the *mai-mai-kaze* (stray wind).

As in the Hiroto village, here again is found a wind den on the W-side of the Toyouke shrine on the summit of Mt. Toyouke. In former times people believed that the wind blew out of this den, and they dedicated this shrine to the God of Wind. Besides this shrine, there are five shrines for wind prevention and one shrine for the love of the wind, though their deities are of unknown origin.

Processes of the *yamaji-kaze* are as follows (Akiyama, 1956): (a) On the lee side, the N–NE wind blows, and the temperature is higher than on the windward side; (b) the *yamaji-kaze* appears in the E-part on the lee side, and the temperature rises due to the föhn effect; (c) the *yamaji-kaze* appears in the western part on the lee side, and the temperature rises; the wind appears and disappears from time to time; (d) The wind changes into a westerly wind, and the temperature desclines. Figure 5.51 shows an example of the wind that blew on February 27, 1954. Incidentally, the *yamaji-kaze* extends some several kilometers over the sea at its peak, and the NE-wind that fishermen call the *Domai* prevails further out at sea.

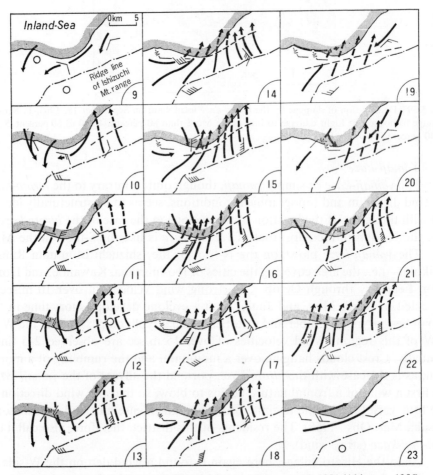

Fig. 5.51 Hourly streamlines of the *yamaji-kaze* on February 27, 1954 (Akiyama, 1956).

Perhaps the organization of the *yamaji-kaze* is, just like that of the *hiroto-kaze*, created by the lee wave, but details are still unsolved. Besides the winds mentioned so far, there are other strong local winds in Japan, such as the *hira-hakko* wind on the W-coast of Lake Biwa and the *rokko-oroshi* that blows exclusively in the vicinity of Kobe Port (Yabuki and Suzuki, 1967).

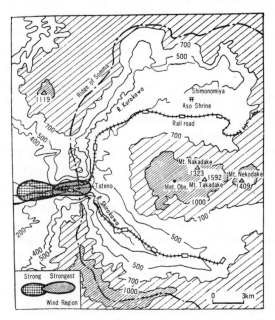

Fig. 5.52   *Matsubori-kaze* region, W of Mt. Aso, Kyushu (Yoshino, 1961a).

Matsubori-kaze

The junction in the crater basin of Mt. Aso, where the Kurokawa River running from the north joins the Shirakawa River running from the south, and from where the confluence flows westward across the somma, is called Tateno. Out of a cold air pool formed in the crater basin, cold air begins to flow outside the somma through Tateno during the night. The topographical situation is shown in Fig. 5.52. Locally, it can be inferred that the strongest wind blows in the area ranging from 1 km E of Tateno to a point about 3 km W, since there are the most developed wind breaks in this area. According to an observation of the *matsubori-kaze* at Tateno and Ohzu from November 10 through 21, 1949, the wind blew very hard when this district came within the range of a migratory anticyclone. The maximum wind velocity appeared twice a day between noon and 16 h and between 19 h and 24 h. On November 15, the maximum appeared during the night; 10.5 m/sec was recorded at 19 h at Tateno and 7.3 m/sec at 20 h 03 m at Ohtsu. The maximum in the afternoon was something like a valley wind, while the midnight maximum seemed to be an outflow of cold air, which is discussed below in Section 5.4.1. There are more wind breaks on the river terraces near Tateno than in the valley basin. From this it can be inferred that the wind blows harder on the valley slopes than in the bottom of the velley.

Wind breaks here are cultivated in order to preserve rice plants from the disease that is supposed to be caused by the *matsubori-kaze*. The wind breaks consist mainly of broad-leaved evergreen trees and also of coniferous trees (Japanese cedar, etc.) or bamboo grass over a small cliff. Many of them are arranged from north to south so as to be at right angles to the contour lines, that is, at right angles to the prevailing wind directions. The word *matsubori* means of surplus. This word is said to have derived from *matsubori-zeni*, which means, in the dialect, a fair amount of money, a fortune that has been made out of many small sums through small trade. The godfathers of this name may not have known the fact that the *matsubori-kaze* is caused in fact by an accumulation of little bits of cold air. Anyway, the name is very appropriate.

Local winds in Hokkaido

Arakawa (1969) studied the local winds in Hokkaido from the standpoint of local climatology as well as dynamic meteorology. In Hokkaido, the following six local winds are unusual: *Hidaka-shimokaze* prevails in the region around Urakawa when a cyclone passes south of Hokkaido in winter or spring. It has a secondary maximum in September when a typhoon passes through the same path. Winds more than 10 m/sec occur about 7 times a year at Urakawa. For example, Fig. 5.53 shows the case on September 27, 1959. The *teine-oroshi*, observed at Otaru, prevails when a cyclone passes through the N-part of Hokkaido in spring. Wind velocity is more than 10 m/sec about 6 times a year. The *tokachi-kaze* prevails on the Tokachi plain when W–NW winds dominate in the upper air layer in spring and winter. Winds more than 10 m/sec occur 8 times a year. The *suttsu-dashikaze* prevails in the Suttsu region when an anticyclone lies in the Okhotsuk Sea after a traveling anticyclone passes over Hokkaido in spring and

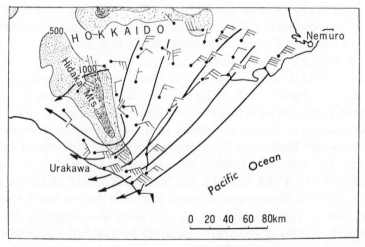

Fig. 5.53 Streamlines in the case of *hidaka-shimokaze* at 9h, September 27, 1959 (Arakawa, 1969).

summer. This wind is more frequent than the others: Winds of over 10 m/sec occur about 48 times per year. The duration of the *suttsu-dashikaze* is longer than that of other local winds in Hokkaido. The *hikata-kaze* prevails when the pressure pattern is similar to that of the *tokachi-kaze* in spring. The frequency of the *hikata-kaze* of more than 10 m/sec is about 27 times a year; that more than 15 m/sec is about twice a year. The data available on the *rausu-kaze* are insufficient to clearly describe its climatological features. The *rausu-kaze* causing great damage occurs mostly in spring associated with the passage of a developed cyclone. An example is shown in Fig. 5.54. A theoretical investigation of these winds is presented by Arakawa (1969) and numerical experiments by the same author (Arakawa, 1973).

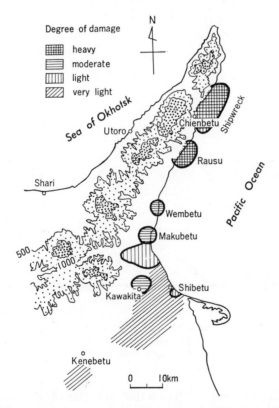

Fig. 5.54 Topography of the Shiretoko Peninsula, NE-Hokkaido, and areas damaged by *rausu-kaze* on April 6, 1959 (Arakawa, 1969).

5.3.3. Classification and Scheme

(A) *Föhn and Bora*

In a comprehensive study on the synoptic problems in the SE-region of the Alps, Čadež (1964, 1967) summarized the effects of mountains on synoptic situations as follows: (a) Generation of a calm region and windy region, (b) formation of a cold air lake on a broader scale, (c) characteristic changes of winds with height and time, (d) characteristic wind distribution with stationary pressure fields, (e) a mountain wave in the lee, and (f) change of stream fields and weather development in relation to the cold air invasion and the general change in pressure fields. Čadež (1967) pointed out that there are three types of föhn, varying according to the relation of the air temperature on both sides of the mountain—$T \doteqdot T_1$, $T \gg T_1$ or $T < T_1$—and the pressure fields, which are shown in Fig. 5.55. He also gives two types of bora as shown in Fig. 5.56.

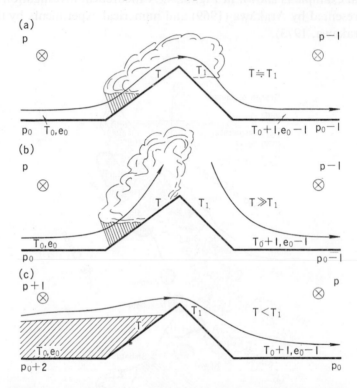

Fig. 5.55 Three types of föhn (Čadež, 1964, 1967). (a) Cyclonic föhn in a stable atmosphere with stronger winds, (b) cyclonic föhn in a less stable atmosphere, and (c) anticyclonic föhn with a damming up of cold air. T: air temperature, p: air pressure, and e: vapour pressure. Suffix 0: values at the ground level, and 1: values on the leeside slope.

It should be clear from this figure that the state of the overflow of cold air depends on the initial condition of pressure fields: the direction of the pressure gradient at higher levels. If the air temperature and the humidity at the lee side mountain-foot are compared with those at the windward mountain-foot the föhn effect occurs within the cold air.

As has been mentioned in Section 5.3.1 (A), (B), (C), and (D), there are many similar characteristics among the föhn, chinook, and bora. It can be said that the fall winds on the lee side of mountains must have the character of the

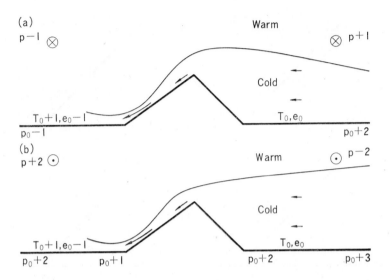

Fig. 5.56 Two types of bora (Čadež, 1964, 1967). The pressure gradient at higher levels is directed (a) in the direction of bora and (b) in the opposite direction of bora.

föhn or the bora, but it is very difficult practically to distinguish between the föhn and the bora in some cases. For instance, if the cold air pool on the windward side in the case of type (c) of Fig. 5.55 becomes thicker and flows to the lee side over the ridge, type (a) of Fig. 5.56 will occur. This means that the föhn blows at first, and then the bora occurs. This is actually what is often observed. Another example may be as follows: If the cold air becomes thicker and thicker in the case of type (a) in Fig. 5.56, that is, if the top of the inversion layer becomes much higher than the ridge level, the föhn effects in the cold air of this type are hard to distinguish from the föhn effects in the case of type (a) in Fig. 5.55. Actually, there are many transistional or combined states of cases shown in Figs. 5.55 and 5.56.

   In my opinion, therefore, the definitions of the föhn and bora should be made in the simplest way as follows: The föhn is a fall wind on the lee side of a mountain range. When it blows, the air temperature becomes higher than before on the lee side slope. The bora is also a fall wind on the lee side of a mountain range, but when it begins, the air temperature becomes lower than before on the lee side slope. This definition can be applied regardless of the vertical structure along the cross section, which is not always easy to observe for every case in the field. The above definitions are explained graphically in Fig. 5.57: The full straight line is a dry adiabat by the föhn or the bora. According to the definitions mentioned above, the air temperature distribution before the föhn on the lee side slope of the mountains must be plotted on the left side of this line as shown by the broken lines. The broken lines A_1–A_3 and B_1–B_3 may also be expressed by curves. On the other hand, the air temperature distribution before the bora on the lee side slope of the mountains must be plotted on the right side of

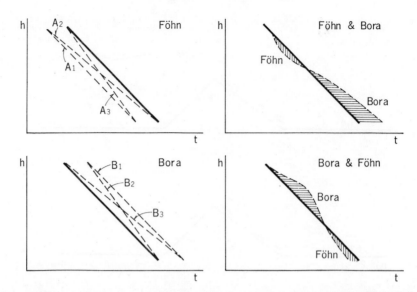

Fig. 5.57 Definition of föhn and bora. Full straight line shows a dry adiabat. Broken lines present the vertical distribution of air temperature on the lee side slope before the föhn or bora. (Yoshino).

this line. Cases like A_1, A_2, B_1, and B_2 can be brought about by the replacement of air masses over the area; by the intensification of the hot summer monsoon, or by the cold polar air outbreaks. Cases A_3 and B_3 can be named pure föhn and pure bora, which originate only by the intensification of a descending current. The complex features of cases A_3 and B_3 are föhn on the upper part and bora on the lower part, or bora on the upper part and föhn on the lower part of the slope. These cases have been discussed by Schmidt (1947). For instance, the *bohorok*, a fall wind in E-Sumatra, is warm on the slope but cold on the plain.

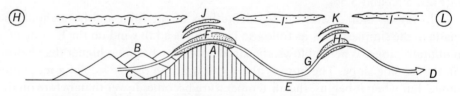

Fig. 5.58 A scheme of structure of fall winds, the föhn and bora, in relation to topographical conditions (Otani, 1956, revised by Yoshino).

(B)   *Scheme of Fall Wind*

As a summary of the structure and the topographical conditions mentioned above a schematic illustration of fall winds, including the föhn and bora as defined above, is given in Fig. 5.58. The fall wind blows from left to right in this figure. A is the mountain range or pass situation with a height of approximately 1,000 m $\pm$ 200 m. This altitude is greater in the region of a great mountain mass like the Alps or the Rockies. The inclinations of the lee side slope of *A* are steep. *B* is the mountainous region, not a plain. *C* is a valley that runs at right angles to the running direction of the mountain range *A*, but is not always an essential factor for the development of the fall wind. *D* is a plain or broad water surface, such as a sea or a great lake. *E* is the mountain foot of the lee side slope, where the fall wind is very gusty. Raindrops often sprinkle from the summit. *F* is a cap cloud on the ridge, called the föhn wall on the Alps, the *crest cloud* on the Rocky Mountains, the *chinook arch* on the Canadian Rockies, the *kapa* on the Dinar Alps and the *helm* on the Crossfell Range, Cumberland. *G* is a jump, a so-called hydraulic jump, of the fall wind. *H* is a cumulus cloud or Rotor cloud, named the *helm bar* on the lee of the *helm* and *tresa* (or *teresa*) on the lee of the *kapa*. Lee waves are mentioned in Section 5.3.4 in some detail. *I* is in most cases the altostratus and, over the top of the wave, it disappears. In some cases, however, no altostratus appears over whole regions. *J* is the lenticular clouds over the crest cloud. *K* is also lenticular clouds on the lee wave. A striking example is the *Bishop wave* formed in the lee of the Sierra Nevada, California.

(C)   *Classification of Local Winds*

Excluding the land and sea breezes and mountain and valley breezes, Schmidt (1947) made an attempt to classify the local winds all over the world. The characteristics of local winds pointed out by Schmidt are: (a) The wind velocity is strong and the frequency is great when the direction of the wind is definite; (b) a type of heavy squall line is present, and the wind occurs suddenly; (c) the air brought by this wind is extremely hot or cold and extremely humid or dry; and (d) the wind is accompanied by precipitations, sand, or dust. Schmidt classified the local winds into four groups, based mainly on the meteorological causes:

Group I is winds caused by a special field of pressure: This is not so important meteorologically, because the special fields of pressure can occur in various cases. For example, a strong wind is the *Elephanta* on the Malabar coast, a wind with a special direction is the Portuguese Trade on the west coast of Portugal, and the frequent wind is the *Etesia* in Greece.

Group II is winds caused by the special relief of the earth's surface in relation to a certain synoptic situation: The warm fall wind is the föhn and the cold fall wind is the bora. They can be defined as described in (A) above. An example of an air current strengthened by the valley topography is *mistral* in the lower Rhône in France.

Group III is winds that bring a characteristic air property from the source region of an air mass: For example, the *blizzard* in Antarctica, the *buran* in Siberia, the *sukhovey* in the Volga River region, the *tramontana* along the N-Mediterranean Sea coast, the *scirocco* in N-Africa or Near East and the *zonda* in Argentina are local winds belonging to this group representing the arctic, polar and tropical air masses.

Group IV winds are caused by the passing of a front or squall line: For example, the *pampero* or *pampeiro* in the La Plata region, Argentina, or *Sumatras* in the Malakka strait.

5.3.4. Air Flow over Mountains and Lee Waves

(A) *Observations*

Glider pilots found in the late 1930s that there were large areas in the neighborhood of mountains where the flow was inclined between 10° and 20° to the horizontal plane, providing the pilots, when the flow was upward, with places where they could soar to around 10 km above the ground (Scorer, 1958). Scorer adds that there was often a very marked inversion of temperature just above the clouds and that the strongest up-currents were a few kilometers down wind of the mountain ridge.

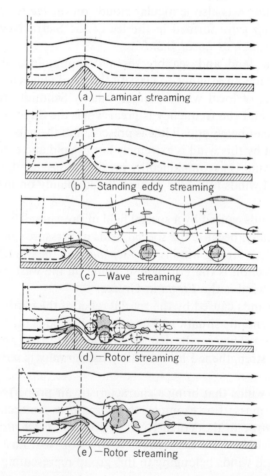

Fig. 5.59 Classification of types of air flow over ridges. The nature of the flow is determined mainly by the wind profile indicated on the left in each case (Förchtgott, 1949).

Prior to these observations, Koschmieder (1921) had described the lee waves, *Moazagotl*, behind the Riesengebirge (Sudety) between Czechoslovakia and Poland. The cylindrically connected cumulus-type clouds lay parallel to the mountain ridge and were 20 km long in some cases and 8–15 km down wind of the ridge. Küettner (1939a, b) reported the lee waves on the Riesengebirge and the Hirschberg valley to the lee of Schneekoppe ridge. Among his results, the following must be noted: (a) There are two cases: first one is a *föhn wave* with a sinking wave and rotor, and second one is *lee eddies* that take place immediately to the lee of the mountain. In the light of recent information, the first case is a kind of cyclonic föhn with relatively strong upper winds, and the second case occurs when the winds are relatively weak. (b) The wavelength at a higher level is longer than at a lower level. This fact has been investigated in great detail observationally, theoretically, and experimentally recently.

Manley (1945) carried out an investigation of the *helm wind* of the Crossfell Range, Cumberland, England, as mentioned above. On the first lee wave appears a slender cigar-shaped rotor cloud, called the *helm bar*, 5–8 km downwind from the ridge and remaining nearly stationary at a height of 1,200–2,000 m above the ground.

Förchtgott (1949) presented an example of a model of air flow over mountains and lee waves, as shown in Fig. 5.59, based on the flight observation by gliders and light aircraft. He points out that there are four types of air currents over mountains according to the distribution of vertical velocity, as shown on the left side in each illustration. There are: (a) Laminar streaming producing no eddies, (b) Standing eddy streaming, a flow producing a large stationary eddy on the lee; above this the air flows smoothly through a shallow wave, (c) Wave streaming that occurs when the stronger winds increasing with altitude produce a chain of waves and eddies. There are lenticular clouds in the crests of the waves and rotor clouds at lower levels. (d) Rotor streaming that occurs when a strong wind is blowing in the lower air layer, comparable to the height of the mountain. Due to the system of quasi-stationary vortices rotating in opposite directions, severe turbulence is experienced. The eddies and waves in the non-laminar case are of importance, because these conditions have as yet not been treated theoretically.

In the Jämtland Mountains, Sweden, Larsson (1954) carefully observed the lee wave phenomena. The main conditions discovered are: (a) The ascending currents on the windward side of wave clouds are much greater than the descending currents on the downwind side, which implies that the structure of the waves is asymmetric. (b) The wavelengths fall in the range of 5–25 km with a majority around 8–10 km, but the length of the first wave (distance from the mountain ridge to the first wave cloud) generally shorter than the length of the second wave. (c) The wave clouds appear in all seasons, but with striking maxima in spring and autumn.

Holmboe et al. (1957) investigated in great detail the lee waves on the Sierra

Nevada, U.S.A. Important results of their study are the observations up to the stratosphere: e.g., the long wavelength reaching about 20 km, the reversal of the wave phase between 6,000 m and 10,500 m, and the single very long wave in the stratosphere at 12,000 m.

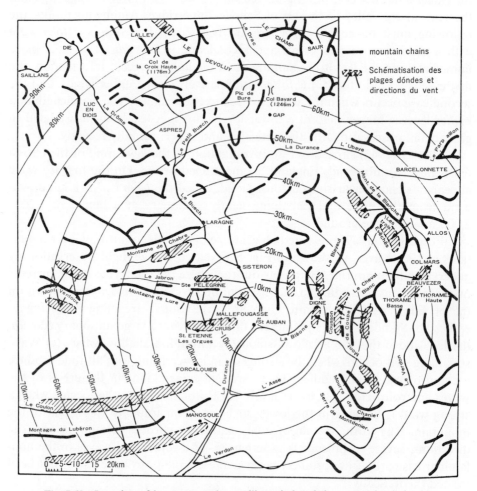

Fig. 5.60 Location of lee waves and prevailing wind and the mountain chains in the French Alps (Gerbier and Bérenger, 1960).

Gerbier and Bérenger (1960, 1961, 1963) and Gerbier and Cachera (1961) reported the wave locations and wind directions in relation to the topography of the French Alps. One of their interesting observations is the change of wavelengths, amplitudes, and position of lift of the wave with height. The wavelength increases and the amplitude decreases with height, as mentioned above. The region of lift of the waves slopes towards the mountain from the first lee wave, but lift in the second lee wave may be vertical or even slope away from the mountain if the wind increase is substantial. The distribution of the lee waves in the French Alps is shown in Fig. 5.60. This figure was drawn through a careful examination of the lee wave by means of airplanes or other methods in January and February from 1956 to 1959. The lee waves appear when the prevailing wind is northerly or westerly. If the relationship between the area where the lee waves rise and the mountain range is examined, the topographical conditions mentioned above (Section 5.3.3 (B)) seem to hold good here also. As a matter of course, the waves become greatest when the amplitude of the lee waves and the distance between the mountain ranges on the lee coincide.

Recently, a very highly sensitive 107 mm radar has been used for studies of lee waves downwind from the Welsh mountains, England, and the results show that in one case a steady-state wave pattern was observable, and in another case an unsteady pattern with waves varying in orientation and wavelength was observed (Starr et al., 1972). The wavelengths were very different from one case to another: e.g., 8–10 km in one case and 15–30 km in another. Viezee and Collis (1973) investigated the mountain waves in the lee of the Sierra Nevada, U.S.A., with lidar (laser radar), and detected the lenticular cloud structure at all times and temporal and spatial variations in the vertical extent of the lower turbulence zone.

For analysis of the wavelength of mountain lee waves, satellite pictures are very useful. Döös (1962) made a theoretical analysis based on the TIROS I pictures; Fritz (1965) showed the relationship between the wavelength observed on the TIROS pictures and the mean wind speed averaged through the troposphere; and Cohen et al. (1967) studied the mountain lee waves along the Jordan rift valley in the Middle East, using the TIROS VI and VIII pictures.

Reynolds et al. (1968) studied a complex lee wave pattern, defined as a series of waves in horizontal succession displaying variable wavelengths and amplitudes. Of over 200 radar plots, 30% showed no detectable wave, 50% showed a simple wave pattern, and 20% showed a complex pattern. Mean altitude was 3.5 km a.s.l., and simultaneously, a very weak wave with a very small amplitude was recorded at 7 km, according to an observation at White Sands Missile Range, U.S.A., on May 6, 1965.

Table 5.15 Wavelength of lee waves observed in various regions
(Compiled by Yoshino).

Mountains	Wavelength	Literature
Riesengebirge (Sudeten)	8–15 km	Koschmieder (1921)
Crossfell, England	5–8	Manley (1945)
Jämtland Mountains, Sweden	8–10	Larsson (1954)
Sierra Nevada, U.S.A.	4.4–28	Holmboe et al. (1957)
French Alps	5–10	Gerbier and Bérenger (1960, 1961)
Rocky Mountains, U.S.A.	11–25	Vergeiner and Lilly (1970)
British Isles	5–13	Stringer (1972a)
Black Mountains, England	{ 8–10 { 15–30	Starr et al. (1972)
Dinar Alps, Yugoslavia	10–12	Yoshimura et al. (1974)

The observed wavelengths of the lee waves in the various regions are summarized in Table 5.15. The values vary in the range from 4 to 30 km with a majority of cases 10 km. The rotor that frequently forms to the lee is as large as 3–6 km in diameter. This rotor is by nature highly turbulent. Generally, the relation between such a rotor cloud formation and the change of wind velocity and gustiness is clearly seen as revealed at the lee of the Little Carpathian ridge (Förchtgott, 1969). The turbulent layers appear to restrict the upward propagation of the larger scale waves, and consequently they absorb much of the drag effect of the mountains on the atmosphere (Lilly, 1971).

There are many comprehensive monographs on the air flow over mountains and lee waves. Among them, the report of a working group of the Commission for Aerology, WMO, is of special importance (Alaka, 1960). Similar problems were also treated by Musaelyan (1964) based on results obtained mainly in the U.S.S.R., with special reference to the formation of lee waves, lee eddies and the orographic cloudiness and turbulence. Crowe (1971) and Stringer (1972a) reviewed the phenomena concerning the lee waves in detail. Nicholls (1973) wrote a well-documented review supplementing the report by Alaka (1960).

(B)   *Theory*

Queney (1947) devised a wave equation, which includes three parameters, k, l_s, and l_f. In it k is the wave-number of the disturbance in the direction of the wind; l_s refers to the effect of stability of the airstream and is expressed by

$$l_s = \sqrt{\frac{g\beta}{U}} \qquad\qquad (5.7)$$

where $\beta = \dfrac{1}{\theta} \dfrac{\partial \theta}{\partial z}$, the coefficient of static stability, and U is the horizontal wind speed.

The effect of the earth's rotation l_f is

$$l_f = \frac{f}{U} \qquad\qquad (5.8)$$

where f equals twice the vertical component of the earth's angular velocity. Queney's pioneer work irrefutably proves that orographic waves are an essential property of the atmosphere, but his model does not agree particularly well with the real atmosphere (Stringer, 1972a).

Lyra (1943) studied lee waves theoretically in a uniform stability and velocity up to infinity. The main difference between Lyra's solution and the real lee waves observed is that the real lee waves have a maximum amplitude at a height of 2–3 km above which they die away. A more realistic model has been provided by Scorer.

He showed that the wavelength, λ, and amplitude, ξ, are expressed by the following equations (Scorer, 1949):

$$\lambda = \frac{2\pi}{l} \qquad\qquad (5.9)$$

where l, named Scorer's parameter, is expressed by;

$$l=\frac{1}{u}\sqrt{\frac{g}{\theta}\frac{\partial\theta}{\delta z}}\qquad(5.10)$$

where u is the wind component (m/sec) perpendicular to the running direction of the mountain ridge at the height of the summit level; g is the acceleration of gravity; θ is a potential temperature. Then, amplitude ξ is expressed by

$$\xi=\frac{u_o}{u_z}\,ab\,\frac{a\cos l_z+x\sin l_z}{a^2+x^2}\qquad(5.11)$$

where u_o is the vertical velocity component (m/sec) at its initial altitude, u_z is the vertical velocity component (m/sec) at a height of z, a is a half of the width of the mountain range (m), b is the height of the mountain range (m), and l_z is the value of l at a height of z. If l decreases with height, waves are likely to occur.

For further theoretical considerations, see Scorer (1953, 1954, 1955) and for a summarized description, see Scorer (1958). The works of Queney and Scorer are reviewed by Corby (1954). A thorough review on the theoretical studies not only of the one-layer model, but also of the two-layer model is found in a report edited by Alaka (1960). Rotor and airstreams around isolated mountains were also treated in this report.

There are many phenomena accompanied by lee waves that cannot be explained by the linear theory, for example, the hydraulic jump, the extraordinarily strong wind in the lee of a mountain. Houghton and Kasahara (1968) used the one-dimensional time-dependent "shallow water" equations that govern the motion of an incompressible, homogeneous, inviscid, and hydrostatic fluid. The

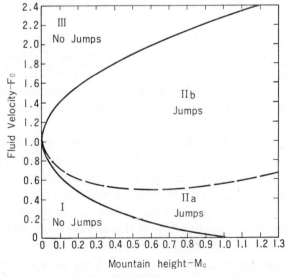

Fig. 5.61 Classification on asymptotic flow conditions as a function of the initial flow speed, F_o and maximum height of the ridge, M_c (Houghton and Kasahara, 1968).

results of the analytical study reveal that there are three classes of motion in the parameter domain of F_o and M_c as shown in Fig. 5.61. F_o is a dimensionless parameter expressed by

$$F_o \equiv \frac{u_o}{\sqrt{gh_o}} \tag{5.12}$$

where u_o and h_o are the velocity and height, respectively, of the approaching flow far from the obstacle. M_c is a ratio of the height of the summit, H_c, over the depth of the approaching fluid, h_o;

$$M_c = \frac{H_c}{h_o} \tag{5.13}$$

In domain I, the motions are subcritical, $F_o < 1$, and steady states exist over the ridge without jumps. In domain III, the motions are supercritical, $F_o > 1$, and steady states exist over the ridge without jumps. In domain II, the steady state solutions exist only with accompanying jumps on both sides of the ridge. The jump on the windward side of the ridge always moves upstream. The jump on the lee side moves downstream in domain IIb and remains stationary over the lee slope of the ridge in domain IIa.

Their model is hydrostatic and therefore cannot produce lee waves. However, it is thought to be applicable for strong down-slope winds. Vergeiner and Lilly (1970) showed a variety of flow patterns that can be explained by hydraulic jump models: (almost) jumps and supercritical and subcritical flows. Vergeiner (1971a) presented a model for linear, stationary mountain waves with arbitrary basic flow and two-dimensional topography. The results of his theoretical study agree well with the observed mountain lee waves. It was shown further that the flow associated with the chinook in Boulder, Colorado, is probably nearly hydrostatic.

Since the model by Houghton and Kasahara (1968) treats a single layer of fluid, it does not permit a study of the vertical extent of jumps associated with the ridge. To correct for this defect, Magata (1969) carried out a numerical experiment with a two-layer model on lee wave clouds, hydraulic jump clouds, and banner clouds around Mt. Fuji, taking into consideration the effects of heating and cooling from the ground surface.

Arakawa (1968, 1969) made a theoretical study on the fall winds in Hokkaido. According to him, the height of the inversion downstream is lower than that upstream. Arakawa and Oobayashi (1968) made a numerical experiment on the one-dimensional unsteady air flow over a mountain ridge, giving the following three initial conditions: (a) subcritical flow for all intervals, (b) subcritical flow for all intervals, but critical flow at the summit level, and (c) supercritical flow for all intervals. In all cases, the height of the inversion increases on the windward side and decreases on the lee side. They have shown that when the initial speed is higher, the jump appears farther downstream, which suggests the possibility of a fall wind.

(C) *Model Experiments*

By observation and a model experiment, Abe (1932, 1941, 1942) has investigated the cloud forms around Mt. Fuji. His experiments were made in a kind of wind channel, where a single conical mountain model was placed. Long (1953a), on the other hand, used a water tank, giving the flow of a three-layer system of immiscible fluids over an obstacle immersed in the lowest liquid. Some features of the motion have turned out to be similar to the flow of air currents over the Sierra Range near Bishop, California. He compared the theoretical results with the experimental facts in a series of studies (Long, 1953b, 1954, 1955).

In making a model experiment, one must take into consideration the dynamical similarity between the model and the actual full-scale problem. This is important first to obtain a reasonable *Reynolds number* in the model on the assumption that molecular viscosity in the model would correspond to eddy viscosity in the full-scale phenomena. Reynolds number, Re, is defined as:

$$\text{Re} = \frac{LU}{\nu} \tag{5.14}$$

where L is a characteristic length, ν is the kinematic viscosity, and U is a characteristic velocity.

Suzuki and Yabuki (1956) made an experimental study using a profile of a mountain range represented by

$$z = \frac{h}{1 + \dfrac{x^2}{a^2}} \tag{5.15}$$

where h is the height of the mountain ridge, z is the height of a point at which the horizontal distance from the mountain top is x, a is a constant. In the experiment, models with $a = 1/2\,h$, h or $2\,h$ were used. A large vortex was observed on the lee side of the mountain in the case of weak velocity and small Re. On the other hand, an eddy train was often observed in the case of medium velocity and small Re.

Assuming that the degree of similarity achieved by equating the "Scorer number" is acceptable for an experiment, the model flow conditions can be calculated for a typical case as follows (Corby, 1954): For a model 3×10^{-5} times actual size, with $U = 10^3$ cm/sec and $\beta = 2 \times 10^{-7}$ cm^{-1} (a lapse rate of about 4°C/km) on the actual scale, then even with a very large temperature gradient of 1°C/cm in the model, the wind tunnel flow speed would have to be as low as 4 cm/sec.

Soma (1969) made a model experiment in a wind tunnel with special reference to the separation in the boundary layer of the air flow, the air flow over the mountains, and so on. Inoue (1948) introduced the *effective Reynolds number* for a similarity between a natural wind and the air flow in the wind tunnel. Namely,

$$\text{Effective } Re = \frac{L \cdot U}{k} = \frac{L \cdot U}{l \cdot \sqrt{\overline{u^2}}} \qquad (5.16)$$

where L is the height of the model, U is the wind speed in the wind tunnel, k is the eddy viscosity coefficient, l is the mean scale of turbulence, $\sqrt{\overline{u^2}}$ is the standard deviation of the turbulent component.

Besides the effective Re, the *Froude number* has generally been used to simulate air flow over mountains. When the stratification of the air flow is considered to be significant, this number must be taken into consideration. Froude number, Fr, is:

$$\text{Fr} = \frac{V^2}{L \cdot g} \qquad (5.17)$$

where V is a characteristic velocity, L is a characteristic length, and g is the acceleration of gravity.

Nemoto (1961a, b, c, 1962) studied the similarity between the wind in the natural atmosphere and the wind in a wind tunnel; his results indicate that the modeling criteria for local wind can be described as $u_{\infty M}/u_{\infty N} = (L_M/L_N)^{1/3}$, where u_∞ and L denote the representative uniform velocity and the representative length (e.g., height of a mountain), respectively. The suffixes M and N mean the winds in the wind tunnel and in the natural atmosphere, respectively.

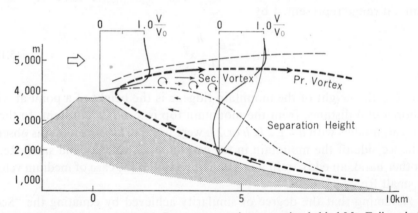

Fig. 5.62 Schematic pattern of the air flow accompanying separation behind Mt. Fuji, an isolated mountain. A thick solid line halfway up the mountainside shows the vertical profile of wind speed, where V_0 is the wind speed of the general flow in the wind tunnel (Soma, 1969, slightly modified by Yoshino).

The height of the separated flow and the wind profile in the lee of Mt. Fuji is schematically illustrated in Fig. 5.62. According to an observation by Soma (1969), there are two vortex systems: one is the large-scale vortex formed by the separation flow and the reverse flow on the lee side slope of the mountain and the other is the small-scale vortex formed in the region of the separated air flow. Soma gives the name "primary vortex" to the smaller vortex. In his experiment, the scale of the topographic model was 1/25,000, the wind speed was about 1 m/sec, and the Reynolds number was of the order of 10^4. An example of the state is shown in Photo 5.3.

Photo 5.3 The small-scale vortices (secondary vortices) in the layer of the separated air flow behind a model of Mt. Fuji (by S. Soma).

Yoshino et al. (to be published) studied the bora on the Adriatic coast by a model experiment in a wind tunnel with a stratified air layer. In this experiment a topographical model with a horizontal scale of $1:50,000$ and a vertical scale approximately $1:40,000$ was used to obtain the state of lee waves along the various topographical profiles. The results can be summarized as follows: (a) When the wind speed is relatively low, the first trough of the lee wave, corresponding to the bora region, appears near the foot of the mountain, and the greater the wind speed, the further downwind the trough moves. (b) The positions of the crest and those of the trough slope from the ground level to the 3 km level. In other words, the wavelength is maximum at the 3 km level. (c) The distance from the first trough to the first crest of a wave becomes longer with an increase of wind velocity. (d) The distance from the first trough to the first crest of a wave is shorter than the distance from the first crest to the second trough. In other words, the wavelength becomes longer down wind. (e) As shown in Table 5.16, the difference between the cross sections is greater in the case of a wind speed of 11.6 m/sec, which is a relatively stronger case. (f) The wavelengths double the values given in Table 5.16, 10–28 km, roughly coincide with the range of the values shown in Table 5.15. An experimental observation shown in Photo 5.4 resembles, qualitatively at least, the case (Photo 5.5) observed at a point 20 km N of Senj at 11 h 30 m on October 16, 1972. In this photograph the cap cloud, *kapa*, and the hydraulic jump cloud, *tresa*, on the right hand, are clearly seen. When this photograph was taken, the wind velocity as a one

Table 5.16 The distance from the trough to the crest of lee waves according to the wind speed along the various topographical cross sections on the Adriatic Coast (Yoshino).

Cross section	Height	Wind speed*				
		3.5	6.0	8.8	11.6	14.0 m/sec
Žrnovrtica	0–1 km	5–8	8–12	10	10	10 km
	3	(9)	8–12	10	10	10
Rt Čardak	0–1	5–8	8–12	10	11	11
	3	(9)	8–12	10	11	12
Senj	0–1	—	9	9	9–10	—
	4	—	10	12	14	—
Cross section	0–1	—	7–8	8–(25)	11–13	—
1.75 km S of Senj	3	—	7–8	8	11–13	—
A cross section	0–1	—	7–8	8	10	—
3 km S of Senj	3	—	8	11	12	—

* Estimated from the wind speed in a wind tunnel, but it seems the wind speed must be stronger in full-scale phenomena.

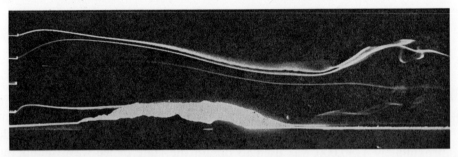

Photo 5.4 Wind tunnel experiment of bora with a topographical model. Smoke streams show a fall wind condition and the hydraulic jump (by M. M. Yoshino).

Photo 5.5 Kapa cloud over the Dinar Alps along the Adriatic Coast associated with the bora situation →

hour mean was 16.2 m/sec, and the instantaneous maximum wind velocity was 25.7 m/sec at Senj.

5.4. Formation and Run-off of Cold Air at Night

This section deals with the inversion layer, cold air lake, and cold air drainage at night. These phenomena are the motion of cold, small air masses and the cold air layer intensified as a result of the motion. But, topography plays an important role in the development of the phenomena and the various topographies cause a great variation of local temperature conditions at night.

5.4.1. Cold Air Drainage at Night

The air layer near the ground grows cold on clear, calm nights. If the ground is sloped, the cold air begins to move to a topographically lower place according to gravity. This is called cold air drainage or cold air run-off. The term *gravity wind* is often seen in meteorological literature for the same cold air motion. Baumgartner (1963) reviewed the cold air formation and movement in flat open land and sloped land at the night.

The cold air masses that flow down the slope have different dimensions from air-drop to air-avalanche. The dimensions of the cold air depend on the slope angle, the relative height between the foot and the top of the mountain, and the thickness of the inversion layer. The movement of the cold air drainage is visible if there is smoke from chimneys. A model experiment can also be an effective means of observing the movement in a complicated mountain region. An example of such an experiment was demonstrated by Teichert et al. (1952). Szász (1964) used the term *microadvection* for this phenomenon on a small scale.

A cold katabatic wind on a mountain slope, a mountain wind in a valley,

→ and cumulus clouds by the hydraulic jump on the right hand (October 16, 1972, by M. M. Yoshino).

and a land breeze in a coastal area are stronger forms of cold air drainage. The cold air masses gathered by the cold air drainage on the respective slopes in upper tributaries make up the first motion of the cold katabatic wind on the mountain slope as a whole; they become the mountain wind in the valley in the middle and lower river courses and then the land breeze on the coastal plain. Lawrence (1954) reviewed these winds, which he called nocturnal winds. According to the results of an observation carried out on the gentle slope in Tsrikovka Village, the Aruik-Baluiksky region of Kokchetavska, U.S.S.R. (Vorontsov, 1958): The cold air drainage begins 40–50 min before sunset, because the inversion of air temperature near the ground surface begins about 1 hr before the sunset in clear calm conditions. The cold air drainage develops on all the mountain slopes and ends 20–30 min after sunrise. Details are mentioned in Section 5.5.5.

The cold airstreams on the slope gather to the valley and then the cold air masses in the valley begin to flow down from the upper part to the lower part of the river as mountain winds. The processes of formation of the mountain wind system can be explained generally by the model illustrated by Defant (1951). There are, however, some cases that deviate from this model such as very large valleys. For instance, Vorontzov and Schelkovnikov (1956) described the aerological structure of the lower air layer in the Azau valley (1,500 m wide) between the Elibrus (3,490 m above the valley bottom) and Caucasus mountain ranges: The nocturnal wind system in the valley was established after 23 h. Davidson and Rao (1958) reported that the mountain wind began to blow at 18 h–20 h and the maximum values of height and velocity of the mountain wind appeared before midnight in a large valley in Vermont, U.S.A.

According to a study on the nocturnal circulation stream on a mountain slope, it was found that the cold air did not flow down continuously, but in bursts, little by little (Nitze, 1936). Küttner (1949) studied such phenomena and came to the conclusion that the instability overcomes the roughness with a rhythmical alternation of the air-avalanche. Aichele (1953) also confirmed the periodicity of the cold air drainage, observing fog as an indicator in the Salemer valley north of Lake Constance.

The air-avalanche from the free atmosphere, which causes a temperature rise on the ground level at night, has been studied by Schmauß (1952), Malsch (1952), and Bleibaum (1952). The air-avalanche from the free atmosphere tends to occur when the winds at the ground level have a tendency to diverge and bring a sudden rise in temperature of more than 3°C within 1 hr in an area with an order of several kilometers in diameter, according to unpublished data observed in the Nasu region, Tochigi Prefecture, Japan. However, details of the air-avalanche from the free atmosphere are omitted here, and those of the air-avalanche on mountain slopes only are dealt with. Berg (1951) observed the periodic temperature change in the mountain region of Hohen Venn. He thought that the cold air had a cell-like structure, which causes the periodic change at night.

Table 5. 17 Occurrence time of cold air run-off at night. The underlined time means the time of minimum air temperature (Yoshino, 1961a).

Observation Region	Date	Literature	Before 20h	20h	21h	22h	23h	24h	1h	2h	3h	4h	5h	6h	7h
Inawashiro, Fukushima Pref., Japan	Monthly mean	Mano (1953)						24:00						<u>6:00</u>	
	Nov., 1948,				21:00				1:00						<u>7:00</u>
	Dec., 1948,			20:00							3:00				<u>7:00</u>
	Jan., 1949,							(24:00)			3:00			6:00	<u>7:00</u>
	Feb., 1949.						23:00					4:00		<u>6:00</u>	
	Dec., 1949.						23:00			2:00					<u>7:00</u>
Funatsu, Yamanashi Pref., Japan	Mar. 4, 1951.	Yoshino (1952, 1953)	17:50	20:20	21:00					2:00	3:40		<u>5:40</u>		
Izu-Nishiura, Shizuoka Pref., Japan	Mar. 21–22, 1952.	Yoshino (1953)		20:00	21:00	22:30			1:50		3:20		<u>5:10</u>		
Sugadaira, Nagano Pref., Japan	Oct. 26–27, 1953.	Yoshino (1957)			21:00			24:30	1:40			<u>4:00</u>			
Bandai, Fukushima Pref., Japan	Oct. 9–10, 1954.	Mano (1956)						24:10	1:40		3:35		<u>5:15</u>		
	Oct. 10–11, 1954.						23:15		1:05	2:55			<u>5:20</u>		

By analyzing the results observed by the low-layer rasonde, regular rasonde, pibal and rabal together with regular meteorological instruments at three places of different localities in Japan, the periodic temperature rise after the cold air run-off was clarified (Mano, 1956). The time when the strongest cold air runs off at night depends on the area, slope angle, shape, and vegetation of the source region of the cold air. Results of some observations in Japan are summarized in Table 5.17. According to this table, cold air drainage occurs mostly once before midnight and two or three times between midnight and sunrise. The last cold air drainage brings the minimum air temperature. Shitara et al. (1973) investigated the temperature fluctuation at night in a small basin and found the relations to the wind velocity in the layer near the ground. Ives (1950a) reported air temperature depressions of as much as 3°C in the Great Basin and Rocky Mountain regions, during the time around sunrise. It was thought that these phenomena are closely related to local air drainages after clear nights with little wind. A sudden rise in temperature during the night occurs after the cold air drainage. Such a rise in temperature was seen in the hours between 20 h and 21 h, 23 h and 24 h, and 2 h and 3 h on clear nights at Inawashiro, Fukushima Prefecture (Mano, 1953). Sudden rises in temperature at night occur also in association with the passage of fronts or with the increase of winds in upper layers. In studies in the Kakuda Basin in the southern part of Miyagi Prefecture, NE-Japan (Kikuchi, 1972), the cases with increase of upper winds were observed almost simultaneously at the stations distributed in small regions; that is, the phenomena were brought about by causes on a larger synoptic scale. Therefore, these are not dealt with here.

Another example of studies on down-slope winds, cold air drainage, came to the conclusions that (a) there is a prevalence of wind directed nearly at right angle to the topographic contour line, (b) the wind developed regularly in a relatively small valley, and (c) in some cases, the cold air flowing down from a small tributary was warmed up by mixing with the warmer air flowing down in the main valley (Sahashi, 1962). The cold air run off from a small side valley is greatly affected by the cold air drainage in a large main valley. An example of such a situation in Odenwald, Germany, was reported by Schneider (1972).

The depth of cold air drainage differs greatly from case to case. A weak case is only 1–2 m, but in the stronger cases, it develops up to 100–200 m. Such a thick drainage might be called a katabatic wind or mountain wind. Where the valley is narrow, the cold air tends to be dammed up and becomes thicker (Schnelle, 1956).

5.4.2. Velocity of Cold Airstream

The velocity of a cold airstream is dynamically the same as that of the katabatic wind on mountain slopes. Heywood (1933) presented an experimental equation of the velocity, v, as a function of the distance from the sea, x, and the angle of the mountain slope, θ:

$$v^2 = \frac{2x}{1.4 + 100 \tan \theta} \qquad (5.18)$$

Another approach to the problem was based on the assumption that the cold airstream flows down in accordance with gravity. Generally, as Geiger (1965) has stated, the following equation on the velocity of a cold airstream (v) fits the actual conditions (Reiher, 1936):

$$v = \sqrt{\frac{2gh\,(T'-T)}{T'}} \qquad (5.19)$$

where, g is the acceleration of gravity, h is difference of height, T' is the absolute temperature of air around the cold air, and T is the absolute temperature of cold air.

Table 5.18 Velocity of cold air drainage obtained from an experiment and the calculated values (Voigts, 1951).

Angle of slope	14°30′	10°51′	6°51′	2°45′
Values calculated from Eq. (5.19)	6.55	5.72	4.51	2.86 cm/sec
Ratio	0.50	0.44	0.343	0.22
Results of experiment on a slope of				
Wood	0.50	0.41	—	0.24 m/min
Sand	0.53	—	0.345	0.22 m/min
Grass	0.50	0.43	0.34	0.20 m/min

An attempt was made to clarify the velocity of the cold air drainage by an experiment (Voigts, 1951). The results, as shown in Table 5.18, coincide quite well with the values calculated by Eq. (5.19). Precisely speaking, however, the initial velocity of the cold air run-off corresponds to the ratio between the calculated values for the various angles of the slope.

If the distance of the slope, l, on which the cold air flows down, and angle of the slope, θ, is taken into consideration, Eq. (5.19) can be rewritten as:

$$v = \sqrt{\frac{2gl \sin \theta\,(T'-T)}{T'}} \qquad (5.20)$$

Equation (5.20) holds quite true in the period when the cold air is developing (Lawrence, 1954) and also on the slopes where θ is large.

Assuming the compressibility of cold air, Fleagle (1950) expressed theoretically the velocity ($\bar{u}$), mean flow velocity of the cooling air layer:

$$\bar{u} = -\frac{c_p C}{gh \tan \theta}\left\{1 - \frac{\sqrt{k^2 + 4w_0^2}}{2w_0}\,e^{-\frac{1}{2}kt} \times \cos\,(w_0 t - \beta)\right\} \qquad (5.21)$$

where $w_0 = \left(-\frac{1}{4}k^2 + g^2\frac{\tan^2 \theta}{c_p T}\right)^{\frac{1}{2}}$

$\beta = \tan^{-1}\frac{1}{2}kw_0^{-1}$

$C = \sigma\,(1-r)\,(T_s^4 - 0.65T_a^4)/\rho c_p$

h: thickness of the air layer, θ: angle of the slope, k: constant expressing friction, 2–$8(10^5\text{sec})^{-1}$, t: time, T: temperature of the air layer, σ: Stefan-Boltzmann constant, T_a: temperature of the top of the air layer, T_s: temperature of the earth's surface, r: proportion of the sky that is obstructed, 0.65: proportion of the earth's radiation that is absorbed by the atmosphere with a clear sky (Hewson and Longley, 1944), and ρ: density of air.

Equation (5.21) means that the velocity is proportional to the net outgoing radiation (C) and is in inverse proportion to the thickness of the cold air layer (h) and the angle of the slope (θ). The velocity changes periodically at first and gradually becomes a constant value.

According to some observations, the thickness of the cold air stream was 100–200 m, as mentioned above, and its velocity was 2–4 m/sec (Defant, 1951). The maximum wind velocity was found at a height of 30 m above the ground and was 0 m/sec at 100 m (Defant, 1949). It seems, however, that this was a developed stronger down-slope wind, a katabatic wind. In such a case, the maximum velocity of the wind appeared at a certain height above the ground corresponding to the theoretical results by Prandtl.

On a longer slope, the mean velocity of the cold air drainage was 1.5–1.8 m/sec with a range of 0.3–3.5 m/sec in the layer 0.5–1.0 m above the ground. The vertical velocity of the cold air downstream was 8–15 cm/sec with a maximum of 35 cm/sec. The upper air was lifted up when the small, cold air masses flowed down on the slope. The vertical velocity of this ascending air frequently reached 50–60 cm/sec, but its mean value was 20–30 cm/sec (Vorontsov, 1958).

Bergen (1969) studied the cold air drainage on a forested mountain slope in the Fraser Experimental Forest, Colorado, U.S.A., and came to the following conclusion. The downslope variation of the wind speed is in general agreement with the variation predicted by the solution and the slope of the hillside. The local mean speed, U_m, is expressed by the equation $U_m \propto \sqrt{\varDelta\theta}$, where $\varDelta\theta$ is the potential temperature drop down the slope, and is also expressed by $U_m/\sqrt{\gamma} \propto x$, where γ is the sine of the angle of the slope to the horizontal along the streamline; x is the downslope distance from the virtual origin of the flow measured along the stream line.

As has been stated, the cold air drainage has some periodicity, and the largest drainage occurs before sunrise in most cases. However, it is interesting that the place where the strongest run-off takes place changes generally from place to place in a small area during a single night, probably under the influence of different topographical effects (Winter, 1956).

5.4.3. Circulation System of Cold Air and Cold Air Lakes

There are several observations on the circulation system of cold air at night. The radiation cooling after sunset begins generally on the mountain foot or the lower part of the slope in the case of a broad slope. In the beginning stage of the cold air drainage, the topography just around the point under consideration has a stronger influence on the drainage, but in a later stage, the topographical situation in a broad area plays a more important role in forming the circulation system of the cold air as well as downslope winds or mountain winds.

The cold air run-off from a higher region in the mountain is collected at a lower place. This is called a cold air lake. In other words, the cold air lake is a clearly developed inversion layer, which is intensified by the topographical situation. This will be discussed in detail in Section 5.5. The term *cold air lake* is used also for the areas covered by the cold air in the regions of the Alps, the Carpathians, and the Balkans from a synoptic meteorological standpoint (Čadež, 1971). However, it might be better to name the phenomena in such a relatively broad area a *cold air sea*, which is not a subject dealt with here. On the other hand, the cold air lake in drainless hollows, such as dolines of limestone regions, was called a *cold air pond* (Wagner, 1970). It was shown that the cold air lake on such a small scale is formed mainly by the cooling effect of nocturnal outgoing radiation of the slopes in the hollow. The development of a cold air pond depends on the depth of the hollow and the vegetation in the hollow. That is, the cold air run-off from the slopes does not play an important role in such a hollow. The cold air run-off has close relation to the formation of a cold air lake in a basin or valley in mountainous and hilly regions. But the cold air drainage on a relatively large scale seems to form its own circulation system, which is independent of the cold air lake formed earlier. Some observations and schematic models follow.

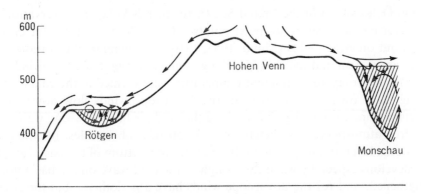

Fig. 5.63 Schema of air circulation and cold air lake on and around the Hohen Venn (Berg, 1951).

Berg (1951) reported the results observed in Hohen Venn and its surroundings and summarized the circulation system as shown in Fig. 5.63. Because the cold air in the basin of Rötgen and in the narrow valley of Monschau cannot easily flow away, it is gradually dammed up there, forming cold air lakes. The dynamic effect of warming has no influence on the descending cold air, because cooling on the ground surface of the mountain slopes is stronger than the warming. Therefore, the ground surface of the table land of Hohen Venn after sunset is always warmer than the valleys in Monschau or Rötgen. According to an observation in summer, 1950, Hohen Venn was 2°C warmer than the cold air lake. The cold air drainage on a relatively greater scale from Hohen Venn flows down over the upper surface of the cold air lake in the basin of Rötgen and in the valley of Monschau, even though part of the cold air flows into the cold air lakes. Plaetschke (1953) supported Berg's opinion that the cold air lake in the valley is formed independently of the cold air advection from the surrounding high plateau.

Lehmann (1952) dealt with a case of a small-scale cold air drainage with a thickness of 3 m. It was shown that, over this flow, there was a windless isotherm layer with a thickness of 2 m and, further, over this layer, an upslope wind with a velocity of 0.2 m/sec.

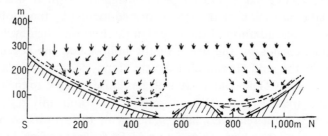

Fig. 5.64   Model of nocturnal air circulation at Ôchô. Broken line shows the top of cold air lake (Mano, 1953).

On Ôchô Island in the Inland Sea (Setouchi), SW-Japan, observations on the nocturnal temperature drops at points 260 m, 130 m, and 20 m a.s.l. were carried out on clear nights (Mano, 1953, 1956). A schematic illustration of the circulation system and the cold air lake was given in Fig. 5.64. It was indicated that (a) cold air layers of different temperatures are formed in the cold air lake, making their own circulation in it, (b) downslope winds flow down on the mountain slopes, (c) to the places where the cold air has run away, warmer air from free atmosphere at a height of 100–250 m a.s.l. invades, (d) there is a transitional layer at a height of 50–100 m from the bottom of the basin, and this layer develops especially when the trough of the cold wave on the basin floor is strengthened, and (e) the downslope winds have a periodicity, but the circulation stated above, item (c), continues from sunset to about 3 in the early morning.

A model of the nocturnal air circulation on the slope of Mt. Bandai, NE-Japan, is presented by Mano (1956) as given in Fig. 5.65. This figure is based on observations of the model given by Wagner (1938). The downslope wind with a velocity less than 1 m/sec from the mountain slope flows into the cold air lake at the mountain foot. Over this, the ascending air current (1–2 m/sec), which is relatively warmer, is seen. Over this up wind layer, the general upper winds are blowing.

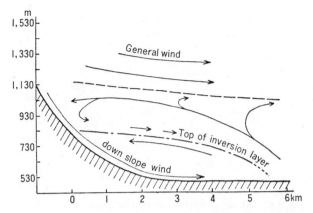

Fig. 5.65 Model of nocturnal air circulation on the slope of Mt. Bandai. Below the top of the inversion layer is a cold air lake (Mano, 1956).

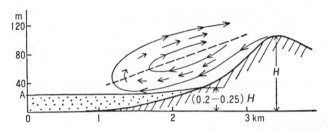

Fig. 5.66 Model of nocturnal air circulation on the gentle slope of Kokchetavska, U.S.S.R., at 22h30m, on July 10, 1956. A shows the upper limit of the cold air lake (Vorontsov, 1958).

From the results obtained in Kokchetavska, U.S.S.R., which are described in details in Section 5.4.1 and in Fig. 5.72, another interesting schematic illustration was given by Vorontsov (1958) as shown in Fig. 5.66. He defined the cold air lake as a *film of cold air*, over which the circulation of downslope winds develops. It was indicated that the thickness (A) of the film is 0.20–0.25 H, where H is the relative height of the mountain. On the lower part of the slope, the inversion, the film of cold air, was very conspicuous, because most of the cold air from the upper part of the slope flowed into that area.

Finally, a schematic illustration summarized from a number of micrometeorological observations on the cold downslope winds in Hakatajima Island, the Setouchi region, SW-Japan, is shown in Fig. 5.67 (Kimura, 1961). The warm upslope wind layer is clearly seen over the cold downslope wind layer on the slope. The asymmetry of this stratification between the southern and northern slopes is also observable, which might be caused by the direction of the upper prevailing winds.

Summarizing the results and schematic illustrations mentioned above, the following conclusions can be drawn, even though the observation areas, seasons, and methods were different. (a) There is a cold air lake in the lowest layer. The

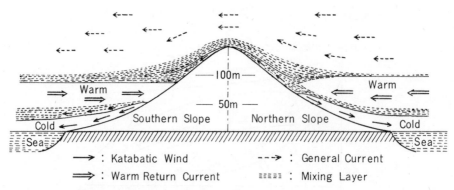

Fig. 5.67 Model of wind circulation on a clear calm night on the small island Hakatajima (Kimura, 1961).

thickness of the cold air lake is 0.2–0.3 of the relative height between the bottom of the valley or basin and the surrounding mountains. (b) Over the cold air lake, a downslope wind and its compensating current, an up wind, blow. These form a circulation system. Some parts of the downslope wind, composed of the cold air drainage, flows into the cold air lake as an important source of cold air. (c) Over this circulation system the general upper winds are blowing.

5.5. INVERSION LAYER

The temperature inversion or stable state is defined here as $\frac{\partial T}{\partial z}>0$. In the case $\frac{\partial T}{\partial z}=0$, it is called the neutral state, and $\frac{\partial T}{\partial z}<0$ is the lapse or unstable state. In some cases, however, it is defined as a weak inversion or weak stable state, when

$$\left(\frac{\partial T}{\partial z}\right)_d \text{ or } \left(\frac{\partial T}{\partial z}\right)_w < \frac{\partial T}{\partial z} < 0 \qquad (5.22)$$

where $\left(\frac{\partial T}{\partial z}\right)_d$ and $\left(\frac{\partial T}{\partial z}\right)_w$ are dry adiabatic lapse rate and wet adiabatic lapse rate, respectively, and a minus sign means that the temperature decreases with height. In this book, the definition of Eq. (5.22) is not taken up, because it is not simple. The layer through which the inversion state occurs is called the inversion layer.

The inversion layer plays an important role in the distribution of weather and climate in a small area. Therefore, the formation and disappearance, intensity, and frequency of occurrence of the inversion layer are dealt with below. Also their relations to the topographical situation are mentioned in detail in this section in order to clarify the distribution patterns of air temperature, wind velocity, and wind direction, as well as air pollution. Schneider-Carius (1953) thoroughly reviewed the inversion layer as an important parameter contributing to the structure of lower troposphere.

5.5.1. Ground Inversion

The air near the ground surface begins to grow cold rapidly about 1 hr before sunset under a windless, clear sky, due primarily to greater radiative loss of heat at and near the ground surface. As a result of this process, an increase of temperature with height begins at the ground level. This is called ground inversion, surface inversion, low-level inversion or radiation inversion. The inversion takes place roughly 1 hr before sunset and continues all night until about 1 hr after sunrise.

According to Austin (1957), the formation and intensification process of the inversion are as follows: The isolation of an air layer near the ground by an inversion results in the lack of vertical motion of mixing between the free atmosphere and the inversion layer. The atmosphere above retains its air mass characteristics, and the isolated layer below is modified by the underlaying ground according to its thermal nature. The modification, the removal of heat, is accomplished by the turbulent transport of sensible heat from the air to the ground and by long-wave radiation from the ground.

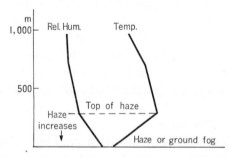

Fig. 5.68   A typical nocturnal ground inversion (Austin, 1957).

A typical example of the ground inversion is shown in Fig. 5.68. The relative humidity, which is close to saturation at the surface, decreases rapidly toward the inversion top. The depth of the inversion layer and the lapse rate (inversion rate) through the layer depend to a great extent on wind velocity, cloud cover, and the radiational properties of the ground surface (Brunt, 1929; Ashburn, 1941; Möller, 1951). The depth of the inversion layer has a great range of values, as described below, but generally is 200–300 m middle latitudes.

Cooling by radiational process alone is restricted to a very shallow layer near the ground because of the length of the night in middle latitudes. Elsasser (1940) stated that heat must be transported from the air to the ground by small-scale eddies and then be radiated into space. The turbulent exchange of heat is the only process rapid enough to permit the formation of ground inversions. According to Mashkova (1965), the depth of the ground inversion in dm, ΔH, and the magnitude of the inversion in °C, Δt, may be estimated simply by the equations $\Delta H = 0.25\,A$ and $\Delta t = 0.5\,A$, where A is the daily range of air temper-

ature (°C) at the 2 m level. Such relations change in accordance with the topographical situations, which is discussed in Section 5.5.5.

Over the Antarctic Continent, a strong inversion is seen during the polar
night over the snow cover. Flohn (1952) called this strong inversion *Nivale
Inversion*. The vertical structure of air temperature up to 3,000 m above the
ground at the Amundsen—Scott base (90°00'S, 2,800 m) was made clear by
Flowers (1960). It was shown that the top of the inversion layer varies within the
500–700 m level above the ground, and a sharp increase (the difference was about
20°C) was found in the lower 200–300 m layer.

The mixing of air by turbulence near the ground plays an important role in
determining the depth and inversion (lapse) rate, if other factors such as roughness of the ground, cloud cover, and state of the ground are constant. The wind
speed in the lower layer is a convenient measure of the turbulence. Turbulence
and the transport of heat by it usually increase with an increase in the wind
speed.

Calm wind conditions under a clear sky produce a very shallow but sharp
inversion. Condensation of water vapour at night in the form of dew or fog
reduces the range of air temperature change near the ground, but in most cases
its reduction is small. The relationships between the occurrence frequency, intensity (inversion rate), depth of the inversion layer, and the wind speed or the
cloud amount are discussed below in detail. The seasonal change of these characteristics of inversion depends on the seasonal change of frequency of anticyclonic air masses, which bring light winds, dry air, and a clear sky.

5.5.2. High-level Inversion

The high-level inversion contains the frontal inversion and the subsidence
inversion. Since the frontal inversion does not play an important role in controlling the local weather, the subsidence inversion is dealt with here.

The subsidence inversion is produced by the adiabatic warming of a layer
of subsiding air. If the subsident air is warmed at a dry adiabatic warming rate,
1°C/100 m, and if the lapse rate before subsidence is smaller than 1°C/100 m,
the inversion is produced in the layer of subsiding air. The subsidence is gener-

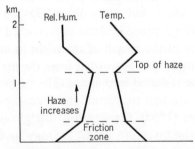

Fig. 5.69 A typical subsidence inversion (Austin, 1957).

ally stronger at a height of 1,000 m a.s.l. in the middle latitudes, and hence the inversion takes place frequently at this level. In lower latitudes its height is about 2,000 m.

The subsidence inversion is characterized in the fully developed stage by a large inversion of temperature together with a sharp decrease of relative humidity with height commencing at the inversion base, as shown in Fig. 5.69 (Austin, 1957). The thickness of the inversion layer is about 300 m.

The inversion generally forms at higher altitudes and gradually intensifies while descending. In a continental air mass, dry clear skies prevail with scattered cumulus or altocumulus clouds. In a maritime air mass, stratus or stratocumulus clouds spread widely beneath the inversion. If the inversion is high enough, drizzle or light rain may occur.

Subsidence inversions are found generally in regions of high pressure and low-level divergence, and their intensity should depend on the intensity of the anticyclonic circulation. Therefore, in the high-pressure belt regions in the middle latitudes in summer, when the temperature of the ocean surface is lower than that of the land, the anticyclonic circulation is at a maximum over the ocean, and hence the inversion intensity is at a maximum. The subsidence inversion is lowest and strongest in the eastern part of the anticyclonic region and highest, or in some cases nonexistent, in the western part.

The California coast in summer is a good example of subsidence inversion. The base of the inversion layer is at a height of 600 m in a typical case, and below this height, stratus develops. At the ground level, haze or fog is seen.

As the subsidence inversion is situated at a higher level, it does not have stronger effects on the behavior of local weather than the ground level inversion. However, its effects on the distribution of wind in a small area must not be disregarded, as discussed below. Also, it is important to understand the development of a haze or smog layer, in which the subsidence inversion plays a definitive role. Further, a subsidence inversion exists even in the daytime and is maintained for several days. On this account, its effects are more important than the ground inversion in considering the distribution of climate and weather in a small area.

5.5.3. Formation and Dissipation of Inversions

First the formation and dissipation of the inversion layer in the case of the ground inversion is dealt with here. The ground inversion takes place at the lowest air layer near the ground, and the inversion layer becomes thicker and thicker. According to the theory of cooling of the air layer, the inversion layer should continue to thicken until the minimum air temperature appears in the early morning. However, observations at a tower 312 m high at Kawaguchi City, north of Tokyo, during anticyclonic weather showed that the inversion layer became thicker until midnight, and then the intensity of the inversion became great (Tohsha, 1953). As shown in Fig. 5.70 (a), the first period of the formation of the inversion layer is the time when the inversion layer grows thicker up to

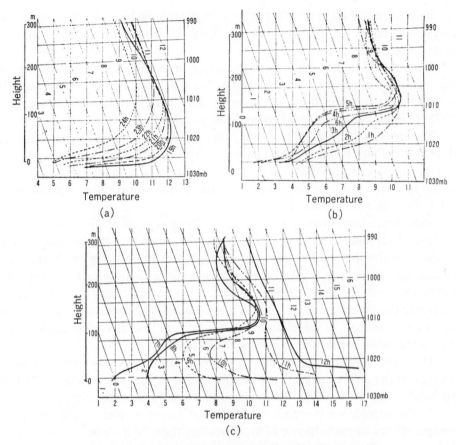

Fig. 5.70 Formation and dissipation of inversion layer at the tower of Kawaguchi, Nov. 26–27, 1952 (Tohsha, 1953).

150 m. The second period, as shown in Fig. 5.70 (b), is characterized by the intensification of the inversion layer. The third period, the dissipation of the inversion, shown in Fig. 5.70 (c), appears after sunrise. These figures were drawn on an aerogram. The oblique lines indicate isolines of potential temperature. A combination of the height and the difference in the potential temperature at the base and top of an inversion layer is a measure of intensity.

A model of inversion development in mountain regions was presented after considering the process of inversion formation, maintenance, and dissipation in the canyon near Salt Lake on the night of September 24–25, 1957 (Thompson and Dickson, 1958): (a) The initial phases of inversion formation occur in the valleys in mountain regions shortly after noon, long before the development will begin over the level ground. (b) About the time when inversion development gets started over the level land, the inversion formation process in the valleys suddenly accelerates almost explosively, so that the inversion reaches its full development within a fairly short time. (c) The effect of motion resulting from the solenoidal field formed during the inversion development process is a quasi-stationary stage for most of the night following the initial short explosive development; this is in contrast with the continued but gradually slowing intensification of inversions over the level land outside of the valley. (d) The mechanism of dissipation is similar to that over the level ground, except that properly exposed locations will be subject to dissipation of the inversion quite early, but properly sheltered locations retain inversions longer.

The relationship between the occurrence frequency of the ground inversion (f in %) and the wind force (F) on the Beaufort wind scale is expressed by

$$f = \frac{193.8}{F+4.8} - 14.8 \qquad (5.23)$$

This experimental equation was obtained from the 2 years' records observed at the mountain foot station (527 m a.s.l.) and the mountain top station (1,142 m) on Mt. Aso, Kyushu, Japan (Masatsuka, 1949).

The inversion between the ground level and a height of 29 m above the ground at Lindenberg, Germany, occurs when the wind velocity is below 5 m/sec and the lapse state occurs when the wind velocity is more than 5 m/sec at the height of 29 m. The strongest inversion occurred in the case when the wind velocity was 1–2 m/sec (Rink, 1953). Similar results were obtained from the observations at a tower at Kawaguchi City, Saitama Prefecture, Japan (Ôta, 1960), as shown in Table 5.19. Some other results at Lindenberg will be given later in Section 5.5.4.

Table 5.19 Relation between the wind velocity* and the inversion (April 1943–March 1944) (Ôta, 1960)

	Lower than 2 m/sec	2–3 m/sec	4–5 m/sec	More than 6 m/sec
Mean lapse rate**	−0.8	+0.1	+0.1	+0.3°C/100 m
Occurrence frequency of inversion	70	31	36	14%

* 10-min mean wind velocity at a height of 10 m above the ground.
** Obtained from the results observed at 293 m and 1.5 m above the ground. Minus means inversion.

The maximum occurrence frequency of the top of the inversion layer in Munich, Germany, appeared in the lowest air layer and the second maximum at 120–180 m when the wind velocity increase between the ground level and the top of the inversion layer was smaller than 1 m/sec. If the wind velocity increased above 2 m/sec, the maximum frequency of the top of the inversion was found at 150–300 m (Feldmann, 1965).

The height of the top of the nocturnal inversion coincides strikingly with the height of the wind maximum, the low-level jet, as studied by Blackadar (1957). There appears to be a slight increase of the height of the nocturnal inversion with an increase of the wind speed at the height of the maximum. This tendency, as Blackadar stated, is no more than would be expected from the fact that the nocturnal inversion is rising during the night and the wind speed is increasing during the same period.

A meteorological study at Douglas Point, Ontario, Canada, on the east shore of Lake Huron, during the period August 13–18, 1961, came to the following conclusion (Munn, 1963). The intensity of the radiation inversion depended on the wind speed at a height of 24 m: When the wind was more than 4.4 m/sec, the temperature difference between 24 and 6 m was rarely more than 0.6°C. The evening maximum of the inversion intensity was followed by a maximum in the wind speed and turbulence. This might be caused by the nocturnal low-level jet discussed by Gifford (1952) and Blackadar (1957).

The existence of inversions in a desert region during the daytime is unexpected, but it has been reported based on observations. A microclimatic observation was carried out under diverse terrain conditions at Yuma, Arizona, U.S.A. (Dodd et al., 1959). There were three stations, (a) Sandy Plains (128 m), (b) Desert Pavement (128 m), and (c) Laguna Top (192 m). At Laguna Top, the temperature profile in the daytime was regular, but at Desert Pavement and Sandy Plains the greatest decrease of temperature occurred at the lowest height of 2.5 cm and showed a slight increase at higher levels. At Sandy Plains, the primary layer of the daytime inversion is between 2.5 and 7.5 cm. Inversions between these levels were often greater than 0.6°C. At Desert Pavement, inversions occurred more frequently than at Sandy Plains and existed between 25 and 50 cm during more than two-thirds of the afternoon hours. At times these inversions were as great as 2.8°C, but were generally less than 1.7°C. It was thought that these daytime inversions may be a function of eddy diffusion or turbulent mixing near the earth's surface. But, a detailed study must be conducted in the future.

In a coastal region, the development of an inversion layer depends not only on the radiation conditions during the night, but also on the conditions of the prevailing wind, which brings cooler air to the surface air layer. A study on the nocturnal inversions at Point Arguello, located about 3 km inland from the Pacific coast, W-U.S.A., shows that they are most frequent in the coldest months and that they develop strongly when the winds from the interior flow as land breezes (Baynton et al., 1965). The wind from the interior is usually cooler than 9°C at night during the colder months and is able to displace the marine layer at the surface level.

5.5.4. Seasonal Change of Occurrence Frequency and Intensity of Inversions

The development of the ground inversion changes seasonally in accordance with the seasonal changes of the characteristics of the prevailing air mass, the surface conditions of the ground, and the hours of night at the point under consideration. The seasonal change of the occurrence frequency and intensity of inversion depends on the seasonal differences of these factors.

Table 5.20 Annual change of the occurrence frequency and the duration of ground inversion in Munich (Feldmann, 1965) and Tübingen (Daubert, 1962).

	Munich				Tübingen	
	Night time (1)	Day time (2)	Total	Mean duration (3)	Occurrence (4)	Mean duration (5)
Jan.	16	8	24 times	19.2 hr	14 times	17.2 hr
Feb.	18	6	24	11.0	17	15.0
Mar.	21	1	22	9.1	20	11.0
Apr.	19	0	19	8.6	17	8.9
May	21	0	21	8.4	20	9.6
June	20	0	20	7.8	18	9.3
July	21	0	21	8.4	20	9.5
Aug.	23	1	24	8.7	23	10.3
Sep.	23	1	24	8.6	23	11.6
Oct.	21	1	22	9.2	22	14.2
Nov.	15	4	19	11.8	17	15.2
Dec.	17	7	24	17.1	16	20.3
Year	235	29	264		227	

(1) Observed at 0 h and 3 h, 1952–1961, including the cases that the bottom of the inversion layer was found at heights between the ground and 100 m above the ground.
(2) Observed at 12 h and 15 h, 1952–1961.
(3) Mean of 1949–1953.
(4) Mean of 1950–1961.
(5) Mean of 1950–1961.

Some examples of the annual change of the ground inversion in Munich (Feldmann, 1965) and Tübingen (Daubert, 1962) are given in Table 5.20. Contrary to expectations, the maximum occurrence appears in the summer months and the minimum in the winter months. The duration, however, is longer in winter and shorter in summer.

According to the observations at 3 h at 1.5 and 150 m height in Budapest, the inversion was the most intense, $0.79°C/100$ m, in August, second in September, and weakest in winter (Szepesi, 1964). The occurrence frequency of the ground inversion also shows its minimum in winter in Detroit, Montreal, and Ottawa and its maximum in summer or fall in these cities (Munn et al., 1963). It can be said, therefore, that the seasonal change of the occurrence frequency of inversion in higher middle latitudes of the northern hemisphere depends on the broad-scale synoptic features. But the duration of the inversion is strongly affected by the duration of the night. Baynton (1962) indicated that the formation, defined as appearance of an inversion between 6 and 90 m, has been related to time of sunset at Detroit, U.S.A. During the winter months, inversions tend to form about 2 hr after sunset and during the summer months they tend to form at sunset.

There are quite clear local differences. For instance, the annual change of the inversion rate in Japan can be grouped into three types. The first type, which appears on the Japan Sea side, has maxima twice a year in autumn or spring,

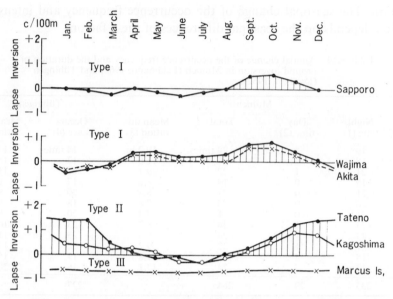

Fig. 5.71   Annual change of lapse and inversion rates in Japan (Yoshino, 1968).

the second type, which appears mostly on the Pacific side, shows a strong inversion from November through February, and the third type on the islands has no definite seasonal change, as shown in Fig. 5.71 (Yoshino, 1968). The temperature difference was obtained from the temperature at the ground level and the 1000 mb level observed by radiosonde at midnight, 1951–1955. These local characteristics shown in Fig. 5.71 depend on the prevailing synoptic conditions. That is, spring and autumn in Japan are seasons with frequent migratory anticyclones, which cause severe ground inversions, winter is the monsoon season with strong winds with cloudy weather on the Japan Sea side and clear, dry whether on the Pacific side, and summer is wet and hot with the summer monsoon. These seasonal changes of various synoptic situations must cause the local difference in the types of annual change of inversion.

   Based on the record at the stations on the mountain top (1,142 m) and the mountain foot (527 m) of Mt. Aso, Kyushu, Japan, the mean duration of inversion was 8–9 h in the winter months and 2–5 h in the summer months, as shown in Table 5.21 (Masatsuka, 1949).

   Very minute measurements of air temperature and other meteorological elements at the 1 and 76 m levels were made at Lindenberg, 56 km SE of Berlin,

Table 5.21   Duration of inversion at Mt. Aso, March 1942 to February 1944
(Masatsuka, 1949).

	Jan.	Feb.	Mar.	Apr.	May	June	July	Aug.	Sep.	Oct.	Nov.	Dec.
Duration*	8.33	8.53	8.00	7.35	8.04	6.10	4.49	2.33	5.06	8.13	9.13	8.57

*   For instance, 8.33 means 8 hours and 33 minutes.

Germany, December 1950–November 1951 (Rink, 1953). Some results, as shown in Table 5.22, show the inversion rate in relation to weather and seasons. From this table, it can be seen that the inversion rate in the case of scattered clouds is about half of the value under a clear sky. If it is cloudy, the inversion will almost disappear both in winter and summer.

Table 5.22   Solar radiation and inversion rate* in the early morning by weather and seasons at Lindenberg (Rink, 1953).

Season	Weather			
Winter (8–9 h)	Clear	{	Solar rad. Inv. rate	2 cal/hr 1.69 °C/100 m
	Scattered cloud	{	Solar rad. Inv. rate	3 0.56
	Cloudy	{	Solar rad. Inv. rate	2 0.00
Summer (5–6 h)	Clear	{	Solar rad. Inv. rate	7 1.13
	Scattered cloud	{	Solar rad. Inv. rate	6 0.00
	Cloudy	{	Solar rad. Inv. rate	5 0.13

* Plus sign means inversion

In the ground inversion layer, two or more inversions can frequently be observed. According to the statistics of occurrence frequency of the height of the inversion top at Munich, 1952–1961, the first main maximum, which appears at lower altitudes, showed no apparent seasonal change, but the second one is higher in summer and autumn (Feldmann, 1965). The occurrence frequency of the first inversion between the ground level and 15 m above the ground at Lindenberg, January 1954–December 1956, was about 80% annually, but the cases with the second inversion over 60 m were 11% annually. The frequency was higher in autumn and winter (Hentschel and Leidreiter, 1960).

The observed results at Klagenfurt, Carinthia, Austria, were analyzed statistically to find the monthly change of occurrence frequency, thickness, intensity, and stability of the inversion layer (Ekhart, 1949). The occurrence frequency showed a winter maximum and summer minimum with a half-year cycle. The thickness and intensity of the inversion layer were also maximum in the winter and minimum in the summer, but small maxima were found in April and August. The gradient of the inversion is larger in May, November, and December. These characteristics might be common in the valleys in the Alps.

5.5.5.   Development of the Inversion Layers and Topography
There are two problems concerning the development of inversion layers in relation to topography. The first concerns how is the temperature distribution on the slope of an isolated mountain located in the ground inversion layer devel-

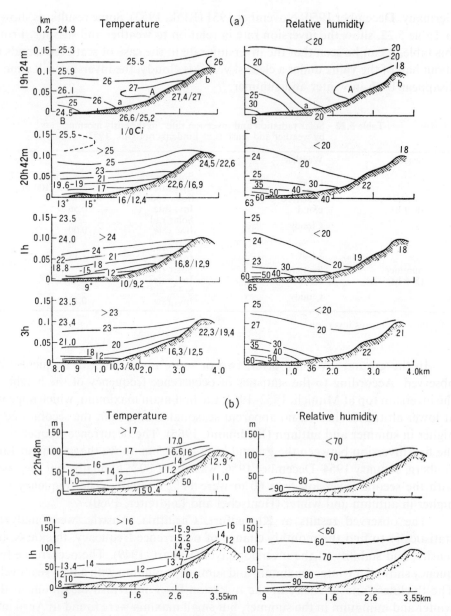

Fig. 5.72 Formation processes of inversion layer on the gentle slope of Kokchetavska, U.S.S.R (a) Dry weather, June 19–20, 1955. (b) Humid weather, August 27–28, 1956 (Vorontsov, 1958).

oped thickly on the surrounding plain. The second is the problem of cold air lakes formed in a basin or valley. The first problem was discussed in Chapter 4 and the second one in Section 5.4 under the title of "the Formation and Run-off of Cold Air at Night." Here, the general structure of the inversion layer on the slope of a mountain will be discussed.

The top of the inversion layer has a tendency to become lower where the influences of the surrounding mountains are strong. For instance, the thickness of the inversion layer in a valley surrounded by mountains was 50–80 m, but on the coast it was 100–200 m, according to observations at Ôchô Island in the Inland Sea, SW-Japan (Mano, 1953). The height of the inversion top changes generally during the night. The behavior of the change is different from that at an open flat surface, which was discussed in Section 5.5.5. That is, the middle of the inversion layer had a height of 200 m at 20 h 00 m, 90 m at 23 h 40 m and 80 m at 5 h 00 m in a valley at Ôchô Island.

The lowering of the inversion top during the night occurs often in fine weather and has a strong influence on the meteorology at the ground level. In some cases, this is related to the low-level jet, as pointed out by Gifford (1952).

Aerological observations of the lower atmosphere were made on the gentle slope in Tsrikovka Village, the Aruik-Baluiksky region of Kokchetavska, U.S.S.R., in summer, 1955 and 1956 (Vorontsov, 1958). The distributions of air temperature and relative humidity observed under dry conditions on the night of June 19–20, 1955, and under wet conditions on the night of August 27–28, 1956, are presented in Figs. 5.72 (a) and (b). At 19 h 24 m, 50 min before sunset, the inversion begins on the lower part of the slope, but the temperature decreases with an increase of height on the middle and upper parts of the slope. At 20 h 42 m after sunset, the inversion is formed over the whole slope. At 3 h before sunrise, the inversion layer develops independently of the shape of the slope; the isotherms run almost horizontally. Therefore, the thickness of the inversion layer is about 60 m on the uppermost part of the slope. The change of relative humidity is similar to that of air temperature mentioned above. With wet but cloudless weather, as shown in Fig. 5.72 (b), the differences of air temperature and relative humidity between the top and the bottom of the inversion layer were smaller than under dry conditions.

If the advection is negligible, in other words, if the inversion is formed mainly by the radiation cooling at a certain point, it can be estimated that the air temperature difference (Δt) between the top and the bottom of the inversion layer is about half of the diurnal range (A) of the air temperature at a height of 1.5–2 m above the ground at this point. From the results obtained in 1955 and 1956, the relation between Δt and A is roughly as follows:

$\Delta t > 0.5A$ \
$\Delta t = (0.7 - 0.5)A$ on the lower part of the slope

$\Delta t \doteqdot 0.5A$ at a relative height of 35–40m on the slope from the foot.

$\Delta t < 0.5A$ on the upper part of the slope

$\Delta t = 0.25A$ at a relative height of 110 m on the slope from the foot under the dry weather.

It can be said, however, that nocturnal radiation cooling is the main cause of the inversion layer, because $\varDelta t \fallingdotseq 0.5\ A$ as a mean state of the relations on the upper, middle, and lower parts of the slope.

Photo 5.6 Observation field of the cold air lake in a doline in the Bükk-Hoscberac region, NE-Hungary, demonstrated by the late Prof. Richard Wagner, Szeged, and his collaborators (August 10, 1966, by M. M. Yoshino).

The reason why the inversion is the strongest on the lower part of the slope is a result of the cold air lake that develops due to the cold air runoff as discussed previously. The cold air lake is thermally very stable, and the inversion rate is 3–4.5°C/100 m along the slope in the cold air lake. Ishikawa et al. (1973) carried out an observation of radiation cooling and formation of the inversion layer in the small basins in Hokkaido. They pointed out that the sensitive heat flux is more than two times greater at the hill top stations than at the basin bottom station under the inversion condition. Thompson (1967) also reported on air temperature in a canyon. MacHattie (1970) concluded that an inversion forms in the valley bottom almost every night in the Kananaskis Valley, Canada. The thickness of the inversion plus the isothermal layer is roughly 300 m. A table on the degree of valley bottom inversions in various regions in Europe and in North America compiled by MacHattie and Table 5.25 on the height of the thermal belt present a general view of the height of the inversions in valleys or basins.

The inversion layer in a basin or valley is called a *cold air lake* as described before. The formation of a cold air lake in a small doline in the Bükk Mountain of NE-Hungary has been studied for many years by the researchers of the Szeged University, Hungary. One of the results, as given in Fig. 5.73,

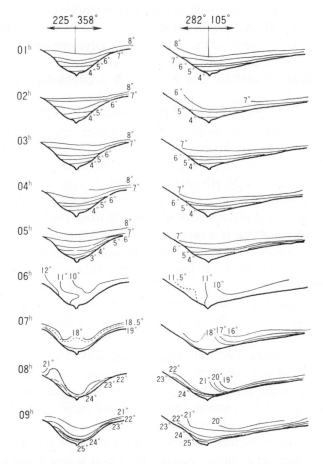

Fig. 5.73 The vertical temperature distribution in the Bükk Mountain on August 22, 1959 (Wagner, 1963).

shows that the cold air lake dissipated suddenly at 6 h (Wagner, 1963). This is fairly striking.

5.5.6. Thermal Belts on Slopes

The thermal belts or thermal zones on mountain slopes are primarily the result of vertical temperature variation. As mentioned above, the lower part of the slope is cold because the cold air lake develops. In contrast, the top of the inversion layer or the upper surface of the cold air lake is relatively warm. Above this height, on the other hand, the temperature decreases at the normal lapse rate and, consequently, the mountain top or the uppermost part of the slope is cold.

Geiger (1965) explained the formation of the thermal belt on the slope of a valley that dissects the flat open surface of a plateau. The valley would contain a cold air lake, but relatively small, individual circulations are built up between the air which is cooling on the valley slope and the reservoir of warmer air above the valley floor. Consequently, the cold air lake develops only near the valley bottom, and the upper part of the valley slope remains as a thermal belt. Over the plateau, a layer of cold air near the ground is established, and hence, it is cold.

It is said that Jefferson described the thermal belt, calling it *verdant zone*, in 1781. The stratification of the air temperature over valleys and the enclosing slopes was brought to the attention of Silas McDowell of Franklin, North Carolina, a learned farmer, in connection with fruit growing. He mentioned the concept *thermal belts* in 1858 (Dunbar, 1966). H. J. Cox (1923) reported thoroughly on thermal belts from the horticultural viewpoint based on extensive observations made in western North Carolina during 1913–1916. It was concluded that the thermal belt is centered more than 360 m above the valley floor on the average. Usually place on a slope having an elevation of 300 m or more above the floor is safety for frost danger (Henry, 1923).

Vertical division into three zones can be recognized from the thermograph records at five heights on the south slope of San Jose Hill in the Pomona Valley, California, during the night December 27–28, 1918 (Young, 1921). The lowest temperature was observed at the base station in the valley bottom. A relatively warm nighttime temperature, indicating the thermal belt, was experienced at heights of 15, 68 and 84 m above the base station, and the warmest was at 68 m. On every hill where the observations were made the top of the hill was colder than points on the hillside some distance below on clear, calm nights. The fall in temperature during the night at the lower stations was fairly steady, but there were larger fluctuations in temperature at the stations on the slope. The duration of the minimum temperature was usually much longer at the base stations than on the slope.

A detailed study on the distribution of the maximum and minimum temperature at sixteen stations around Mt. Tsukuba (876 m), 70 km NNE of Tokyo, was made from July 1953 to December 1956 (Ibaraki-ken and Mito Weather Observatory, 1955, 1957; Gunj, 1958). According to the results, the number of

Table 5.23   Number of frost days at the stations on the slopes of Mt. Tsukuba (Yoshino, 1961a).

	Height a.s.l.						
	870	240	230	70	35	35	35 m
Oct. 1953–Mar. 1954	84	40	51	67	99	73	62 days
Oct. 1954–Mar. 1955	83	38	48	65	85	80	74
Oct. 1955–Mar. 1956	76	45	57	69	82	75	71
Mean	81	41	52	67	89	76	69 days

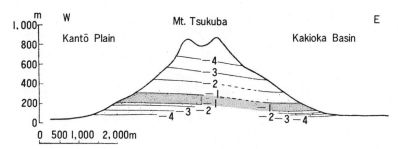

Fig. 5.74 Vertical distribution of monthly mean minimum temperature along the W–E cross section of Mt. Tsukuba, January 1955 (Yoshino, 1961a).

frost days was smallest at heights of 230 and 240 m a.s.l., as shown in Table 5.23.

A cross section of the distribution of the minimum temperature is given in Fig. 5.74. From this figure and the table, the thermal belt is clearly seen at heights of 200–300 m a.s.l. The height of the thermal belt is different for slopes in different directions on Mt. Tsukuba. On the eastern slope facing the Kakioka Basin, the height is slightly lower than on the western slope facing the broad Kanto plain.

An investigation was made on the temperature distribution in the orchard region of the Fukushima Basin, Japan, during the night from April 6–7, 1955 (Fukushima Weather Station, 1956). It was found that the region at a height of 100–200 m above the bottom of the basin was warmer. Further examination of the results from the Fukushima Basin revealed that the absolute heights of the thermal belt differ according to the slopes, but the relative heights are an average of 100–200 m above the foot of the slope (Umeda, 1958). The thermal belt was found at a height of 250 m above the Kôfu Basin, Yamanashi Prefecture, Japan, according to the results of the minimum air temperature observed in the early morning of frosty nights (Sekiya, 1958).

The thermal belt on the slope depends on the shape of the side slopes of the valley. A detailed study on this problem came to the conclusion that the profile of the terrain must be a main factor in deciding the height of the thermal belt in the valley (Koch, 1961). The thermal belt on a steep valley slope appears in accordance with the angle of the slope. That is, the thermal belt is located where the slope is steepest. It was also pointed out that below the thermal belt there was a isothermal zone with a thickness of 30–60 m, which is a transitional zone from the top of the cold air lake in the valley to the thermal belt on the upper part of the valley slope. In the case of a valley slope with some terraces the cold zone is observed on the surface of the terraces. In other words, the height of the thermal belt has no relation to the absolute or relative height, but to the shape of the valley slope in such a case. Finally, a schematic illustration was presented, as shown in Fig. 5.75.

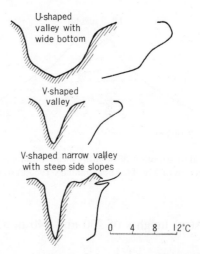

Fig. 5.75 Schematic temperature profile (right hand), showing the position of the thermal belt on the slope in relation to the shape of the valley cross section (Koch, 1961).

Urfer-Henneberger (1969) reported on the thermal belt in Dischma valley near Davos, Switzerland. The height of center of the thermal belt is 100–300 m above the valley floor and is slightly higher on the NE-exposed slope than on the SW-exposed slope.

Relations between the thermal belt and weather are dealt with next. A comprehensive study was made on the temperature distribution on Mt. Gr. Falkenstein, Germany (49°05′N, 13°17′E), by Baumgartner (1960, 1961, 1962). The average of the monthly mean of the nocturnal minimum temperature was the highest at the station at the relative height of 228 m above the mountain foot. It was shown, however, that the thermal belt disappeared on cloudy and windy nights. The relative inversion height from the mountain foot decreases with an increase of the amount of clouds, as is shown in Table 5.24. Of course, the height of the thermal belt must follow this rule. The vertical profile of the average of the monthly mean minimum temperature is given in relation to the amount of clouds in winter and summer in Fig. 5.76.

The thermal belt was located higher in winter than in summer, because of the stable situation of the cold air layer at the lower level. Under overcast conditions in winter the temperature profile is not in good order, probably due to the winter synoptic situation with the warm air advection at the higher level resulting in a heavy overcast.

Table 5.24 Cloud amount and the relative inversion height in Mt. Gr. Falkenstein (Baumgartner, 1962).

Cloud amount	0	2	4	6	8	10/10
Relative inversion height	300	220	160	130	100	70 m

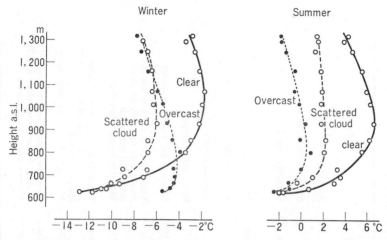

Fig. 5.76 Average of monthly mean minimum temperature on the western slope of Mt. Gr. Falkenstein, Germany, in winter (Nov. 1953–Mar. 1954) and in summer (April–July 1954) (Baumgartner, 1962).

The seasonal change in the height of the thermal belt on the WNW-slope of a valley running in the N–S direction in Obergurgl (46°53′N, 11°02′E), Ötz valley in Tyrol, Austria, was studied based on the detailed observations carried out in 1954–1962 (Aulitzky, 1967). According to the results from stations on the slope from 1,820 to 2,440 m a.s.l., the position of the thermal belt was much higher in winter months than in summer months. In addition, the thermal belt was found not only on the vertical profiles of the minimum temperature, but also on those of the daily mean and maximum temperature. Therefore, the thermal belt in the high altitudes near the timber line has particular significance for the plant life. Phenological observations in relation to the thermal belt will be discussed elsewhere. Further information can be obtained from the description by Geiger (1961, 1965). Recently, Obrebska-Starkel (1970) summarized the previous studies on the temperature inversion in basins and valleys and presented a table (Table 4.18) showing the average height of the inversion tops in the various regions, mainly in Europe. The average height of the inversion top might correspond to the height of the center of the thermal belt under consideration. According to her study, the average height is 150–200 m.

As a summary of the knowledge concerning thermal belts: (a) The height of the center of the thermal belt on a slope appears at a height of 100–400 m above the valley floor or mountain foot, as shown in Table 5.25. In most cases, it is 200–300 m. (b) The position of the thermal belt depends on the cross section shape of the valley. (c) The thermal belt is higher and sharper during clear, calm nights. (d) The thermal belt is higher in winter than in summer.

5.5.7. Inversion Layers and Local Weather

The distributions of meteorological variables in a small area under the in-

Table 5.25 Relative height of the center of the thermal belt (Compiled by Yoshino).

Relative height	Topography	Surround-ing peak	Region	Observation period	Literature	Notes
360m	Valley floor, 680m a.s.l.	1,200m a.s.l.	W-North Carolina, U.S.A.	1913–1916	Henry, 1923	Average
68m	Broad valley floor, 130m a.s.l.	600m a.s.l.	S-California, U.S.A.	Dec. 27–28, 1928	Young, 1921	Clear, calm night
230m	Mountain foot, 30m a.s.l.	870m a.s.l.	Kanto district, Japan	July 1953–Dec. 1956	Mito Wea. Obs., 1955, 1957 Gunji, 1958	Monthly mean
180m	Mountain foot, 100m a.s.l.	1,000m a.s.l.	Fukushima Basin, Japan	April 6–7, 1955	Fukushima Wea. Obs., 1956	Frosty night
250m	Bottom of basin, 260m a.s.l.	1,500m a.s.l.	Kōfu, Yamanashi, Japan	April, May 1956, 1957	Sekiya, 1958	Frosty night
150–160m	Valley bottom, 170m a.s.l.	340m a.s.l.	Jägerberg, Jena, Germany	1945–1959	Koch, 1961	Various nights
210–220m	Valley bottom, 140m a.s.l.	350–390m a.s.l.	Gleissberge, Jena, Germany	?	ditto	Steep valley slope
440m	Mountain foot, 620m a.s.l.	1,320m a.s.l.	Gr. Falkenstein, Germany	Nov. 1953–Mar. 1954	Baumgartner, 1962	Mean of clear days in winter
300m	ditto	ditto	ditto	ditto	ditto	Mean of clear days in summer
230m	ditto	ditto	ditto	April–Nov. 1955	ditto	Average of monthly mean of nocturnal minimum
100–200m	Valley bottom, 1,820m a.s.l.	More than 3,000m a.s.l.	Obergurgl, Austria	June 1954–May 1955	Aulitzky, 1967	Monthly mean minimum
120–370m	Valley bottom, 1,700m a.s.l.	2,500m a.s.l.	Dischma Valley, Switzerland	Summers, 1960–1965	Urffer-Henneberger, 1969	Monthly mean minimum and total of hours of air temperature lower than 5°C
100–240m	Valley bottom, 1,300m a.s.l.	3,000m a.s.l.	Kananaskis Valley, Canada	Summer, 1960	MacHattie, 1970	Mean of 20 days for each month
100–120m	Valley bottom, 302m a.s.l.	611m a.s.l.	Czarna Woda Valley, Poland	July 18–30, 1971	Niedźwiedź et al., 1973	Mean minimum temperature (5 cm), Anticyclonic clear weather
100m	Valley bottom, 251m a.s.l.	400m a.s.l.	Raba Valley, Poland	1967–1968	Niedźwiedź, 1973	Mean annual temperature

version condition are quite different from those under the lapse condition, because the local air flows that definitively influence the distributions are affected by the presence of the inversion layer.

Excluding the cases of frontal inversion, the air masses, amount of clouds and wind speed over the area are different in the cases with inversion from the cases without inversion. Therefore, in some cases, the inversion layer is a factor of such meteorological conditions prevailing in the area.

Table 5.26 Classification of gustiness by the fluctuation of wind direction (Smith, 1951; Singer and Smith, 1953).

Type	Range of wind direction fluctuation	Taype of turbulence	Stability (Lapse rate)*	Mean wind velocity and standard deviation	Period Occurred	Season
A	$>90°$	Large eddies produced thermally	Unstable (-1.25)	1.8 m/sec (±1.1)	9–15 h only	Abnormal type in winter
B_2	45–90°	Large eddies produced thermally	Unstable (-1.6)	3.8 (±1.8)	9–15 h only	Mostly in summer
B_1	15–45°	Eddies produced thermally and mechanically	Unstable (-1.2)	7.0 (±3.1)	6–18 h occasionally at night with strong lapse condition	Every season
C	$>15°$	Eddies produced mechanically	Stable (-0.64)	10.4 (±3.1)	At night and daytime with heavy cloud deck	Every season
D	0–15°	none	Stable $(+2.0)$	6.4 (±2.6)	Mostly at night	Every season

* Lapse rate is the mean temperature difference (°C) between 120 m and 11 m.

Among all the parameters that influence the diffusion, the most important ones are wind direction, wind speed, and turbulence, whose characteristics differ according to the gradient of temperature up to 120 m in height (Smith, 1951; Singer and Smith, 1953). For the purpose of forecasting, a classification of gustiness was developed from the types of traces found on gust recorders. Five types are distinguished based on the results observed at Brookhaven, New York, U.S.A., as shown in Table 5.26. Types A, B_2, and B_1 are distinguished by the fluctuation range of the wind direction. Type C is distinguished by the unbroken solid core of the trace, through which a straight line can be drawn for the entire hour, without touching "open space." By Type D, the trace approximates a line —short term fluctuations do not exceed 15°. These definitions are based entirely on the wind direction traces recorded by a Bendix Friez Aerovane. During April 1950–March 1952 the occurrence frequency of Type A was 1%; Type B_2, 42%; Type B_1, 3%; Type C, 14%; and Type D, 40%. Type A appeared generally before Type B. The greater the gradient, the higher the wind speed in the cases of Types A and B_2. Type B_1 was observed when the wind was strong under the normal temperature gradient. Type C was the cases with overcast weather and the normal lapse rate. Type D is seen under the conditions of the typical inversion layer. As mentioned above, the frequency of Type D was 40% at Brookhaven, because the inversion state occurred very frequently due to the surrounding hilly topography. It can be understood from Table 5.26 that the thermal condition of the air layer, stability, plays much more important role for turbulence. The concentration measurements during these five types revealed also that the relations between the cross wind or vertical standard deviation (gustiness) of plumes from continuous point sources and the down wind distance are expressed by linear lines for each type. This proved the usefulness of this classification and the importance of the vertical gradient in making definitions of meteorological phenomena (Singer and Smith, 1966).

From the distribution of wind direction at the ground level in Tokyo, it was supposed that the formation of eddies in the city region might depend on the height of the inversion layer (Yamashita, 1954). Further investigation on this problem based on the wind direction observed at about 200 stations in Tokyo revealed that peculiar phenomena were found when the inversion layer develops (Yoshino, 1955b). Namely, there are two cases; the formation of eddies estimated from the regional fluctuation of wind direction were strong or weak when the inversion was located between 200 and 1,000 m a.s.l.

There are, in general, as observed at Oak Ridge, Tennessee, two regimes in which the hilly terrain influences the wind condition as a factor of roughness (Holland, 1952). (a) In inversion periods, Richardson number $Ri \geq 1$, surface wind roses are dependent largely on the local slope and exposure, and (b) in lapse periods with turbulent flow, $Ri \leq 1$, the wind losses are less dependent on the terrain situation and more sensitive to the gradient wind direction. Another study that observed low-level air trajectories at the same place reported that they do not depend primarily on large surface obstacles such as ridges, but are characteristic properties of low-level air flows (Gifford, 1953).

Air pollution in the urban and industrial areas has been studied intensively since World War II. These problems were discussed in Section 3.2.7. The elements of atmospheric electricity change clearly with the passage of the temperature inversion. A comprehensive review was given by Reiter (1964).

Chapter 6

LOCAL AND MICROCLIMATE AND NATURE

6.1 PLANT ECOLOGY

6.1.1. Forest Structure and Microclimate

The local climate is a constituent of nature, but this does not mean that the local climate described in the preceeding chapters constitutes directly the environment of plant life. The relationship between the local climate and the environment should be viewed from the standpoint of the eco-system. Furthermore, in considering the eco-system the "scale" classes are important; they should correspond to the scale divisions of climate as given in Table 1.1. The level of the environment, and orders of climate and soil, in relation to the level of life are given in Table 6.1 (Numata, 1958).

Table 6.1 Level of plants and animals and its relation to the orders of climate and soil phenomena (Numata, 1958).

Level of plants and animals	Order of climate	Order of soil
Formation Association	} Macroclimate	Zonal soil
Interstand Intrastand	Mesoclimate Microclimate	} Intrazonal soil

The factors contributing to a single ecotope were discussed by Troll (1963) in relation to the landscape distribution. A distinction may be made between the climatic range of factors above the ground surface and the edaphic range of factors under the ground. The vegetation is affected by the conditions of climate, soil, and ground water, on the one hand, and the vegetation affects edaphic factors such as soil formation, soil climates, and the ground water table on the other; further, it is the major element in microclimate differentiations.

In a study on the environmental factors for a bamboo forest (*Phyllostachys bambusoides*), Numata et al. (1957) pointed out that three levels of integration of the environment can be recognized (see Table 6.2). This table was compiled by analyzing the extremes of air temperature, wind velocity and the number of stormy days, amount of rainfall, deviation of temperature from the annual mean relative humidity, light intensity, ground temperature, etc. It should be noted that the primary factors that control the position of the borderline of the distribution of a bamboo species are not always the same as those factors that

Table 6.2 Environmental physical factors controlling the distribution
and growth of a Japanese bamboo forest from the stand-
point of the level of integration (Numata et al., 1957).

Organismic level	Environmental scale	Primary factors	Secondary factors
Association (Specific synusia)	Macro-	Minimum temperature	Wind velocity Stormy days Amount of rainfall
Interpopulation	Local	Wind velocity Stormy days % Amount of rainfall	Grain size composition of soil Water content of soil Draining
Stand (Interpopulation)	Micro-	Depth of surface soil Draining	Direct solar radiation

control the local distribution. For instance, the wind is considered to be the primary factor on the local environmental scale, whereas it plays only a secondary role on the macro-environmental scale (Schmithüsen, 1968).

Detailed research into the forest composition and minute observation of the local climate has been made in Harvard Forest, Massachusetts, U.S.A. (Spurr, 1956, 1957; Rasche, 1958). On the whole, the forest composition has a close relationship to the state of soil moisture, but a closer observation shows that the micro-topography, concave or convex, plays an important role in its composition. In a concave topography, the minimum temperature is usually low, hence the predominance of northern-type trees, such as red spruce (*Picea rubens* Sarg.) or black spruce (*Picea mariana* B.S.P.), whose southern limit of distribution lies around this area, or a kind of tamarack (*Larix laricina* K. Loch). On the other hand, mature trees of white ash (*Fraxinus americana* L.) are seldom found here. On a convex slope where minimum air temperature is usually high, it might be assumed that southern-type trees should dominate, but there is no distinctive feature of tree distribution like that found in a concave topography. The local difference in the average temperature is 4–5°C within small areas in this forest. Difference amounts to the same value by which it is supposed to have been warmer in the climatic optimum of the postglacial age than today. Since the tableland in this area was not covered with glaciers, the difference in temperature between a ridge and a valley immediately after the glacier age was similar to that today. The present distribution of the forest is supposed to have been determined directly by the forest type based on the local climate at that time through the forest succession during the period of the climatic optimum, and indirectly by forest soil.

An example of a major influence of the local climate on the plants is found at a small valley at Neotoma, Ohio, as stated in Section 4.2.2(D), where the distributions of various kinds of species depend largely on the local climate in this district (Wolfe et al., 1949; Wolfe, 1951).

Shanks and Norris (1950) observed a microclimatic variation in a small valley in E-Tennessee, U.S.A. According to them, *Quercus coccinea*, *Q. alba*,

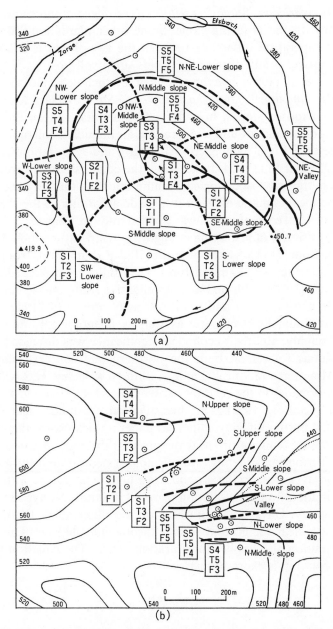

Fig. 6.1 Local climatic division of Staufenberg (upper) and Wurmberg (lower) in the Harz Mountains, Germany. S: Insolation, T: temperture, and F: humidity. 5 means the highest or greatest and 1 the lowest or smallest values (Hartmann et al., 1959).

Carya glabra, C. tomentosa, Diospyros virginiana, Vaccinium vacillans, etc., are found only on the S-facing slope, whereas *Fagus grandifolia, Liriodendron tulipifera, Fraxinus americana, Staphylea trifolia, Hydrangea arborescens, Rhus radicans*, etc., only on the N-facing slope. Some other plants are found on both slopes. The temperature difference between the two slopes is 1–2°C, but the difference of species is still observable in this small valley.

Hartmann et al. (1959) give the results of microclimatic measurements carried out in the *"montane"*- and *"hochmontane"*-*Stufe* of Harz Mountains; on the differently exposed slopes, in valleys, on summits, and on the ridge. The measurements of air and ground temperature, air humidity, soil moisture, radiation, brightness (cf. Fig. 3.39 in Section 3.3.1(B)), and evaporation were made in September 1953 and June 1954. In addition, the composition of forest trees and plants, the resulting phytosociological communities and the soil conditions were analyzed. The local climatic division was made according to the three meteorological parameters, each of which was divided into five degrees: S (insolation), from S1–S5 by priority of insolation strength; T (temperature, including ground temperature), from T1–T5 by temperature; and F (humidity), from F1–F5 by air moisture (together with soil moisture). The results are presented in Fig. 6.1. Hartmann et al. (1970) published a comprehensive study on the climatic background of the natural forest and the forest association in Mittelgebirge, Germany. In this study, local climatic as well as phenological conditions were analized thoroughly, and their ecological significance was discussed.

In deciduous forests on both the northern and southern slopes of the Cushetunk Mountains (about 250 m a. s. l.) in New Jersey, U.S.A., a close investigation was made on the relation between the vegetation and the local and microclimate (Cantlon, 1953). This investigation has made clear that in the foliation period on the S-facing slope: (a) under a thick crown there is a slight inversion of the air temperature near the ground; (b) under a moderate crown, the vertical distribution of temperature is almost neutral, and (c) under a thin crown, the air layer close to the ground has a high temperature with a complex vertical distribution. In the defoliation period it is almost always similar to case (c) during the day. On a N-facing slope, on the other hand, the temperature decreases a little near the ground surface throughout the year. This pattern in vertical distribution also applies to saturation deficit. The equation for the difference in vegetation is expressed as:

$$D = \frac{s}{S} \times 100 \qquad (6.1)$$

where D stands for the degree of vegetation difference, S is the sum of the values for the character on both slopes for all species, and s is the sum of the differences in the character between the slopes for all species. Such calculations were made for the tree layers using the character density and basal area, for the shrub and herb layers using density and frequency, and for the bryophyte layer using fre-

Table 6.3 Degree of vegetation difference between the S and N slopes for the various layers in the Cushetunk Mountains, New Jersey (Cantlon, 1953).

Layer and the character used to obtain the degree of difference	Degree of difference* between the slopes
Tree layer	
Density	45 %
Basal area	46
Lower tree layer	
Density	74
Basal area	75
Shrub layer	
Frequency	53
Density	40
Herb layer	
Frequency	52
Density	72
Bryophyte layer	
Frequency	92
Cover	95
Average difference	64

* expressed by Eq. (6.1)

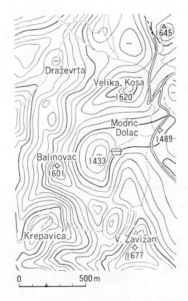

Fig. 6.2 Topography of a doline in Modrić dolac near the Mt. Veliki Zavižan, Croatia, Yugoslavia (Kušan, 1970).

quency and cover. The results are presented in Table 6.3. This table shows that the difference in vegetation between the northern and southern slopes becomes most conspicuous in the layers next to the ground in a forest.

Whittaker et al. (1965) studied the vegetation on the SW-slope of Santa

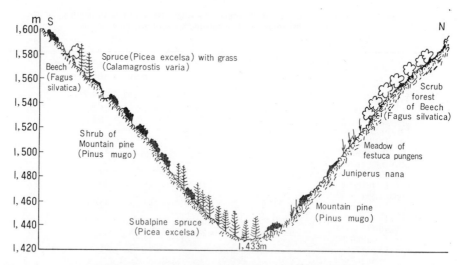

Fig. 6.3 Vegetation distribution along the profile of a doline in Modrić dolac (Kušan, 1970).

Photo 6.1 Vegetation inversion on the slope of a doline, the south-facing slope of Mt. Velika Kosa (1,620 m) near Zavižan, Yugoslavia (November, 1972, by M. M. Yoshino).

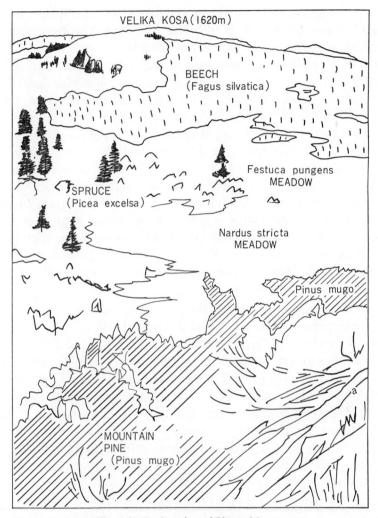

Fig. 6.4 Explanation of Photo 6.1.

Catalina Mountain, SE-Arizona, by preparing transects for 305 m elevation belts on granite and gneiss soils from the summit forests (2,440–2,750 m) to the base of the mountain (900 m). They showed in a mosaic chart the relations of communities to elevation and topographic moisture gradients. Zoller (1954) investigated the *Bromus erectus* meadow in the Swiss Jura. In his floristic-statistic analysis in relation to local climate, it is pointed out that the xerobromion is distributed in an extremely dry, warm region, and the mesobromion appears independently of environmental factors.

The primary productivity of Japanese beech (*Fagus crenata*) forests on the Naeba Mountains was estimated by means of the summation method and also by the photosynthetic model method (Maruyama, 1971). The primary production of the climax beech forest is approximately 40 t·ha⁻¹·year⁻¹ but it is 20 t·ha⁻¹·year⁻¹ at the upper limit on the mountain slope. The decreasing rate was calculated to be 2 t·ha⁻¹·year⁻¹ per 100 m altitude.

Based on his ecological, physiological, and climatological studies, Aulitzky (1963 a, 1965) discusses the forest management in the subalpine zone of the Alps with particular reference to reforestation and stand improvement. Because of the great importance of the snow cover in these high regions, all climatological elements have to be taken with regard to the thickness of the snow cover as well as its duration. The protection by the snow cover is more effective than the damage caused by the snow cover. For reforestation on the present timber line, the so-called "wind-snow ecogram" serves as a planting instruction that pays particular attention to site selection. Aulitzky made the ecogram based on the zonation of characteristic plants, which indicate various degrees of wind- and sun-exposure as well as thickness and duration of snow cover. With the help of these indicators the distribution of the snow cover during the winter months can be estimated. This is of great importance for reforestation as well as for the control and prevention of avalanches.

In the Velebit Mountain Range along the Adriatic Coast, Yugoslavia, there are two famous mountains, Mt. Veliki Zavizan (1,667 m) and Mt. Balinovac (1,601 m). Near them an alpine garden is opened in Modrić dolac, including a doline, which has an extremely interesting phenomenon of *inversion of vegetation*. The doline, shown in Fig. 6.2, is located at the subalpine zone; the bottom has a height of 1,433 m a.s.l., and the surrounding fringe of this doline is 1,500–1,600 m. It is said that air temperature falls to $-7°C$ in the bottom of the doline even in summer. Such conditions strongly affect the vegetation distribution (Kušan, 1970). The normal vertical distribution from lower to higher altitude, beech-spruce-mountain pine, appears inversely. As shown in Fig. 6.3, on the slope of the doline the following vertical distribution of vegetation is seen, mountain pine (*Pinus mugo*) and spruce (*Picea excelsa*) near the bottom and beech (*Fagus silvatica*) at higher altitude indicating warmer conditions. The landscape of the slope of the doline, the right hand slope in Fig. 6.3, is shown in Photo 6.1 and Fig. 6.4. The temperature inversion phenomena is discussed in Section 4.2.2, and the formation of cold air lakes in Sections 5.4 and 5.5, but the effect of the phenomena on vegetation can be seen here clearly.

6.1.2. Relicts and Local and Microclimate

We sometimes find a relict, a remainder of the former vegetation, among the dominant species or the indicator plants of the present vegetation pattern in an area. Occasionally a climax association itself remains, and in other cases only a fragment of it is found. Weaver and Clements (1938) made a short general review of the relicts in the United States.

On main cause of the existence of such a relict flora may be the local climatic condition. A typical example is the hemlock forest at 35°43′N, 78°47′W in

North Carolina, U.S.A. (Oosting and Hess, 1956). The hemlock (*Tsuga cana-densis*) there stands on the N-facing cliff with a gradient of 35–40°, about 300 km apart from the normal horizontal distribution limit. A year's observation has made it clear that sunshine, temperature, and evaporation are all little or low, and soil moisture is high, where the hemlock trees grow. This isolated hemlock forest can be regarded as the relict of what had once extended very widely during the diluvial epoch but retreated thereafter to Appalachia in the postglacial period. The climatological conditions of low temperature and high moisture in this area must have been a help toward the isolation of this hemlock.

Sugar maples (*Acer saccharum*) in the arid area of Oklahoma, U.S.A., afford another example of such a relict (Rice, 1960). Sugar maples are distributed mainly in a humid climate. Therefore the limit of the distribution of these maples is located at Oklahoma, 300 km afar. An observation was made from April to October, 1958, on the difference in climate between the areas inside Devils Canyon and outside the Canyon. By analyzing the observations of air temperature, moisture, wind velocity, amount of evaporation, precipitation, ground temperature, and soil moisture, it was concluded that the humid climate in this region contributed to the survival of the maple trees.

As stated above concerning Harvard Forest, some southern elements have extended beyond the present northern limits of their distribution during the climatic optimum and now remain isolated in the forests of the northern elements.

These climatic remnants are classified into five groups by their geological ages and origin: remnants from the pre-Tertiary period, Tertiary period, glacial period, interglacial period, and postglacial period. A relict, as a result of the influence of the local climate, usually has its origin in the postglacial period.

From fossil evidence the local climate of the past can be estimated. Perring (1962) noted that W-Ireland, where fossil evidence suggests that forest development was very much slower than in Great Britain and where high winds and exposure limit tree growth even today, and where large areas of hard limestones occur, was a very likely area in which species of open habitats, intolerant of shade and low pH, could survive.

6.1.3. Wind-shaped Trees

Deformed trees develop under the influence of severe environmental conditions. One factor in forming wind-shaped trees is the prevailing winds. Although the physiological processes are not yet fully understood, the effects of winds on tree growth can be recognized in many regions of the world: From the tropics to the polar regions and from the coasts to the alpine zones of mountains.

Using the relation inversely, the wind conditions can be inferred by observing the wind-shaped trees. Wind-shaped trees are a powerful tool in the study of ecological climatography on wind conditions. Among the studies on wind distribution using wind-shaped trees, are some very early works, such as by Früh

(1902). Since then, many studies, reports, descriptions, and photographs have been published concerning this subject (Yoshino, 1973a).

.   (A) *Classification of Wind-shaped Trees*

According to the studies by Walter (1951) and Yoshino (1967) there are four types of wind-shaped trees, based mainly on the shapes of tree trunks and branches: Type 1——Trunks are vertical, but branches are bent drastically to the leeward of the tree by the prevailing winds during their growing period. Examples are shown in Photos 6.2 and 4.10 (right). Type 2——Trunks are veritcal, but branches on the windward side of the tree are severed by the effects of winter winds carrying snow or frozen rain. Type 3——Trunks and branches are both deformed drastically by the prevailing winds during their growing season. Examples are shown in Photos 6.3 and 4.10 (left). Type 4——Trunks are declining due to occasional strong winds such as typhoons, but the shape of the canopy is almost symmetrical. This type is not formed by the prevailing winds.

Photo. 6.2 An example of a wind-shaped tree, Grade 3, of Type 1. Larch (*Larix leptolepis*) on Mt. Kirigamine, central Japan (April, 1973, by M. M. Yoshino).

Photo 6.3 An example of a wind-shaped tree, Grade 4, of Type 2. *Abies Mariesii* on Mt. Aumza, central Japan. a, b, c and d are defined as given in text (August, 1959, by M. M. Yoshino)

Yoshino collected examples from all over the world for the purpose of classification. A list of the references has been published elsewhere (Yoshino, 1973a).

Type 1 trees are mainly conifers such as larches. The difference between Types 1 and 2 is clearly seen in the shape of branches. If observed from the top, the branches around the trunk are bent in Type 1 by the prevailing wind during the growing season, as shown in Fig. 6.5 (a), and in Type 2 by the prevailing

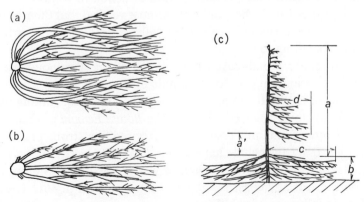

Fig. 6.5 Schematic section of wind-shaped tree of (a) Type 1 and (b) Type 2. (c) Typical tree form of Type 2, the wind-shaped tree deformed during winter (Yoshino, 1973 a and b).

winter winds carrying snow and frozen rain, as shown in Fig. 6.5 (b). Type 1 is found in mountainous regions in the subalpine zone. Yoshino (1964) gives a few examples from the upper course of the Rhône valley, Switzerland. Examples of Type 1 occur in the region near Pru del Vent and others, Switzerland (Holtmeier, 1971), and in the eastern part of the Columbia River gorge (Lawrence, 1939). Cypress trees and pine trees are deformed by the mistral in the lower Rhône valley, France, and on the coast of Riviera, Italy (Runge, 1957). A more detailed classification was made of the wind-shaped trees of Type 1 in a fog prevention forest area near Atsukeshi in Hokkaido, Japan (Tatewaki, 1953). In this report, *Alnus maximowiczii*, *Betula ermani*, and *Quercus crispula* are classified into five patterns considering tree height, diameter at breast height, and crown width. When the trees of Type 3 grow densely, their crowns, if viewed from high above, form a smooth surface with little unevenness due to the wind. Some examples of *cushion-shaped trees* are often found near a cliff facing strong prevailing winds.

Type 2 is found most commonly in the subalpine zones all over the world. The tree form of this type can be measured by Parts *a*, *b*, *c*, and *d* shown in Fig. 6.5. (c). Part *a* stands for the part of the tree above snow cover in winter; the height of which decreases with the increase of the wind velocity. This is also the case with the value of Part *d*, which shows the part of branches shooting out toward the lee. The section of this Part *a* as given in Fig. 6.5 (b) is quite different from one deformed by the winds during the growing season. Occasionally, there is a tree whose branches never spread leeward above Part *b*. It is thought that Part *a'* of a trunk is formed when a trunk has been deprived of every sprout by small particles of snow blown off from the snow cover due to a strong wind (Daubenmire, 1950). Consequently, where there is no strong prevailing wind, the trees exhibit no such peculiar type as Part *a'*. The height of Part *b*, in which the branches spread in all directions under the protection of the winter snow, may be interpreted as the average snow depth. Values of Part *d* also suggest the strength of the prevailing wind in winter and the branches of Part *d* develop only leeward, which indicates the prevailing wind direction in winter.

In either Types 1 or 2, the trunks are almost vertical, but their branches are bent down by winds. Therefore, they are called *flag-shaped trees* when all the branches are spreading to one side.

Type 3 trees are broad-leaved trees. Some of the deciduous broad-leaved trees show a deformation due to the winds during the growing season in the subalpine zone or mountain zone as well as in the lowlands. Among the trees of Type 3 in the subalpine zone, the shape of the *Betula Ermani* is often most striking. Type 3 is often found in a coastal area exposed to a strong prevailing wind, as is exemplified by the wind-shaped trees, mostly *Quercus petracea*, on the western coast of England (Oliver, 1960) or western Scotland (Stamp, 1950). An example on the Adriatic coast is shown in Photo 6.4 on p. 461. They develop also under the influence of a NW-monsoon, for example, along the Pacific or Sea of Japan coasts. Type 3 of *Alnus glutinosa*, *Betula alba*, *B. pubescens*, etc., are also observable in the sand dune region of the East Frisian Islands in the North Sea (Runge, 1955). In general, the trees of Type 3 are expected to develop in the temperate zone, but often appear in the tropical region under the influence of trade winds and in the polar region where a strong wind prevails.

Type 4 includes trees whose trunks alone bend to the lee, often seen among pine trees (*Pinus thumbergii*) on the Pacific coast of Japan. Unusually strong but rare winds (for example, in a typhoon) have acted on the trees of Type 4, because the neighboring trees in a small area do not show similar inclination gradients or trunk directions, and their crowns do not present a conspicuous deformation.

(B)   *Scale of Wind-shaped Trees*

In his very interesting book *Power from the Wind*, Putnam (1948) classified the deformation of trees into 5 grades: (a) brushing, (b) flagging, (c) throwing, (d) wind clipping, and (e) tree carpets. Putnam's investigation is based on observations in various regions of the world. Barsch (1963) made an intensive study on the wind-shaped trees in the lower Rhône valley in France using six deformation grades, as shown in Fig. 6.6. Yoshino (1964) studied the wind con-

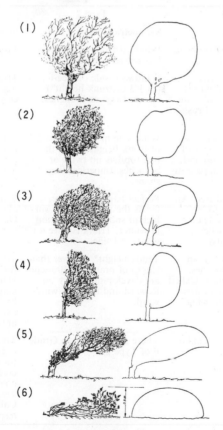

Fig. 6.6   Deformation grades of wind-shaped trees (Barsch, 1963).

ditions in the Rhône valley in Switzerland based on the degrees of deformation by Barsch and presented an illustration on the grades of *Prunus avium, Populus alba, P. nigra* var *italica, Pinus silvestris, Larix decidua*, etc. In this region no extreme shapes were found, so that only four grades were enough, but they coincided with the ones from Grades 1 to 4 by Barsch (1963) and Weischet (1963). Scultetus (1969) introduced the scale by Weischet (1963) as a useful tool for estimating the wind conditions in an area.

Based on the grades mentioned above, a standard scale of grading has been made as shown in Table 6.4 and Fig. 6.7. Several surveys using this standard scale were carried out in Hokkaido (Owada et al. 1971; Yoshino et al., 1972) and in Slovenia (Yoshino et al., 1973c), in the coastal region of the Adriatic Sea, Croatia (Yoshino et al., 1974), and they proved its applicability.

Owada (1973) clarified the relationship between the grade of the larches (*Larix leptolepsis*), the wind-shaped trees of Type 3, and the mean wind velocity of the prevailing wind in summer in the Ishikari plain, the Shari-Abashiri region,

Table 6.4 Scale of deformation grade of wind-shaped trees (Yoshino, 1973a).

Grade	Type 1	Type 2	Type 3
0	Symmetrical form	Symmetrical form	Symmetrical form
1	Tree-top and twigs are bent slightly to leeward.	Windward side of tree-top has severed slightly.	Trunk or branches are bent slightly to leeward.
2	Tree-top (ca 1/3 of tree height) and twigs are bent apparently to leeward. This form can be called "Brushing."	Windward side of tree-top (ca 1/3 of trunk height*) has no twigs or branches.	Trunk and branches in ca 1/3 tree height are bent markedly to leeward.
3	Imperfect flag-shaped tree. About 2/3 of the height of tree from the tree-top and branches are bent apparently to leeward.	Wind side of trunk in ca 2/3 of trunk height* from the tree-top has no twigs or branches. Imperfect flag-shape.	Trunk and branches in ca 2/3 tree height are bent drastically to leeward. Imperfect flag-shape.
4	Perfect flag-shaped tree. All branches are bent to leeward. Windward side of trunk has no branches. Trunk leans slightly.	Perfect flag-shaped tree, as far as the part over the surface of snow accumulation in winter. Trunk is bare on windward side.	Trunk and branches are bent almost perfectly to leeward. On windward side, trunk is bare.
5	Tree height is lower than length of main branches. Trunk leans leewards. Called *Krüppelform, Windflüchter* (Weischet) or "Throwing" (Putnum).	Trunk height* is lower than length of branches. Branches are developing poorly on leeward and none on windward.	Trunk and branches are deformed so as to lean away. Height of tree is lower than length of main trunk or branches. Branches on the extreme leeward part only are living.
6	The shape becomes similar to that of Type 3.	There are no trees in Grade 6 of Type 2.	Trunk and branches grow only leeward very low on the ground level, look like creeping. Called *wind creeping* (Putnum) or *Teppich-form* (Barsch). Called *cushion shape, tree-carpet*, and so on.

* Trunk height of Type 2 is measured from the estimated surface of snow accumulation in winter.

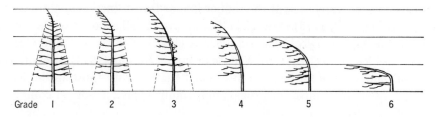

Fig. 6.7   Scale of wind-shaped broad-leaved trees of Type 3 (Yoshino et al., 1972).

and Konsen-Genya in Hokkaido. The relation is expressed by the following equation:

$$W_s = 1.6 + 0.95G_{SL} \qquad\qquad (6.2)$$

where W_s is the mean wind velocity (m/sec) observed at the meteorological stations in summer and G_{SL} is the grade of wind-shaped trees of *Larix leptolepis* in summer. This means that the mean wind velocity in summer is approximately equal to the grade value of wind-shaped larch trees plus 1.4–1.6.

By observing the distribution of the grade of wind-shaped larch trees, the distribution of the wind velocity can be estimated using Eq. (6.2). Such a relationship has so far been obtained only for the larch trees, but it demonstrates the usefulness of wind-shaped trees as an indicator of wind conditions.

(C)   *Wind and Tree Deformation*

The wind is the key factor affecting deformation directly and indirectly. The direct influence is due to a mechanical effect (a) the wind pressure, generally during the growing period of a tree and (b) the breaking off of the branches and top twigs by the winds. The indirect influences can be divided into: (a) strong winds on the windward side promote transpiration, causing branches to wither due to low temperature and dryness during the growing season. Consequently, there are no branches on the windward side. (b) Snowflakes or glaze drops carried by winds stick to the branches on the windward side, under the burden of which the young branches are broken, so that the branches on the windward side are missing.

According to the classification of wind-shaped trees stated in Section 6.1.3 (A), Type 1 deformity is the result of the indirect influence (a) of the wind, but in the subapline regions (e.g., Yoshino, 1973b) the indirect influence (b) is sometimes at work. Given with a deformed needle tree, it is sometimes difficult to decide whether it is to Type 1 or Type 2; the latter type is the result of a phenomenon called *storm-pruning* caused by winter glaze storms. A reliable decision can be reached by observing the branches shown in Fig. 6.5 (a) and (b), as mentioned above.

Type 2 wind-shaped trees are the results of the direct influence (b). Type 3 is caused mainly by the indirect influence (b) and in some cases by the direct influences (a) and (b). Type 4 is caused by the direct influence (a) at any period.

Type 2 is generally seen at the timber line or forest line in subalpine zones all over world (Troll, 1973). Aulitzky (1963b) gives a full account of the trees in Type 2 in the Alps in Tyrol, Austria, Plesnik (1971) of those in Tatra, Czechoslovakia, and Yoshino (1973a, b) of those in Japan. Plesnik (1973) compared the trees of Type 2 on the timber line in Central Europe with those in the Rocky Mountains. The mechanism or process in forming Type 2 is not yet completely known; for instance, snow polish or snow blower (*Schneeschliff*) effects on Parts *a* and *a'* of Fig. 6.5 (C) have been discussed by Turner (1968) and Holtmeier (1968). It seems that further experimental field observations are needed before any definite conclusions can be reached.

Degrees of residence vary according to species of trees. The influence of winds accompanied by snow or glaze storms in winter is obvious, as a matter of course, among evergreen coniferous trees, whose leaves do not fall even in winter, in the subalpine or polar regions but not so apparent among deciduous broad-leaved trees.

The influence of wind during the growing period shows a wide variation among various kinds of trees. Among the trees whose crowns indicate clearly the direction of a prevailing wind are poplars, fruit trees, and medlar trees, but with trees such as pine trees, birches, oaks, and beeches, it is very difficult to estimate the wind direction. Trees such as willows and spruces are the borderline cases between the two groups (Walter, 1951). According to another investigation made on the basins of the lower Rhein region (Weischet, 1951, 1953, 1955), the following kinds of trees are arranged according to their degree of residence in sequential order from the weakest to the strongest: sweet cherry (*Prunus avium*), pine (*Pinus silvestris*), mountain ash (*Sorbus aucuparia*), Silver poplar (*Populus alba*), apple (*Prunus malus*), pear (*Pirus communis*), beech (*Fagus silvatica*), horse chestnut (*Aesculum hippocastanum*), poplar (*Populus nigra var italica*), linden (*Tilia platyphyllos*, *T. cordata*), maple (*Acer platanoises*), spruce (*Picea excelsa*). Likewise, the resisting power against transpiration and damage by a strong wind increases in the following order: hazels, birches, ashes, beeches maples, firs and spruces.

The difference of opinions concerning the deformation sensibility may arise from lack of any definite standard, since one observer pays attention to crowns and another to trunks, without classifying deformed trees into groups. Further study is needed to analyze the influences of winds according to the Types mentioned above.

Growth of deformed trees and their relation to the local climate was studied by a careful observation carried out on Cornwallis Island, Canada (75°N, 98°W), in 1954 (Wilson, 1959). The subject for this experiment was the young shoots of willows (*Salix arctica*). The branches with buds were pulled upward with a rope, so that the new buds originally 1 cm above the ground were raised to 5–19 cm on July 5, when they were expected to begin blooming. Twelve of the twenty-eight buds on the branches were found dead on August 24, the end of their growing period. Besides, the leaves alive on these branches were only half the size of the other leaves on other branches lowered by the wind (but not pulled up by ropes). The pulled-up branches that escaped death had not grown as much as one-fourth of the ordinary branches bent down by the wind as shown in Table 6.5. At another island (71°N) the effects of the wind on the rate of assimilation of leaves of *Oxyria digyna* were studied. It was found that the rate for three days of the last decade of August was 0.34 g·dm^{-2}·week^{-1} on the mountain-top exposed to a stronger wind, and 0.46 g·dm^{-2}·week^{-1} in the mountain cove less exposed to the wind.

Table 6.5 The growth of branches in natural and experimental conditions (Wilson, 1959).

	Experimental (Branches made higher up by rope)	Natural (4 Branches lowered by wind)			
Dead, young buds (%)	43	10	28	7	23
Mean length of living, young branches (mm)	0.6	2.2	2.1	2.1	3.1

Wind-shaped Douglas fir (*Pseudotsuga taxifolia*) were studied in the Columbia River gorge in the boundary between Oregon and Washington states, U.S.A. (Lawrence, 1939). In the western part of the gorge, the prevailing wind blows westward, since the glaze storm that develops in this region in winter forms a silver thaw on the windward side, breaking off the branches (Type 2). On the eastern part of the gorge, the prevailing west wind tends to bend the branches of Douglas fir trees toward the east (Type 1). These inflected branches are indicative of direct influence (a) mentioned above by the west wind that develops during the summer. Thus two different types of wind action are observable within tens of kilometers from the eastern part of the gorge to the western part, due to the seasonal change of the prevailing wind direction.

(D) *Local Distribution of Wind-shaped Trees: Examples*

An observation was made using wind-shaped trees as indicators of wind conditions in the bora region, Ajdovščina in Slovenia, Yugoslavia (Yoshino et al., 1973). It was intended to make clear the local distribution of a strong fall wind zone and to test the validity of the deformation grades. The prevailing wind-shaped trees were those of Type 3; they are *Quercus pubescens*, *Q. petraea*, *Q. robur*, *Robinia pseudacacia*, *Ostrya carpinifolia*, *Alnus glutinosa*, *Prunus avium*, *P. Armeniaca*, *P. domestica*, *Fraxinus ornus*, *Gleditschia triacanthas*, *Tilia cordata*, *Yuglans regia*, *Populus* sp., and *Ulmus campestris*. Among these, *Quercus*, *Prunus*, *Populus*, *Alnus*, and *Ulmus* were most frequent. We ascertained the deformation grade for this region, as shown in Fig. 6.7. There were no trees of Grades 5 and 6 in this region. About 100 points were observed but three or four trees were observed at each point and an average was calculated for that point. The distribution of grades of wind-shaped trees and the wind direction

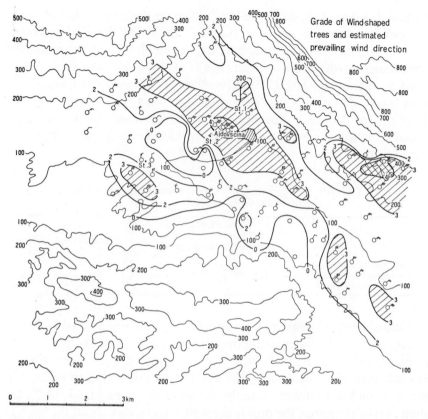

Fig. 6.8 Distribution of grades of wind-shaped trees and the direction of prevailing wind estimated from the wind-shaped trees in the Ajdovščina region, Slovenia, Yugoslavia (Yoshino et al., 1973).

estimated from the wind-shaped trees are shown in Fig. 6.8. This figure shows the following facts.

A wind maximum zone runs from NW to SE along the foot of the mountain, the NE-border of the Ajdovščina basin. On the lee of this zone, there is a weak wind region; in other words, the grade is 0. Further leeward, a strong wind region is seen, though small in area. For comparison, the results of wind observation by anemometers on November 24, 1970, are given in Fig. 6.9. The wind maximum zone, caused by the strong fall winds from the NE-mountains as bora, runs also from NW to SE. In this area comparatively weak wind appears on the immediate leeward of the maximum zone. There is an arca situated about 2.5 km leeward where a stronger wind blows. Concerning the prevailing wind direction, NNE–NE winds are most frequent within the limits of N to E. Grade 4 prevails in the region of the NNE–NE wind region. There are some minor differences between Figs. 6.8 and 6.9. This is thought to be because of the fact that the wind-shaped trees reflect a long-year average, whereas the observed results in Fig. 6.9 present only the condition on the day measured. From the facts

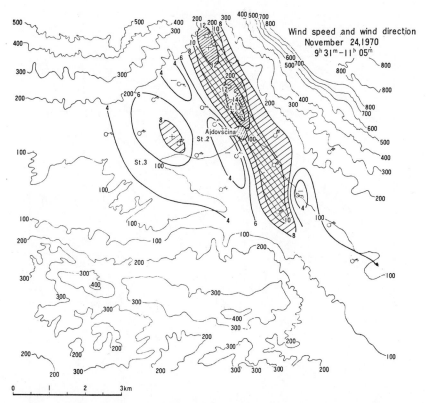

Fig. 6.9 Distribution of the wind speed (m/sec) and direction observed on November 24, 1970, in the same region given in Fig. 6.8 (Yoshino et al., 1973).

mentioned above, it can be concluded that the wind-shaped trees express the climatological state of wind conditions.

The second example is the results observed in the Ishikari plain, Hokkaido (Owada et al., 1971). In this region W–NW winds from the Japan Sea prevail during the winter monsoon season and S–SE winds from the Pacific Ocean during the summer monsoon season. To study the detailed distributions of these prevailing winds, wind-shaped trees were observed at 180 points in the plain. The wind-shaped trees observed were: *Fraxinus mandshurica*, *Alnus japonica*, *Larix leptolepis*, and *Populus* sp. The distribution of wind-shaped Type 3 larch trees shows the following facts (see Figs. 6.10 and 6.11): (a) The southerly winds from the Pacific Ocean go up to the north in the central part of the plain and then gradually change direction to the northeast. (b) One branch of the southerly winds flows northwestward from the central part of the plain to the Japan Sea. (c) The strongest wind region is found in the area around the City of Ebetsu, the central part of the plain. There are other smaller regions where strong winds caused by the local topographical conditions blow.

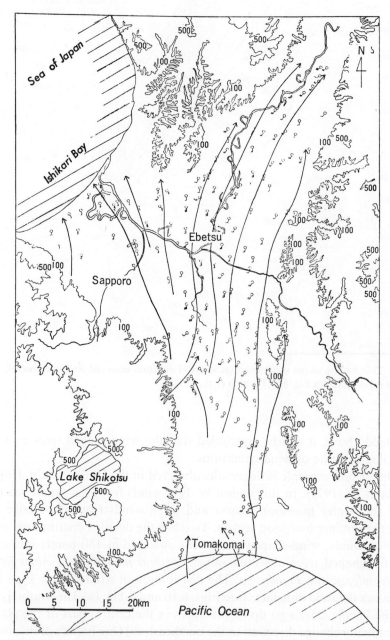

Fig. 6.10 The distribution of wind-shaped trees caused by the prevailing wind in summer and the estimated streamlines (Owada et al., 1971).

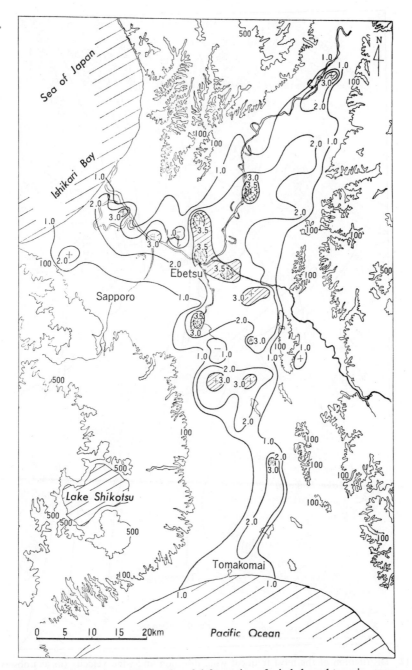

Fig. 6.11 Distribution of the grades of deformation of wind-shaped trees in summer (Owada et al., 1971).

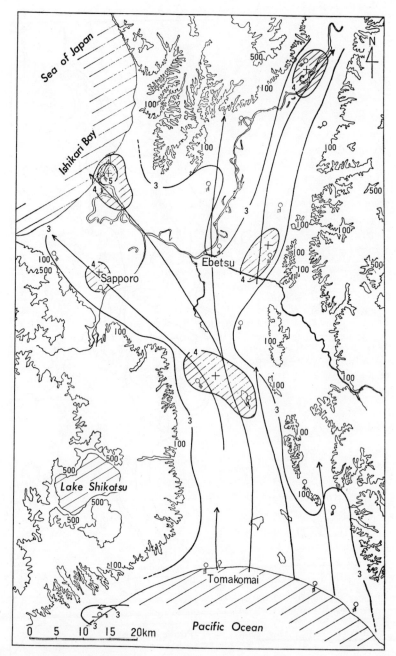

Fig. 6.12 Distribution of the most frequent wind direction and wind velocity (m/sec) observed at the meteorological stations in summer. Compare the distribution pattern with those given in Figs. 6.10 and 6.11 (Owada et al., 1971).

The results mentioned above are compared with the results of an instrumental observation at the 20 agrometeorological stations distributed in the area studied. The results during the summer (Fig. 6.12) demonstrate a striking coincidence with the conditions of southerly winds given in Figs. 6.10 and 6.11. Of course, the distribution map obtained by observing the wind-shaped trees is more detailed because of the number of observation points. A comparison for the winter conditions is given elswhere (Yoshino, 1973a).

In conclusion, wind-shaped trees indicate very well the climatological conditions of winds. As has been discussed by Primault (1971), ecological methods together with orthodox climatological methods are useful for the study of climatic conditions. Wind-shaped trees may be an outstanding example of powerful ecological methods.

6.1.4. Local Plant Ecology and Local Climate

(A) *Wind-shaped Trees and Lichen in Mountainous Regions*

Because wind conditions are well reflected in wind-shaped trees, an elaborate study of the distribution of wind-shaped trees will bring forth a distribution map of local wind direction and velocity. This is especially useful for mountain regions, because the results of instrumental observations are quite limited.

Obrebska-Starklowa (1969) showed wind conditions in great detail in her comprehensive mesoclimatic study in the Jaszcze and Jamne valleys and the surrounding mountain regions in the Gorce Mountains, Poland. A distribution map showing the wind directions and velocity under the condition of NW general wind on June 27 and 28, 1962, has made clear that the *flag-shaped trees* are seen on the ridges and the slopes exposed to the prevailing winds. Although the definition of flag-shaped trees is not shown in the study, the situations of those trees are topographically striking.

Usnea, a kind of lichen that grows in places with low temperature and high humidity, is often found in the subalpine zone and the upper montane zone at the altitude of 1,000–2,500 m in middle latitudes. It grows thickly under favorable conditions, but does not grow locally where a strong wind promotes evaporation making the humidity is relatively low. *Usnea diffracta* is one of the typical inland or alpine species of *Usnea* in Japan (Sato, 1960). The "*U. diffracta* line" of Sato might be a border line between a maritime and an inland region or between a lowland and a highland region. In central Japan, Yoshino (1967) studied the distributions of the wind-shaped trees, which indicate the strong wind regions and of the *Usnea*, which indicates the weak wind or calm regions. Such coniferous varieties as *Abies Veitchii, Abies firma, Abies homolepis, Picea jezoensis*, of Type 2 mentioned in Section 6.1.3 (A) were surveyed. In addition, the *Usnea* that grows on branches or trunks of the trees was also investigated. The results are shown in Fig. 6.13. The distribution of *Usnea* is uneven from the effect of microclimatic wind conditions which apparently precluded its growth in windier, less humid areas. The marked effect of topography on microclimatic

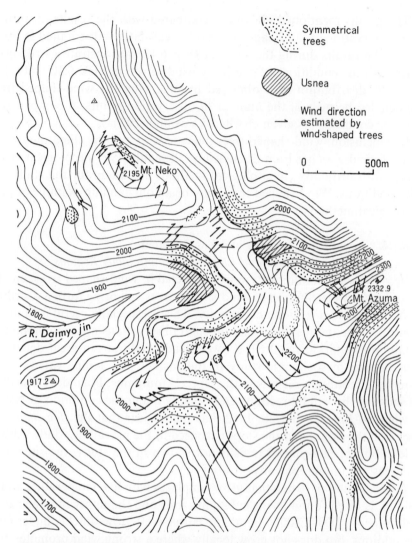

Fig. 6.13 Distribution of wind-shaped trees and *Usnea* in the subalpine zone of the Mt. Azuma and Mt. Neko region, central Japan (Yoshino, 1967).

wind movements in summit level areas is shown in this figure in the wind directions, as revealed by wind-shaped trees, and the occurrence of symmetrical trees. Figure 6.13 shows that (a) the wind-shaped trees are spread over areas higher than 2,020 m a.s.l. on the wind-facing slopes, (b) they are seen at points 40–70 m lower in relative height on the windward side of a ridge, and (c) they are also found at points 20–70 m lower in relative height on the leeward side of a ridge. But below these points wind velocity decreases rapidly, and hence there is no distribution of wind-shaped trees. It is worth noting that, due to an eddy formed topographically on the leeward side of a narrow ridge, a wind blows from the

Photo. 6.4 An example of a wind-shaped tree, Grade 5, of Type 3. *Fraxinus ornus* at the central part of Cres Island on the Adriatic Coast, Yugoslavia (October, 1972, by M. M. Yoshino)

opposite direction to that of the wind prevailing on the other side. Such a situation is shown in Photo 4.6.

A similar zonal distribution is observable on the region around Mt. Yake and the Nakao Pass in Nagano Prefecture, central Japan. The wind-shaped trees of Type 2 range from the altitude of 2,200 to 1,900 m. The col runs from SW to NE around Nakao Pass, which consequently reinforces the NW-wind. The wind in the upper layer in central Japan blows from NW–W in winter, so that wind-shaped trees develop in the areas 80–100 m lower in relative height on the leeward side. The col between Mt. Azuma and Mt. Neko, on the other hand, runs from NW to SE. As a result, the trees become wind-shaped within the areas only 30–40 m below on the leeward side. Thus the running direction of the col affects the local distribution of the wind-shaped trees. *Usnea* ranges from the altitude of 1,800–1,850 m on the eastern slope of Nakao Pass to the altitude of 1,600–1,700 m, i.e., 100–150 m in relative height.

The wind distribution in a small area in mountainous regions is thus well reflected in the distributions of *Usnea* and wind-shaped trees, or alternatively, the locality of the wind in the region may be inferred from an investigation into the state of their distributions.

(B) *Wind and Plant Distribution*

The wind exerts a strong influence on the distribution of those plants whose seeds are scattered by the wind. Species having light seeds grow generally in broad and open areas. Those with large seeds, on the other hand, tend to grow in shaded places.

About 75% of the trees in the forests in Europe need the wind to scatter their seeds. The wind velocity required for the flight of seeds ranges in the following order of strength (Domes, 1950): *Fraxinus*, *Acer*, *Carpinus*, *Tilia*, *Picea*, and *Pinus*.

An investigation in a tropical rain forest in Nigeria, Africa, showed an obvious difference in the species of trees with seeds spread by the wind, a difference attributable to the difference of layers of the forest (Keay, 1957). He has shown that the number of species of trees whose seeds are dispersed by the wind increases with height. Thus, the relation in the temperate zone remains unchanged in the tropical rain forest.

Roughness parameter z_0 (cf. Section 3.1.4) for wind velocity affords an important clue to the relationship between the wind in mountainous regions and plant ecology (Whitehead, 1954, 1957). Whitehead observed the height of plants, dry weight, the number of growth forms, and the number of species in the four places in Mt. Maiella, Italy, the places having the values z_0, 0.07, 0.15, 0.3, and 0.5. A positive relationship, although not a linear one, was found: As the value of z_0 rises from 0.1 to 0.5 cm, the height of the plant increases two and a half times, the dry weight four times, the number of growth forms two times, and the number of species three and a half times. In other words, the number of species and the density of growth forms are large in a sheltered community, and on the other hand, they are limited in an exposed community. Kalium and phosphorus in dry weight decrease in percentage with the increase of z_0 values.

The dry conditions in valleys, where valley breezes develop well, is observable in the vegetation distribution. This phenomenon was first studied by Troll (1952) along the mountain belts of the tropical Andes from Bolivia to Venezuela and of East Africa from Natal to Northern Ethiopia and by Schweinfurth (1956) in the humid zones of the Himalayan system (Sutlej valley, Karnali valley, six valleys in Bhutan, the Lohit valley in Zayul). This was discussed in Section 4.2.4 (C) above. Braun-Blanquet (1961) described the characteristics of dry valleys in the Alps from the standpoint of geobotany. The cause of dry conditions can also be attributed to the valley breeze conditions.

(C)   *Light and Radiation*

Light is as important as temperature for plant growth. The greatest local difference of light in a small area is found between the inside and the outside of a forest. However, since other climatic elements relevant for plant growth also have local distribution, light presents extremely complicated aspects. When the growth of a plant is affected by some unusual light conditions, the influence of the other environmental factors will be different from normal light conditions. For instance, a decrease in photosynthesis due to a deficiency of light prevents the normal growth of roots. Underdeveloped roots may lead to the death of plants in the event of a very slight environmental deterioration.

Since 1959, experimental plantations of different tree species have been established in the upper tree limit zone of the Dischma valley near Davos, Switzerland. Growth rates of larch (*Larix decidua*) and spruce (*Picea abies*) are closely related to wind speeds and to solar irradiation on the slopes (Turner, 1971). The surviving trees show greatest height increments on strongly irradiated and wind-sheltered positions in the micro-topographical situations. On the sunny slopes, the height of trees decreases sharply with increasing wind velocity. As shown in Fig. 6.14, the relationships between the growth of larch and wind velocity differ largely according to radiation conditions.

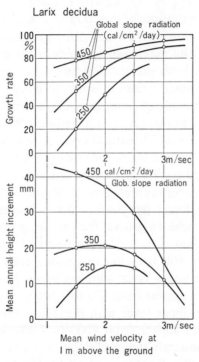

Fig. 6.14 The relationship between the growth rate or the mean annual height increment of larch (*Larix decidua*) in relation to global slope radiation during the period May 15 to September 30 in Stillberg, Switzerland, 1960–1967 (Turner, 1971).

To study the relationship between plant yields on slopes and the amount of solar radiation, Sato and Kira (1960) first made an experimental study and then a theoretical one. As a conclusion, they showed that the yield per unit slope surface is proportional to the amount of total solar radiation received.

(D)   *Temperature and Humidity*

One of the basic factors influencing plant growth is temperature. The temperature influences the effectiveness of precipitation and the rate of transpiration through its effect on relative humidity and evaporating power. As discussed above, the air temperature varies strikingly from place to place, and hence there is a local difference of plant life in a small area and its relation to the temperature conditions is of special academic importance.

Balchin and Pye (1950) observed plant responses in relation to local temperature variations, which were mentioned in Section 4.2.2 (B). The results discussed in Sections 6.1.1 and 6.1.2 have also contributed to the study of this problem.

Yoshino and Yoshino (1963) studied the local differences of air temperature, precipitation, and wind conditions in the Mt. Kirishima region, S-Kyushu, Japan, together with the vegetation distribution. The summit areas above 1,400 m in this region have Köppen's D-climate. The distribution patterns of precipitation do not coincide with those of the 500 and 1,000 m contour lines, but differ according to synoptic situations as described in Sections 4.1.7 (B) and 5.1.2 (B). The lapse rate of the maximum temperature is greater than that of the minimum temperature. With greater cloudiness, the lapse rate is greater. The lapse rate in summer is smaller than in winter as discussed in Section 4.1.3 (A). The results of wind observation from the microclimatological viewpoint were shown in Section 4.1.10. In this region, vegetation is classified into the following twelve types: 1. laurel forest; 2. natural coniferous forest; 3. deciduous scrub forest; 4. grassland; 5. bare land; 6. forest mixed with big deciduous trees; 7. afforested area of mixed conifers; 8. afforested area of Japanese cedar forest; 9. afforested area of Japanese cypress forest; 10. afforested area of Japanese red pine forest; 11. young forest (species are not distinguished on aerial photographs); 12. cultivated field. In general the following vertical zonation of vegetation can be distinguished: 1. laurel forests up to a height of 800 m; 2. natural coniferous forests from 800 to 1,400 m (among them, forests consisting of coniferous trees in the upper layer with evergreen broad-leaved trees in the middle and lower layers appear from 800 to 1,000 m, and forests consisting of coniferous trees in the upper layer with deciduous broad-leaved trees in the middle and lower layers appear from 1000 to 1,400 m); 3. deciduous scrub forests near the steep mountain tops and ridges or surrounding the active cones; 4. grassland, including the *Bambusaceae*, and other bush areas on the windy summits. Along the cross-section on the SW–NE line, the profiles of climatic elements and vegetation are shown in Fig. 6.15. The relation of vegetation to the temperature is especially striking.

Chung (1962) reports some interesting observations on the growth of bamboos in Cholla-nam-do, Korea. He states that the internodal length of bamboos is longer when the density of bamboo in the groves is higher. Therefore, the internodal length is longer on the slopes facing N, NE, and E than on the slopes facing SE, S, and SW. The temperature conditions should be favorable to the growth of bamboos on the slopes facing SE, S, and SW, but the results are the reverse of what is expected because of the characteristics of bamboos concerning the density and growth.

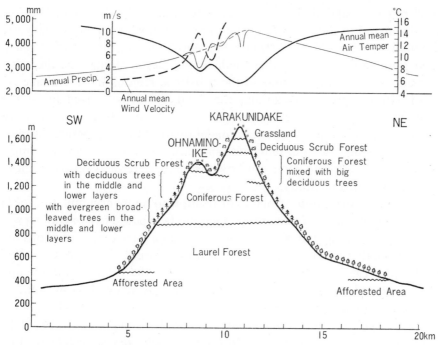

Fig. 6.15 The profiles of air temperature, precipitation, and wind and vertical zonation of vegetation in the Mt. Kirishima region (Yoshino et al., 1963).

Sakharov (1952) studied the effects of forest types on the temperature of spruce trunks. This is especially important when the moss and lichen development on the trees is considered in relation to an indicator of moisture conditions in the subalpine zone, as mentioned in Section 6.1.4 (A). The temperature of spruce trunks with moss and lichen was higher than that of tree trunks with *Vaccinium* sp. The former was 16.1°C as an average at 13 h during the growing period, and the latter is 13.9°C.

Perring (1960) studied the climatic gradients of the Chalk grassland. He examined the climatic factors, such as temperature, rainfall, and relative humidity, and came to the conclusion that the most significant difference, as far as the local distribution pattern of species is concerned, is the relatively low humidities in the spring in the Rouen area of France and in the Cambridge area in England compared with the relatively high humidities of north Dorset and East Riding in England.

(E) *Snow and Plant Distribution*

Since snow cover protects plants from an extremely low temperature during the winter, plants of southern element are distributed more often in a snowy area, the average annual temperature being equal. In Japan, the normal vertical distribution of forest zones in the areas less covered with snow is: an evergreen broad-leaved forest zone, deciduous broad-leaved forest zone, evergreen coniferous forest zone, and alpine zone. On the other hand, on the Japan Sea Coast, especially in and to the north of Niigata Prefecture, where there is heavy snow accumulation, there seldom exists an evergreen coniferous forest zone. From the Echigo Mountains to the southern half of the Dewa Mountains, there is no evergreen coniferous forest zone in the subalpine region, and the alpine zone covered with creeping pines adjoins the deciduous broad-leaved forest zone. This difference in the vertical distribution may be ascribed to the following causes (a) pressure of heavy snow accumulation, (b) low temperature during the growth period due to a lot of remaining snow, or (c) strong monsoon during the winter (Oota, 1956). The details of how the snow cover acts on the plant life, however, await further study.

The local distribution of plants in relation to snow accumulation occurs also on a high mountain. In an area of a mountain exposed to a prevailing wind, the snow does not stay but is blown away, so that a distribution of plants quite different from that in the areas covered with snow appears in these areas. Poore and McVean (1957) studied the difference between the vegetation in the upper exposed area and the vegetation in the lower area covered with snow in the subalpine zone in Norway. In the former area such vegetation as lichen, stunted shrub, or a kind of heath (*Calluna*) are commonly found, whereas in the area covered with snow, *Vaccinium myrtillus*, *Phyllodoce*, *Cassiope* etc. grow. In the zone between them, at a height of 800–900 m a.s.l., *eye-brow*-shaped birch trees are developed by snow. The same type of distribution by height can be found in the subalpine zones in Scotland and in Japan.

In the Japanese Northern Alps, the distribution of vegetation has a close relation to the snow accumulation period (Ogasawara, 1964): (a) Creeping pine (*Pinus pumila*) and *Vaccinium uliginosum* appear in the region where the snow disappears toward the end of May or the beginning of June, (b) *Abies Mariesii*, *Betula ermani*, *Alnus Maximowiczii*, and *Sorbus* during the middle to end of June, and (c) the alpine sedge meadow (*Faurieto-caricentum blepharicarpae*) from the beginning to the middle of July. The mean depth of snow accumulation estimated was to be 70–155 cm in the region of (a) but 460–660 cm in the region of (c). Sugiyama and Saeki (1965) investigated the occurrence frequency of avalanches in relation to vegetation distribution in Niigata and Fukui Prefectures by means of aerial photographs and field surveys. The relation between the distribution of avalanches and of vegetation types was analyzed by comparing the aerial photographs taken during the winter of 1963 and 1964 with those taken during the summer of these two years. No avalanches occurred in the mature forests, whereas many avalanche occurred in the grass or bush lands. An example from Shiozawa, Niigata Prefecture, is shown in Fig. 6.16.

The vegetation types in accumulated snow can be classified into the following four types, A: stems appear on the snow surface, B: crowns appear on the snow surface, C: tree-tops are scattered over the snow surface, and D: the vegetation is completely buried in the snow. In the cases of Types A and B, the

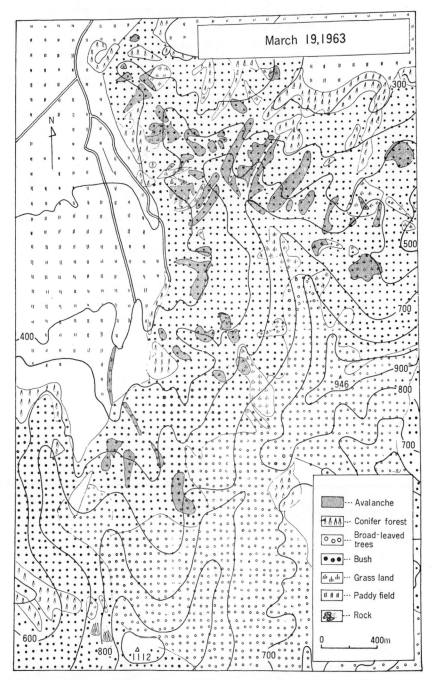

Fig. 6.16 Distribution of vegetation and avalanches in Shiozawa, Niigata Prefecture, Japan, on March 19, 1963 (Sugiyama et al., 1965).

accumulated snow on the slopes is stable, although in Type B sometimes the balance is lost. In the cases of Types C and D, on the other hand, the accumulated snow on the slopes is unstable or apt to slide. However, the effect on vegetation is changeable according to the amount of snow. Consequently it is necessary to ascertain what conditions of vegetation (height, diameter, density, etc.) can break snow movement on the slopes for various snow depths. Where avalanches occur frequently, the vegetation is distinctly different from places having no avalanches. By surveying species, inclination of stems and cracks on stems, which are caused by the snow movement on the slopes, the dangerous sites for avalanches can be forecast. Where the maximum snow depth is more than 1.5 m and there are no trees with diameters at breast height of more than 6 cm, there is no place with stable snow accumulation on the slopes. If the maximum snow depth exceeds 3 m, the density of trees with diameter at breast height of more than 6 cm, must be more than 7 trees/10 m² for the snow accumulation on the slopes to be stable.

Sugiyama and Saeki (1965) also studied the relation between the inclination of stems, measured as shown in Fig. 6.17 (a), and the height of the trees. The results are shown in Fig. 6.17 (b). Examples are shown in Photos 6.5 and in 4.1.

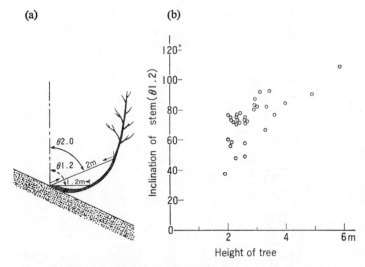

Fig. 6.17 (a) Measurement of stem inclination.
          (b) Relation $\theta_{1.2}$ and the height of trees. (Sugiyama et al., 1965).

Similar phenomena concerning the stem inclination are seen in the regions with soil creeping, especially under severe frost action. The angle is small in the younger stage of trees, but it becomes larger in accordance with the development of roots (Ishikawa et al., 1965).

Photo 6.5 Effect of creeping of snow accumulation on the slope of Japanese cedar (*Cryptomeria japonica*) on Mt. Kurohime, central Japan (April, 1965, by M. M. Yoshino).

6.2   EFFECTS OF CLIMATE ON MICRO-TOPOGRAPHY

Climatic geomorphology aims at elucidation of the process of topography development and especially of the characteristics of erosion power and its role; but it is not so developed as to deal with the differences in a small area in the process of development. Only a few cases will be discussed here.

6.2.1.   The Asymmetry of Valley Topography and Microclimate

As Troll (1969) points out, valley asymmetry is one of the most striking phenomena in geomorphology in relation to the local climate and microclimate. In valleys running in the east-west direction, the cross sections of eroded valley slopes may be of the following three types: (a) the north-facing slope has a steeper gradient, (b) the south-facing slope has a steeper gradient, and (c) both slopes have the same gradient. A brief sketch of the investigations on these three types is given below.

(A)   *On a Steep North-facing Slope*

In arid regions in the temperature zone in the northern hemisphere, the north-facing slopes have a steeper gradient, other conditions being equal. Most geomorphologists agree on this point.

This phenomenon can be accounted for by the following facts. The amount of runoff and erosion on the north-facing slopes are less than those on the south-facing slopes, since (a) the vegetation density is high on the north-facing slopes, (b) the soil has developed enough to promote infiltration into the ground, and (c) the surface resistance against the eroding stress by runoff is large. The south-facing slopes, on the other hand, are subject to faster erosion, so that the gradient of the slopes decreases. This phenomenon is called Gilbert's *law of divides* (Gilbert, 1904).

As a result of comparatively rapid erosion of the south-facing slopes, the amount of debris increases on the foot of the south-facing slope and hence the course of the river at the valley bottom is pushed away a little to the south. The foot of the north-facing slope is eroded and becomes steeper. Russell (1931) stresses the effect of this mechanism in an arid region or subhumid region, where the differences of local climatic conditions between the sides of the valley are comparatively great.

In areas where the monthly mean temperature in winter is below 0°C, the north-facing slopes are covered with snow or ice for a longer period. This means a faster erosion on the south-facing slopes.

According to a measurement of the slope gradients of the eroded slopes of the valleys at the Laramie Range in Wyoming, U.S.A., where the river flows eastward with a slight inclination, the north-facing slope is steeper by 4.42 degrees than the south-facing slope (Melton, 1960). However, another measurement of the gradient of forty-seven slopes in northern Arizona shows that where the inclination of the river flow at the bottom is above eight degrees the difference in gradient between the north-facing and the south-facing slopes disappears, and that, as the inclination decreases below eight degrees, the north-facing slopes become steeper than the south-facing slopes and the difference in gradient becomes greater.

From this it may be clear that, in a V-shaped valley with a steeply inclined river at its bottom, the valley slopes may have a symmetrical form, despite the fact that Gilbert's law is at work here and local climate and vegetation on both slopes differ very noticeably from each other. The north-facing slope becomes steeper when the river at its bottom flows with a slight inclination. Most cases of valley asymmetry are attributed to the asymmetry of basal corrasion caused by the changes of conditions in the valley.

The south-facing slopes generally have less vegetation, especially grass and shrubs, due to higher insolation and evapotranspiration. An observable difference of vegetation and slope morphology between the sunny slope and the shady slope was reported by Chiba (1949) for the Dauria region in the upper course of the Amur River. He also studied the valley asymmetry on the border of the forest zone and the grassland zone in the regions along the Argun River, Manchuria (Chiba, 1950), where the erosion occurs only during the period from May to September, and arrived at the same conclusion as that mentioned above.

(B)   *On a Steep South-facing Slope*

An instance of the steep south-facing slope is observable in tundras. In the periglacial regions the south-facing slopes get more heat from insolation and are less subject to freezing than the north-facing slopes, so that the higher frequency of denudation caused by surface runoff makes the south-facing slopes steeper than the opposite ones. The valley asymmetry seen on the slopes today serves as an indicator of the past climate, if there was no counter-action against such an asymmetry during the recent erosion period (Smith, 1949).

Kennedy and Melton (1972) investigated statistically the valley asymmetry and slope forms of a permafrost area in the Mackenzie River area, Canada. They have come to the interesting conclusion that the asymmetry in maximum slope angles is opposite between (a) areas with more severe climate and low available relief, where the north-facing slopes are steeper than those facing south, and (b) areas with milder climate and deeper valleys, where the south-facing slopes are steeper. They pointed out that there is no one form of asymmetry characteristic of permafrost areas, because of the co-existence of classes of valleys exhibiting steeper north- and south-facing slopes within one small area, where it seems reasonable to suppose that both types of asymmetry correspond with modern conditions.

The measurement of gradients on 105 slopes of hills in California did not show that the difference of $\pm 1.6°$ in gradient between two slopes is statistically significant (Strahler, 1950).

Karrasch (1970) made a comprehensive study on the relief asymmetry produced by the climatic conditions in central Europe. Full literature concerning the valley asymmetry in this region can be found in his paper. He defines an asymmetry index, L, of slopes as follows:

$$L = 1 - \frac{\text{length of steep slope}}{\text{length of gentle slope}} \tag{6.3}$$

The mean values and maximums of L measured for the Swabian Alb and upper Pfälz Alb region in South Germany are shown in Table 6.6. The index is larger on the NE- and E-facing than on the other slopes. As for causalities, the solifluction, nival and glacial processes, and fluvial lateral erosion are considered to be effective for this phenomena.

Table 6.6 Mean and maximum asymmetry index, L, on the various slopes (Karrasch, 1970).

Study region	Summit height		NW	N	NE	E
Swabian Alb	750–950 m	mean	0.21	0.34	0.37	0.35
		max.	0.23	0.67	0.63	0.69
Upper Pfälz Alb	500–630 m	mean	—	0.40	0.51	0.41
		max.	—	0.44	0.64	0.49

In many cases the glacial conditions are most intense on shaded slopes, and hence the north-facing slopes are most eroded and become gentle. On the other hand, under periglacial conditions insolation promotes breeze-thaw and solifluction on the south-facing slopes in high latitudes, and hence the south-facing slope becomes gentle. However, in middle latitudes, a deep and longer snow cover acts as protector from erosion on the north-facing slopes so that the north-facing slope is gentle.

In some cases, valley asymmetry is found between the west- and east-facing slopes. For example, on a cross section of the Little Hampden valley, the west-facing slope is steeper (Ollier et al., 1957). The most probable explanation for this phenomenon is that the asymmetry developed at the same time as the down-cutting. The west-facing slope experienced considerable creep erosion under conditions that did not effect the east-facing slope. Büdel (1944) thought that the prevailing west wind deposited a thick loess layer and snow on the east-facing slopes of the valleys. The thick, wet soil layer caused by melted snow tends to increase the runoff with materials on the east-facing slopes and, accordingly, the river course is shifted eastward, resulting in the steep west-facing slopes. According to Büdel's study, these conditions must have occurred in middle Europe in the last Pleistocene, when the climate was cold with a stronger west wind.

Such complicated conditions on the micro-scale as well as macro-scale must be precisely defined in discussing the problems. Furthermore, in order to clarify the effect of local climatic conditions on the formation of small-scale topography, it is necessary to make a more detailed quantitative analysis on the climatic factors and the erosion processes of topography. In the case where both slopes do not stand in a rectilinear line, a strict definition of an asymmetrical shape is required. No consistent investigation is possible in cases where one reporter is talking about the gradient of the lower part of a valley slope, while another is talking about that of the upper part.

6.2.2. Landslides, Landslips, and Erosion by Rainfall

In Japan, landslides and landslips often occur accompanied by heavy rain during the typhoon periods or at the end of the bai-u period, causing damage by floods on the lower courses of a river. This problem is treated here as a process of erosion in mountains, with special reference to the local pattern of rainfall distribution. On September 26 and 27, 1958, a typhoon, having passed the southern tip of the Izu Peninsula, went across the eastern part of the Kanto district and went beyond the offing of the Sanriku district. Kawamura (1960) drew a detailed distribution map of rainfall from 9 h, September 25, to 9 h, September 26, using the coaxial method, since the rainfall distribution in a mountainous region is greatly affected by the complexity of the topography. The map in Fig. 6.18 reveals that the maximum amount of over 700 mm appears on the northern slope of the Amagi Mountains in the central part of the Izu Peninsula. The areal distribution of density of landslides in the central part of the landslides does not correspond to the distribution pattern of the degradation density classified by rocks, but is closely correlated to the rainfall distribution (Ichikawa, 1960). The areas where the total amount of rainfall attains 500–750 mm overlap those areas where landslides occur very frequently.

When the amount of rainfall surpasses a certain limit, any slope with a gradient of 30–50° is subject to landslides, regardless of the difference in geological features (Iwatsuka, 1954). The limit of the maximum rainfall amount varies

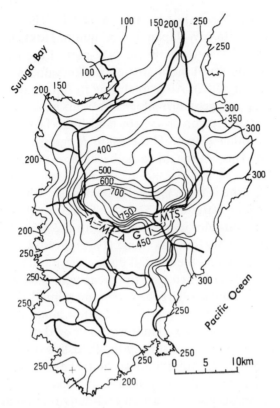

Fig. 6.18 Rainfall distribution in the Izu Peninsula, Japan, from 9h, Sept. 25 to 9h, Sept. 27, 1958 (Kawamura, 1960). Thick line presents the border line of the Kano River basin and corresponds to the area given in Fig. 6.19.

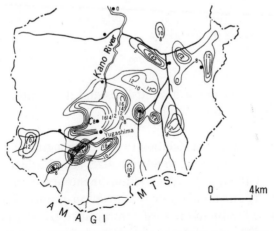

Fig. 6.19 Areal distribution of density of landslides in the upper course of Kano River in the central part of Izu Peninsula after the typhoon on September 26–27, 1958 (Ichikawa, 1960).

according to the form of rain just before a landslide. The limit is estimated at approximately 100 mm in NW-Japan and 50 mm in Hokkaido, N-Japan.

On the volcano of Mt. Akagi, in the NW-part of the Kanto plain, Japan, 1,126 landslips were observed in an area of 558 km^2 after a typhoon hit in autumn 1949 (Murakami, 1956). After analyzing in full detail the distribution in relation to geological and geomorphological condition, scales and types, Murakami concluded that the erosion process by landslips is faster on the north-facing slopes than on the other slopes. The erosion progresses most slowly on the south-facing slopes. This is thought to be because of the difference in climatic conditions.

Shirasu in S-Kyushu, Japan, is a volcanic deposit, which has a porosity of 40–50% and can contain water up to 20–25%. Therefore, the regions covered by Shirasu often suffer landslides in association with the heavy rainfall during the Bai-u season or typhoon season. Sakanoue (1972) investigated the landslides in the Shirasu region caused by the rainfall during the Bai-u season from June 28 to July 8, 1969. It was confirmed that the areas with most frequent landslides nearly coincide with the areas where the maximum rainfall exceeds 10–15 mm/10 min or 30–50 mm/hr. The Shirasu areas where few landslides occur in spite of rainfall exceeding 50 mm/hr have a landform composed mainly of gentle slopes.

Takeshita (1971) studied statistically the landslide occurrence in Fukuoka Prefecture, Kyushu, Japan, taking into account the geomorphological factors such as slope types, slope angles, slope length, and densities of slope creases and meteorological factors such as air current and daily rainfall amount. The relation between the percentage of landslides, Y_i, and the daily maximum rainfall, P in mm, is expressed as

$$Y_i = 4 \left(\frac{P-50}{100} \right)^2 \left(\frac{100-x}{100} \right)^{1.5} \tag{6.4}$$

where x is the percentage of past landslides occurrence and its average value in the area under investigation is approximately 10%. The areal density of landslides, Y in ha/100 ha, is obtained by

$$Y = \frac{Y_{100} \times Y_i}{100} \tag{6.5}$$

where Y_{100} is the standard area of landslides and its average value in the area studied is approximately 6 ha/100 ha.

6.2.3. Snow and Frost in Relation to Micro-topography

As has been discussed above in Section 4.2.5 (C), the distribution of snow is affected strongly by the topography and this phenomenon in turn causes peculiar topography such as nivation hollows or nivation cirques. Temple (1965) deals with the cirque distribution in the west-central Lake District in N-England and shows the relation between altitude of cirques and position (direction). Of the 73 cirques in the district 53.4% appear from N to E, 19.2% from S to E, 24.7%

from N to W and 2.7% from S to W. This means that the cirques develop most frequently in the quadrant from N to E.

Snow patches, the remaining snow on the upper part of the leeward slope, develops in the quadrant from N to E in middle latitudes in most cases, because of the westerly upper winds and insolation effects. The snow patches, which pass into the new year amounted to 200 in number and cover 300 ha in total area in the N-Alps in central Japan (Higuchi and Iozawa, 1971). Most of them are located on the east-facing slopes.

Soons and Rainer (1968) reported observations of the total amount of sediment yielded by the various plots in the S-Alps in New Zealand. They have explained in some detail the energy exchange in a mountain environment and the effect of snow cover on net radiation and soil temperatures. Another interesting observation is the effects of needle ice (1.5–2.5 cm long) on the movement of stones (1.25–6.5 cm in diameter in the experiment). They moved 5–63.5 cm in distance due to the eleven freeze-thaw cycles during the period from May 27 to June 6, 1966. This occurred on the west-facing slope, which receives the maximum possible insolation.

According to an investigation at Gaik-Brzezowa, the Wieliczka foothills in Poland, there were three phases of frost action on the soil movement (Olecki et al., 1970). The first phase is characterized by movements of all the soil layers to the depth of 50 cm with an amplitude of 1 mm and on very rare occasions of 2 mm. The height of curves never attains 2 mm above the initial state of the soil. The movements take place when the mean diurnal temperature falls to $-5°C$ to $-8°C$. The second phase is the period in which the soil strata rise decidedly, even up to 2.5 cm. This happens when minus temperatures continue for a long time at depths of 5, 10, 15, and 20 cm. The third phase begins on the day in the autumn when the curve of the soil movements returns to its original state or below. This is connected with a warming to higher than 0°C. For instance, the phases of soil movements at Gaik-Brzezowa in the winter of 1966 to 1967 were: first phase, until January 15; second phase, from January 16 to February 20, and third phase, from February 21 at a depth of 20 cm. The snow cover greatly influences the state and intensity of the soil movements: The third phase began on April 10 in 1969.

Kim (1967) studied the structural soils in Iceland and presented a model of development of the great stone rings, as shown in Fig. 6.20. This model was obtained from observations at the N-border of a col of Fjardarheidi in E-Iceland. Stage a of this figure shows the ground surface covered by grass. Fine materials fill the soil layer and debris weathered from the mother rocks fills the wet layer. At Stage b, frost penetrates into the top of the mounds more frequently than the other parts where the snow cover prevents short period frost action. At Stage c, the mounds are gradually destroyed, but the frost action (freeze-thaw effect) is stronger in the center of the mound, and hence the center is drier. The fine materials of the dry surface are eroded by the wind, and as a result, a crater-

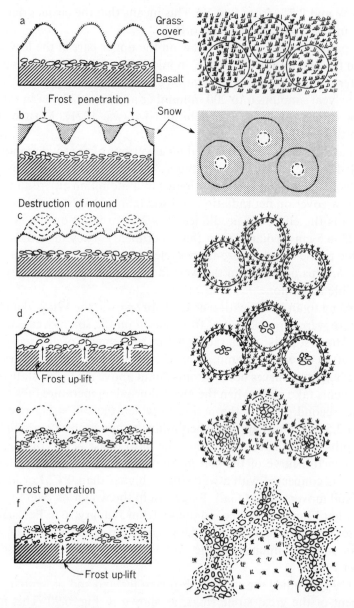

Fig. 6.20 A model of the development of stone rings at Fjardarheidi, E-Iceland (Kim, 1967).

formed circle develops. At Stage d, the frost lifts the debris, because the layer of the fine materials becomes so thin that the frost action reaches the layer of debris. The stones from the layer of the debris appear on the surface in the center of the crater-formed circle. In Stage e, all stones from the debris layer come up to the surface by front action, except the parts under the grass. At Stage f, frost action reaches gradually to the parts between the circles and lifts the stones to the surface, but some other parts remain under the grass. The stones surround the grass surface, making a great ring or polygons finally.

As has been studied by Thorarinsson (1951), patterned ground in Iceland can be classified according to the altitude: Those with a diameter of 20–80 cm appear in the lower coastal plains and valley bottoms, those of 60–200 cm on the higher inland plateaus with a height of 400–800 m above sea level, and the stone rings with a diameter of 2–10 m on the basalt plateaus with an altitude of 600–1,000 m above sea level.

Washburn (1956) classifies patterned ground and reflects on its origins. Mino-Ishikawa (1972) discussed problem concerning patterned ground in Japan and presented some observations from Mt. Kusatsu-Shirane and Mt. Gassan. If the macroclimatological situation, the material compositions of the ground strata, and the micro-topographical conditions such as surface inclination are equal, the distribution must be determined by the local and microclimatological conditions. The suggested local and microclimatological factors influencing the local distribution of patterned ground are soil moisture, insolation, snow accumulation, wind velocity, and the intensity and occurrence frequency of frost. Every factor is connected with the local differences in frost actions, that is, freeze-thaw actions in the soil in a small area. Troll (1944) states that the conditions of the frost action are defined especially by the frequency of freeze-thaw phenomena and the penetrating depth of the frost action. As these climatic conditions change widely from place to place in a small area, further study must be done in various regions from the viewpoint of local and microclimatology.

6.2.4. Wind and Micro-topography

(A) *Sand Dunes and Related Phenomena*

Micrometeorological information about the sand and soil transported by wind is useful to elucidate the influence of the wind on dune formation and wind-erosion processes. Nakano (1956) gives results of an observation on sand ripples and fresh sand surfaces on the Kashima dune in Ibaraki Prefecture on February 23 to 24, 1943. Most of the wavelength of sand ripples ranges from 8 to 12 cm, but none are below 6 cm. The measurements at some points show that the wind direction is usually at right angles to sand ripples.

Naruse (1971) observed the coastal sand dunes of Tanegashima Island, S-Kyushu. He concluded that the distribution of aeolian sand is dominant on the west coast and that the prevailing wind directions can be estimated by analyzing the distribution of aeolian sand. The distribution of sand dunes and wind roses is shown in Fig. 6.21. The ridge directions of sand dunes on the west coast roughly coincide with the prevailing wind direction, that is, between NNW and W.

In Hungary there are large areas (20% of the total country's area) with aeolian sand, most of which developed on Pleistocene alluvial fans and has a thickness of a few cm to 25–32 m. In cases like this, the problem of wind erosion

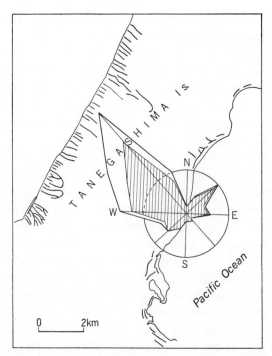

Fig. 6.21 Distribution of sand dunes (position of ridges) and the wind roses in the middle part of Tanegashima Island. The shaded area presents the cases of 4.5–10m/sec and the white area the cases stronger than 10m/sec (Naruse, 1971).

Photo 6.6 Stripes on the slope by wind erosion under periglacial conditions on Mt. Washigamine, Kirigamine, central Japan. Prevailing wind flows from right to left (April, 1973, by M. M. Yoshino).

is of special importance (Borsy, 1972). In the areas where the height of dunes is less than 8 m, blankets of loessy sand and sandy loess are formed on the dunes. In the case of dunes covered by thernozemic soil, the height of the dunes decreases by 60–100 cm as a result of agricultural cultivation, and the eroded humid soil accumulates in the depressions between the dunes.

Hastings (1971) observed the interesting phenomenon of *sand streets*. During a flight over a *ghibli* between El Adem in Libya and the coast on the morning of Feb. 14, 1969, he noticed that lines were visible below in the rising sand in a number of places. The lines were analogous to cloud streets aligned in the direction of the winds. They were the process of longitudinal dune formation in action. The diameter of helical eddies would have been about 300 m, and the spacing between them 600 to 900 m.

(B)   *Periglacial Wind Conditions and Micro-topography*

Svensson (1972) studied palaeoclimatological wind conditions on the Laholm Plain on the W-coast of Sweden. In this region, the aeolian sand deposits, although low and smooth, show parabolic contours representing old dune fields. Another feature of the wind activity in this region is the ventifacts sculptured in an early period of the deglaciation of this area. Namely, intensely blasted and facetted stones are found on the upper part of glacifluvial deposits and below a cover of wind-blown sand. The facets blasted by the most active winds are in the E–SE, they are considered to be older. This blasting must have been caused by the wind accompanied by sand as well as by snow. The facets caused by the WSW–SW winds must be younger, because of the reshaping features. The older facets are considered to have been formed by katabatic winds from the nearby ice-cap during the early period of the last glacial period, whereas the younger facets belong probably to the Alleröd period. The fossil parabolic dunes were formed by W-winds, which began to prevail in the Alleröd period.

In order to investigate the vegetation destruction in the Cairngorm Mountains, Scotland, King (1971) measured the orientation of turf scarps and sand ripples and compared them with the wind roses. It was confirmed that the turf scarps are primarily due to needle ice and that the denuded surfaces are caused by both needle ice erosion and deflation. These processes reflect wind direction sharply. Small hillslope terraces, previously described as solifluction terraces, are considered here to be caused also by wind action, possibly in the 18th century.

Rudberg (1968) studied the distributions of winds eroding the ground surface in the two mountain areas in Sweden and one on Axel Heiberg Island in the Canadian Arctic. The most important features are the scars in the vegetation, distribution of lichens on boulders, and shrub vegetation on lee sides. It was proved that these phenomena as indicators of prevailing wind directions in winter were useful, like the wind-shaped trees discussed in Section 6.1.3, but the variations in the force of winds in connection with grades of the phenomena are as yet not known fully.

(C) *Oriented Lakes*

Wind action causes the orientation of lakes. According to Cabot (1947), the lake orientation in the Point Barrow region, Alaska, is attributable to summer winds blowing from the Cordilleran ice-cap, which existed some thousands of years ago, parallel with the long axes of the lakes. Black and Barksdale (1949) have come to the conclusion that the prevailing wind direction is parallel with the long axes of lakes. However, Livingstone (1954) suggests that the Alaskan lakes owe their elongation to modern processes, primarily that of longshore rip currents on the lake ends, resulting from winds blowing across the lakes. Mackey (1963) states that the oriented lakes of the Tuktoyaktuk Peninsula-Liverpool Bay area are of Pleistocene and recent sand and silt. In both areas, the prevailing winds blow across the minor lake axes, and longitudinal sand dunes parallel the minor axes. As this area is situated in the tundra, neither trees nor bushes break the sweep of the wind. The oriented lakes have four shapes: egg-shape, ellipse, triangle, and lemniscate. Washburn (1973) says also that the oriented lakes lengthened at right angles to the present predominant winds as the result of bars forming on the lee shores. As a basin enlarges, mainly by thawing, these bars tend to protect the shores from erosion by waves, currents, and thawing. As has been illustrated by Carson and Hussey (1962), the distribution maps of present lakes, old shorelines of present lakes, recently drained lakes, and ancient drained lakes show a strikingly impressive parallel pattern. Further study from the standpoint of local and microclimatology must be done on the precise mechanism.

REFERENCES

Chapter 1

Barry, R. G. (1970): A framework for climatological research with particular reference to scale concepts. *Trans. Papers, Inst. Brit. Geog.*, **49**, 61–70.

Barry, R. G., Perry, A. H. (1973): "Synoptic Climatology, Methods and Applications." Methuen, London, 1-555.

Baum, W. A., Court, A. (1949): Research status and needs in microclimatology. *Trans. Amer. Geoph. Union*, **30**, 488–493.

Berényi, D. (1967): "Mikroklimatologie". Gustav Fischer Verlag, Stuttgart, 1–328.

Bjelanović, M. M. (1967): Mesoklimatische Studien im Rhein- und Moselgebiet.——Ein Beitrag zur Problematik der Landesklimaaufnahme——Inaug. Diss. Doktor. Mathem.-naturw. Bonn, 1–232

Böer, W. (1959): Zum Begriff des Lokalklimas. *Zeitsch. Met.*, **13**, 1–6.

Böer, W. (1964): Einige Überlegungen zur raumzeitlichen Struktur des Geländeklimas und den Möglichkeiten seiner Darstellung. *Ang. Met.*, **5**, 34–36.

Bogdan, O. (1972): Efectuarea observatiilor microclimatice si topoclimatice cu elevii pe itinerarii geografice. *Terra*, (5), 72–79.

Boyko, H. (1962): Old and new principles of phytobiological climatic classification. *Biometeorology*, 113–127.

Brunt, D. (1953): Bases of micro-climatology. *Nature*, **171**, 322–323.

Byers, H. R. (1959): "General meteorology" 3rd ed. McGraw-Hill, N.Y., 363–364.

Chromow, S. P. (1952): Klimat, makroklimat, mestniiklimat, i mikroklimat. *Vses. Geograf. Obsh., Izv.*, **84** (3), 289–298.

van Eimern, J. (1968): The topoclimate and its mapping for agricultural purposes. *Proceeding WMO Regional Training Seminar on Agrometeorology. Wageningen, Netherlands*, 213–219.

Endrödi, G. (1965): Basic research methods of topoclimatology. *Időjárás*, **69** (2), 90–94.

Eriksen, W. (1964): Das Stadtklima, seine Stellung in der Klimatologie und Beiträge zu einer witterungsklimatologischen Betrachtungsweise. *Erdkunde*, **18** (4), 257–266.

Eriksen, W. (1967): Das Klima des mittelrheinischen Raumes in seiner zeitlichen und räumlichen Differenzierung. In: Die Mittelrheinlande. Festschrift zum 34. Deutsch. Geographentag in Bad Godesberg. Wiesbaden, 16–30.

Fiedler, F., Panofsky, H. A. (1970): Atmospheric scale and spectral gaps. *Bull. Amer. Met. Soc.*, **51**, 1114–1119.

Flemming, G. (1971): Studie zu den Einteilungsprinzipien der Meteorologie. *Veröf. Met. Dienst. DDR.*, (21), 1–59.

Flohn, H. (1959): Bemerkungen zum Problem der globalen Klimaschwankungen. *Arch. Met. Geoph. Biokl.* (B) **9**, 1–13.

Fukui, E. (1938): "Kikogaku (Climatology)." Kokonshoin, Tokyo, 1–566.

Fukui, E. (1962): Meaning of climate and three-dimensional division of climatology. *Jour. Geogr.* (Tokyo) (730), 232–236.

Geiger, R. (1927): "Das Klima der Bodennahen Luftschicht." Friedr. Vieweg, Braunschweig, 1–246.

Geiger, R. (1929): Die vier Stufen der Klimatologie. *Met. Zeitsch.*, **46**, 7–10.

Geiger, R., Schmidt, W. (1934): Einheitliche Bezeichnungen in kleinklimatischer und mikroklimatischer Forschung. *Biokl. B.*, **1**, 153–156.

Geiger, R. (1950): "Das Klima der bodennahen Luftschicht." 3 Aufl. Friedr. Vieweg, Braunschweig, 1–460.

Geiger, R. (1961): "Das Klima der bodennahen Luftschicht." 4 Aufl. Friedr. Vieweg, Braunschweig, 1–646.

Geiger, R. (1965): "Climate near the ground." Harvard Univ. Press, Cambridge, 1–611.

Gol'tsberg, I. A. (1969): "Microclimate of the USSR." Israel Program for Scientific Translations, Jerusalem, 1–236.

Gol'tsberg, I. A. (1970): Research in agro- and microclimatology. In: Voeikov Main Geophysical Observatory 1917–1967. ed. by M. I. Budyko, IPST, Jerusalem, 84–97.

Haase, G. (1964): Landschaftsökologische Detailuntersuchung und naturräumliche Gliederung. *Pet. Geog. Mitt.*, **108**, 8–30.

Hall, W. F., Holloway, L. (1955): System analysis of data provision and processing. Progress Rep. No. 2, Contract No. CSO & A 54–24.

Hess, M. (1971): Studien über die quantitative Differenzierung der makro-, meso- und mikroklimatischen Verhältnisse der Gebirge als Grundlage zur Konstruierung detaillierten Klimakarten. *Prace Geograficzne*, **26**, 265–278.

Holmes, R. M., Dingle, A. N. (1965): The relationship between the macro- and microclimate. *Agricul. Met.*, **2**, 127–133.

Huschke, R. E. (1959): "Glossary of meteorology." *Amer. Met. Soc.*, Boston, 1–365.

Iudin, M. I. (1955): Invariantnye velichiny v krupnomasshtabnykh atmosfernykh protsessakh. (Invariable values of the large scale atmosphere processes.) *GGO, Trudy*, **55** (117), 3–12.

Kayane, I. (1966): Meso-climatological research on the temperature distribution in the Kanto Plain, Japan. *Sci. Rep. T.K.D. Sec. C*, **9** (87), 125–187.

Keil, K. (1950): "Handwörterbuch der Meteorologie." Verl. Fritz Knapp, Frankfurt a. M., 1–604, see 286.

Knoch, K. (1942): Weltklimatologie und Heimatklimakunde. *Met. Zeit.*, **59**, 245–249.

Knoch, K. (1949): Die Geländeklimatologie, ein wichtiger Zweig der angewandten Klimatologie. *Ber. z deutsch. Landeskunde*, **7**, 115–123.

Kurashima, A. (1968): The scale and predictability of meteorological phenomena. *Tenki*, **15** (3), 79–88.

Landsberg, H. (1960): "Physical climatology." (2nd Rev. Ed.) DuBois, Penn., 1–446.

Mäde, A. (1964): Zur Methodik der meteorologischen Geländeaufnahme. *Ang. Met.*, **5**, 1–2.

Mason, B. J. (1970): Future developments in meteorology: an outlook to the year 2000. *Q.J.R.M.S.*, **96**, 349–368.

Mattsson, J. O. (1964): Mikroklimat och mikroklimatologi. *Särtryck ur Geografiska Notiser Årg.*, **22** (4), 179–183.

Miháilescu, V., Seitan, O., Neamu, Gh. (1965): Microclimat et topoclimat. *Rev. Roum. Géol., Géoph. et Géog., Ser. Géog.*, **9** (2), 173–177.

Monin, A. S. (1972): Weather forecasting as a problem in physics. MIT Press, Cambridge, 1–199.

Mörikofer, W. (1947): Die Bedeutung lokalklimatischer Einflüsse für die Kurortplannung. Ann. schweiz. Gesell. f. Balneologie und Klimatol. (38): 31–38.

Neef, E. (1963): Topologische und chorologische Arbeitsweisen in der Landschaftsforschung. *Pet. Geog. Mitt.*, **107**, 249–259.

Neef, E. (1964): Zur grossmassstäbigen landschaftsökologischen Forschung. *Pet. Geog. Mitt.*, **108**, 1–7.

Okołowicz, W. (1960): Macro-, meso- and microclimate. *Przeg. Geograficzny*, **32** (Supplement) 97–102.

Paffen, K. H. (1948): Ökologische Landschaftsgliederung. *Erdkunde*, **2**, 167–173.

Prosek, P. (1970): Soucasny stav a problematika clenění kategorií klimatu. *Sborník Československé Společnosti Zeměpisné*, **75**, (2) 126–141.

Raethjen, P. (1960): Über den "Scale"—Unterschied vertikaler und horizontaler Störbewegungen. *Beitr. z. Phy. Atm.*, **32**, 257–264.

Sapozhnikova, S. A. (1950): "Mikroklimat I mestnyn klimat." Gidromet. Izdat., Leningrad, 1–241.

Sasakura, K. (1950): "Shôkikôgaku (Kleinklima)." Kokonshoin, Tokyo, 1–169.

Scaëtta, M. H. (1935); Terminologie climatique, bioclimatique et microclimatique. *La Met.*, **11**, 342–347.

Schmithüsen, J. (1948): "Fliesengefüge der Landschaft" und "Ökotop." Vorschläge zur begrifflichen Ordnung und zur Nomenklatur in der Landschaftsforschung. *Ber. z. Deutsch. Landeskunde*, **5**, 74–83.

Schmithüsen, J. (1949): Grundsätze für die Untersuchung und Darstellung der naturräumlichen Gliedelung von Deutschland. *Ber. z. Deutsch. Landeskunde*, **6**, 8–19.

Schnelle, F. (1968): Agrotopoclimatology. Symposium on Methods in Agroclimatology. UNESCO, Paris, 251–260.

Sekiguti, T. (1950): Introduction to local climatology. *Geoph. Mag.*, **22**, 29–33.

Sekiguti, T. (1952): Kikôgaku no taikei ni tsuite (On the system of climatology). Monogr. ded. to

Prof. K. Uchida's 60th Birthday, No. 2, 269–280.

Shcherban, M. I. (1968): "Mikroklimatologiia (Microclimatology)." Kiev, Izdatelśvo Kievskogo Universiteta, 1–210.

Stone, K. H. (1968): Scale, scale, scale. *Economic Geography*, **44** (2) Guest Editorial facing 55p.

Stringer, E. T. (1958): Geographical meteorology. *Weather*, **13** (11) 377–384.

Suzuki, S. (1944): On the new classification of climatology by R. Geiger and W. Schmidt. *Jour. Agricul. Met.*, **1**, 175–178.

Suzuki, S. (1951): "Nôgyô-kishôgaku (Agricultural Meteorology)." Yôkendo, Tokyo, 1–267.

Takahashi, K. (1969): " Sôkan kishôgaku. (Synlptic Meteorology)" Iwanami, Tokyo, 1–385.

Teodoreanu, E. (1971): Microclima si topoclima. *Progresele stiintei*, **7** (4), 186–191.

Tepper, M. (1959): Mesometeorology——the link between macroscale atmospheric motions and local weather. *Bull. Amer. Met. Soc.*, **40**, 56–72.

Thornthwaite, C. W. (1953): Topoclimatology. Proc. Toronto Met. Conf., 227–232.

Troll, C. (1950): Die geographische Landschaft und ihre Erforschung. *Studium Generale*, **3**, 163–181.

Ukita, N. (1970): Chirigaku ni okeru chiiki no scale. (Scale of region in geography). *Jinbunchiri*, (Kyoto), **22** (4), 405–419.

Ukita, N. (1972): Nôgyôteki tochi riyôno kenkyû ni okeru scale. (Scale in studying agricultural land use). Essays of Geographical Sciences, Hiroshima, 34–40.

Wagner, R. (1956): Microclimatic spaces and their mapping. *Földrajzi Közlemények*, **80**, 201–206.

Watanabe, K. (1960): Ranges of meso- and micro- scale. *Kishô-kenkyû Note*, **11**, 76–77.

Weischet, W. (1956): Die räumliche Differenzierung klimatologische Betrachtungsweisen. Ein Vorschlag zum Gliederung der Klimatologie und zu ihrer Nomenklatur. *Erdkunde*, **10**, 109–122.

Yoshino, M. M. (1961): "Shôkikô (Microclimate an introduction to local Meteorology)." Chijinshokan, Tokyo, 1–274.

Chapter 2

Budyko, M. I. (1956): "Teplovoi balans zemnoi poverkhnosti." Gidromet. Izdat., Leningrad, 1–254.

Fels, E. (1935): "Der Mensch als Gestalter der Erde." Leipzig l. Aufl.

Fujiwara, S. (1951): "History of meteorology in Japan." Iwanami, Tokyo, 40–52.

Geiger, R. (1927): "Das Klima der bodennahen Luftschicht." Friedr. Vieweg, Braunschweig, 1–246.

Götz, G. (1964): Über die Methode und einige Ergebnisse der geländeklimatologischen Forschungen in Ungarn. *Ang. Met.*, **5**, 27–30.

Gol'tsberg, I. A. (1969): "Microclimate of the USSR". IPST, Jerusalem, 1–236.

Gol'tsberg, I. A. (1970): Research in agro- and microclimatology. In: Voeikov Main Geophysical Observatory 1917–1967. ed. by M. I. Budyko, (Trudy GGO No. 218) IPST, Jerusalem, 84–97.

Hann, J. (1908): "Handbuch der Klimatologie". 2.Aufl., 3.Aufl., Stuttgart.

Howard, L. (1833): Climate of London deduced from meteorological observations. 3Vols. 3rd ed.

Kerner, F. (1891): Die Änderung der Bodentemperatur mit der Exposition. Sitzungsb. d. Wiener Akad.

Knoch, K. (1951): Über das Wesen einer Landesklimaaufnahme. *Zeitsch. Met.*, **5**, 173–177.

Knoch, K. (1961): Methodische Erfahrungen zur Durchführung einer Landesklimaaufnahme. *Zeitsch. Met.*, **15**, 1–6.

Knoch, K. (1963): Die Landesklimaaufnahme——Wesen und Methodik. *Ber. Deutsch. Wett.*, (85) 1–64.

Kratzer, A. (1937): "Das Stadtklima." 1 Aufl. Friedr. Vieweg, Braunschweig, 1–144.

Leighly, J. (1949): Climatology since the year 1800. *Trans. Amer. Geoph. Union*, **30**, 658–672.

Miller, D. H. (1968): Development of the heat budget concept. In: Eclectic climatology. ed. by A. Court, Oregon State Univ. Press, Corvallis, 123–144.

Munn, R. E. (1966): "Descriptive micrometeorology." Academic Press, N.Y., 1–245.

Sammadar, J. N. (1912): Indian meteorology of the 4th century B.C. *Q.J.R.M.S.*, **38**, 65–66.

Sapozhnikova, S. A. (1950): "Mikroklima i mestnyii klimat." Gidromet. Izdat., Leningrad, 1–241.

Shitara, H. (1966): Meso- and Microclimatology in Japan. In: Japanese Geography 1966. Ass. Japanese Geog., 52–56.

Tanaka, Y. (1887): Revised report on plant distribution in Japan. 1–176.

Voeikov (Woeikof), A. (1887): "Die Klimate der Erde." 2Bd, Jena, 1–396, 1–422.

Chapter 3

Aichele, H. (1957): Kleinklimatische Untersuchungen, eine Voraussetzung für Windschutzplanungen. Veröff. d. Landesst. f. Naturschutz und Landschaftspflege Baden-Württemberg und der Württembergischen Bezirksstellen in Stuttgart und Tübingen, Ludwigsburg, 115–124.

Aizenshtat, B. A. (1958): Teplovoi balans i mikroklimat nekotorykh landshaftoy peschanoi pustyni. In: Sovremennye problemy meteorologii prizemnogo sloia vozdukha. ed. by M. I. Budyko, Gidrometeoizdat, Leningrad, 67–130.

Aldrich, J. H. (1970): Convergence zones in southern California, *Weather*, **25** (4), 140–146.

Alissow, B. P., Drosdow, O. A., Rubinstein, E. S. (1956): "Lehrbuch der Klimatologie" Deutsch Verl. der Wissenschaften, Berlin, 1–536.

Andersson, T. (1969): Small-scale variations of the contamination of rain caused by washout from the low layers of the atmosphere. *Tellus*, **21** (5), 685–692.

Angell, J. K., Pack, D. H., Holzworth, G. C., Dickson, C. R. (1966): Tetroon trajectories in an urban atmosphere. *Jour. Appl. Met.*, **5**, 565–572.

Angell, J. K., Pack, D. H., Dickson, C. R., Hoecker, W. H. (1971): Urban influence on nighttime airflow estimated from tetroon flights. *Jour. Appl. Met.*, **10** (2), 194–204.

Angell, J. K., Hoecker, W. H., Dickson, C. R., Pack, D. H. (1973): Urban influence on a strong daytime air flow as determined from tetroon flights. *Jour. Appl. Met.*, **12**, 924–936.

Arakawa, H. (1937): Increasing air temperature in large, developing cities. *Gerl. Beitr. Geoph.*, **50**, 3–6.

Arakawa, H. (1938): The effect of coasts on winds. *Geoph. Mag.*, **11**, 193–196.

Arakawa, H., Tsutsumi, K. (1967): Strong gusts in the lowest 250-m layer over the city of Tokyo. *Jour. Appl. Met.*, **6** (5), 848–851.

Asai, T. (1940): Temperature increase of sea breeze on the sandy coast. *Kagaku*, **10**, 11.

Asai, T. (1952): Low temperature zone along the coast of Tohoku District caused by sea breeze in summer. Monogr. ded. to Prof. K. Uchida's 60th Birthday, No. 2, 215–233.

Asai, T. (1964): On the water temperature rising and its microclimatic environment, in summer at Shirait. Waterfall, Shizuoka Prefecture, Japan. *Misc. Rep. Research Inst. Natural Resources.* (62), 1–12.

Asakawa, T. (1950): Fog of the river mouth of Saigawa, Kawanakajima. *Shinano*, **2** (11) 742–757.

Asakuno, K. et al. (1970): Analysis of sub-micron particles in air pollution at Tokyo Tower. *Ann Rep. Tokyo Metropolitan Res. Inst. Environmental Protection*, **1** (Sec. 1), 132–135.

Aslyng, H. C. (1960): Evaporation and radiation heat balance at the soil surface. *Arch. Met. Geoph. Biokl.* (B), **10**, 359–375.

Aslyng, H. C., Nielsen, B. F. (1960): The radiation balance at Copenhagen. *Arch. Met. Geoph. Biokl.* (B), **10**, 342–358.

Atlas, D. (1960): Radar detection of the sea breeze. *Jour. Met.*, **17**, 244–258.

Atkinson, B. W. (1968): A preliminary investigation of the possible effect of London's urban area on the distribution of thunder rainfall, 1951–60. *Trans. Inst. Brit. Geog.*, **44**, 97–118.

Atkinson, B. W. (1969): A further examination of the urban maximum of thunder rainfall in London, 1951–60. *Trans. Inst. Brit. Geog.*, **48**, 97–120.

Atkinson, B. W. (1970): The reality of the urban effect on precipitation. A case-study approach. *W.M.O. Tech. Note* (108), 342–362.

Atkinson, B. W. (1971): The effect of an urban area on the precipitation from a moving thunderstorm. *Jour. Appl. Met.*, **10** (1), 47–55.

Azuma, S. (1956): Micrometeorological observations on natural and artificially-blackened snow surfaces. *Sci. Rep. Saikyo Univ. Japan*, (A), (3), 147–156.

Azuma, S. (1958): A micrometeorological study on the diurnal variations of air and earth temperatures. *Sci. Rep. Saikyo Univ. Japan*, (A), **2** (5), 55–92.

Bach, W. (1970): An urban circulation model. *Arch. Met. Geoph. Biokl.*, (B), **18**, 155–168.

Barnes, F. A. (1960): The intense thunder rains of 1st July, 1952, in the northern Midlands. *East Midland Geographers, Nottingham*, (14) 11–26.

Baumgartner, A. (1956): Untersuchungen über den Wärme- und Wasserhaushalt eines jungen Waldes. *Ber. Deutsch. Wett.*, **5** (28), 1–53.

Baumgartner, A., Kleinlein, G., Waldmann, G. (1958): Klimatische Standortsfaktoren am Gr. Falkenstein. *Forstw. Cbl.*, **75**, 290–303; **77**, 230–237; **78**, 98–109.

Baumgartner, A. (1961): Baum und Wald im Windfeld. *Allgemeine Forstzeitschrift. München*, (13/14) 2p.

Baumgartner, A. (1963): Wärmeumsätze des Bodens und den Pflanze. In: Frostschutz im Pflanzenbau. ed. by Schnelle, 84–150.

Baumgartner, A. (1967 a): Entwicklungslinien der forstlichen Meteorologie. *Forstw. Cbl.*, **86** (3), 156–175, (4), 201–220.

Baumgartner, A. (1967 b): The balance of radiation in the forest and its biological function. *Biometeorology*, **2**, 743–754.

Baumgartner, A. (1967 c): Ermittlung der tatsächlichen Verdunstung aus Messungen des vertikalen Wasserdampfaustausches und der Energiebilanz. *Deutsche Gewässerkdl. Mitteilgn.* Sonderheft 1967, 192–195.

Baynton, H. W., Hamilton, H. C., Sherr, P. E., Worth, J. J. B. (1965): Temperature structure in and above a tropical forest. *Q.J.R.M.S.*, **91**, 225–232.

van Bemmelen, W. (1922): Land- und Seebrise in Batavia. *Beitr. zur Phys. d. fr. Atm.*, **10**, 169–177.

Berényi, D. (1965): Mikroklimatologische Beobachtungen und Wärmehaushaltsmessungen auf der Hortobágy Puszta (Heide) bei Debrecen. *Ang. Met.*, **5**, (3/4) 87–91.

Bergeron, T. (1949): The coastal orographic maxima of precipitation in autumn and winter, I and II. *Tellus*, **1**, (1), 32–43, (3), 15–32.

Berlyant, T. G. (1956): Teplovoi balans atmosfery severnogo polushariya. In: A. I. Voeikov i sovremennye problemy klimatologii. Gidromet. Izdat., Leningrad, 57–83.

Berson, F. A. (1958): Some measurements on undercutting cold air. *Q.J.R.M.S.*, **84**, 1–16.

Best, A. C. (1935): Transfer of heat and momentum in the lowest layers of the atmosphere. Geoph. Mem. No. 65 (London).

Biggs, W. G., Graves, M. E. (1962): A lake breeze index. *Jour. Appl. Met.*, **1** (4), 474–480.

Blackadar, A. K. (1962): The vertical distribution of wind and turbulent exchange in a neutral atmosphere. *Jour. Geoph. Res.*, **67** (8), 3095–3102.

Blom, J., Wartena, L. (1969): The influence of changes in suface roughness on the development of the turbulent boundary layer in the lower layers of the atmosphere. *Jour. Atm. Sci.*, **26**, 255–265.

Bornstein, R. D. (1968): Observations of the urban heat island effect in New York City. *Jour. Appl. Met.*, **7** (4), 575–582.

Boros, J. (1966): Temperaturverhältnisse auf Bergwiese und in Tannenwaldbestanden sonnigen Sommertagen. *Acta Climat.*, (6), 53–72.

Bowa, N. W. (1950): Die Niederschläge an den Bergufern der Flüße. Sowjetwiss. Naturwiss. Abt. (2), 124–126.

Bradley, E. F. (1968): A micrometeorological study of velocity profiles and surface drag in the region modified by a change in surface roughness. *Q.J.R.M.S.*, **94**, 361–379.

Brooks, C. E. P. (1954): "The English climate." London, 1–214.

Brown, K. W., Rosenberg, N. J. (1972): Shelter-effects on microclimate, growth and water use by irrigated sugar beets in the Great Plains. *Agricul. Met.*, **9**, 241–263.

Brunt, D. (1932): Notes on radiation in the atmosphere. *Q.J.R.M.S.*, **58**, 389–420.

Brunt, D. (1945): Some factors in microclimatology. *Q. J.R.M.S.*, **71**, 1–10.

Brunt, D. (1953): Basis of micro-climatology. *Nature*, **171**, 322–323.

Budyko, M. I. (1956): "Teplovoi balans zemnoi poverkhnosti." Gidromet. Izdat., Leningrad, 1–254.

Burgos, J. J., Cagliolo, A. Y., Santos, M. C. (1951): Exploracion microclimatica en la selva Tucumano-Oranense. *Meteoros*, **1** (4), 314–341.

Busch, N. E., Panofsky, H. A. (1968): Recent spectra of atmospheric turbulence. *Q.J.R.M.S.*, **94**, 132–148.

Byzova, N. L. (1963): (Some measurement findings on the horizontal diffusion of admixture in the bottom atmospheric layer) Nekotorye rezul'taty izmerenii gorizontal'noi diffuzii primesi v nizhnem sloe atmosfery. In: (Investigation of the bottom 300-meter layer of the atmosphere) Issledovanie nizhnego 300-metrovogo sloya atmosfery. ed. by N. L. Byzova. Moskva. Akad. Nauk SSSR. IPST, Jerusalem, 1965, 26–35.

Caborn, J. M. (1957): Shelterbelts and microclimate. *Forestry Commission Bulletin, Edinburgh*, (29), 1–135.

Cantlon, J. E. (1953): Vegetation and microclimates on north and south slopes of Cushetunk Mountain, New Jersey. *Ecol. Monogr.*, **23**, 241–270.

Chandler, T. J. (1965): "The climate of London." Hutchinson, London, 1–292.

Chander, T. J. (1970): Urban climatology—inventory and prospect. In: Urban Climates, *WMO Tech. Note*, (108), 1–14.

Chang, Jen-hu. (1957a): Global distribution of the annual range in soil temperature. *Trans. Amer. Geoph. Union*, **38**, 718–723.

Chang, Jen-hu (1957 b): World patterns of monthly soil temperature distribution. *Ann. Ass. Amer. Geog.*, **47**, 241–249.

Chang, Jen-hu (1958): Ground temperature (Vols. I and II.) Blue Hill Met. Observ. Milton, 1–300, and 1–196.

Chang, Jen-hu (1961): Microclimate of sugar cane. *Hawaiian Planters' Record*, **56**, 195–225.

Changnon, S. A. Jr. (1961 a): A climatological evaluation of precipitation patterns over an urban area. Symposium on "Air over Cities." Public Health Service, Washington D.C., 37–67.

Changnon, S. A., Jr. (1961 b): Precipitation contrasts between the Chicago urban area and an off-shore station in Southern Lake Michigan. *Bull. Amer. Met. Soc.*, **42** (1), 1–10.

Changnon, S. A., Jr. (1968): The La Porte weather anomaly——Fact or fiction? *Bull. Amer. Met. Soc.*, **49** (1), 4–11.

Changnon, S. A., Jr. (1970): Recent studies of urban effects on precipitation in the United States. *W.M.O. Tech. Note*, (108), 325–341.

Chia, L. S. (1967): Albedos of natural surfaces in Barbados. *Q.J.R.M.S.*, **93** (1), 116–120.

Clarke, J. F. (1969): Nocturnal urban boundary layer over Cincinnati, Ohio. *Mon. Wea. Rev.*, **97**, 582–589.

Clarke, J. F., Peterson, J. T. (1973): An empirical model using eigenvectors to calculate the temporal and spatial variations of the St. Louis heat island. *Jour. Appl. Met.*, **12** (1), 195–210.

Cramer, H. E., Record, F. A. (1953): The variation with height of the vertical flux of heat and momentum. *Jour. Met.*, **10**, 219–226.

Daigo, Y., Nagao, T. (1972): Toshikikô (Urban climate). Asakura Shoten, Tokyo, 1–214.

Daubenmire, R. F. (1950): "Plants and environment." 285–287.

Davis, N. E. (1951): Fog at London Airport. *Met. Mag.*, **80**, 9–14.

Deacon, E. L. (1949): Vertical diffusion in the lowest layers of the atmosphere. *Q.J.R.M.S.*, **75**, 89–103.

Deacon, E. L. (1953 a): Vertical profiles of mean wind in the surface layers of the atmosphere. *Geoph. Mem. London*, **11** (91), 1–68.

Deacon, E. L. (1965): Wind gust speed: averaging time relationship. *Aust. Met. Mag.*, **51**, 11–14.

Defant, F. (1951): Local winds. *Comp. Met.*, 655–672.

DeMarrais, G. A. (1961): Vertical temperature difference observed over an urban area. *Bull. Amer. Met. Soc.*, **42**, 548–554.

Devyatova, V. A. (1957): Microaerological investigations of the lower kilometer layer of the atmosphere. Transl. from "Mikroaerologicheskie issledovaniya nizhnego kilometrovogo sloya atmosfery. Leningrad, 1–163.

Dirmhirn, I. (1953): Einiges über die Reflexion der Sonnen- und Himmelsstrahlung an verschiedenen Oberflächen. *Wett. Leben*, **5**, 86–94.

Dirmhirn, I. (1961): Light intensity at different levels. *Trans. Roy. Entomol. Soc. London*, **113** Pt. 11, 270–274.

Djunin, A. K. (1963): "Mechanika metelej." Novosibirsk, 1–378.

Domrös, M. (1966): Luftverunreinigung und Stadtklima im Rheinisch-Westfälischen Industriegebiet und ihre Auswirkung auf den Flechtenbewuchs der Bäume. Arbeiten zur Rheinischen Landeskunde. Bonn, (23) 1–132.

Druyan, L. M. (1968): A comparison of low-level trajectories in an urban atmosphere. *Jour. Appl. Met.*, **7** (4), 583–590.

Druyan, L., Pack, D. H., Holzworth, G. C., Dickson, C. R. (1966): Tetroon trajectories in an urban atmosphere. *Jour. Appl. Met.*, **5**, 565–572.

Duckworth, F. S. et al. (1954): The effect of cities upon horizontal and vertical temperature gradients. *Bull. Amer. Met. Soc.*, **35**, 198–207.

Edwards, R. S., Claxton, S. M. (1964): The distribution of wind-borne salt of marine origin in some western areas of Wales. In: Major weather hazards affecting British agriculture. ed. by J. A. Taylor. Memorandum Univ. Coll. Wales, Aberystwyth, (7), 25–26.

Egli, E. (1951): Die neue Stadt in Landschaft und Klima. Erlenbach, Zürich, 1–155.

van Eimern, J. (1957): Über die Veränderlichkeit der Windshutzwirkung einer Doppelbaumreihe bei verschiedenen meteorologischen Bedingungen. *Ber. Deutsch. Wett.*, **5** (32), 1–21.

van Eimern, J. et al. (1964): Windbreaks and shelterbelts (Report of a working group of the Commission for Agricultural Meteorology). *WMO Tech. Note*, (59), 1–188.

Emonds, H. (1954): Das Bonner Stadtklima. Arbeiten zur Rheinischen Landeskunde. Bonn, (7), 1–79.

Eriksen, W. (1964 a): Beiträge zum Stadtklima von Kiel. *Schriften des Geographischen Instituts der Universität Kiel.*, **22** (1), 1–218.

Eriksen, W. (1964 b): Das Stadtklima, seine Stellung in der Klimatologie und Beiträge zu einer witterungsklimatologischen Betrachtungsweise. *Erdkunde*, **18** (4), 257–266.

Eskuche, U. (1957): Über Windschutzuntersuchungen an der Donau bei Herbertingen. Veröff. Landesstelle f. Naturschutz und Landschaftspflege Baden-Württemberg und der Württembergischen Bezirksstellen in Stuttgart und Tübingen, Ludwigsburg, (25), 57–114.

Estoque, M. A. (1961): A theoretical investigation of the sea breeze. *Q.J.R.M.S.*, **87** (372), 136–146.

Estoque, M. A. (1962): The sea breeze as a function of the prevailing synoptic situation. *Jour. Atm. Sci.*, **19**, 244–250.

Fedorov, S. F. (1965): Evaporation from forest and field in years with different amounts of moisture. *Soviet Hydrology*, (4), 337–348. (Trudy GGO, 1965, (123), 22–35)

Fiedler, F., Panofsky, H. A. (1972): The geostrophic drag coefficient and the "effective" roughness length. *Q.J.R.M.S.*, **98**, 213–220.

Fisher, E. L. (1960): An observational study of the sea breeze. *Jour. Met.*, **17**, 645–660.

Fisher, E. L. (1961): A theoretical study of the sea breeze. *Jour. Met.*, **18**, 216–233.

Flemming, G. (1967): Die drei klimatischen Grundfunktionen der Kronenschicht des Waldes. *Arch. Forstwes*, **16** (6/9), 573–577.

Flemming, G. (1968 a): Waldmeteorologie und Forstmeteorologie——Definition- und Einteilungsfragen. *Arch. Forstwes*, **17** (3), 303–311.

Flemming, G. (1968 b): Die Windgeschwindigkeit auf waldungebenen Freiflächen. *Arch. Forstwes*, **17** (1), 5–16.

Flemming, G. (1969): Zur Umlagerung der Schneedecke durch den Wind. *Arch. Naturschutz u. Landschaftsforsch*, **9** (2), 175–194.

Flohn, H., Fraedrich, K. (1966): Tagesperiodische Zirkulation und Niederschlagsverteilung am Victoria-see (Ostafrika). *Met. Rdsch.*, **19** (6), 157–165.

Flower, W. D. (1937): An investigtion into the variation of the lapse rate of temperature in the atmosphere near the ground at Ismailia, Egypt. Geoph. Mem. No. 71 (London).

Forest Section, Hokkaido Government Office, (1951, 1952, 1953, 1954): Studies on the fog prevention forests; I, II, III and IV.

Forschungsstelle für Lawinenvorbeugung, Innsbruck (1961, 1963): Ökologische Untersuchugen in der subalpinen Stufe zum Zwecke der Hochlagenaufforstung. I, II. Mitteilungen der Forstlichen Bundes-Versuchsanstalt Mariabrunn, (59), 1–430, (60), 433–887.

Frankenberger, E. (1962): Beiträge zum internationalen Geophysikalischen Jahr 1957/58. I, Mess-Ergebnisse und Berechnungen zum Wärmehaushalt der Erdoberfläche. *Ber. Deutsch. Wett.*, **10** (73), 1–41.

Frenkiel, J. (1962): Wind profiles over hills (in relation to windpower utilization). *Q.J.R.M.S.*, **88**. 156–169.

Frimescu, M., Romanof, N., Moroianu, I. (1973): Meteorologische Aspecte der Luftverunreinigung in einer Industriezone mit komplexem Relief. *Időjárás*, **77** (5), 284–289.

Frizzola, J. A., Fisher, E. L. (1963): A series of sea breeze observations in the New York City area. *Jour. Appl. Met.*, **2**, 722–739.

Fugglc, R. F. (1972): Rural/urban variations in atmospheric counter radiation. *S. African Geog. Jour.*, **54**, 138–144.

Fukaishi, K. (1971): On summer fogs of eastern part of Hokkaido (1). *Kushiro-ronshû*, (2), 17–55.

Fukaishi, K. (1973): On the summer fog in Kushiro district. *Geog. Rev. Japan*, **46** (11), 741–754.

Fukui, E. (1939): Distribution of UV-radiation around the great cities in Japan. *Geog. Rev. Japan*, **15** (5), 377–394.

Fukui, E., Wada, N. (1941): Air temperature distribution in the great cities in Japan. *Geog. Rev.*

Japan, **17**, 354–372.

Fukui, E. (1943): Secular change of climate at the great cities in Japan. *Jour. Met. Soc. Japan*, **21**, 428–434.

Fukui, E. (1957): Increasing temperature due to the expansion of urban areas in Japan. *75th Anniversary Vol. Jour. Met. Soc. Japan*, 336–341.

Fukui, E. (1969): Recent rise of temperature in Japan. *Sci. Rep., Tokyo Kyoiku Daigaku, Sec. C.*, **10** (97), 145–164.

Fukuoka, S., Itoo, M., Odaira, T. (1972): Analysis of oxidant pollution in a broad area. Ann. Rep. Tokyo Metropolitan Res. Inst. Environmental Protection, 87–104.

Fukuoka, Y. (1965): The effect of the sea and land water on the soil temperature. *Ann. Tôhoku Geog, Ass.*, **17** (4), 221p.

Fukuoka, Y. (1966): Soil temperature observations in western suburbs of Tokyo and their statistical studies. *Geog. Rev. Japan*, **39** (2), 103–117.

Fukuoka, Y., Yamashita, S. (1970): Air pollution in Japanese cities. In: Japanese Cities. Ass. Japanese Geog., 227–236.

Fukutomi, T. et al. (1951): On the distribution of fog water contents in the surroundings of the forest at Ochiishi, 1950. In: Studies on the fog prevention forests, I. 67–73.

Funatsu, Y. (1964): On the salty wind damage by typhoon in Kanto district. *Jour. Met. Res.*, **16** (2), 103–117.

Futi, H. (1933): The effects of coast and mountain ranges on the direction and velocity of wind in vicinity. *Jour. Met. Soc. Japan*, **11** (11), 492–496.

Galahkov, N. N. (1955): Mikro-klimaticheskie nabliudenüa v raionakh srednego Priangar'ia i basseina verkhnei Leny. *Trudy Inst. Geog. Akad. Nauk S.S.S.R.* (64), 173–192.

Garnett, A., Bach, W. (1965): An estimation of the ratio of artificial heat generation to natural radiation heat in Sheffield. *Mon. Wea. Rev.*, **93**, 383–385.

Garnett, A., Bach, W. (1967): An investigation of urban temperature variations by traverses in Sheffield (1962–1963). *Biometeorology*, (2), 601–607.

Gavrilova, M. K. (1966): Radiation climate of the arctic. IPST, Jerusalem, 1–178.

Gavrilova, M. K. (1972): Radiation and heat balances, thermal regime of an icing. A paper presented to the International Symposium "Snow and Ice", Banff, Canada, 1972.

Geiger, R. (1961): "Das Klima der bodennahen Luftschicht." Friedr. Vieweg & Sohn. Braunschweig, 1–646.

Geiger, R. (1965): "The climate near the ground". Harvard Univ. Press, 1–611.

Geisler, J. E., Bretherton, F. P. (1969): The sea-breeze forerunner. *Jour. Atm. Sci.*, **26** (1), 82–95.

Georgii, H. W. (1970): The effects of air pollution on urban climates. *W.M.O. Tech. Note* (108), 214–237.

Gifford, F. A., Hanna, S. R. (1973): Modelling urban air pollution. *Atmosph. Env.*, **7**, 131–136.

Gilbert, O. L. (1965): Lichens as indicators of air pollution in the Tyne Valley. In: Ecology and the industrial society. Brit. Ecol. Soc. Sym. No. 5, ed. by Goodmann, G. T. et al., Oxford, 35–47.

Grah, R. F., Wilson, C. C. (1944): Some components of rainfall interception. *Jour. Forestry*, **42**, 890–898.

Gregory, S., Smith, K. (1967): Local temperature and humidity contrasts around small lakes and reservoirs. *Weather*, **22** (12), 497–505.

Grillo, J. N., Spar, J. (1971): Rain-snow mesoclimatology of the New York metropolitan area. *Jour. Appl. Met.*, **10** (1), 56–61.

Grunow, J. (1961): Die relative Globalstrahlung, eine Masszahl der vergleichenden Strahlungsklimatologie. *Wett. Leben.*, **13** (3–4), 47–56.

Grunow, J. (1965): Die Niederschlagszurückhaltung in einem Fichtenbestand am Hohenpeißenberg und ihre meßtechnische Erfassung. *Forstw. Cbl.*, **84** (7/8), 201–229.

Hamm, J. M. (1969): Untersuchungen zum Stadtklima von Stuttgart. *Tübinger Geog. Studien. Univ. Tübingen*, (29), 1–150.

Hanna, S. R. (1969): Urban meteorology. ATDL Contrib. (35), Air Resources Lab., Oak Ridge, Tenn. 1–8.

Hanna, S. R. (1971): Simple methods of calculating dispersion from urban area sources. *Jour. Air Poll. Control Ass.*, **21**, 774–777.

Hare, F. K., Ritchie, J. C. (1972): The boreal bioclimates. *Geog. Rev.*, **62** (3), 333–365.

Hartmann, F. K., Eimern, J. V., Jahn, G. (1959): Untersuchungen reliefbedingter kleinklimatischer Fragen in Geländequerschnitten der hochmontanen und montanen Stufe des Mittel-und Südwestharzes. *Ber. Deutsch. Wett.*, **7**, 1–39.

Hartmann, F., Schnelle, F. (1970): "Klimagrundlagen natürlicher Waldstufen und ihrer Waldgesellschaften in deutschen Mittelgebirgen." Gustav Fischer Verl., Stuttgart, 1–176.

Heckert, L. (1959): Die klimatischen Verhältnisse in Laubwäldern. *Zeitsch. Met.*, **13**, 211–233.

Heigel, K. (1964): Minimum-Temperaturen auf und über der Schneedecke auf dem Hohenpeissenberg während des Winters 1962/63. *Met. Rdsch.*, **17** (1), 25–28.

Hellmann, G. (1915, 1917): Über die Bewegung der Luft in den untersten Schichten der Atmosphäre. *Met. Zeitsch.*, **32**, 1–16, (Zweite Mitteilung) *Met. Zeitsch.*, **34**, 273–285.

Hirt, M. S., Shaw, R. W. (1973): The passage of a lake-breeze front at Toronto—A comparison between the city and the suburbs. *Atmosph. Env.*, **7**, 63–73.

Holmes, R. M. (1970): "Oasis effects" caused by the Cypress hills. Proc. 3rd forest microclimate symposium, Seebe, Alberta, 159–180.

Holmgren, B. (1971): On the katabatic winds over the north-west slope of the ice cap. Variations of the surface roughness. In: Climate and energy exchange on a sub-polar ice cap in summer. *Uppsala Univ. Met. Inst. Meddelande*, (109), 1–43.

Hopkins, A. D. (1938): Bioclimatics, a science of life and climate relations. *U.S. Dept. Agr. Misc. Publ.*, (280), 1–188.

Höschele, K. (1966): Der zeitliche Verlauf und die örtliche Verteilung der SO_2- Konzentrationen in einem Stadtgebiet mit einer Analyse der Einflussgrössen. *Met. Rdsch.*, **19** (1), 14–22.

Huff, F. A., Changnon, S. A. (1972): Climatological assessment of urban effects on precipitation. Final Rep. Pt. I. Ill. State Water Survey, Urbana. NSF-GA 18781, 34p.

Hufty, A. (1970): Les conditions du payonnement en ville. *W.M.O. Tech. Note* (108), 65–69.

Huss, E., Stranz, D. (1970): Die Windverhältnisse am Bodensee. *Pure Appl. Geoph.*, **81** (4), 323–356.

Iizuka, H. (1952 a): On the width of windbreak. *Bull. Gov. Forest. Exp. Sta., Meguro*, (56), 1–218.

Iizuka, H. (1952 b): Research on so called "blow-in" at the side end of a windbreak. *Bull. Gov. Forest Exp. Sta., Meguro*, (56), 219–223.

Iizuka, H. (1952 c): Research on the correlation between the turbulence intensity and the wind velocity on the leeward of windbreak as one method to judge the function of windbreak. *Bull. Gov. Forest Exp. Sta., Meguro*, (56), 225–231.

Imahori, K. (1952): On the vanishing mechanism of advection fog and the effect of woods upon fog reduction. In: Studies on the fog prevention forests, II. 121–142.

Inoue, E. (1963): Micrometeorological research in Japan. *Sci. Progress*, **51** (201), 69–80.

Inoue, K. (1972): Numerical experiments of humidity change of air mass moving from seashore into inland. *Jour. Agricul. Met.*, **28** (2), 103–113.

Ishihara, Y., Kobatake, S. (1970): Initial storage of rain-water in runoff process. *Ann. Disaster Prevention Res. Inst. Kyoto Univ.*, (13 B), 69–81.

Ishihara, Y., Kobatake, S. (1971, 1972): On the water balance in Ara experimental basin (1), (2). *Ann. Disaster Prevention Res. Inst., Kyoto Univ.*, (14) B, 131–141, (15) B, 321–331.

Ishikawa, M., Kawaguchi, T., Sato, S. (1970): Estimation of snow depths by tree forms of Japanese cedar and Japanese larch. *Seppyô*, **32** (1), 10–17.

Itoo, K., Moriguchi, M., Naruse, H. (1955): Atmospheric SO_2 concentration observed in Tokyo and its relation to meteorological elements. *Pap. Met. Geoph.*, **6** (1), 19–25.

James, D. (1953): Fluctuations of temperature below cumulus clouds. *Q.J.R.M.S.*, **79**, 425–428.

Jones, P. M., Larrinaga, M. A. B.de, Wilson, C. B. (1971): The urban wind velocity profile. *Atmosph. Env.*, **5** (2), 89–102.

Johnson, A., Jr., O'Braien, J. J. (1973): A study of an Oregon sea breeze event. *Jour. Appl. Met.*, **12**, 1267–1283.

Junghaus, I. H. (1961): Zur Temperaturmessung in verschiedenen Bodenarten. *Ang. Met.*, **4** (3), 83–90.

Kalb, M. (1962): Einige Beiträge zum Stadtklima von Köln. *Met. Rdsch.*, **15** (4), 92–99.

Kalma, J. D., Stanhill, G. (1972): The climate of an orange orchard: physical characteristics and microclimate relationships. *Agricul. Met.*, **10**, 185–201.

Kaps, E. (1953/54): Die Temperaturverhältnisse an der Elbe zwischen Ufer und Deich. *Ann. Met.*, **6**, 15–25.

Karapiperis, P. P. (1953): Characteristics of local winds at Athens, Greece. *Geof. Pura Appl.*, **25**, 203–206.

Kataoka, K., Sato, S. (1959, 1964): On the bend of lower part of the afforested *Cryptomeria* by snow (1) (2). *Seppyô*, **21** (4), 13–19; **26** (2), 39–45.

Katsumi, S. (1956): Variation of flow in the period before and after cutting at Kamikawa Experimental Forest (1). *Rep. Hokkaido Branch, Gov. Forest Exp. Sta., Sp. Rep.*, (5) 139–149.

Kauper, E. K. (1960): The zone of discontinuity between the land and sea breezes and its importance to southern California air-pollution studies. *Bull. Amer. Met. Soc.*, **41**, 410–422.

Kawamura, T., Mizukoshi, M. (1957, 1959): Micro-variation of air temperature (1) (2). *Tenki*, (75th Ann. Volume) 134–137, **6**, 159–161.

Kawamura, T. (1963): The geographical distribution of solar radiation amount. *Tenki*, **10** (1), 18–20.

Kawamura, T. (1964 a): Analysis of temperature distribution in Kumagaya City. *Geog. Rev. Japan*, **37** (5), 243–254.

Kawamura, T. (1964 b): Some considerations on the cause of city temperature at Kumagaya City. *Geog. Rev. Japan*, **37** (10), 560–565.

Kawamura, T. (1966): Urban climatology in Japan. "Japanese Geography 1966——its recent trends." Special Publ. (2) Ass. Japanese Geographers, Tokyo, 61–65.

Kawamura, T. (1974): Change of air temperature in Tokyo. (Mimeograph) 7p+18figs.

Kayane, I. (1960 a): Temperature increase due to the expansion of urban area in Tokyo. *Tenki*, **7** (9), 269–274.

Kayane, I. (1960 b): The distribution and climatological analysis of the daily minimum temperature in and around the Tokyo metropolitan area. *Geog. Rev. Japan*, **33** (11), 564–572.

Kayane, I. (1961): Temperature distribution and its meso-climatological analysis in the Kanto Plain. *Geog. Rev. Japan*, **34**, 438–449.

Kayane, I. (1966): Meso-climatological research on the temperature distribution in the Kanto Plain, Japan. *Sci. Rep. Tokyo Kyoiku Daigaku Sec. C.*, **9** (87), 125–187.

Keller, Hs. M. (1968): Der heutige Stand der Forschung über den Einfluß des Waldes auf den Wasserhaushalt. *Schweizerischen Zeitschr. f. Forstwesen*, (4/5), 353–363.

Khemani, L. T., Ramana Murty, Bh. V. (1973): Rainfall variations in an urban industrial region. *Jour. Appl. Met.*, **12**, 187–194.

Kitazawa, Y. et al. (1960): Distribution of species populations of warm-temperate broad-leaved trees along the elevation gradient at Ôsumi Peninsula, the southernmost portion of Kyûshû, Japan. *Misc. Rep. Res. Inst. Natural Resources*, (52–53), 24–35.

Klinov, F. Ya., Poltavskii, V. V. (1963): Ob izmerenii skorosti vetra v nizhnem 300-metrovom sloe atmosfery na vysotnoi meteorologicheskoi machte. (The measurement of the wind velocity in the bottom 300-meter atmospheric layer on an upper-air meteorological mast.) In: Issledovanie nizhnego 300-metrovogo sloya atmosfery, ed. by N. L. Byzova. Izdatel'stvo Akad. Nauk SSSR, Moskva 1963. IPST, Jerusalem, 57–63.

Kluge, M., Krawczyk, B. (1964): Mapa albeda okolic Wojcieszówa Górnego. (The albedo map of the region of Wojcieszów Górny, Silesia) *Przegl. Geog.*, **36** (1), 131–141.

Knudsen, J. (1941): Über Windmessungen an Hindernissen. Berg. Mus. Årbok 1941, Naturvit. rekke, (10), 1–35.

Knudsen, J. (1963): Vindforholdene på Bergen Hamn. *Det Norske Met. Inst. Tec. Rep.*, (4), 1–16.

Koch, H. C. (1934): Temperaturverhaltnisse und Windsystem eines geschlossenen Waldgebietes. *Veröff. Geoph. Inst. Univ. Leipzig.*, **6** (3), 121–175.

Kohlbach, W. (1942): Seewindbeobachtungen an der Südostküste der Danziger Bucht. *Das Wetter*, **59**, 6–12.

Konda, K. (1962): Frost height as a microclimatic phenomenon. *Res. Bull. Coll. Exp. Forests, Hokkaido Univ.*, **21**, 177–185.

Kondo, J. (1967): Analysis of solar radiation and downward long-wave radiation data in Japan. *Sci. Rep. Tôhoku Univ., Ser. 5, Geoph.*, **18** (3), 91–124.

Kopec, R. J. (1965): Continentality around the Great Lakes. *Bull. Amer. Met. Soc.*, **26** (2), 54–57.

Kopec, R. J. (1966): Effects of the Great Lakes' thermal influence on freezefree dates in spring and fall as determined by Hopkins' bioclimatic law. *Agrical. Met.*, **4**, 241–253.

Kopec, R. J. (1967): Areal patterns of seasonal temperature anomalies in the vicinity of the Great Lakes. *Bull. Amer. Met. Soc.*, **48** (12), 884–889.

Kopec, R. J. (1973): Daily spatial and secular variations of atmospheric humidity in a small city. *Jour. Appl. Met.*, **12**, 639–648.

Kratzer, A. (1956): "Das Stadtklima." Friedr. Vieweg & Sohn, Braunschweig, 1–184.

Kreutz, W. (1943): Die Jahresgang der Temperatur unter gleich Witterungsverhältnisse. *Zeitsch. f. angew. Met.*, **60**, 65–76.

Kreutz, W. (1952): "Der Windschutz: Windschutzmethodik, Klima und Bodenertrag" Dortmund, 1–167.

Kristensen, K. J. (1959): Temperature and heat balance of soil. *Oikos*, **10**, 103–120.

Kubo, T. (1943): An example of the effect of the river on its neighbouring air temperature. *Jour. Met. Soc. Japan*, **21**, 494–501.

Kurashige, K. (1943): Air temperature distribution in the mediumsized cities. *Mem. Central Met. Obs.*, (19), 495–498.

Laboratory of Snow Damage, Government Forest Experimental Station, Japan (1952): Study of the fallen snow on the forest trees. *Bull. Gov. Forest Exp. Sta.*, Meguro, (54), 115–164.

Laykhtman, D. L., Chudnovskii, A. P. (1949): "Fizika prizemnogo sloia atmosfery. (Physics of the layer of air near the ground)." Gosudarst. Izdatel'stvo Tekhniko-Teoret. Literatury, Leningrad, 1–254.

Laitinen, E. (1970): Energy balance of the earth's surface in Finland. *Finnish Met. Inst. Cont.*, (76), 1–16.

Landsberg, H. (1951): Climatology and its part in pollution. *Met. Monogr.*, **1** (4), 7–8.

Landsberg, H. E. (1956): The climate of towns. In: Man's role in changing the face of the earth. ed. by W. L. Thomas, Chicago, 584–606.

Landsberg, H. (1957): Review of climatology, 1951–1955. *Met. Monogr.*, **3**, 27–29.

Landsberg, H. E. (1967): Air pollution and urban climate. *Biometeorology*, (2), 648–656.

Lauscher, F. (1934): Wärmeausstrahlung und Horizonteinengung. *Sitzungsberichte Wien Akad.*, **143**, 503–519.

Lauscher, F. et al. (1959): Witterung und Klima von Linz. *Wett. Leben, Sonderheft*, (6), 1–235.

Lavoie, R. L. (1972): A mesoscale numerical model of lake-effect storms. *Jour. Atm. Sci.*, **29** (6), 1025–1040.

Lawrence, E. N. (1952): Frost investigation. *Met. Mag.*, **81**, 65–74.

Lawrence, E. N. (1971): Urban climate and day of the week. *Atmosph. Env.* **5** (11), 935–948.

Leopold, L. B. (1949): The interaction of trade wind and sea breeze, Hawaii. *Jour. Met.*, **6**, 312–320.

Lettau, H. H., Davidson, B. (1957): Exploring the atmosphere's first mile, Vol. I & II. Pergamon Press, N.Y., 1–376, 377–578.

Lettau, H. H. (1957): Computation of Richardson mumbers, classification of wind profiles, and determination of roughness parameters. In: "Exploring the atmosphere's first mile, Vol. I" ed. by H. H. Lettau et al. 328–336.

Lindqvist, S. (1968): Studies on the local climate in Lund and its environs. *Geog. Ann.*, **50** (A), 79–93.

Lingova, St. (1965): Über die Strahlungsverhältnisse Bulgariens. *Ang. Met.*, **5**, 73–79.

Löbner, A. (1935): Horizontale und vertikale Staubverteilung in einer Großstadt. *Veröff. Geoph. Inst. Univ. Leipzig*, 2 Ser., **7** (2), 53–99.

Lomas, J., Gat, Z. (1967): The effect of windborne salt on citrus production near the sea in Israel. *Agricul. Met.*, **4**, 415–425.

Long, I. F. et al. (1964): The plant and its environment. *Met. Rdsch.*, **17** (4), 97–101.

Louw, W. J., Meyer, J. A. (1965): Near-surface nocturnal winter temperatures in Pretoria. *Notos*, **14**, 49–56.

Lowry, W. P. (1956): Evaporation from forest soils near Donner summit, California, and a proposed field method for estimating evaporation. *Ecology*, **37**, 419–430.

Lüdi, W., Zoller, H. (1949): Über den Einfluss der Waldnähe auf das Lokalklima. *Ber. Geobot. Foschungsinst. Rübel in Zürich, Jahr* 1948, 85–108.

Ludwig, F. L. (1970): Urban temperature fields. *W.M.O. Tech. Note*, (108), 80–107.

Lumb, F. E. (1970): Topographic influences on thunderstorm activity near Lake Victoria. *Weather*, **25** (9), 404–410.

Lyons, W. A. (1972): The climatology and prediction of the Chicago lake breeze. *Jour. Appl. Met.*, **11** (8), 1259–1270.

MacHattie, L. B., McCormack, R. J. (1961): Forest microclimate—a topographic study in Ontario.

Jour. Ecol., **49**, 301–323.

Mäde, A. (1970): Zur Methodik landwirtschaftlicher Windschutzuntersuchungen. *Idöjárás*, **74**, 200–208.

Magata, M. (1965): A study of the sea breeze by the numerical experiment. *Pap. Met. Geoph.*, **14**, 23–27.

Mano, H., Okazaki, G. (1954): The effect of coast on precipitation. *Jour. Met. Res.*, **6** (10/11), 473–481.

Marsh, K. J., Foster, M. D. (1967 a): An experimental study of the emissions from chimneys in Reading——1: The study of longterm average concentration of sulphur dioxide. *Atmosph. Env.*, **1**, 527–550.

Marsh, K. J., Bishop, K. A., Foster, M. D. (1967 b): An experimental study of the dispersion of the emissions from chimneys in Reading——2: A description of the instruments used. *Atmosph. Env.*, **1**, 551–559.

Martin, H. C. (1971): Average winds above and within a forest. *Jour. Appl. Met.*, **10** (6), 1132–1137.

Maruyama, I., Inose, T. (1952): First report on the forest hydrology at Kamabuchi. *Bull. Gov. Forest. Exp. Sta., Meguro*, (53), 1–46.

Mattsson, J. O. (1961): Microclimate observations in and above cultivated crops. *Lund Stud. Geog.*, (A) (16), 1–117.

Mattsson, J. O. (1965): Brief fluctuations of temperature in and over a potato field and over follow ground. *Lund Stud. Geog.*, (A) (31), 1–69.

Mattsson, J. O. (1966 a): The temperature climate of potato crops. *Lund Stud. Geog.*, (A) (35), 1–193.

Mattsson, J. O. (1966 b): Soil temperatures and light, wind and air humidity conditions of potato crops. *Lund Stud. Geog.*, (A) (36), 7–30.

Mattsson, J. O. (1971): Dagg som klimatindikator (Dew as a climatic indicator). *Svensk Geografisk Årsbok*, **47**, 29–52.

Mayama, T. (1923): Fog precipitation on the top of Mt. Oodaigahara. *Shinrin-chisui-kishô-ihô* (2), 9–19.

McCormick, R. A. (1969): Meteorology and urban air pollution. *WMO Bulletin*, **18** (3), 155–165.

Miara, K. (1969): Wyznaczenie turbulencyjnego strumienia ciepla metoda gradientowa wedlug Kazanskiego i Monina (Determination of turbulent heat flux with Kazanskij and Monin's gradient method). *Przegl. Geof.*, **14** (2), 181–189.

Miller, D. H. (1955): Snow cover and climate in the Sierra Nevada, California. *Univ. of Calif. Publication in Geog.*, (11), 1–218.

Miller, D. H. (1956 a): The influence of open pine forest on daytime temperature in the Sierra Nevada. *Geog. Rev.*, **46** (2), 209–218.

Miller, D. H. (1956 b): The influence of snow cover on local climate in Greenland. *Jour. Met.*, **13** (1), 112–120.

Miller, D. H. (1965): The heat and water budget of the earth's surface. *Advances in Geophysics*, (11), 175–302.

Miller, D. H. (1971): Kontseptsiia energo——i massoobmena v prirodnoi srede kak method analiza iavlenii, obuslavlivaemykh deiatel'nost'iu cheloveka. *Akad. Nauk. S.S.S.R., Izv. Ser. Geog.*, (2), 118–133.

Mitchell, J. M. (1953): On the causes of instrumentally observed secular temperature trends. *Jour. Met.*, **10** (4), 244–261.

Mizukoshi, M. (1965): A contribution to the relationships between the distribution of air temperature and the wind in an urban area. *Geog. Rev. Japan*, **38** (2), 92–102.

Mizukoshi, M. (1968): The geographical distribution of solar radiation in Yokkaichi City, Mie Prefecture. *Tenki*, **15** (1), 5–8.

Mizukoshi, M., Nakagawa, Y., Morioka, Y. (1968): The regional distribution of daily minimum temperature in local cities. *Miedaigaku Kyôikukenkyûjo-Kenkyû Kiyô*, (39), 17–29.

Monin, A. S., Obukhov, A. M. (1954): Osnovnye zakonomernosti turbulentnogo peremeshivaniia v prizemnom sloe atmosfery. *Trudy Geog. Inst. Akad. Nauk SSSR.*, (24), 163–187.

Monteith, J. L. (1961): An empirical method for estimating long-wave radiation exchanges in the British Isles. *Q.J.R.M.S.*, **87**, 171–179.

Monteith, J. L. (1965): Evaporation and environment. *Symp. Soc. Exp. Biol.*, **19**, 205–234.

Monteith, J. L. (1966): Local differences in the attenuation of solar radiation over Britain. *Q.J.R.M.*

S., **92**, 254–262.

Morita, Y., Tabata, T. et al. (1952): On the sea fog invasion over Nemuro and Kushiro District (1st Rept). In: Studies on fog prevention forests II, 71–74.

Moroz, W. J., Hewson, E. W. (1966): The mesoscale interaction of a lake breeze and low level outflow from a thunderstorm. *Jour. Appl. Met.*, **5** (2), 148–155.

Moroz, W. J. (1967): A lake breeze on the eastern shore of lake Michigan: Observations and model. *Jour. Atm. Sci.*, **24**, (4) 337–355.

Muller, R. A. (1966): Snowbelts of the Great Lakes. *Weatherwise*, **19**, 248–255.

Munn, R. E., Stewart, I. M. (1967): The use of meteorological towers in urban air pollution programs. *Jour. Air Poll. Control Ass.*, **17**, 98–101.

Munn, R. E., Hirt, M. S., Findley, B. F. (1969): A climatological study of the urban temperature anomaly in the lakeshore environment at Toronto. *Jour. Appl. Met.*, **8** (3), 411–422.

Munn, R. E. (1970): Airflow in urban areas. *W.M.O. Tech. Note*, (108), 15–39.

Munn, R. E. (1973): Urban meteorology: some selected topics. *Bull. Amer. Met. Soc.*, **54** (2), 90–93.

Myrup, L. O. (1969): A numerical model of the urban heat island. *Jour. Appl. Met.*, **8** (6), 908–918.

Nägeli, W. (1943): Untersuchungen über die Windverhältnisse im Bereich von Windschutzstreifen. *Mitt. Schweiz. Anst. forstl. Vers'wes.*, **23** (1), 221–276.

Nägeli, W. (1946): Weitere Untersuchungen über die Windverhältnisse im Bereich von Windschutzstreifen. *Mitt. Schweiz. Anst. forstl. Vers'wes.*, **24** (2), 659–737.

Nägeli, W. (1965): Über die Windverhältnisse im Bereich gestaffelter Windschutzstreifen. *Mitt. Schweiz. Anst. forstl. Vers'wes.*, **41** (5), 221–300.

Nakamura, S. (1964): Study on the heat balance at a snow surface in the thawing season. *Bull. Hokuriku Exp. St.*, (7), 1–28.

Neiburger, M., Johnson, D. S., Chien, C-w. (1961): Studies of the structure of the atmosphere over the Eastern Pacific Ocean in summer. I. The inversion over the Eastern North Pacific Ocean. *Univ. Calif. Pub. Met.*, **1** (1), 1–94.

Neuber, E. (1970): Einige Aspekte des Einflusses der Ostsee auf das Klima Mecklenburgs. *Veröff. Geoph. Inst., Univ. Leipzig*, **19** (4), 413–424.

Nishizawa, T. (1958 a): The influence of buildings on the urban temperature. *Misc. Rep. Res. Inst. Natural Resources*, (48), 40–48.

Nishizawa, T. (1958 b): On the air temperature distribution in a small area and a method of areal division, with special regard to the urban area and its surroundings. *Geog. Rev. Japan*, **31** (3), 168–175.

Nishizawa, T. (1966): The sensible heat transfer coefficient at the earth's surface. *Misc. Rep. Res. Inst. Natural Resources*, (66), 9–14.

Nishizawa, T. (1973): Urban climate, with special reference to heat island. *Kagaku*, **43** (8), 487–494.

Nishizawa, T., Hasegawa, T. (1969): Soil temperature in Japan. *Geog. Rev. Japan*, **42** (12), 775–777.

Nitzschke, A. (1970): Zum Verhalten der Lufttemperatur in der Kontaktzone zwischen Land und Meer bei Zingst. *Veröff. Geoph. Inst., Univ. Leipzig*, **19** (4), 425–433.

Nkemdirim, L. C., Yamashita, S. (1972): Energy balance over prairie grass. *Canadian Jour. Plant Sci.*, **52**, 215–225.

Nordlund, G. (1971): The sea breeze in South Finland according to a numerical model. *Finnish Met. Inst. Contr.*, (77), 1–28.

Nyberg, A. (1938): Temperature measurements in an air layer very closed to a snow surface. *Geog. Ann.*, **20**, 234–275.

Ôdaira, T., Fukuoka, S. (1970): Study on relationship between weather elements and SO_2 concentration surveyed at Tokyo Tower. *Ann. Rep. Tokyo Metropolitan Res. Inst. f. Environmental Protection*, **1** (Sec. 1), 65–69.

Ogden, T. L. (1969): The effect on rainfall of a large steelworks. *Jour. Appl. Met.*, **8** (4), 585–591.

Oguntoyinbo, J. S. (1970): Reflection coefficient of natural vegetation, crops and urban surfaces in Nigeria. *Q.J.R.M.S.*, **96** (409), 430–441.

Ogura, Y. (1952): Note on the wind velocity profile in the non-adiabatic atmosphere. *Jour. Met. Soc. Japan*, **31**, 1–5.

Ohta, Y. (1965): Studies on the meteorological analysis of air pollution. *Jour. Met. Res.*, **17** (11), 661–726.

Okanoue, M. (1950): Anomalous temperature in Fukuoka. *Mem. Ind. Met.*, **14** (2), 81–91.

Oke, T. R., East, C. (1971): The urban boundary layer in Montreal. *Bound. Layer Met.*, **1**, 411–437.

Oke, T. R., Hannell, F. G. (1970): The form of the urban heat island in Hamilton, Canada. *W.M.O. Tech. Note*, (108), 113–126.

Okita, T. (1960): Estimation of direction of air flow from observation of rime ice. *Jour. Met. Soc. Japan*, **38**, 207–209.

Olecki, Z. (1973): Wplyw miasta na niektóre elementy bilansu radiacyjnego na przykladzie Krakowa (Influence of a town upon certain elements of the radiation balance as exemplified by Cracow). *Prace Geog.*, **32**, 105–118.

Oliver, H. R. (1971): Wind profiles in and above a forest canopy. *Q.J.R.M.S.*, **97**, 548–553.

Oliver, J. (1959): The climate of the Dale Peninsula, Pembrokeshire. *Field Studies*, **1** (1), 1–17.

Oliver, J. (1962): The thermal regime of upland peat soils in a maritime temperate climate. *Geog. Ann.*, **44** (3–4), 293–302.

Oliver, J. (1966): Soil temperatures in the arid tropics with reference to Khartoum. *Jour. Trop. Geog.*, **23**, 47–54.

Onodera, S. et al. (1954): Microclimate of the small unsheltered openings in fog preventing forests. In: Studies on the fog prevention forest, IV, 152–165.

Ooura, H. (1952): The capture of fog particles by the forest. In: Studies on the fog prevention forest II. 239–258.

Oosawa, S. (1950): Climate in a mixed forest. *Mem. Ind. Met.*, **14** (3), 36–38.

Ozawa, Y. (1966): On the influences of low temperature of sea water on the inland air temperature. *Bull. Nat. Res. Center Disaster Prevention (Tokyo)*, (6), 59–64.

Pack, D. H., Angell, J. K. (1963): A preliminary study of air trajectories in the Los Angeles basin as derived from tetroon flights. *Mon. Wea. Rev.*, **91**, 583–604.

Paeschke, W. (1937 a): Experimentelle Untersuchungen zum Rauhigkeits und Stabilitätsproblem in der bodennahen Luftschicht. *Beitr. zur Phys. d. fr. Atm.*, **24**, 163–189.

Paeschke, W. (1937 b): Mikroklimatische Untersuchungen innerhalb und dicht über verschiedenartigen Bestand. *Biokl. B.*, **4**, 155–163.

Panofsky, H. A., Petersen, E. L. (1972): Wind profiles and change of terrain roughness at Risφ. *Q.J. R.M.S.*, **98**, 845–854.

Parry, M. (1956 a): Local temperature variations in the Reading area. *Q.J.R.M.S.*, **82**, 45–57.

Parry, M. (1956 b): An "urban rainstorm" in the Reading area. *Weather*, **11**, 41–48.

Parry, M. (1967): Air pollution patterns in the Reading area. *Biometeorology*, (2), 675–667.

Parry, M. (1970): Sources of Reading's air pollution. *W.M.O. Tech. Note*, (108), 295–305.

Paszyński, J. (1960): Transparence de l'atmosphére comme élément du climate local des régions industrielles. *Przegl. Geog.*, **32** (Supplement), 103–107.

Paszyński, J. (1964 a): Mikroklimatische Untersuchungen über den Wärmehaushalt der Erdoberfläche. *Ang. Met.*, **5**, 55–59.

Paszyński, J. (1964 b): Topoclimatological investigations on heat balance. *Polonica*, **2**, 70–77.

Paszyński, J. (1965): The distribution of short-wave net radiation in Poland. *Idöjárás*, **69** (3), 129–134.

Paszyński, J. (1966): Die Strahlungsbilanz Polens. *Zeitsch. Met.*, **17** (9–12), 321–327.

Paszyński, J. (1970): Études topoclimatologiques du bilan thermique dans les montagnes moyennes. In: L'Amenagement de la Montagne. III Colloque franco-polonais de géographie. Acad. Polonaise des Sci., Centre sci. à Paris Conf., Warszawa, **87**, 47–70.

Paszyński, J. (1972): Studies on the heat balance and on evaporation. *Geographia Polonica*, **22**, 35–51.

Patton, C. P. (1956): Climatology of summer fogs in the San Francisco Bay area. *Univ. Calif. Publ. Geog.*, **10** (3), 113–200.

Paulsen, H. S. (1948): Investigations carried through at the station of forest meteorology at Os. *Univ. i Bergen Årbok 1948, Naturvit. rekke*, (7), 1–44.

Peace, R. L., Sykes, R. B. Jr. (1966): Mesoscale study of a lake effect snowstorm. *Mon. Wea. Rev.*, **94**, 495–507.

Pearce, R. P. (1955): The calculation of a sea-breeze circulation in terms of the differential heating across the coastline. *Q.J.R.M.S.*, **81**, 351–381.

Pearce, R. P. (1962): A simplified theory of the generation of sea breeze. *Q.J.R.M.S.*, **88**, 20–29.

Peattie, R. (1936): "Mountain geography." Harvard Univ. Press, Cambridge, Mass., 1–271.

Peppler, W. (1936): Über die Windverhältnisse in der untersten Luftschicht über dem Bodensee und dem Ufer bei Friedrichshafen. *Beitr. zur Phys. d. fr. Atm.*, **23**, 289–309.

Peterson, J. T. (1969): The climate of cities: A survey of recent literature. Publ. No. AP-59, U.S. Public Health Service, Nat. Air Pollution Control Admin., Raleigh, N.C., 1–48.

Pooler, F., Jr. (1963): Airflow over a city in terrain of moderate relief. *Jour. Appl. Met.*, **2** (4), 446–456.

Preston-Whyte, R. A. (1968): Some observations of land breezes and katabatic winds in the Durban area. *Jour. for Geogr.*, **3**, 337.

Preston-Whyte, R. A. (1970 a): Land breezes and rainfall on the Natal Coast. *S. African Geog. Jour.*, **52**, 38–43.

Preston-Whyte, R. A. (1970 b): A spatial model of an urban heat island. *Jour. Appl. Met.*, **9** (4), 571–573.

Priestley, C. H. B. (1959): "Turbulent transfer in the lower atmosphere." The Univ. of Chicago Press, Chicago, 1–130.

Quitt, E. (1964): Method of the establishment of mesoclimatic regions in towns. *Sbornik Československé Společnosti Zkměpisné*, **69** (Supplement), 105–110.

Rassow, L. (1959): Zunahme der Immissionen in einem mitteldeutschen Industrieort. *Zeitsch. Met.*, **13** (1/6), 68–69.

Reid, D. C. (1957): Evening wind surge in South Australia. *Aust. Met. Mag.*, **10**, 23–32.

Rider, N. E. (1954): Evaporation from an oat field. *Q.J.R.M.S.*, **80**, 198–211.

Reifsnyder, W. E., Furnival, G. M., Horowitz, J. L. (1971): Spatial and temporal distribution of solar radiation beneath forest canopies. *Agricul. Met.*, **9**, 21–37.

Ross, P. (1958): Microclimatic and vegetational studies in a cold-wet deciduous forest. *Black Rock Forest Papers, New York*, (24), 1–89.

Rossi, V. (1957): Land-und Seewind an dem Finnischen Küsten. *Mitt. Met. Zentralans. Helsinki*, (41), 1–17.

Rubner, K. (1932): Der Nebelniederschlag im Wald und seine Messung. *Tharandter Forstliches Jb.*, **83** (3), 121–149.

Rutkovskii, V. I. (1948): Climatic and hydrological influences of forests. *Trudy Vses. Geog.*, **2**, 387–405.

Saito, M. (1963): Meso-scale analysis on fog in the Yufutsu-Plain. *Kishô-kenkyû-nôto*, **14** (1), 15–23.

Sakagami, J. (1960): On the relations between the diffusion parameters and meteorological conditions. *Nat. Sci. Rep. Ochanomizu Univ. Japan*, **11** (2), 127–159.

Sanderson, M., Kumanan, I., Tanguay, T., Schertzer, W. (1973): Three aspects of the urban climate of Detroit-Windsor. *Jour. Appl. Met.*, **12**, 629–638.

Sapozhnikova, S. A. (1950): "Mikroklimat i mestnyi klimat." Gidromet. Izdat., Leningrad, 1–241.

Sarson, P. B. (1962): Validity of soil-temperature records. *Memorandum, Univ. Coll. of Wales, Aberystwyth* (4), 8–13.

Sasaki, N., Bokura, T. (1970): A study on relation between "Yamase" wind and solar irradiation. *Jour. Agricul. Met.*, **26** (3), 143–146.

Sasaki, N., Bokura, T. (1972): Local climate characters of Yamase wind. *Jour. Agricul. Met.*, **27** (4), 159–163.

Sasaki, Y. et al. (1969): Distribution of the air temperature and the phase of its diurnal variation in the urban area of Sendai. *Ann. Tôhoku Geog. Ass.*, **21** (4), 198–202.

Sasakura, K. (1950): "Shôkikôgaku (Kleinklimatologie)." Kokonshoin, Tokyo, 1–169.

Sasakura, K., Yamada, K. (1956): Investigation into the forestless zone in the southern part of the Izu Peninsula. *Bull. Education Fac., Shizuoka Univ. Japan*, (7), 183–185.

Sauberer, F. (1952): Beiträge zur Kenntnis des Strahlungsklimas von Wien. *Wett. Leben*, **4**, 187–192.

Schmidt, G. (1952): Zur Nutzbarmachung staubklimatischer Untersuchungen für die städtebauliche Praxis. *Ber. Deutsch. Wett. U.S.-Zone*, **6** (38), 201–205.

Schmitz, H. P. (1962): Über die Interpretation bodennaher vertikaler Geschwindigkeitsprofile in Ozean und Atmosphäre und die Windschubspannung auf Wasseroberflächen. *Deutschen Hydrographischen Zeitschrift.*, **15** (2), 45–72.

Schneider, K., Sonntag, K. (1936): Die meteorologischen Verhältnisse von München und Umgebung in ihrer Bedeutung für das Flugwesen. Veröff. d. Wett. München.

Schöne, V. (1962): Geländemeteorologische Untersuchungen im Havelländischen Luch. *Aug. Met.*,

4 (6), 166–172.

Schöne, V. (1964): Geländemeteorologische Untersuchungen im Klützer Winkel. *Ang. Met.*, **5**, 31–33.

Schroeder, M. J. et al. (1967): Marine air invasion of the Pacific Coast: A problem analysis. *Bull. Amer. Met. Soc.*, **48**, 802–808.

Sekiguti, T. (1951): Studies on local climatology (VI). Humidity distribution and surface covers. *Pap. Met. Geoph.*, **2** (2), 180–188.

Sekiguti, T. (1960): The geographical distribution of spring-time city temperatures in and around Yonezawa, Yamagata, in Northern Japan. *Tokyo Geog. Papers*, (4), 17–40.

Sekiguti, T. et al. (1960): Geographical distribution of solar radiation in Tokyo. *Geog. Rev. Japan*, **33**, 269–277.

Sekiguti, T. (1963 a): City climate distribution——in and around Ogaki, a medium-sized city in Japan. *Tokyo Geog. Papers*, (7), 193–240.

Sekiguti, T. (1963 b): Geographical distribution of air temperature up to 30m above the city areas. *Geog. Rev. Japan*, **36**, 577–586.

Sekine, T. (1956): Shigaichi jôkû no kaze no kansoku (An observation of the winds over the city region). *Bull. Fire Prev. Society Japan*, **5** (2), 33–39.

Selleck, G. W., Schuppert, K. (1957): Some aspects of microclimate in a pine forest and in an adjacent prairie. *Ecology*, **38**, 650–653.

Sellers, W. D. (1965): "Physical climatology." Univ. Chicago Press, Chicago, 1–272.

Seppänen, M. (1961): On the accumulation and the decreasing of snow in pine dominated forest in Finland. *Fennia*, **86** (1), 1–51.

Shanks, R. E. (1956): Altitudinal and microclimatic relationships of soil temperature under natural vegetation. *Ecology*, **37**, 1–7.

Shaposhnikova, M., Lykosov, V., Gutman, L. (1968): Nonstationary nonlinear problem of a breeze in a stably stratified atmosphere. *Akad. Nauk S.S.S.R., Izv. Fiz. Atm. i Okeana*, **4**, 79–89.

Sharon, D., Koplowitz, R. (1972): Observations of the heat island of a small town. *Met. Rdsch.*, **25** (5), 143–146.

Shieh, L. J., Halpern, P. K., Clemens, B. A., Wang, H. H., Abraham, F. F. (1972): Air quality diffusion model; application to New York City. *IBM Jour. Res. Develop.*, **16** (2), 162–170.

Shitara, H. (1952): On the temperature of a coastal plain. *Sci. Rep. Tôhoku Univ. 7th Ser. Geog.* (1), 43–55.

Shitara, H. (1955): An analysis of the distribution of the nocturnal air temperature in a coastal area. *Geog. Rev. Japan*, **28**, 609–620.

Shitara, H. (1957): On the discontinuous distribution of the air temperature during the summer season in the Sanbongi Plain. *Ann. Tôhoku Geog. Ass.*, **9**, 67–71.

Shitara, H. (1963): Meso-climatic divide seen from the discontinuity of the weather. *Sci. Rep. Tôhoku Univ. 7 th Ser. Geog.* (12), 21–34.

Shitara, H. (1964 a): Sea-breeze air-mass boundary in a coastal plain as an example of meso-climatic divide. *Sci. Rep. Tôhoku Univ. 7th Ser. Geog.*, (13), 37–50.

Shitara, H. (1964 b): High temperature area over a lake due to Lake Breezes. *Ann. Tôhoku Geog. Ass.*, **16** (3), 150.

Shitara, H. (1967): Air temperature in summer on and around Lake Inawashiro, Fukushima Prefecture. *Sci. Rep. Tôhoku Univ. 7th Ser. Geog.*, (16), 69–84.

Shitara, H. (1969): Thermal influence of Lake Inawashiro on the local climate in summer daytime. *Sci. Rep. Tôhoku Univ. 7th Ser. Geog.*, **18** (2), 213–220.

Simpson, J. E. (1965): Sea breeze front in Hampshire. *Met. Mag.*, **19**, 208–220.

Simpson, J. E. (1967): Aerial and radar observations of some sea-breeze fronts. *Weather*, **22** (8), 306–316.

Skoczek, J. (1970): Wpływ podłoza atmosfery na przebieg dobowy bilansu cieplnego powierzchni cynnej (Influence of local factors upon daily variation of the heat balance on the active surface). *Prace Geog. IG PAN*, **84**.

Slatyer, R. O., McIlroy, I. C. (1961): Practical microclimatology with special reference to the water factor in soil-plant-atmosphere relationships. UNESCO. 7 sects. Separately paged +10 app.

Stanhill, G., Hofstede, G. J., Kalma, J. D. (1966): Radiation balance of natural and agricultural vegetation. *Q.J.R.M.S.*, **92**, 128–140.

Steinhauser, F., Eckel, O., Sauberer, F. (1955, 1957, 1959): Klima und Bioklima von Wien. I, II, III. *Wett. Leben*, (3), 1–120, (5), 1–136, (7), 1–136.

Stewart, J. B. (1971): The albedo of a pine forest. *Q.J.R.M.S.*, **97**, 561–564.

Stewart, J. B., Thom, A. S. (1973): Energy budgets in pine forest. *Q.J.R.M.S.*, **99**, 154–170.

Sundborg, Å. (1951): Climatological studies in Uppsala, with special regard to the temperature conditions in the urban area. *Geographica* (22), 1–111.

Sutton, O. G. (1953): "Micrometeorology." McGraw-Hill, New York, 1–333.

Tabata, T. et al. (1952): On the distribution of fog water content in the surrounding of the forest at Ochiishi, 1951. In: Studies on the fog prevention forests, II. 203–209.

Takahashi, K. (1955): "Dôkikôgaku (Dynamic climatology)." Iwanami-shoten, Tokyo, 1–316.

Takahashi, M. (1959): Relation between the air temperature distribution and the density of houses, in small cities of Japan. *Geog. Rev. Japan*, **32**, 305–313.

Takasu, K. (1957): Microclimatic study. The Ohara Inst. for Agricul. Biology., Okayama Univ., Kurashiki (Coll. of 4 Papers). Separately paged.

Takasu, K., Kimura, K. (1970): Microclimate in the field. (1) Diurnal variations of air temperature, humidity, soil temperature and CO_2 concentration in the sweet potato field. *Nôgaku Kenkyû*, **53** (3), 167–179.

Takasu, K., Kimura, K. (1971): Microclimate in the field. (2) Diurnal variations of air temperature, humidity and CO_2 concentration in the soybean field. *Nôgaku Kenkyû*, **53** (4), 205–213.

Tamiya, H. (1968): Night temperature distribution in a new-town, western suburbs of Tokyo. *Geog. Rev. Japan*, **41** (11), 695–703.

Tanaka, M., Toba, Y. (1969): Basic study on salt damage (3). *Ann. Disaster Prev. Res. Inst.* (12) B, 201–212.

Tanaka, M. (1970, 1971, 1972): Basic study on salt damage (4), (5), (6). *Ann. Disaster Prev. Res. Inst.* (13) B, 445–456, (14) B, 499–510, (15) B, 295–304.

Tani, N., Inoue, E., Imai, K. (1955): Some measurements of wind over the cultivated field (3). *Jour. Agricul. Met.*, **10**, 105–108.

Tani, N., Inoue, E., Imai, K. (1956): Some measurements of wind over the cultivated field (5). *Jour. Agricul. Met.*, **12**, 17–20.

Tani, N. (1958): On the wind tunnel test of the model shelter-hedge. *Bull. Nat. Inst. Agricul Sci. Japan*, (A) (6), 1–80.

Tani, N. (1963): The wind over the cultivated field. *Bull. Nat. Inst. Agricul. Sci. Japan*, (A) (10), 1–99.

Tanner, J. T. (1963): Mountain temperatures in the southeastern and southwestern United States during late spring and early summer. *Jour. Appl. Met.*, **2** (4), 473–483.

Taylor, J. A. (1962): Soil climate: its definition and measurement. *Memorandum, Univ. Coll. of Wales, Aberystwyth*, (4), 1–7.

Terjung, W. H. et al. (1970 a): Energy balance climatology of a city-man system. *Ann. Ass. Amer. Geog.*, **60** (3), 467–492.

Terjung, W. H., et al. (1970 b): A nighttime energy and moisture budget in Death Valley, California, in mid-August. *Geog. Ann.*, **52** (A) (3–4), 160–173.

Thiele, A. (1974): Luftverunreinigung und Stadtklima in Grossraum München. *Bonner Geog. Abhandlungen*, (49), 1–175.

Thom, A. S. (1972): Momentum, mass and heat exchange of vegetation. *Q.J.R.M.S.*, **98**, 124–134.

Thornthwaite, C. W., Mather, J. R. (1951): The role of evapotranspiration in climate. *Arch. Met. Geoph. Biokl.*, (B) **3**, 16–39.

Thornthwaite, C. W. (1953): Topoclimatology. *Johns Hopkins Univ. Lab. Clim.*, 1–13.

Toba, Y., Tanaka, M. (1967): Basic study on salt damage (1). *Ann. Disaster Prev. Res. Inst.* (10 B), 331–342.

Toyama Local Met. Obs. (1967): Report of winds in Toyama Prefecture. *Tech. Rep. Japan Met. Agency*, (58), 1–101.

Tsuchiya, I. et al. (1973): A preliminary report on the experiment of thermal convective circulation under the inversion layer in the stratified wind tunnel. *Pap. Met. Geoph*, **24** (1), 55–60.

Turnage, W. V. (1939): Desert subsoil temperature. *Soil Science*, **47**, 195–199.

Turner, D. B. (1964): Diffusion model for an urban area. *Jour. Appl. Met.*, **3** (1), 83–91.

Turner, D. B. (1970): Workbook of atmospheric dispersion estimates. U.S. Dept. of Health, Edu. and Welfare, Public Health Serv. Publ. (999-AP-26), 1–84.

Turner, H. (1968): Der heutige Stand der Forschung über den Einfluß des Waldes auf das Klima. *Schweizerischen Zeistchrift für Forstwesen*, (4/5), 364–379.

Uchijima, Z. (1959): Turbulent transport of the momentum, heat and water vapour in the internal layer over the warming pond. *Bull. Nat. Inst. Agricul. Sci. Japan*, (A) (7), 69–97.

Uchijima, Z. et al. (1967): Studies of energy and gas exchange within crop canopies (1). *Jour. Agricul. Met.*, **23** (3), 99–108.

Uchijima, Z. (1974): Micrometeorology of cultivated fields. In: Agricultural meteorology of Japan. ed. by Y. Mihara, Univ. of Tokyo Press, Tokyo, 41–79.

Undt, W., Zawadil, R., Roller, M. (1970): Wetter und Klima im Raum von Wien. In: Naturgeschichte Wiens, Bd. I, F. Starmühlner, Jugend und Volk Wien-München, Wien, 287–391.

Utsumi, T. (1944): On the cause of sea fog in the northern Korean region. *Chuôkishôdai-himitsu-kishô-hôkoku*, **4** (2), 93–107.

Verber, J. L. (1955): The climates of South Bass island, western Lake Erie. *Ecology*, **36**, 388–400.

Visher, S. S. (1943): Some climatic influences of the Great Lakes, latitude and mountains; an analysis of climatic charts in Climate and Man (1941). *Bull. Amer. Met. Soc.*, **24**, 205–210.

Voeikov, A. (1885): On the influence of accumulation of snow on climate. *Q.J.R.M.S.*, **11**, 299–309.

Voeikov (Woeikow), A. (1889): Der Einfluss einer Schneedecke auf Boden, Klima und Wetter. *Geog. Abh. Wien* III, **3**, 1–115.

Volkovitskaya, Z. I., Mashkova, G. B. (1963): O profilyakh vetra i kharakteristikakh turbulentnogo rezhima v nizhnem 300-metrovom sloe atmosfery (The wind profiles and the turbulence characteristics in the bottom 300-meter layer of the atmosphere). In: Issledovanie nizhnego 300-metrovogo sloya atmosfery. ed. by N. L. Byzova. Izdatelśtro Akad. Nauk SSSR, Moskva, IPST 1965, 13–25.

Vorontsov, P. A. (1956): Ob urovne rarvirrliia konvektivnykh dvizhenii v atmosfere (On the level of the development of convective movements in the atmosphere). *Met. Gidr.*, (9), 26–29.

Vorontsov, P. A. (1958 a): O brizakh Ladozhskago ozera. *Trudy GGO*, (73), 87–106.

Vorontsov, P. A. (1958 b): Nekotorie voprosi aerologicheskih issledovanii pogranichnogo sloia atmosferi. In: Sovremennie prob. meteorol. prizem. voz. 157–179.

Vorontsov, P. A. (1960): "Aerologicheskikh issledovaniia pogranichnogo sloia atmosfery (Aerological studies of the boundary layer of the atmosphere)." Gidrometeoizdat, Leningrad, 1–451.

Vorontsov, P. A. (1961): "Metody aerologicheskikh issledovaniia progranichnogo sloia atmosfery (Methods of aerological studies of the boundary layer of the atmosphere)." Gidrometeoizdat, Leningrad, 1–220.

Wagner, R. (1966): Die Temperatur des Bodens, des Wassers und der Luft in Kopáncs. *Acta Climat.*, (6), 3–52.

Warner, J. (1972): The structure and intensity of turbulence in air over the sea. *Q.J.R.M.S.*, **98**, 175–186.

Watanabe, S., Ozeki, Y. (1964): Study of fallen snow on forest trees, II. Experiment on the snow crown of the Japanese cedar. *Bull. Gov. Forest Exp. Sta., Meguro*, (169), 121–139.

Webb, E. K. (1965): Aerial microclimate. *Met. Monogr.*, **6** (28), 27–58.

Weickmann, H. (1972): Man-made weather patterns in the Great Lakes Basin. *Weatherwise*, **25** (6), 260–267.

van Wijk, W. R., de Wilde, J. (1962): Microclimate, the problems of the arid zone. *UNESCO, Arid. Zone Research Series*, **18**, 83–113.

van Wijk, W. R. (1965): Soil microclimate, its creation, observation and modification. *Met. Monogr.*, **6** (28), 59–73.

Wilkins, E. T. (1954): Air pollution aspects of the London fog of December 1952. *Q.J.R.M.S.*, **80**, 267–271.

Woelfle, M. (1939, 1942): Windverhältnisse im Walde, I, II, III. *Forstw. Cbl.*, **61**, 65–75, 461–475; **64**, 169–182.

Woelfle, M. (1950): "Waldbau und Forstmeteorologie." 2. Aufl. Bayrischer Landwirtschaftsverlag, München, 1–68.

Wolfe, J. N., Wareham, R. T., Scofield, H. T. (1943): The microclimates of a small valley in Central Ohio. *Trans. Amer. Geoph. Union*, **24**, 154–166.

Wolfe, J. N., Wareham, R. T., Scofield, H. T. (1949): Macroclimates and microclimates of Neotoma, a small valley in Central Ohio. *Bull. Ohio Biol. Survey*, **8** (1), 1–267.

Yamamoto, G. (1957): Estimation of additional downward radiation from aerosols over large cities. *75th Ann. Vol. Jour. Met. Soc. Japan*, 1–4.

Yamamoto, G., Kondo, J. (1964): Evaporation from Lake Towada. *Jour. Met. Soc. Japan*, **42** (2), 85–96.

Yamamoto, G., Yasuda, N., Shimanuki, A. (1968): Effect of thermal stratification on the Ekman layer. *Jour. Met. Soc. Japan*, **46** (6), 442–455.

Yamashita, I. (1953): Lake wind on Lake Suwa. *Jour. Met. Res.*, **5**, 701–702.

Yamashita, S. (1973): Air pollution study from measurements of solar radiation. *Arch. Met. Geoph. Biokl.*, (B) **21**, 243–253.

Yamashita, S. (1975): A comparative study of turbidity in an urban and a rural environment at Toronto. *Atmosph. Env.* (to be published)

Yokoi, S. (1953): Report on the experiment of wind velocity distribution in Shizuoka City. *Kasai-no-kenkyû*, (2), 47–49.

Yokoyama, O. (1971): An experimental study on the structure of turbulence in the lowest 500m of the atmosphere and diffusion in it. *Rep. Nat. Res. Inst. for Pollution and Res.* (2).

Yoshino, M. M. (1953): Local climate of snow accumulation and snowfall on the coastal region. *Geog. Rev. Japan*, **26**, 475–485.

Yoshino, M. M. (1955): Wind direction distribution in the city area of Tokyo. *Tenki*, **2**, 203–207.

Yoshino, M. M. (1957 a): Distribution of rainfall and increase of number of days with drizzle in the Tokyo area. *75th Ann. Vol. Tenki*, 121–125.

Yoshino, M. M. (1957 b): The structure of surface winds over a small valley. *Jour. Met. Soc. Japan*, **35**, 184–195.

Yoshino, M. M. (1957 c): Some aspects on the distribution of the surface winds within a small area. *75th Ann. Vol. Jour. Met. Soc. Japan*, 365–371.

Yoshino, M. M. (1958): Wind speed profiles of the lowest air layer under influences of micro-topography. *Jour. Met. Soc. Japan*, **36**, 174–186.

Yoshino, M. M. (1961 a): "Shôkikô (microclimate)." Chijin- shokan, Tokyo, 1–274.

Yoshino, M. M. (1961 b): Local climate in the Mt. Kirishima region. *Tokyo Geog. Papers*, **5**, 79–109.

Yoshino, M. M. (1968): A climatological estimation of distribution of air pollution in Japan. *Tenki*, **15** (1), 9–14.

Yoshino, M. M., Urushibara, K., Owada, M. (1970): Lake-breeze of Lake Suwa, Nagano Pref., Central Japan. *Tenki*, **17** (2), 55–62.

Yoshino, M. M., Hoshino, M., Owada, M. (1972 a): On the local wind distribution and the wind break density in the Shari-Abashiri region, Hokkaido. *Jour. Agricul. Met.*, **27** (4), 145–152.

Yoshino, M. M., Kai, K. (1973): Change of air temperature in Japanese cities in the recent years and its relation to the synoptic patterns and population. *Tenki*, **20** (9), 489–497.

Yoshino, M. M., Kudo, T., Hoshino, M. (1973): On the land and sea breeze on the Japan Sea Coast. *Geog. Rev. Japan*, **46** (3), 205–210.

Zimmermann, H. (1953): Kleinklimatische Ozonmessungen in Bad Kissingen. *Mitt. D.W.*, (1), 1–13.

Chapter 4

Abe, M. (1937): Distribution and movement of clouds around Mt. Fuji studied through photographs. *Bull. Centr. Met. Obs. Japan*, **6** (1), 1–466.

Ageta, Y. (1971): Some aspects of the weather conditions in the vicinity of the Mizuho Plateau, East Antarctica. *Antarctic Rec., Tokyo*, (41), 42–61.

Ageta, Y. (1972): Some aspects of ablation of sastrugi and drifts in the area subjected to katabatic winds in Antarctica. *Antarctic Rec., Tokyo*, (43), 8–19.

Aichele, H. (1951): Frostgefährdete Gebiete in der Baar, eine kleinklimatische Geländekartierung. *Erdkunde*, **5**, 70–73.

Aichele, H. (1953 a): Kaltluftpulsationen. *Met. Rdsch.*, **6**, 53–54.

Aichele, H. (1953 b): Lokalklimatische Froststudien am westlichen Bodensee. *Met. Rdsch.*, **6**, 126–130.

Aichele, H. (1953 c): Der Beginn der Apfelblüte 1953 am westlichen Bodensee als Hilfsmittel der kleinklimatischen Gelädekartierung. *Met. Rdsch.*, **6**, 204–206.

Aichele, H. (1964): Beitrag zur Festlegung phänologischer Jahreszeiten und deren Höhenabhängigkeit in Rheinland-Pfalz. *Met. Rdsch.*, **17** (2), 42–46.

Aigner, S. (1952): Die Temperaturminima im Gstettnerboden bei Lunz am See, Niederösterreich. *Wett. Leben*, Sonderheft 1, 34–37.

Albrecht, F. (1962): Die Berechnung der natürlichen Verdunstung (Evapotranspiration) der Erdoberfläche aus klimatologischen Daten. *Ber. Deutsch. Wett.*, (83), 1–19.

Alissow, B. P., Drosdow, O. A., Rubinstein, E. S. (1956): Lehrbnch der Klimatologie. Deutscher Verlag der Wissenschaft, Berlin, 1–536.

Arai, T., Sekine, K. (1973): Study on the formation of perennial snow patches in Japan. *Geog. Rev. Japan*, **46** (9), 569–582.

Arakawa, H. (1955): Three unusual depths of snowfall in Japan. *Q.J.R.M.S.*, **81**, 99–101.

Asakawa, T. (1950): Fog at the mouth of the Sai River, Kawanakajima. *Shinano*, **2**, 742–757.

Aubert de la Rüe, E. (1955): "Man and the winds." Hutchinson's Sci. Tech. Publ., London, 1–206.

Aulitzky, H. (1955): "Über die lokalen Windverhältnisse einer zentralalpinen Hochgebirgs-Hangstation. *Arch. Met. Geoph. Biokl.* (B), **6**, 353–373.

Aulitzky, H. (1961, 1962a, 1962b): Die Bodentemperaturvehältnisse an einer zentralalpinen Hanglage beiderseits der Waldgrenze (I), (II), (III). *Arch. Met. Geoph. Biokl.* (B), **10**, 445–532; **11**, 301–362; **11**, 363–376.

Aulitzky, H. (1963): Grundlagen und Anwendung des vorläufigen Wind-Schnee-Ökogrammes. *Mitt. d. Forstl. Bundes-Vers. Mariabrunn*, (60), 763–834.

Aulitzky, H. (1967): Lage und Ausmass der "warmen Hangzone" in einem Quertal der Innenalpen. *Ann. Met.*, (3), 159–165.

Aulitzky, H. (1968): Die Lufttemperaturverhältnisse einer zentralalpinen Hanglage. *Arch. Met. Geoph. Biokl.* (B), **16**, 18–69.

Balchin, W. G. V., Pye, N. (1947): A micro-climatological investigation of Bath and the surrounding district. *Q.J.R.M.S.*, **73**, 297–319.

Balchin, W. G. V., Pye, N. (1948): Local rainfall variation in Bath and the surrounding district. *Q.J.R.M.S.*, **74**, 361–378.

Balchin, W. G. V., Pye, N. (1950): Observations on local temperature variations and plant responses. *Jour. Ecol.*, **38** (2), 345–353.

Barry, R. G. (1972): Climatic environment of the east slope of the Colorado Front Range. Inst. Arctic and Alpine Res. Occas. Paper, (3), 1–206.

Baumgartner, A. (1958): Nebel und Nebelniederschlag als Standortsfaktoren am Großen Falkenstein (Bayerischer Wald). *Fortsw. Cbl.*, **77** (9/10), 257–272.

Baumgartner, A. (1960): Gelände und Sonnenstrahlung als Standortsfaktor am Gr. Falkenstein (Bayerischer Wald). *Forstw. Cbl.*, **79**, 286–297.

Baumgartner, A. (1960, 1961, 1962): Die Lufttemperatur als Standortsfaktor am G. Falkenstein. I, II, III. *Forstw. Cbl.*, **79** (11/12), 362–373, **80** (3/4), 107–120; **81** (1/2), 17–47.

Baumgartner, A. (1963 a): Wärmeumsätze des Bodens und -der Pflanze. In: Frostschutz im Pflanzenbau. Herg. F. Schnelle, BLV., München, 82–150.

Baumgartner, A. (1963 b): Einfluss des Geländes auf Lagerung und Bewegung der nachtlichen Kaltluft. In: Frostschutz und Pflanzenbau. Herg. F. Schnelle, BLV., München, 151–194.

Bergeron, T. (1949): The problem of artificial control of rainfall on the globe. II, The coastal orographic maxima of precipitation in autumn and winter. *Tellus*, **1** (1), 15–32.

Bernot, F. (1959): Temperaturinversion im unteren Teil des Ljubljana-Beckens. *Ber. Deutsch. Wett.*, (54), 194–196.

Best, A. C. (1951): Drop-size distribution in cloud and fog. *Q.J.R.M.S.*, **77**, 418–426.

Billwiller, R. (1914): Der Walliser Talwind. *Ann. Schweiz. Met. Zentr. Anst.*, **50**, 1913 (7), 1–13.

Bleibaum, I. (1953): Studien zur Meteorologie der südlichen Rhön. *Ber. Deutsch. Wett.*, **4**, 1–15.

Bluskowa, D. (1965): Über den Einfluss des Reliefs auf die Lufttemperatur. *Ang. Met.*, **5**, 65–72.

Böer, W. (1952): Über den Einfluss des Lokalklimas auf die Mitteltemperatur. *Ber. Deutsch. Wett.*, (42), 217–220.

Böhm, H. (1966): Die geländeklimatische Bedeutung des Bergschattens und der Exposition für das Gefüge der Natur-und kulturlandschaft. *Erdkunde*, **20** (2), 81–93.

Boros, J. (1971): Angaben zur Bordentemperatur E- und W-exponierter Hänge. *Acta Climat.*, **10** (1–4), 47–56.

Bouët, M. (1961): Le vent en Valais (Suisse). *Mémoires Soc. Vaudoise, Sci. Nat.*, (79), **12** (7), 277–

352.

Buettner, K. J. K., Thyer, N. (1966): Valley winds in the Mount Rainier area. *Arch. Met. Geoph. Boikl.* (B), **14** (2), 125–147.

Burckhardt, H. (1963 a): Meteorologische Voraussetzungen der Nachtfröste. In: Frostschutz im Pflanzenbau. Herg. F. Schnelle, BLV., München, 13–81.

Burckhardt, H. (1963 b): Kleinklimatische Kartierung nach Frostgefährdung und Frostschaden. In: Frostschutz im Pflanzenbau. Herg. F. Schnelle, BLV., München, 195–268.

Cagliolo, A. (1951): Estudio microclimatico de pendientes en el sudeste de la provincia de Buenos Aires (Dionisia). *Meteoros*, **1**, 134–149.

Cantlon, J. E. (1953): Vegetation and microclimates on north and south slopes of Cushetunk Mountain, New Jersey. *Ecol. Monogr.*, **23**, 241–270.

Central Meteorological Observatory (1952): Iso hara chiku uryôchosa. (Investigations of rainfall in the Isohara region). 1–69.

Collinge, V. K., Jamieson, D. G. (1968): The spacial distribution of storm rainfall. *Jour. Hyd.*, **6** (1), 45–57.

Conrad, V. (1935): Beitrage zur Kenntnis der Schneedeckenverhältnisse (3). *Gerl. Beitr. Geoph.*, **45**, 225–236.

Conrad, V. (1936): Die klimatologischen Elemente und ihre Abhängichkeit von terrestrischen Einflüssen. Handbuch der Klimatologie. Bd. Ib., 143–144.

Czelnai, R. (1972): Confidence levels of monthly areal rainfall depths estimated from sample-points in Hungary. Központi Met. Intézet, Budapest, 1–83.

Dammann, W. (1960): Die Windverhältnisse in Rhein-Main Gebiet, Eine Studie zur dynamischen Klimatologie der Mittelgebirge. *Erdkunde*, **14**, 10–29.

Defant, F. (1949): Zur Theorie der Hangwinde, nebst Bemerkungen zur Theorie der Berg-und Talwinde. *Arch. Met. Geoph. Biokl.* (A), **1**, 421–250.

Defant, F. (1951): Local winds. *Comp. Met.*, 655–672.

DeMarrais, A., Downing, G. L., Meyer, H. E. (1968): Transport and diffusion of an aerosolized insecticide in mountainous terrain. *ESSA Res. Lab. Tech. Mem. ARL*, (6), 1–46.

Diamond, M. (1960): Air temperature and precipitation on the Greenland Ice Sheet. *Jour. Glac.*, **3**, 558–567.

Dimitrov, D., Vekilska, B. (1966/1967/1968): The influence of the Balkan Mountain on the genesis regime and distribution of rainfall in the sub-Balkan valleys. Godišvnik na Sofijskija universitet, Geologo-geografski fakultet, geografia, **61** (2), 95–115.

Domrös, M. (1968): Über die Beziehung zwischen äquatorialen Konvektionsregen und der Meereshöhe auf Ceylon. *Arch. Met. Geoph. Biokl.* (B), **16**, 164–173.

Domrös, M. (1970): Frost in Ceylon. *Arch. Met. Geoph. Biokl.* (B), **18**, 43–52.

Drewes, W. U., Drewes, A. T. (1957): Climate and related phenomena of the eastern Andean slopes of Central Peru. Syracuse Univ. Res. Inst., New York, 1–85.

Doumani, G. A. (1967): Surface structures in snow. In: Physics of snow and ice. Sapporo Conf. 1966. ed. by H. Oura, Inst. Low Temp. Sci., Hokkaido Univ., 1119–1136.

Ebara, S. et al. (1915, 1917): Meteorological observation on Mt. Nantai. *Shinrin-sokkôjo-tokubetsu-hôkoku*, 2:79–116, 4:107–131.

Eimern, J. v. (1951): Kleinklimatische Geländeaufnahme in Quickborn Holstein. *Ann. Met.*, **4**, 259–269.

Eimern, J. v., Kaps, E. (1954): Lokalklimatische Untersuchungen im Raum der Harburger Berge und der benachbarten Elbniederung. *Landwirtschaft-Angewandte Wissenschaft*, 1–111.

Eimern, J. v. (1955): Über eine Windbeeinflussung durch die Randhöhen des Elbtals bei Hamburg. *Met. Rdsch.*, **8**, 97–99.

Eimern, J. v. (1964): Über den jahreszeitlichen Gang der geländeklimatisch bedingten Differenzen der nächtlichen Minimumtemperatur. *Agricul. Met.*, **1**, 149–153.

Eimern, J. v. (1968): Some experiences obtained in mapping the local wind pattern. Proc. WMO Reg. Training Sem. on Agrometeorology, Wageningen, Netherland, 303–317.

Ekhart, E. (1950): Zur Bewölkungsklimatologie der Alpen. *Geog. Ann.*, **32**, 21–36.

Elomaa, E. (1970): Pinnanmuotojen vaikutus lömpötilaoloihin Lammin Untulanharjulla kesällä 1968 (The influence of topography on the air temperature on an eskar at Lammi (S-Finland) in the summer of 1968). *Terra*, **82** (3), 97–107.

Fleagle, R. G. (1950): A theory of air drainage. *Jour. Met.*, 7, 227–232.

Flemming, G. (1964): Zur Windbeständigkeit in Sachsen. *Zeitsch. Met.*, 17, 264–247.

Flemming, G. (1966 a): Wind, Richtungsböigkeit und Rauchausbreitung in Tal-und Hochflächenlage. *Zeitsch. Met.*, 18 (8–10), 330–351.

Flemming, G. (1966 b): Zum Einfluss unterschiedlicher grossräumiger Strömungsrichtung und thermischer Stabilität auf die lokalen Windrichtungsverhältnisse in Sachsen. *Zeitsch. Met.*, 19 (1/2), 34–43.

Flemming, G. (1967 a): Die morphogene Schicht der Atmosphäre. *Zeitsch. Met.*, 19 (7/8), 232–238.

Flemming, G. (1967 b): Concerning the effect of terrain configuration on smoke dispersal. *Atmosph. Env.*, 1, 239–252.

Fletcher, R. D. (1951): Hydrometeorology in the United States. *Comp. Met.*, 1033–1049.

Flint, R. F. (1957): "Glacial and Pleistocene geology." John Wiley, N.Y., 1–533.

Fliri, F. (1967 a): Beitäge zur Kenntnis der Zeit-Raum-Struktur des Niederschlags in den Alpen. *Wett. Leben*, 19, 241–268.

Fliri, F. (1967 b): Beiträge zur Kenntnis zeitlichen und räumlichen Verteilung des Niederschages in den Alpen in der Periode 1931–1960. *Int. Tag. Alpine Met.*, (4), 72–79.

Fliri, F. (1967 c): Über die klimatologische Bedeutung der Kondensationshöhe im Gebirge. *Die Erde.*, 98 (3), 203–210.

Fliri, F. (1971): Neue klimatologische Querprofile der Alpen—ein Energiehaushalt. *Ann. Met.*, (5), 93–97.

Flohn, H. (1968): Ein Klimaprofil durch die Sierra Nevada de Mérida (Venezuela). *Wett. Leben*, 20, 181–191.

Flohn, H. (1969): Local wind systems. World Survey of Climatology. II. General Climatology, (2), 139–171.

Flohn, H. (1970): Comments on water budget investigations, especially in tropical and subtropical mountain regions. World Water Balance (Proc. Reading Symp.), 251–262.

Flohn, H. (1971): Beiträge zur vergleichenden Meteorologie der Hochgebirge. *Ann. Met.*, (5), 9–16.

Förchtgott, J. (1969): Winter thunderstorms in Central Europe. *Weather*, 24 (11), 448–452.

Franken, E. (1955/56): Unterschiedliche Frostgefährdung im norden Hamburg. *Ann. Met.*, 7, 135–148.

Franken, E. (1959): Über eine Abhängigkeit der Temperaturverteilung in Strahlungsnächten von Geländeformung und Windrichtung. *Met. Rdsch.*, 12, 25–31.

Franken, E. (1962): Über den Geländeeinfluss auf Windrichtung und Windgeschwindigkeit im Raum Hamburg. *Deutscher Wetterdienst Seewetteramt*, (34), 1–15. +28Abb.

Frenkiel, J. (1962): Wind profiles over hills (in relation to windpower utilization). *Q.J.R.M.S.*, 88, 156–169.

Friedel, H. (1961): Schneedeckendauer und Vegetationsverteilung im Gelände. Ökologische Untersuchungen in der subalpinen. *Mitt. d. Forstl. Bundes-Vers. Mariabrunn*, (59), 317–369.

Fuh, B.-p. (1962): The influence of slope orientation on the microclimate. *Acta Met. Sinica*, 32 (1), 71–86.

Fuh, B.-p. (1963): The wind speed in valleys. *Acta Met. Sinica*, 33 (4), 518–526.

Fujimoto, J. et al. (1971): Geography of Ohzu basin. Ehime-kôkô-chiri-kyôdô-chôsa-hôkoku, 1–76.

Fujimura (Huzimura), I. (1948): On the relation between rainfall and altitude on Mt. Fuji and in neighbourhood of it. *Geoph. Mag.*, 20, 113–126.

Fujimura, I. (1950): On the relation between rainfall and altitude on Mt. Fuji and in the neighbourhood of it. *Geoph. Mag.*, 21, 190–198.

Fujimura, I. (1952): Altitude and precipitation. *Tokyo-kanku-kishô-kenkyûkaishi*, (11), 299–302.

Fujimura, I. (1971): Meteorology of Mt. Fuji. In: Fujisan-sôgô-gakujutsuchôsa-hôkoku. Fujikyu, 215–304.

Fujita, T. et al. (1968): Aerial measurement of radiation temperatures over Mt. Fuji and Tokyo areas and their application to the determination of ground- and water-surface temperatures. *Jour. Appl. Met.*, 7 (5), 801–816.

Fukuoka, Y. (1966): Lapse rate of soil temperature with altitude. *Jour. Agricul. Met.*, 21 (4), 145–147.

Garnett, A. (1935): Insolation, topography and settlement in the Alps. *Geog. Rev.*, 25, 601–617.

Garnett, A. (1937): Insolation and relief——Their bearing on the human geography of the alpine regions. London.

Garnier, B. J. (1968): Estimating the topographic variation of direct solar radiation: A contribution to geographical microclimatology. *Canad. Geog.*, **12** (4), 241–248.

Garnier, B. J., Ohmura, A. (1968): A method of calculating the direct short wave radiation income of slopes. *Jour. Appl. Met.*, **7** (5), 796–800.

Garnier, B. J., Ohmura, A. (1970): The evaluation of surface variations in solar radiation income. *Solar Energy*, **13**, 21–34.

Gavrilova, M. K. (1966): Radiation climate of the Arctic. Trans. from Russian National Sci. Foundation, Washington, D.C., 1–178.

Gavrilović, D. (1970): Die Überschwemmungen im Wadi Bardagué im Jahr 1968 (Tibesti, Rep. du Tchad). *Zeitsch. Geomorph.*, (2), 202–218.

Geiger, R., Woelfle, M., Seip, L. P. (1933–1934): Höhenlage und Spätfrostgefährdung. *Forstw. Cbl.*, **55**, 579–592; 737–746, **56**, 141–151; 221–230, 253–260, 357–374, 465–484.

Geiger, R. (1961): "Das Klima der bodennahen Luftschicht." Friedr. Vieweg & Sohn, Braunschweig, 1–646.

Geiger, R. (1965): "The climate near the ground." Harvard University Press, Cambridge, 1–611.

Gelmgelts, N. F. (1963): Gorno-dolinnaia tzirkuliatsiia severnykh sklonov Tian-Shania (Mountain-valley circulation of the northern slopes of Tien Shan). Gidrometeoizdat, Leningrad, 1–328.

Georgii, W. (1923): Die Luftströmung über Gebirgen. *Met. Zeitsch.*, (4), 108–112.

Gifford, F., Jr. (1953): A study of low level air trajectories at Oak Ridge, Tenn. *Mon. Wea. Rev.*, **81**, 179–192.

Gleeson, T. A. (1953): Effects of various factors on valley winds. *Jour. Met.*, **10**, 262–269.

Gocho, Y., Nakajima, C. (1971): On the rainfall around the Suzuka Mountains. *Ann. Disaster Prev. Res. Inst.*, (14)B, 103–118.

Godske, C. L. (1953): Studies in local meteorology and representativeness. *Univ. Bergen Årbok 1952, Nat. rekke*, (10), 1–99.

Goldreich, Y. (1971): Influence of topography on Johannesburg's temperature distribution. *S. African Geog. Jour.*, **53**, 84–88.

Grunow, J. (1960): The productiveness of fog precipitation in relation to the cloud droplet spectrum. *Amer. Geoph. Union Monogr.*, (5), 110–117.

Grunow, J. (1963): Weltweite Messungen des Nebelniederschlags nach der Hohenpeissenberger Methode. Int. Hyd. Conf. IUGG, IAHS, Berkeley, 1–17.

Grunow, J., Tollner, H. (1969): Nebelniederschlag im Hochgebirge. *Arch. Met. Geoph. Biokl*, **17** (B), 201–228.

Gugiuman, I. et al. (1970): Profils topoclimatiques dans les Carpates orientales. *Prace Geog.*, **26**, 279–287.

Hann, J. v., Knoch, K. (1932): "Handbuch der Klimatologie." I Bd. 252–256.

Harrison, A. A. (1971): A discussion of the temperatures of inland Kent with particular reference to night minima in the lowlands. *Met. Mag.*, **100** (1185), 97–111.

Hartmann, F. K., Eimern, J. v., Jahn, G. (1959): Untersuchungen reliefbedingter kleinklimatisther Fragen in Geländequerschnitten der hochmontanen und montanen Stufe des Mittel-und Südwestharzes. *Ber. Deutsch. Wett.*, (50), 1–39.

Hastenrath, S. (1967): Rainfall distribution and regime in Central America. *Arch. Met. Geoph. Biokl.*, **15** (B), 202–241.

Hawke, E. L. (1944): Thermal characteristics of a Hertfordshire frost hollow. *Q.J.R.M.S.*, **64**, 23–48.

Heigel, K. (1955): Exposition und Höhenlage in ihrer Wirkung auf die Pflanzenentwicklung. *Met. Rdsch.*, **8** (9/10), 146–148.

Heigel, K. (1967): Talnebel und Taldunst aus der Sicht des Hohenpeissenbergs als Beitrag zur Klimatologie des Alpenvorlandes. *Met. Rdsch.*, **20** (3), 77–83.

Hendl, M. (1966): Grundriss einer Klimakunde der Deutschen Landschaften. B. G. Teubner, Leipzig 1–95.

Henning, D. (1967): Mt. Waialeale. *Wett. Leben*, **19**, 93–100.

Henning, I. und D. (1967): Abbildung lokaler Zirkulationen durch Wolkenfelder auf Hawaii. *Met. Rdsch.*, **20** (4), 109–114.

Henry, A. J. (1919): Increase of precipitation with altitude. *Mon. Wea. Rev.*, **47**, 33–41.

Henz, J. F. (1972): An operational technique of forecasting thunderstorms along the lee slopes of mountain range. *Jour. Appl. Met.*, **11** (8), 1284–1292.

Hess, M. (1962): Influence of snow and ice cover upon the radiation balance and the microclimate of mountains. *Prace Geog.*, (5), 1–154.

Hess, M. (1965): Occurrence of different daily mean temperature values in particular vertical climatic zones of the west Polish Carpathian Mts. *Przegl. Geof.*, **10**, 257–270.

Hess, M. (1967): Wpływ lodowców górskich na klimat na przykładzie lodowca Fedczenki w Pamirze. *Przegl. Geog.*, **39** (4), 743–774.

Hess, M. (1971): Studien über die quantitative Differenzierung der makro-, meso-, mikro-klimatischen Verhältnisse der Gebirge als Grundlage zur Konstr···ierung detaillierten Klimakarten. *Prace Geog.*, **26**, 265–277.

Higuchi, K., Fujii, Y. (1971): Permafrost at the summit of Mount Fuji, Japan. *Nature*, **230** (5295), 521.

Hinds, W. T. (1970): Diffusion over coastal mountains of Southern California. *Atmosph. Env.*, **4** (2), 107–124.

Hirata, T., Tamate, M. (1922): Results of meteorological observation on Mt. Nantai. *Shinrin-sokkôjo-tokubetsu-hôkoku*, (8), 1–108.

Hočevar, A. (1961): Die Grundzüge der frostfreien Periode des südöstlichen Alpenrandes und ihre Abhängigkeit von der Seehöhe. *Met. Tagung, Jubil. 75-jähr. Best. Observator. Sonnblick*, 128–131.

Höhndorf, F. (1952): Zur Böigkeit auf dem Hohenpeissenberg. *Ber. Deutsch. Wett. US-Zone*, (38), 394–396.

Hoinkes, H., Untersteiner, N. (1952): Wärmeumsatz und Ablation auf Alpengletschern, I. *Geog. Ann.*, **24**, 99–158.

Hoinkes, H. (1953): Wärmeumsatz und Ablation auf Alpengletschern II. *Geog. Ann.*, **25**, 116–123.

Hoinkes, H. (1955): Measurements of ablation and heat balance on Alpine glaciers. *Jour. Glac.*, **2**, 497–501.

Holland, J. Z. (1952): The diffusion problem in hilly terrain. In: Air Pollution. ed. by McCabe, McGraw-Hill, New York, 815–821.

Holmgren, B. (1971): Wind- and temperature-field in the low layer on the top plateau of the ice cap. In: Climate and energy exchange on a sub-polar ice cap in summer. *Uppsala Universitet, Met. Inst., Meddel.*, (108), 1–43.

Holtmeier, F.-K. (1966): Die "Malojaschlange" und die Verbreitung der Fichte. *Wett. Leben*, **18**, 105–108.

Hoshiai, M., Kobayashi, K. (1957): A theoretical discussion on the so-called "snow-line" with reference to the temperature reduction during the last glacial age in Japan. *Japanese Jour. Geol. Geog.*, **28**, 62–75.

Huff, F. A., Stout, G. E. (1952): Area-depth studies for thunderstorm rainfall in Illinois. *Trans. Amer. Geoph. Union*, **33**, 495–498.

Huttenlocher, F. (1923): Sonner- und Schattenlage. Ihr Klima und ihr Einfluss in den Alpen sowie im Schwaben-und Frankenland. *Erdgeschichte and landeskundl. Abh. aus Schwaben u. Franken. Herausg. Geol. Geog. Inst. Univ. Tübingen*, (7), 1–62.

Investigation Committee of the Matsumoto Fukashi Senior High School (1969): An accident caused by lightning on Doppyo Peak at Mt. Nishihotaka in the North Japan Alps. Matsumoto, 1–336.

Ishida, T., Yamamoto, S. (1960): On the snow avalanche on Mt. Fuji. *Jour. Japanese Soc. Snow and Ice*, **22**, 28–36.

Ishihara, Y., Kobatake, S. (1971, 1972): On the water balance in Ara experimental basin (1), (2). *Ann. Disaster Prev. Res. Inst.* (14) B, 131–142; (15) B, 321–331.

Ishimaru, Y. (1952): "Nephology." Hokuryukan, Tokyo, 129–137.

Ives, R. L. (1941): Colorado Front Range crest clouds and related phenomena. *Geog. Rev.*, **31**, 23–45.

Kaiser, H. (1954): Über die Stromungsverhältnisse im Bergland. *Met. Rdsch.*, **7**, 214–217.

Kaps, E. (1955): Zur Frage der Durchlüftung von Tälern im Mittelgebirge. *Met. Rdsch.*, **8**, 61–65.

Kato, T. (1922): Results of rainfall observation on Mt. Haruna. *Shinrin-sokkôjo-tokubetsu-hôkoku*, (8), 109–136.

Kauf, H. (1950): Die Einwirkung der Orographie des mittleren Saaletales auf die Niederschlagsverteilung. Teil II. *Mitt. Thüring. Landeswetterwarte*, (10), 35–62.

Kern, H. (1951 a): Untersuchung lokaler Beeinflussungen der nächtlichen Tiefsttemperaturen in einem Moorgebiet. *Landwirtsch. Jahrb. f. Bayern*, **28** (1/2), 3–13.

Kern, H. (1951 b): Kleinklimakartierung zur Erfassung der bodennahen nächtlichen Tiefsttemperaturen im Donaumoos. *Mitt. f. Moor-und Torfwirtschft. Beilage z. Landwirtsch. Jahrb. f. Bayern*, **28** (4), 53–58.

Kinoshita, S. (1952): The drop size distribution of a mountain fog. In: Studies on fog prevention forests, (2), 235–237.

Knudsen, J. (1961): Mountain and valley winds from a synoptical point of view. In: Local and synoptic meteorological investigations of the mountain and valley wind system. Forsvarets Forskningsinstitutt, Norw. Defence Res. Establ., Intern rapport K-242, 75–139.

Kobayashi, T. (1960): Errors of average areal rainfall in relation to number of raingauge and calculation method. *Denryoku-kishô-renrakukaihô*, **11**, 13–25.

Koch, H. G. (1955): Hanggewitter, Hangerwarmung und Hangströmung am Thüringerwald. *Pet. Geog. Mitt.*, **99**, 107–109.

Kodama, R. (1954): On the örographic rain on the slopes of Mt. Ibuki (I). *Jour. Met. Res.*, **6**, 333–338.

Konček, M. (1959): Schneeverhältnisse der Hohen Tatra. *Ber. Deutsch. Wett.*, (54), 132–133.

Konček, M. (1960): Zur Frage der Nebelfrostablagerungen im Gebirge. *Studia Geoph. Geod.*, **4**, 69–84.

Konček, M., Briedon, V. (1964): "Sneh a snehová pokrývka na Slovensku." Vydavatelstvo Slovenskeg akadémie vied, Bratislava, 1–71.

Kondratyev, K. Ya. (1965): "Radiative heat exchange in the atmosphere." Trans. from Russian by O. Tedder, Pergamon Press, Oxford, 1–411.

Kosiba, A. (1954): The snowfalls in Silesia. *Acta Geoph. Polonica*, **2**, 136–139.

Kreutz, W., Schubach, K. (1952): Lokalklimatische Geländekartierung der südlichen Bergstrasse unter besonderer Berücksichtigung der Gemarkung Heidelberg. *Mitt. D. W. US-Zone*, (13), 1–11.

Kreutz, W. (1955): Tätigkeitsberichte der agrarmeteorologischen Dienststellen, Giessen. *Mitt. D. W.*, (14), 40–65.

Kreutz, W., Schubach, K., Walter, W. (1955): Untersuchungen über die geländeklimatischen Verhältnisse der Gemarkung Dieburg und über Wind- und Temperaturverhältnisse in den Gemarkungen Hausen und Langenseifen im Untertaunuskreis mit der Zielsetzung ihrer Verbesserung durch Windschutzanlagen. *Arch. f. Raumf. Hessen.*, (1), 1–46.

Kreutz, W., Schubach, K. (1960): Lokalklimatische Untersuchungen im Ohmbecken und die Frage der Windschutzbedürftigkeit. *Inst. f. Naturschutz Darmstadt*, **4**, 104–110.

Kronfuss, H. (1972): Kleinklimatische Vergleichsmessungen an zwei subalpinen Standorten. *Mitt. Forstl. Bundes-Vers. Wien*, (96), 159–176.

Lauscher, F. et al. (1932): Ein Profil der Sonnenstrahlungsintensität durch die steirisch-niederösterreichischen Kalkalpen. *Met. Zeitsch.*, (8), 300–306.

Lauscher, F. (1934): Weitere Studien über die Sonnenstrahlungsintensität in den steirisch-niederösterreichischen Kalkalpen. *Met. Zeitsch.*, (9), 336–341.

Lauscher, F. (1962–1964): Die Tagesschwankung der Lufttemperatur auf Höhenstationen in allen Erdteilen. 60–62. Jahrb. d. Sonnblick-Vereines für 1962–1964, 3–17.

Lawrence, E. N. (1956): Minimum temperatures and topography in a Herefordshire valley. *Met. Mag.*, **85**, 79–83.

Lee, R. (1962): Theory of the "equivalent slope". *Mon. Wea. Rev.*, **90**, 165–166.

Lee, R., Baumgartner, A. (1966): The topography and insolation climate of a mountainous forest area. *Forest Sci.*, **12** (3), 258–267.

Linsley, R. K., Kohler, M. A. (1951): Variations in storm rainfall over small area. *Trans. Amer. Geoph. Union*, **32**, 245–250.

Loewe, F. (1974): Polar dry deserts. *Bonner Met. Abh.*, (17), 195–208.

Lomas, J. (1968): The effect of the climate on agricultural production in the upper Jordan Valley. Agroclimatological Methods (Proc. Reading Symp.) UNESCO, 261–268.

MacHattie, L. B., McCormack, R. J. (1961): Forest microclimate: a topographic study in Ontario. *Jour. Ecol.*, **49**, 301–323.

MacHattie, L. B. (1968): Kananaskis valley winds in summer. *Jour. Appl. Met.*, **7** (3), 348–352.

Mäde, A. (1956): Über die Methodik der meteorologischen Geländevermessung. *Deutsch. Akad. d. Landwirtschaftswissenschaften zu Berlin, Sitz. Ber.*, **5**, (5), 1–25.

Maisel, Ch. (1931): Der Einfluss der kontinentalen Lage auf die Jahresschwankung der Monatsmittel der Lufttemperatur im Deutschen Reich. *Heimatk. Arbeiten. Geog. Inst. Univ. Erlangen*, (5), 1–50.

Manley, G. (1944): Topographical features and the climate of Britain: A review of some outstanding effects. *Geog. Jour.*, **103**, 241–258.

Maruyasu, T. et al. (1968): Analysis of snow depth distribution on aerial photographs. *Doboku-gakkai-ronbunshû*, (153), 41–54.

Maruyasu, T. et al. (1969): Analysis of snow melt and runoff on aerial photographs. *Doboku-gakkai-ronbunshû*, (164), 59–69.

Masatsuka, A. (1960, 1962): A comparative study of meteorological and hydrological rainfalls on the windy slope near the summit (1) (2). *Jour. Met. Res.*, **12**, 681–690, **14**, 65–71.

Masatsuka, A. (1964): Some hydrometeorological studies on rainfall and runoff in mountainous drainage area. *Jour. Met. Res.*, **16** (1), 1–50.

Masatsuka, A. (1965): Studies on the analysis of water balance in the upper stream basins. *Kishô-kenkyû-nôto*, **16** (2), 358–369.

Mendonca, B. G. (1969): Local wind circulation on the slopes of Mauna Loa. *Jour. Appl. Met.*, **8** (4), 533–541.

Michalczewski, J. (1962): Dlugotrwale zastoiska mrozowe Kotliny Podhalanskiej. *Acta Geog. Lodziendzia*, (13), 27–70.

Miller, D. H. (1956): Landscape and climate: Topographic influences on heat supply at the earth's surface. *Ann. Ass. Amer. Geog.*, **46**, 265.

Miller, D. H. (1965): The heat and water budget of the earth's surface. *Advances in Geoph.*, **11**, 175–301.

Milosavljević, M. and K. (1961): Luftströmungen im Gebiet des Gebirges Maljen und ihr Einfluss auf die Lufttemperatur, relative Luftfeuchte, Bewölkung und den Niederschlag. "Met. Tagung in Verbindung mit der Jubiläumsfeier, 75-jahr. Bestehens der Obs. Sonnblick", 90–102.

Misawa, K., Kawasumi, H. (1940): Measurements of local geographical characteristics by phenology. *Chirigaku*, **8**, 35–48, 299–309.

Montefinale, A. C., Petriconi, G. L., Gori, E. G., Papee, H. M. (1970): On ground-level visibilities in the Po-Valley during winter, and on their intercorrelations. *Pure Appl. Geoph.*, **83**, 201–221.

Morawetz, S. (1952): Kleinklimatische Beobachtungen in der Weststeiermark bei St. Stefan ob Stainz. *Ang. Met.*, **1**, 146–150.

Morawetz, S. (1961): Schneegrenze, Gletscherablation, Temperatur und Sonnenstrahlung in den Ostalpen. *Pet. Geog. Mitt.*, (2), 93–104.

Müller, W. (1964): Zur Verdunstung im Gebirge. *Carinthia II*. **24**. *Sonderheft*, 232–236.

Nagel, F. W. (1956): Fog precipitation on Table Mountain. *Q.J.R.M.S.*, **82**, 452–460.

Nägeli, W. (1971): Der Wind als Standortsfaktor bei Aufforstungen in der subalpinen stufe. (Stillbergralp im Dischmatal, Kanton Graubünden). *Mitt. Schweiz. Anst. Forstl. Vers'wes*, (47), 33–147.

Nakahara, M. (1944): On the progression of plant isophenes. *Jour. Agricul. Met.*, **1**, 153–158.

Nakajima, S. (1973): Observed profile of wind velocity in the lower atmospheric layer. *Chôkôsô-Tower-Kishôshiryô*, (5), 17–32.

Nakaya, U. (1949): Snow of Mt. Daisetsu. *Shizen*, (1), 15–20.

Niedźwiedź, T. (1970): Primer kartograficeskogo predstaulenija zamorozkov v uslovijach pogornogo reliefa Karpat. *Prace Geog.* (26).

Nogami, M. (1970): The snowline: its definition and determination. *The Quaternary Research, Tokyo*, **9** (1), 7–16.

Nose, A. et al. (1953): On application of observed temperature in a screen. *Mem. Ind. Met., Tokyo*, **17**, 25–31.

Obrebska-Starkel, B. (1969 a): Mezoklimat zlewni potoków Jaszcze i Jamne. *Studia Naturae, S. A., Kraków*, (3), 7–99.

Obrebska-Starklowa, B. (1969 b): Przebieg dobowy temperatury powietrza jako podstawa wudzielania regionów mezoklimatycznych w beskidach (Diurnal course of air temperature as basis for delimiting mesoclimatic regions in the Beskids). *Prace Geog.*, **25**, 49–61.

Obrebska-Starklowa, B. (1969 c): Stosunki mikroklimatyczne na pograniczu pieter leśnych i pól uprawnych w Gorcach. (Microclimatic conditions on the border of forest belts and arable land in the Gorce Mts.) *Prace Geog.*, **23**, 1–142.

Obrebska-Starkel, B. (1970): Über die thermische Temperaturschichtung in Bergtälern. *Acta Climat.*, **9** (1–4), 33–47.

Okanoue, M. (1957): On the intensity of the solar radiation on any slope. *Nippon-ringakkaishi*, **39**, 435–437.

Okanoue, M., Makimura, H. (1958): Calculation method for solar radiation receiving a sloped surface. *Nippon-ringakkaishi*, **40**, 40–41.

Olecki, Z. (1969): Wpływ zachmwzzenia na przebieg dobowy temperatury gleby w okresie letnim w pietrze pogórskim Karpat. (Effect of cloudiness on the diurnal course of soil temperature in the summer season in the Carpathian submontane zone). *Prace Geog.*, **25**, 99–116.

Oliver, J. (1964): A study of upland temperatures and humidities in South Wales. *Trans. Papers Inst. Brit. Geog.*, (35), 37–54.

Oliver, J. (1969): Problems of determining evapotranspiration in the Semi-Arid tropics illustrated with reference to the Sudan. *Jour. Trop. Geog.*, **28**, 64–74.

Ono, S. (1925): On orographic precipitation. *Phil. Mag.*, **49**, 144–164.

Onuma, T. (1955): Morphology of snow accumulation in mountainous regions. *Seppyô-no-kenkyû*, (2), 129–138.

Onuma, T. (1956): Studies on the mountain snow cover (1). *Studies on Snow* (4), 1–45.

Ortolani, M. (1965): Osservazioni sul clima delle Ande centrali. *Riv. Geog. Ital.*, **72**, 217–235.

Owada, M., Okura, M. (1972): Height division according to the daily change of climatic elements on the SE slope of Mt. Fuji. *Geog. Rev. Japan*, **45** (5), 372–380.

Ozawa, Y., Tsuboi, Y. (1952): On the distribution of air temperature in winter at Aoki village. *Jour. Agricul. Met.*, **7**, 131–133.

Ozawa, Y., Tsuboi, Y. (1953): On the distribution of air temperature in summer at Aoki village. *Jour. Agricul. Met.*, **8**, 143–145.

Peattie, R. (1936): "Mountain geography." Harvard Univ. Press, Cambridge, Mass., 1–271.

Peterson, P. (1964): The rainfall of Hong Kong: Royal Obs. Hong Kong Tech. Note, (17), 2nd Ed.

Petkovšek, Z. (1969): Frequency of fog in the lowlands of Slovenia. *Slovenian Met. Soc., Papers*, (11), 57–89.

Pleiss, H. (1951): Die Windverhältnisse in Sachsen. *Abh. Met. Dienst. DDR*, (6), 1–127.

Pockels, F. (1901): Zur Theorie der Niederschlagsbildung an Gebirgen. *Ann. d. Physik*, **4**, 459.

Poggi, A. (1959): Contribution a la connaissance de la distribution altimétrique de la durée de lénneigement dans les Alpes françaises du Nord. *Ber. Deutsh. Wett.*, (54), 134–149.

Prutzer, E. (1961): Die Verdunstungsverhältnisse einiger subalpiner Standorte. *Mitt. d. Forstl. Bundes-Vers. Mariabrunn*, (59), 231–256.

Quitt, E. (1961): Mesoklimatische Untersuchung des mittleren Teils von Dyjskosvratecký úval. *Práce, Brnenske Zakladny Československé Akademie Ved*, **33**, 77–112.

Rao, P. K. (1960): Observation and theory of local wind systems. New York Univ., Dept. of Met. Ocean. Coll. of Eng., Res. Div. Contract DA-36-039-Sc-78127, 1–19.

Reiter, R. (1966): Luftverunreinigung und Kleinionendichte in Abhängigkeit von Windströmung und Austausch. *Arch. Met. Geoph. Biokl.* (B), **14**, (1) 53–80.

Rosenkranz, F. (1951): Grundzüge der Phänologie mit besonderer Berücksichtigung von Österreich. Wien, 1–69.

Roux, L. (1951): Über den Anteil des Schnees am Gesamtniederschlag im Harz. *Ann. Met.*, **4**, 252–258.

Rouse, W. R., Wilson, R. G. (1969): Time and space variations in the radiant energy fluxes over sloping forested terrain and their influence on seasonal heat and water balances at a middle latitude site. *Geog. Ann.*, **51**A (3), 160–175.

Russler, B. H., Spreen, W. C. (1947): Topographically adjusted normal isohyetal maps for Western Colorado. *U.S. Dept. of Comm. Wea. Bur., Tech. Paper*, (4), 1–27.

Sarker, R. P. (1966): A dynamic model of orographic rainfall. *Mon. Wea. Rev.*, **94** (9), 555–572.

Sasaki, N. (1962): Studies on frost-injury of planted forest in Hokkaido with special references to micro-topography and applicable methods against damage. *Res. Bull. Coll. Exp. Forests, Hokkaido Univ.*, **21**, 377–395.

Sauberer, F. (1948): Kleinklimatische Niederschlagsuntersuchungen im Lunzer Gebiet. *Umwelt*, **1**, 410–415.

Sauberer, F. (1953): Der Strahlungshaushalt der Alpen-Gletscher. *Wett. Leben*, **5**, 38–42.

Sauberer, F., Dirmhirn, I. (1950): Die Bedeutung des Strahlungsfaktors für den Gletscherhaushalt. *Wett. Leben*, **2** (11/12), 248–261.

Sauberer, F., Dirmhirn, I. (1952): Der Strahlungshaushalt horizontaler Gletscherflächen auf dem Hohen Sonnblick. *Geog. Ann.*, **24**, 261–290.

Sauberer, F., Dirmhirn, I. (1953): Über die Entstehung der extremen Temperaturminima in der Doline Gstettner-Alm. *Arch. Met. Geoph. Biokl.* (B), **5**, 307–326.

Sauberer, F., Dirmhirn, I. (1956): Weitere Untersuchungen über die Kaltluftansammlungen in der Doline Gstettner-Alm bei Lunz in Niederösterreich. *Wett. Leben*, **8** (8/11), 187–196.

Schindler, G. (1961): Autofahrten als Hilfsmittel zur Feststellung Lokalklimatischer Besonderheiten im Nebelvorkommen. *Met. Rdsch.*, **14**, 123–125.

Schirmer, H. (1951): Umstrittenen Niederschlagsmessungen im Hochgebirge. *Mitt. D. W. US-Zone*, (11), 1–14.

Schmidt, W. (1930): Die tiefsten Minimumtemperaturen in Mitteleuropa. *Naturw.*, **18**, 367–369.

Schneider, M. (1963): Begriff und Einteilung des Frostes. In: Frostschutz im Pflanzenbau. Herg. F. Schnelle, BLV., München, 3–12.

Schnelle, F. (1950 a): Kleinklimatische Geländeaufnahme am Beispiel der Frostschäden in Obstbau. *Ber. Deutsch. Wett. US-Zone*, (12), 99–104.

Schnelle, F. (1950 b): Kleinklimatische Gelandeaufnahme zur Feststellung der Frostlagen im Obstbau. *Arch. Wissenshaftl. Gesellschaft für Land-und Forstwirtschaft*, (2), 25–27.

Schnelle, F. (1950 c): Feststellung der Frostlagen durch kleinklimatische Geländeaufnahme. *Gartenjahr (Geisenheim)*, **5** (11), 1–4.

Schnelle, F. (1955): "Pflanzenphänologie." Akademische Verl. Geest & Portig K.-G. Leipzig. 1–299.

Schnelle, F. (1956): Ein Hilfsmittel zur Feststellung der Höhe von Forstlagen in Mittelgebirgstälern. *Met. Rdsch.*, **9**, 180–182.

Schnelle, F. (1963): "Frostschutz im Pflanzenbau." Bd. 1, BLV., München, 1–488.

Schubach, K. (1960): Das Spätfrostphänomen, dargestellt am Beispiel oberes Modautal. Inst. f. Naturschutz Darmstadt, (V4), 111–116.

Schüepp, M. et al. (1963): Die Windverhältnisse im Davoser Hochtal. *Arch. Met. Geoph. Biokl.*, (B), **12**, 337–349.

Schweinfurth, U. (1956): Über klimatische Trockentäler im Himalaya. *Erdkunde*, **10**, 287–302.

Schweinfurth, U. (1972): The eastern marches of High Asia and the river gorge country. *Erdwissenschatliche Forschung*, (4), 276–287.

Schltetus, H. R. (1959): Geländeausformung und Bewindung in Abhängigkeit von der Austauschgrösse, *Met. Rdsch.*, **12**, 73–80.

Scultetus, H. R. (1964): Auswirkungen eines 12m hohen Dammes auf des Kleinklima. *Ang. Met.*, **5**, 13–26.

Sekiguti, T. (1950): Local climatological study of the 80% flowering dates of cherry blossom (*Prunus subhirtella*) at the Akaho Fan, Kami-ina, Nagano Pref. *Geoph. Mag.*, **22**, 89–97.

Sekiguti, T. (1951 a): Studies on local climatology (III). On the prevailing wind in early summer judged by bending shapes of top-twigs of persimmon trees at the Akaho Fan, Nagano Pref., Japan. *Pap. Met. Geoph.*, **2** (2), 168–179.

Sekiguti, T. (1951 b): Studies on local climatology (VI). Humidity distribution and surface covers. *Pap. Met. Geoph.*, **2** (2), 180–188.

Sekiguti, T. (1951 c): Humidity and temperature distribution and surface covers (2). *Geog. Rev. Japan*, **24**, 404–411.

Shitara, H. (1965): The influence of hills on the distribution of nocturnal cooling. *Sci. Rep. Tôhoku Univ. 7th Ser.*, (14), 27–38.

Simonis, R. (1960): Modifizierung phänologischer Daten durch lokalklimatische Sonderverhältnisse. *Ang. Met.*, **3**, 339–345.

Smith, L. P. (1954): Length of a frost-free period. *Met. Mag.*, **83**, 81–83.

Stanev, S., Simeonov, P. (1970): The distribution of the snow cover in mountain regions. *Hidrologija i meteorologija*, **19** (2), 33–40.

Steinhauser, F. (1938): Die Meteorologie des Sonnblicks. 1. Teil. Springer, Wien, 1–180.

Steinhauser, F. (1948): Die Schneehöhen in den Ostalpen und die Bedeutung der winterlichen Temperaturinversion. *Arch. Met. Geoph. Biokl.*, (B), **1**, 63–74.

Steinhauser, F. (1950): Über die Windverstärkung an Gebirgszügen. Ein Beitrag zur Frage der Beeinflussung der Luftströmungen durch Gebirge. *Arch. Met. Geoph. Biokl.* (B), **2**, 39–64.

Steinhauser, F. (1956): Über die kartographische Darstellung der Sonnenscheindauer. *Wett. Leben*, **8**, 1–12.

Steinhauser, F. (1968): Mithodische Bemerkungen zur Bearbeitung von Berg-und Talwinden im Gebirgsland. In: Einfluss der Karpathen auf die Witterungserscheinungen. ed. by M. Čadež, Prirodno Mathematicki Fak., Met. Zavod, Beograd, 17–48.

Sterten, A. K. (1961): Local meteorological investigations of the mountain and valley wind systen in South-eastern Norway. In: Local and synoptic meteorological investigations of the mountain and valley wind system. *Forsvarets Forskningsinstitutt Norw. Defence Res. Establ., Intern rapport K-* 242, 9–73.

Sterten, A. K. (1965): Alte und neue Berg- und Talwindstudien. *Carinthia II*, 24. Sonderheft, Wien, 186–194.

Stoenescu, S. M. (1963): Über die Temperaturverhältnisse im Karpatengebirge. "Die II Konferenz für Karpatenmeteorologie. Budapest, 13–15, Nov. 1961", 209–215.

Sugawara, Y. et al. (1939): On the air temperature lapse rate between Mt. Fuji and Funatsu. *Kôkûki-sho-hôkoku*, **2**, 125–138.

Sugaya, J. (1949 a, 1950): On the precipitation distribution in the mountainous regions in summer (1) (2). *Suigaino-sôgôteki-kenkyû*, (2), 15–53, (3), 29–40.

Sugaya, J. (1949 b): A quantitative investigation on the water content of snow accumulation. In: Water content of snow accumulation and runoff in the Mt. Daisetsu region by Keizai-antei-honbu, Tokyo, 1–42.

Sugaya, J. et al. (1949): Water content of snow accumulation and runoff in the Mt. Daisetsu region. Keizai-antei-honbu and Hokkaido-cho, Tokyo and Sapporo, 1–90.

Sugaya, J. (1953): Water resources in the area of Towada-lake and Mt. Hakkoda. Sugaya-mizushigen-ken, (3), 1–70.

Suzuki, T. (1969): Preliminary report on the distribution of the lingering snows on some strato-volcanic cones in Japan. *Bull. Facul. Sci. & Eng., Chuô Univ.*, **12**, 160–173.

Sverdrup, H. U. (1936): The eddy conductivity of the air over a smooth snow field. Results of the Norwegian-Swedish Spitsbergen expedition in 1934. *Geof. Publ.*, **11** (7), 1–69.

Szepesi, D. (1960): Die orographische niederschlagsbildende Wirkung der Gebirge des Karpatenbeckens. *Idöjárás*, **3**, 144–152.

Takahashi, K. (1953): Properties of snow deposited on mountain slopes. *Research on Snow and Ice* (1), 207–210.

Takasaki, M. et al. (1964): Snow survey by means of aerial photographs. *Jour. Japan Soc. Photogrammetry, Spec. Vol.*, **1**, 43–47.

Takeda, K. (1960): On the mountain rainfall. *Shinrin-ritchi*, **1** (2), 36–39.

Takeuchi, Y. (1964): Statistical investigation on the movement of thunderstorm in the Kanto District. *Jour. Met. Res.*, **16** (7), 403–408.

Tateishi, Y. (1969): Effects of wind velocity and microtopography on the snow accumulation. *Geog. Rev. Japan*, **42** (8), 527–532.

Thyer, N., Buettner, K. J. K. (1962): On valley and mountain winds III. Dept. of Atmosph. Sci. Univ. of Washington. Final Rep. AF Contract, **19** (604) -7201, 1–106.

Thyer, N. H. (1966): A theoretical explanation of mountain and valley winds by a numerical method. *Arch. Met. Geoph. Biokl.* (A), **15** (3–4), 318–348.

Tiszkov, Ch. (1963): Temperaturnite inversji prez Studenoto polugodije v srednija Predbalkan mezdu rekite Rosica i Belica. *Izvestija na Geografskija Institut, T.* 8.

Troll, C. (1942): Büsserschnee (Nieve de los penitentes) in den Hochgebirgen der Erde. *Pet. Geog. Mitt., Ergäzht*, (240), 1–103.

Troll, C. (1943): Thermische Klimatypen der Erde. *Pet. Geog. Mitt.*, (89), 81 89.

Troll, C. (1952): Die Lokalwinde der Tropengebirge und ihr Einfluss auf Niederschlag und Vegetation. *Bonner Geog. Abh.*, (9), 124–182.

Troll, C., Paffen, K-H. (1964): Karte der Jareszeiten-Klimate der Erde. *Erdkunde*, **18**, 5–28.

Troll, C. (1972): The three-dimensional zonation of the Himalayan system. *Erdwissenschaftl. Forschung*, (4), 264–275.

Turner, H. (1958): Über das Licht- und Strahlungsklima einer Hanglage der Ötztaler Alpen bei

Obergurgl und seine Auswirkung auf das Mikroklima und auf die Vegetation. *Arch. Met. Geoph. Biokl.* (B), **8**, 273–325.

Turner, H. (1961): Die Niederschlags- und Schneeverhältnisse. *Mitt. d. Forstl. Bundes-Vers. Mariabrunn*, (59), 265–315.

Turner, H. (1966): Die globale Hangbestrahlung als Standortsfaktor bei Aufforstungen in der subalpinen Stufe. *Mitt. Schweiz. Anst. forstl. Vers'wes*, **42** (3), 111–168.

Tyson, P. D. (1968): Velocity fluctuations in the mountain wind. *Jour. Atm. Sci.*, **25** (1–3), 381–384.

Tyson, P. D., Keen, C. S. (1970): Some observations of velocity spectra in mountain and valley winds. *S. African Geog. Jour.*, **52**, 58–66.

Uehara, M. (1949, 1950 a, b): On the micrometeorology at the slope farm (1), (2), (3). *Tech. Bull. Kagawa Agricul. College*, **1** (1), 46–57; **1** (1), 1–13; **2** (1), 31–38.

Ueno, T. (1938): The flowering date of cherry blossoms (*Prunus yedoensis*) on Mt. Tsukuba. *Tenki-to-kikô*, **5**, 270–282.

Ueno, T. (1941): The flowering date of cherry blossoms (*Prunus yedoensis*) in Miyagi Pref. in 1941. *Mem. Ind. Met., Tokyo*, **10**, 119–130.

Umeda, S. (1958): On the characteristic features of the temperature distribution in small areas at frosty night. *Mem. Ind. Met., Tokyo*, **22**, 15–21.

Urfer-Henneberger, Ch. (1964): Wind- und Temperaturverhältnisse an ungestörten Schönwettertagen im Dischmatal bei Davos. *Mitt. Schweiz. Anst. Forstl. Vers'wes.*, **40** (6), 389–441.

Urfer-Henneberger, Ch. (1970): Neuere Beobachtungen über die Entwicklung des Schönwetterwindsystems in einem V-förmigen Alpental (Dischmatal bei Davos). *Arch. Met. Geoph. Biokl.*, (B), **18**, 21–42.

U. S. Weather Bureau and Corps of Eng. (1947): Thunderstorm rainfall. *Hydromet. Rep.* (5).

Vorontsov, P. A., Schklkovnikov, M. S. (1956): Opyt aerologicheskogo issledovanniia nizhnego sloia atmosfery v doline Azau. *Trudy GGO*, (63), 138–167.

Vorontsov, P. A. (1958 a): O brizakh Ladozhskago ozera. *Trudy GGO*, (73), 87–106.

Vorontsov, P. A. (1958 b): Nekotorie voprosi aerologicheskih issledovanii pofranichnogo sloia atmosferi. In: Sovremennie prob. meteorol. prizem. voz. 157–179.

Wagner, A. (1937): Gibt es im Gebirge eine Höhenzone maximalen Niederschlages? *Gerl. Beitr. Geoph.*, **50**, 150–155.

Wagner, A. (1938): Theorie und Beobachtungen der periodischen Gebirgswinde. *Gerl. Beitr. Geoph.*, **52**, 408–449.

Wagner, R. (1963): Der Tagesgang der Lufttemperatur einer Doline im Bükk-Gebirge. *Acta Climat.*, (2–3), 49–79.

Wagner, R. (1965): Lufttemperaturmessungen in einer Doline des Bükk-Gebirges. *Ang. Met.*, **5**, 92–99.

Wagner, R. (1970): Kaltluftseen in den Dolinen. *Acta Climat.*, **9** (1–4), 23–32.

Weischet, W. (1965): Der Tropisch-Konvektive und Aussertropisch-Advektive Typ der vertikalen Niederschlagsverteilung. *Erdkunde*, **19** (1), 6–14.

Weischet, W. (1969): Klimatologische Regeln zur Vertikalverteilung der Niederschläge in Tropengebirgen. *Erde*, **100** (2–4), 287–306.

Wendler, G. (1971): An estimate of the heat balance of a valley and hill station in Central Alaska. *Jour. Appl. Met.*, **10** (4), 684–693.

Williams, L. D., et al. (1972): Application of computed global radiation for areas of high relief. *Jour. Appl. Met.*, **11** (3), 526–533.

Williams, P., Jr., Peck, E. L. (1962): Terrain influences on precipitation in the intermountain west as related to synoptic situations. *Jour. Appl. Met.*, **1** (3), 343–347.

Williams, P., Jr., Heck, W. J. (1972): Areal coverage of precipitation in Northwestern Utah. *Jour. Appl. Met.*, **11** (3), 509–516.

Winiger, M. (1972): Die Bewölkungsverhältnisse der zentralsaharischen Gebirge aus Wettersatellitenbildern. *Hochgebirgsforschung*, (2), 87–120.

Wolfe, J. N., Wareham, R. T., Scofield, H. T. (1943): The microclimates of a small valley in Central Ohio. *Trans. Amer. Geoph. Union*, **24**, 154–166.

Wolfe, J. N., Wareham, R. T., Scofield, H. T. (1949): Macroclimates and microclimates of Neotoma, a small valley in Central Ohio. *Bull. Ohio Biol. Survey*, **8** (1), 1–267.

Yoshida, S. (1951) Relation between rainfall and altitude on Iwate. *Jour. Met. Res.*, **3**, 364–367.

Yoshida, Y. (1960): Snow survey of the catchment basin of the projected Lake Yonezawa. *Geog. Rev. Japan*, **33**, 26–43.

Yoshida, Y., Moriwaki, K. (1972): Some characteristics of the climate of Wright Valley, Victoria Land, Antarctica. In: Essays of Geographical Sciences. Commemoration Vol. Prof. Funakoshi, 218–233.

Yoshino, M. M. (1952): On the influences of the topography which affects the wind directions. *Geog. Rev. Japan*, **25**, 100–110.

Yoshino, M. M. (1957 a): Local characteristics of surface winds in a small valley. *Sci. Rep. Tokyo Kyoiku Daigaku*, (C), **5** (46), 129–151.

Yoshino, M. M. (1957 b): The structure of surface winds over a small valley. *Jour. Met. Soc. Japan*, **35** (3), 184–195.

Yoshino, M. M. (1958): Wind speed profiles of the lowest air layer under influences of micro-topography. *Jour. Met. Soc. Japan*, **36** (5), 174–186.

Yoshino, M. M. (1959): A micro-climatological study of surface winds affected by micro-topography. *Proc. IGU Reg. Conf. Japan 1957*, 243–249.

Yoshino, M. M. (1960): The distribution of maximum observed rainfall and the characteristics of depth-duration and intensity-duration curves in Japan. *Jour. Met. Soc. Japan*, **38**, 27–46.

Yoshino, M. M. (1961 a): "Shôkikô (Microclimate——An introduction to local meteorology)." Chijinshokan, Tokyo, 1–274.

Yoshino, M. M. (1961 b): Local climate in the Mt. Kirishima region. *Tokyo Geography Papers*, (5), 79–109.

Yoshino, M. M., Yoshino, M. T. (1963): Lokalklima und Vegetation im Kirishima-Gebirge im südlichen Kyushu, Japan. *Erdkunde*, **17**, 148–162.

Yoshino, M. M. (1964): Some local characteristics of the winds as revealed by wind-shaped trees in Rhône valley in Switzerland. *Erdkunde*, **18**, 28–38.

Yoshino, M. M. (1965): Some aspects of air temperature climate of the high mountains in Japan. *Carinthia II*, 24. Sonderheft 147–153.

Yoshino, M. M. (1973): Studies on wind-shaped trees: Their classification, distribution and significance as a climatic indicator. *Climat. Notes, Hosei Univ.*, (12), 1–52.

Yuyama, Y. (1972): A climatological study of orographic wave clouds on and leeside of Mt. Fuji. *Jour. Met. Res.*, **24** (9), 415–418.

Chapter 5

Abe, M. (1932): The formation of cloud by the obstruction of Mount Fuji. *Geoph. Mag.* (Tokyo), **6**, 1–10.

Abe, M. (1941): Mountain clouds, their forms and connected air current, Pt. II. *Bull. Centr. Met. Obs. Japan, VII*, **3**, 93–145.

Abe, M. (1942): An attempt to make visible the mountain air current. *Jour. Met. Soc. Japan*, **20**, 69–76.

Aichele, H. (1953): Kaltluftpulsationen. *Met. Rdsch.*, **8**, 53–54.

Akademiia Nauk SSSR, Inst. Geografii, Inst. Lesa, (1953): Mikroklimaticheskie i klimaticheskie issledovania v Prikaspiĭskoĭ Nizmennosti. (Microclimatic and climatic investigations in the Caspian Plain). Moscow, 1–167.

Akademiia Nauk SSSR, Institut Geografii, (1957): Sukhovei, ikh proiskhozhdenie i bor'ba s nimi (Sukhovey, their origin and control). Moscow, 1–370.

Akiyama, T. (1956): On the occurrence of the local severe wind "Yamaji" (2). *Jour. Met. Res.*, **8**, 627–641.

Alaka, M. A. (1960): The airflow over mountains. *W. M. O. Tech. Note*, (34), 1–135.

Alissow, B. P. et al. (1956): "Lehrbuch der Klimatologie." VEB Deutsch. Verl. d. Wissensch., Berlin, 1–536.

Arakawa, S. (1968): A proposed mechanism of fall winds and Dashikaze. *Pap. Met. Geophy.*, **19** (1), 69–99.

Arakawa, S., Oobayashi, T. (1968): On the numerical experiments by the method of characteristics of one-dimensional unsteady airflow over the mountain ridge. *Pap. Met. Geoph.*, **19** (3), 341–361.

Arakawa, S. (1969): Climatological and dynamical studies on the local strong winds, mainly in Hokkaidô, Japan. *Geoph. Mag.*, **34** (4), 349–425.

Arakawa, S. (1973): Numerical experiments of the local strong winds, Bora and Föhn. *Climat. Notes, Hosei Univ.*, (14), 1–20.

Asai, T., Nishizawa, T. (1959): An analysis of interdiurnal change of air temperature on "Yamase" weather by synoptic pattern. *Misc. Rep. Res. Inst. Natural Resources.*, (50), 3–10.

Ashburn, E. V. (1941): The vertical transfer of radiation through atmospheric temperature inversions. *Bull. Amer. Met. Soc.*, **22**, 239–242.

Ashwell, I. Y., Marsh, J. S. (1967): Moisture loss under chinook conditions. *Proc. 1st Canadian Conf. Micrometeorology Toronto*, 307–310.

Aubert de la Rüe, E. (1955): "Man and the winds." Hutchinson's Sci. and Tech. Pub., London, 1–206.

Aulitzky, H. (1967): Lage und Ausmaß der "warmen Hangzone" in einem Quertal der Innenalpen. *Ann. Met.*, (3), 159–165.

Austin, J. M. (1957): Low-level inversions. Final Report. Quart. Res. Eng. Command DA 19–129-QM-377, 1–153.

Baumgartner, A. (1960, 1961, 1962): Die Lufttemperatur als Standortsfaktor am Gr. Falkenstein. 1, *Mitt. Forstw. Cbl.*, **79**, 362–373; 2, *Mitt. Forstw. Cbl.*, **80**, 107–120; 3, *Mitt. Forstw. Cbl.*, **81**, 17–47.

Baumgartner, A. (1963): Einfluss des Geländes auf Lagerung und Bewegung der nächtlichen Kaltluft. In: Frostschutz im Pflanzenbau. Bd. 1. ed. by F. Schnelle, BLV, München, 151–194.

Baynton, H. W. (1962): Time of formation and burn off of nocturnal inversions at Detroit. *Jour. Appl. Met.*, **1** (2), 244–250.

Baynton, H. W., Bidwell, J. M., Beran, D. W. (1965): The association of low-level inversions with surface wind and temperature at Point Arguells. *Jour. Appl. Met.*, **4** (4), 509–516.

Berau, D. W. (1967): Large amplitude lee waves and chinook winds. *Jour. Appl. Met.*, **6** (5), 865–877.

Berg, H. (1951): Kleinmeteorologische Messungen im Hohen Venn. *Zeitsch. Met.*, **5**, 229–235.

Bergen, J. D. (1969): Cold air drainage on a forested mountain slope. *Jour. Appl. Met.*, **8** (6), 884–895.

Blackadar, A. K. (1957): Boundary layer wind maxima and their significance for the growth of nocturnal inversions. *Bull. Amer. Met. Soc.*, **38** (5), 283–290.

Bleibaum, I. (1952): Abendliche und nächtliche Temperaturanstiege im Thüringer Wald und in der Rhön. *Ber. Deutsch. Wett. US-Zone*, (38), 110–113.

Bleibaum, I. (1953): Studien zur Meteorologie der südlichen Rhön. *Ber. Deutsch. Wett.*, (4), 1–15.

Blüthgen, J. (1966): "Lehrbuch der allgemeinen Klimageographie." Walter de Grayter & co. Berlin, 1–720.

Bosart, L. F., Vaudo, C. J., Helsdon, J. H. J. (1972): Coastal frontogenesis. *Jour. Appl. Met.*, **11** (8), 1236–1258.

Brinkmann, W., Ashwell, I. Y. (1968): The structure and movement of the chinook in Alberta. *Atmosphere*, **6**, 41–50.

Brinkmann, W. A. R. (1970): The chinook at Calgary (Canada). *Arch. Met. Geoph. Biokl.* (B), **18**, 279–286.

Brinkmann, W. A. R. (1971): What is a foehn? *Weather*, **26** (6), 230–239.

Brunt, D. (1929): The transfer of heat by radiation and turbulence in the lower atmosphere. *Proc. Roy. Met. Soc.*, (124), 201–218.

Čadež, M. (1954): Über einige Einflüsse orographische Hindernisse auf die Luftbewegung. *Arch. Met. Geoph. Biokl.* (A), **6**, 403–416.

Čadež, M. (1963): Entstehung der thermischen Antizyklonen. *Geof. Met.*, **12** (3/4), 1–4.

Čadež, M. (1964): Vreme u Jugoslavia (Weather in Yugoslavia). *Prirodno-Mathematicki Fak. u Beograd-Met. Zavod, Rasprave*, (4), 1–83.

Čadež, M. (1967): Über synoptische Probleme im Südostalpinen Raum. *Veroff. Schweiz. Met. Zentralanstalt*, (4), 155–175.

Čadež, M. (1971): Einige Resultate der internationalen Zusammenarbeit hinsichtlich der Analyse der Wetterentwicklung im Alpengebiet während der Alpentage. *Ann. Met.*, (5), 79–87.

Chopra, K. P., Hubert, L. F. (1965): Mesoscale eddies in wakes of islands. *Jour. Atm. Sci.*, **22**, 652–657.

Chopra, K. P. (1972): Velocity field in vortices leeward of island. *Jour. Atm. Sci.*, **29**, 396–399.

Clarke, R. H. (1972): The morning glory: An atmospheric hydraulic jump. *Jour. Appl. Met.*, **11** (2), 304–311.

Cohen, A., Doron, E. (1967): Mountain lee waves in the Middle East: theoretical calculations compared with satellite pictures. *Jour. Appl. Met.*, **6** (4), 669–673.

Corby, G. A. (1954): The airflow over mountains. *Q.J.R.M.S.*, **80**, 491–521.

Cox, H. J. (1923): Thermal belts and fruit growing in North Carolina. *Mon. Wea. Rev., Suppl.* (19), 1–106.

Crowe, P. R. (1971): "Concepts in climatology." Longman, London, 1–589.

Daubert, K. (1962): Ein Beitrag zur Kenntnis der Bodeninversionen. *Met. Rdsch.*, **15**, 121–130.

Davidson, B., Rao, P. K. (1958): Preliminary report on valley wind studies in Vermont, 1957. Final Report Under Contract No. AF, 19 (604), 1971, 1–54.

Defant, F. (1949): Zur Theorie der Hangwinde, nebst Bemerkungen zur Theorie der Berg- und Talwinde. *Arch. Met. Geoph. Biokl.* (A) **1** (3/4), 421–450.

Defant, F. (1951): Local winds. *Comp. Met.*, 655–672.

Dodd, A. V., McPhilimy, H. S. (1959): Yuma summer microclimate. Environmental Protection Research Division, Mass., *Tech. Rep. Ep.*, **120**, 1–34.

Domrös, M. (1969): Die Niederschlagsverhältnisse im Uva-Becken auf Ceylon. Eine geländeklimatologische Untersuchungen. *Erdkunde*, **23**, 117–127.

Domrös, M. (1971): Der Monsun im Klima der Insel Ceylon. *Erde*, **102** (2/3), 118–140.

Döös, B. R. (1962): Theoretical analysis of lee wave clouds observed by TIROS I. *Tellus*, **14**, 301–309.

Drinkwater, T. A. et al. (1969): Atlas of Alberta. Gov. of Alberta and Univ. Alberta, Edmonton, 158 maps.

Dunbar, G. S. (1966): Thermal belts in North Carolina. *Geog. Rev.*, **56** (4), 516–526.

Dzerdzervski, B. L. (1963): Sukhoveis and draught control. *Israel Progr. Sci. Transl.*, Jerusalem, 1–366.

Edagawa, H. (1971): Studies on heavy rainfall (III)——On the heavy rainfall in the Central Kinki District——*Ann. Disas. Prev. Res. Inst.*, (14B), 119–129.

Ehrlich, A. (1953): Note on local winds near Big Delta, Alaska. *Bull. Amer. Met. Soc.*, **34**, 181–182.

Ekhart, E. (1949): Über Inversionen in den Alpen. *Met. Rdsch*, **2** (5/8), 153–159.

Elsasser, W. (1940): Radiative cooling in the lower atmosphere. *Mon. Wea. Rev.*, **68**, 185–188.

Eriksen, W. (1966): Über den Einfluss der Stadt auf die Temperaturverteilung bei Frontdurchgang. *Met. Rdsch.*, **19** (1), 23–24.

Evseev, P. K. (1957): O prirode letnik sukhoveev na iugo-vostoke Evpopečskoč chasti Sovetskogo Soiuza (Nature of summer sukhovey in the SE-part of the European USSR). In: Sukhovei, ikh proiskhozhdenie i bor'ba s nimi. Akad. Nauk S.S.S.R., Moscow, 106–116.

Feldmann, G. (1965): Bodeninversionen über München-Riem, ihre Häufigkeit und Entwichkung im Tagesgang. *Met. Rdsch.*, **18** (1), 3–13.

Ficker, H. v., de Rudder B. (1948): Föhn und Föhnwirkungen. Probl. d. Bioklimatologie Bd. I, Leipzig, 1–114.

Findlater, J. (1971): The strange winds of Ras Asir (formerly Cape Guardafui). *Met. Mag.*, **100**, 46–54.

Fleagle, R. G. (1950): A theory of air drainage. *Jour. Met.*, **7**, 227–232.

Flemming, G. (1966 a): Das Verhalten der Temperatur am Talgrund bei Luftströmung parallel und quer zum Tal. *Zeitsch. Met.*, **18** (11–12), 445–448.

Flemming, G. (1966 b): Zum Einfluß unterschiedlicher großräumiger Strömungsrichtung und thermischer Stabilität auf die lokalen Windrichtungsverhältnisse in Sachsen. *Zeitsch. Met.*, **19** (1–2), 34–43.

Fliri, F. (1972): Statistische Untersuchung über den Zusammenhang von Südföhn und Gesamtklima in Innsbruck (1906–1972). In: Beiträge zur Klimatologie, Meteorologie und Klimamorphologie. ed. by E. Lendl and H. Riedl. Geog. Inst. Univ. Salzburg, 45–57.

Flohn, H. (1952): Zur Aerologie der Polargebiete. *Met. Rdsch.*, **5** (5/6) 81–87, (7/8) 121–128.

Flowers, E. (1960): "Antarctic meteorology" Proc. of Symp. Melbourne, Feb. 1959, 453–462.

Förchtgott, J. (1949): Vlnové proudění v závětr i horských hřebenů (Wave streaming in the lee of mountain ridges). *Met. Zprávy*, **3**, 49–51.

Förchtgott, J. (1969): Evidence for mountainsized lee eddies. *Weather*, **24** (7), 255–260.

Fritz, S. (1965): The significance of mountain lee waves as seen from satellite pictures. *Jour. Appl. Met.*, **4** (1), 31–37.

Fukaishi, K. (1961): Distribution of snowfall in Niigata Prefecture. *Tenki*, **8**, 395–402.

Fukaishi, K. (1963, 1964): Distribution of snowfall in Niigata Prefecture. *Suion-no-kenkyû*, **7** (4), 185–191; (5), 236–238, **8** (1), 295–299.

Fukuda, K. (1965): Synoptic study on the mechanism of heavy snowfall. *Geoph. Mag.*, **32** (4), 317–359.

Fukui, E. (1966): Further studies on the climatic boundary dividing the Japan Sea Coast and the Pacific side of Japan. *Geog. Rev. Japan*, **39** (10), 643–655.

Fukushima Weather Station (1956): On the nocturnal temperature variation surrounding the orchards on a frosty nights (1) (2). *Mem. Ind. Met., Tokyo*, **19** (1), 19–30, **19** (2), 91–99.

Geiger, R. (1961): "Das Klima der bodennahen Luftschicht." Friedr. Vieweg, Braunschweig, 1–646.

Geiger, R. (1965): "The climate near the ground." Harvard Univ. Press, Cambridge, 1–611.

Gerbier, N., Bérenger, M. (1960): Études expérimentales das ondes dues au relief. *Monogr. Météorol. Nationale*, (20), 1–137.

Gerbier, N., Bérenger, M. (1961): Experimental studies of lee waves in the French Alps. *Q.J.R.M.S.*, **87**, 13–23.

Gerbier, N., Bérenger, M. (1963): Mesures aérologiques dans les Alpes Françaises du Sud. *Geof. e Met.* (11), 110–118.

Gerbier, N., Cachera, P. (1961): Considerations generales theoriquest et experimentales sur les mouvements ondulatoires dus au Relief. *Notice d'Inf. Tech. Sec.* 23, (8) *Météorol. Nationale*, 1–52.

Gifford, F., Jr. (1952): The breakdown of a low-level inversion studied by means of detailed soundings with a modified radiosonde. *Bull. Amer. Met. Soc.*, **33**, 373–379.

Gifford, F., Jr. (1953): A study of low level air trajectories at Oak Ridge, Tenn. *Mon. Wea. Rev.*, **81**, 179–192.

Glenn, C. L. (1961): The chinook. *Weatherwise*, **14** (5), 175–182.

Gunji, T. (1958): On temperature inversion on the foot of Mt. Tsukuba. *Mem. Ind. Met., Tokyo*, **21**, 89–91.

Gutermann, T. (1970): Vergleicheude Untersuchumgen zur Föhnhäufigkeit im Rheintal zwischen Chur und Bodensee. *Veröff. Schweizer. Met. Zeutralanstalt*, (18), 1–69.

Henning, I., Henning, D. (1967): Abbildung lokaler Zirkulation durch Wolkenfelder auf Hawaii. *Met. Rdsch.*, **20** (4), 109–114.

Henry, A. J. (1923): Cox on thermal belts and fruit growing in North Carolina. *Mon. Wea. Rev.*, **51**, 199–207.

Hentschel, G., Leidreiter, W. (1960): Die Häufigkeit von Inversionen im bodennahen Luftraum (15–76m über Grund) in Abhängigkeit von Jahresseit, Tageszeit und Windrichtung. *Ang. Met.*, **3**, 353–362.

Heywood, G. S. P. (1933): Katabatic winds in a valley. *Q.J.R.M.S.*, **59** (248), 47–85.

Higuchi, K. (1963): The band structure of snowfalls. *Jour. Met. Soc. Japan*, **41** (1), 53–70.

Holland, J. Z. (1952): The diffusion problem in hilly terrain. In: Air Pollution. ed. by McCabe, McGraw-Hill, New York, 815–821.

Holmboe, J., Klieforth, H. (1957): Investigations of mountain lee waves and the air flow over the Sierra Nevada. Final Report, Contract No. AF 19 (604)–728, 1–290.

Holmes, R. M., Hage, K. D. (1971): Airborne observations of three chinook-type situations in Southern Alberta. *Jour. Appl. Met.*, **10** (6), 1138–1153.

Holmgren, B. (1971): On the katabatic winds over the north-west slope of the ice cap: Variations of the surface roughness. In: Climate and energy exchange on a sub-polar ice cap in summer. Pt. C. *Uppsala Univ. Met. Inst. Meddel.*, (109), 1–43.

Houghton, D. D., Kasahara, A. (1968): Nonlinear shallow fluid flow over an isolated ridge. *Communications on Pure and Appl. Math.*, **21**, 1–23.

Ibaragi-ken, Mito Weather Observatory (1955, 1957): Researches on the climatic environment of fruit growing on the mountain slopes of Mt. Tsukuba. (1), 1–31, (2), 1–69.

Ino, H. (1965): On discontinuous line formed in the Wakasa Bay in winter. *Kishô-kenkyû Nôto.*, **16** (2), 329–332.

Inoue, E. (1948): Similarity between natural wind and airflow in wind tunnel. *Jour. Soc. Appl. Mech.*

Japan, **1** (1), 15–21.

Ishihara, K. et al. (1957): Forecasting of small scale precipitation (1st Report). *Jour. Met. Res.*, **9** (9), 615–632.

Ishikawa, N., Ishida, T. (1973): Observations of radiative cooling at basins in midwinter and snow-melting season. *Jour. Met. Soc. Japan*, **51** (3), 197–203.

Ives, R. L. (1950 a): Sunrise temperature dips. *Trans. Amer. Geoph. Union*, **31** (4), 536–538.

Ives, R. L. (1950 b): Frequency and physical effects of chinook winds in the Colorado high plains region. *Ann. Ass. Amer. Geog.*, **40**, 293–327.

Jacobs, W. C. (1946): Synoptic climatology. *Bull. Amer. Met. Soc.*, **27**, 306–311.

Jacobs, W. C. (1947): Wartime developments in applied climatology. *Met. Monogr.*, **1** (1), 1–52.

Japan Meteorological Agency (1968): Heavy snowfalls in the Hokuriku District. *Tech. Rep. J. M. A.*, (66), 1–481.

Julian, L. T., Julian, P. R. (1969): Boulder's winds. *Weatherwise*, **22**, 1–8 and 112.

Kabasawa, M. (1950): An example of rainfall by the local front produced topographically. *Jour. Met. Res.*, **2** (3), 65–69.

Kamiko, T. (1955): Precipitation associated with the Bôsô front. *Jour. Met. Res.*, **6**, 609–614.

Katsuya, M. (1925 a): On the rainfall in the Izumo-Yokota region. *Shinrin-chisui-kishô-ihô* (6), 54–89.

Katsuya, M. (1925 b): On the snow in Chizu, Tottori Prefecture. *Shinrin-chisui-kishô-ihô* (7), 49–77.

Katsuya, M. (1925 c): On the rainfall in Chizu, Tottori Prefecture. *Shinrin-sokkôjô-tokubetsu hôkoku* (10), 77–98.

Katsuya, M. (1928): Topography and snow accumulation on the coast of Japan Sea. *Shinrin-chisui-kishô-ihô* (10), 91–110.

Kawamura, T. (1961): The synoptic climatological consideration on the winter precipitation in Hokkaido. *Geog. Rev. Japan*, **34** (11), 583–595.

Kawamura, T. (1965): The distribution of surface wind over Hokkaido in winter. *Japanese Jour. Geol. Geog.*, **36** (2–4), 135–141.

Kawamura, T. (1966): Surface wind systems over Central Japan in the winter season. *Geog. Rev. Japan*, **39** (8), 538–554.

Kendrew, W. G., Currie, B. W. (1955): "The climate of Central Canada." Ottawa, 1–194.

Kendrew, W. G., Kerr, D. (1955): "The climate of British Columbia and the Yukon Territory." Ottawa, 1–222.

Kikuchi, R. (1972): Distribution of nocturnal temperature in the Kakuda Basin area. *Ann. Tôhoku Geog. Ass.*, **24** (4), 242–248.

Kimura, H. (1961): A micrometeorological study on small islands. Micrometeorological surveys at Hakatajima and Ôshima. *Met. Notes, Met. Res. Inst., Kyoto Univ. Ser.*, **2** (22), 1–60, 1–140.

Koch, H. G. (1960): Zum Begriff des Mittelgebirgsföhns. *Zeitsch. Met.*, **14**, 29–46.

Koch, H. G. (1961): Die warme Hangzone. Neue Anschauungen zur nächtlichen Kaltluftschichtung in Tälern und an Hängen. *Zeitsch. Met.*, **15**, 151–171.

Koschmieder, H. (1921): Zwei bemerkenswerte Beispiele horizontaler Wolken-Schläuche. *Beitr. zur Phys. d. fr. Atm.*, **9**, 176–180.

Kreutz, W. (1952): Lokalklimatische Studie im oberen Vogelsberg. *Ber. Deutsch. Wett. US-Zone*, (42), 171–176.

Küttner, J. (1939 a): Moazagotl und Föhnwelle. *Beitr. zur Phys. d. fr. Atm.*, **25**, 79–114.

Küttner, J. (1939 b): Zur Entstehung der Föhnwelle. *Beitr. zur Phys. d. fr. Atm.*, **25**, 251–299.

Küttner, J. (1949): Periodische Luftlawinen. *Met. Rdsch.*, **2**, 183–184.

Küttner, J. (1959): The band structure of the atmosphere. *Tellus*, **11** (3), 267–294.

Kusano, K. (1960): On the streamline pattern in Miyagi Prefecture. *Jour. Met. Res.*, **12**, 709–718.

Larsson, L. (1954): Observations of lee wave clouds in the Jämtland Mountains, Sweden. *Tellus*, **6**, 124–138.

Lauer, W. (1973): The altitudinal belts of the vegetation in the Central Mexican highlands and their climatic conditions. *Arc. Alp. Res.*, **5** (3, pt. 2), A 99–113.

Lauscher, F. (1956): Gibt es bioklimatisch kalten Föhn? *Wett. Leben*, **8**, 168–173.

Lauscher, F. (1958): Bemerkungen zur Bioklimatologie des Föhns. *Zeitsch. Met.*, **12** (4–6), 130–131.

Lawrence, E. N. (1954): Nocturnal winds. *Prof. Notes*, (111), 1–13.

Lee, Ch. W., Kikuchi, K., Magono, Ch. (1972): The horizontal distribution of snowfalls on the Ishikari Plain, Hokkaido, I, II. *Geoph. Bull. Hokkaido Univ.*, **27**, 13–23, **28**, 1–12.

Lehmann, P. (1952): Abkühlung und Erwärmung im nächtlichen Kaltluftfluß. *Ber. Deutsch. Wett. US-Zone* (38), 113–116.

Leopold, L. B. (1951): Hawaiian climate: its relation to human and plant geography. *Met. Monogr.*, **1** (3), 1–6.

Levi, M. (1965): Local winds around the Mediterranean Sea. *Misc. Papers. Israel Met. Serv. Ser. C* (13), 1–6.

Lied, N. T. (1964): Stationary hydraulic jumps in a katabatic flow near Davis, Antarctica, 1961. *Aust. Met. Mag.* (47), 40–51.

Lilly, D. K. (1971): Observations of mountain induced turbulence. *Jour. Geoph. Res.*, **76** (27), 6585–6588.

Long, R. R. (1953 a): A laboratory model resembling the "Bishop Wave" phenomenon. *Bull. Amer. Met. Soc.*, **34** (5), 205–211.

Long, R. R. (1953 b): Some aspects of the flow of stratified fluids. I: A theoretical investigation. *Tellus*, **5** (1), 42–58.

Long, R. R. (1954): Some aspects of the flow of stratified fluids. II: Experiments of a two-fluid system. *Tellus*, **6** (2), 97–115.

Long, R. R. (1955): Some aspects of the flow of stratified fluids. III: Continuous density gradients. *Tellus*, **7**, 341–357.

Longley, R. W. (1966–7): The frequency of chinooks in Alberta. *Albertan Geog.*, **3**, 20–22.

Longley, R. W. (1967): The frequency of winter chinooks in Alberta. *Atmosphere*, **5** (4), 4–16.

López, M. E., Howell, W. E. (1967): Katabatic winds in the equatorial Andes. *Jour. Atmos. Sci.*, **24**, 29–35.

Lowry, W. P. (1963): Observations of atmospheric structure during summer in a coastal mountain range in northwest Oregon. *Jour. Appl. Met.*, **2** (6), 713–721.

Lyra, G. (1943): Theorie der stationären Leewellenströmung in freier Atmosphäre. *Z. angew. Math. Mech.*, **23**, 1–28.

MacHattie, L. B. (1970): Kananaskis valley temperatures in summer. *Jour. Appl. Met.*, **9** (4), 574–582.

Magata, M. (1969): On the study of the airflow over mountains by the numerical experiment. *Pap. Met. Geoph.*, **20** (2), 91–110.

Magono, C., Yamazaki, T. (1965): A laboratory model experiment of the cloud bands over the sea of Japan. Proc. Autumnal Meet., Met. Soc. Japan, 82–83.

Magono, C. (1971): On the localization phenomena of snowfall. *Jour. Met. Soc. Japan*, **46** (Sp. Iss.), 824–836.

Makino, Y., Hitsuma, M. (1973): A statistical study of line echo as observed by Mt. Fuji radar. *Tenki*, **20** (1), 29–38.

Malsch, W. (1952): Luftlawinen aus der freien Atmosphäre. *Met. Rdsch.*, **5** (1/2), 20–21.

Manley, G. (1945): The Helm wind of Crossfell, 1937–1939. *Q.J.R.M.S.*, **71**, 197–219.

Mano, H. (1953): Sudden increase of nocturnal temperature in valleys or basins. *Jour. Met. Res.*, **5** (6), 525–545.

Mano, H. (1956): A study on the sudden nocturnal temperature rise in the valley and the basin. *Geoph. Mag.*, **27**, 169–204.

Marsh, J. S. (1968–9): Early references to the chinook. *Albertan Geog.*, **5**, 61–64.

Masatsuka, A. (1949): Investigation of temperature lapse rate and inversion on the Mt. Aso. (1) (2), *Jour. Met. Res.*, **1** (12), 395–401, 402–405.

Masatsuka, A. (1960): A comparative study of meteorological and hydrologic rainfalls on the windy slope near the summit (1st Report). *Jour. Met. Res.*, **12**, 681–690.

Mashkova, G. B. (1965): Atmospheric stratification characteristics in inversions. In: Investigation of the bottom 300 meter layer of the atmosphere. ed. by N. L. Byrova, Israel Progr. Sci. Transl., Jerusalem, 43–47.

Matsumoto, S., Ninomiya, K., Akiyama, T. (1967): A synoptic and dynamic study on the three dimensional structure of meso-scale disturbances observed in the vicinity of a cold vortex center. *Jour. Met. Soc. Japan*, **45**, 64–82.

Matsumoto, S., Tsuneoka, Y. (1969): Some characteristic features of the heavy rainfalls observed over Western Japan on July 9, 1967. (Pt. 2). *Jour. Met. Soc. Japan*, **47** (4), 267–278.

Mather, K. B., Miller, G. S. (1967): Notes on topographic factors affecting the surface wind in Antarctica, with special reference to katabatic winds; and bibliography. *Univ. Alaska, Tech. Rept. UAG-R-189*, 1–125.

Meiklejohn, J. (1955): The local wind at Milfield, Northumberland. *Q.J.R.M.S.*, **81**, 468–474.

Meteorological Office (1963): Weather in the Black Sea. Her Majesty's Stationery Office, London, 1–294.

Milosavljević, M. (1959): Einfluss der Transsylvaner Alpen auf die Struktur des östlichen Windes in der Pannonischen Ebene. *Ber. Deutsch. Wett.*, (54), 236–240.

Mitchell, J. M. (1956): Strong surface winds at Big Delta Alaska. *Mon. Wea. Rev.*, **84**, 15–24.

Miyazawa, S. (1960): On the heavy snowfall by the local front in the Hokuriku District on December 29–31, 1957. *Jour. Met. Res.*, **12**, 370–401.

Miyazawa, S. (1962, 1964, 1965, 1966): Study on the heavy snowfall in Hokuriku District. *Jour. Met. Res.*, **14**, 703–718; **16**, 491–497; **17**, 751–758; **18**, 22–29.

Miyazawa, S. (1968): A mesoclimatological study on heavy snowfall. *Pap. Met. Geoph.*, **19** (4), 487–550.

Möller, F. (1951): Long-wave radiation. *Comp. Met.*, 34–49.

Munn, R. E. (1963): Micrometeorology of Douglas Point. *Canadian Met. Mem.*, (12), 1–39+30 Figs.

Munn, R. E., Emslie, J. H., Wilson, H. J. (1963): A preliminary analysis of the inversion climatology of Southern Ontario. A Paper presented at the 56th Ann. Meeting APCA, June 12, 1963. 13p. +Figs, Tables.

Musaelyan, Sh. A. (1964): "Barrier waves in the atmosphere." (Transl. from Russian.) I.P.S.T., Jerusalem, 1–112.

Nakajima, C., Gocho, Y. (1968): On the heavy rainfall in the Kinki district, Western Japan. *Ann. Disaster Prev. Res. Inst.*, **17**, Pt. 3, (128) 29–44.

Nakajima, C., Yoshioka, H. (1971): On air-sea interaction in the Kii Channel. *Ann. Disaster Prev. Res. Inst.*, **20**, 217–226.

Nakamura, K. (1966): Prevailing southerly winds in winter in the Hokuriku District, Japan. *Geog. Rep. Tokyo Met. Univ.*, (1), 149–162.

Nakayama, A. (1953 a): On the so-called secondary orographic frontal rainfall. (1) (2) (3). *Jour. Met. Res.*, **4**, 913–917, 919–923, 975–979.

Nakayama, A. (1953 b): On the cold front passing over the mountainous region in the central part of Japan (2nd report) (On the strong wind near Fuji-machi). *Jour. Met. Res.*, **4**, 963–974.

Nakayama, A. (1960): Small scale analysis of heavy rainfall in Chubu District on 26 August 1959. *Jour. Met. Res.*, **12** (7), 437–445.

Nemoto, S. (1961 a, b, c, 1962): Similarity between natural wind in the atmosphere and model wind in wind tunnel (I, II, III, IV). *Pap. Met. Geoph.*, **12** (1), 30–52; **12** (2), 117–128; **12** (2), 129–154, **13** (2), 171–195.

Nicholls, J. M. (1973): The airflow over mountains. Research 1958–1972. *WMO Tech. Note* (127), 1–74.

Niedźweidź, T. (1973): Temperatura i wilgotność powietrza w warunkach rzeźby pogórskiej karpat (Temperature and air humidity of a hilly landscape). *Prace Geog.*, (32), 7–88.

Niedźwiedź, T., Obrebska-Starklowa, B., Olecki, Z. (1973): Stosunki termiczno-wilgotnościowe wybranych zbiorowisk roślinnych w zachodniej cześci Świetokrzyskiego parku narodowego (The hygrothermic conditions of chosen vegetational associations in western part of the Świetokrzyski national park). *Folia Geogr., Ser. Geographica-physica*, **7**, 27–75.

Nishimura, D. (1932): On the rainfall in Japan. *Jour. Met. Soc. Japan.*, **10**, 128–215.

Nitze, F. W. (1936): Untersuchung der nächtlichen Zirkulationsströmung am Berghang durch stereophotogrammetrisch vermessene Ballonbahnen. *Biokl. B.*, **3**, 125–127.

Nosek, M. (1964): Dynamic aspects of the urban climate. *Sbornik Československé Společnosti Zeměpisné*, **69**, 77–86.

Obrebska-Starkel, B. (1970): Über die thermische Temperaturschichtung in Bergtälern. *Acta Clima.*, **9** (1–4), 33–47.

Okabayashi, T. (1972): Kishôeisei kara mita yukigumo to kôsetsu ni tsuite no kenkyû eno riyô. (Clouds with snowfall taken from the meteorological satellites and their application to studying snowfall.) *Kishô Kenkyû Nôto*, (113), 74–106.

Ooi, S. (1951): Synoptic researches of winter weather in Japan (1).——Inland high and local fronts.

——*Jour. Met. Res.*, **3** (11), 386–400.

Orlicz, M. (1954): A stosunkach anemometrycznych na szcytach tatraza skich (Wind conditions at the summit of Tatra Mountains). *Hydrologiczny i Meteorologiczney*, **3** (4), 316–337.

Osaka Meteorological Observatory (1956): "Hirota-kaze." 1–58.

Osaka Meteorological Observatory (1958): "Yamaji-kaze." 1–57.

Ôta, M. (1960): Gyakutensô to kemuri no kakusan (Inversion layer and diffusion of smoke). *Kishô Kenkyû Nôto*, **11**, 303–332.

Ôtani, T. (1956): Oroshi which causes poverty. *Tenki*, **3**, 65–68.

Ozaki, K. (1973): A study of forecasting methods for meteorological damage to agriculture (Pt. 3). *Jour. Met. Res.*, **25** (1), 1–20.

Paradiž, B. (1959): Bura u Sloveniji (Bora in Slovenia). *Zagreb, Hidrometeorološki Zavod NR Hrvatske, Rasprave i Prikazi*, **8** (4), 165–169.

Plaetschke, J. (1953): Zur Bildung der Kälteseen in Tälern und Mulden. *Zeitsch. Met.*, **7** 346–347.

Polli, S. (1970): Valori normali del clima di Trieste (Normal values of climate of Trieste). Istituto Sperimentale Talassografico "Francesco Vercelli" Pubbl., (460), 1–9.

Preston-Whyte, R. A. (1971): Instability line thunderstorms in Natal. *S. African Geog. Jour.*, **53**, 70–77.

Prügel, H. (1949): Wolkenstrassen bei schwachen Winden. *Ann. Met.*, **2** (3/4), 99–104.

Prügel, H. (1950): Wolkenstrassen. *Met. Rdsch.*, **3** (5/6), 131–133.

Putnins, P. (1970): The climate of Greenland. In: Climates of the polar regions. *World Survey of climatology*, **14**, 3–128.

Queney, P. (1947): Theory of perturbations in stratified currents with application to airflow over mountain barriers. *Pub. Dept. Met., Univ. Chicago, Misc. Rep.*, (23), 1–81.

Reiher, M. (1936): Nächtlicher Kaltluftfluss an Hindernissen. *Biokl. B.*, **3**, 152–163.

Reiter, R. (1964): "Felder, Ströme und Aerosole in der unteren Troposphäre." Dr. Dietrich Steinkopff Verl., Darmstadt, 1–603.

Reynolds, R. D., Lamberth, R. L., Wurtele, M. G. (1968): Investigation of a complex mountain wave situation. *Jour. Appl. Met.*, **7** (3), 353–358.

Riehl, H. (1974): On the climatology and mechanisms of Colorado chinook winds. *Bonner Met. Abh.* (17), 493–504.

Rink, J. (1953): Über das Verhalten des mittleren vertikalen Temperaturgradienten der bodennahen Luftschicht (1–76m) und seine Abhängigkeit von speziellen Witterungsfaktoren und Wetterlagen. *Abh. Met. Dienst.*, (18), 1–43.

Rodewald, M. (1950): Wolkenstrassen über Holstein. *Ann. Met.*, **3**, (7/8) 202–213.

Rossi, V. (1948): On the effect of the Scandinavian mountains on the precipitation fronts approaching from the sea. *Fennia*, (4), 1–23.

Sahashi, K. (1962): A study of the down-slope wind (I). *Met. Notes, Met. Res. Inst., Kyoto Univ. Ser.*, **2** (26), 1–72.

Saito, R. et al. (1973): Shûchû-gôu (A heavy rain concentration). Nippon Hôsô Shuppan Kyokai, Tokyo, 1–285.

Samokhvalov, N. F. (1957): Klimaticheskaia kharakteristika sukhoveev Kazakhstana (Climatic characteristics of sukhovey in Kazakhstan). In: Sukhovei, ikh proiskhozhdenie i bor'ba s nimi. Akad. Nauk SSSR, Moscow, 51–58.

Sakanoue, T. (1969): Studies on the mountain precipitation. *Bull. Agr., Kyushu Univ.*, **24**, 29–113.

Schamp, H. (1964): Die Winde der Erde und Ihre Namen. *Erdkundliches Wissen* (8), 1–94.

Schirmer, H. (1951): Niederschlagsstreifen——Spurlinien von Wolkenstrassen——. *Met. Rdsch.*, **4**, 97–99.

Schirmer, H. (1952): Über die räumliche Struktur der Niederschlagsverteilung. *Ann. Met.*, **5**, 248–253.

Schmauß, A. (1952): Über Luftlawinen. *Ber. Deutsch. Wett. US-Zone*, **4**, (31), 14–16.

Schmidt, F. H. (1947): Remarks on some classes of local winds, a contribution to dynamical climatology. *Mededel. Verh. B, Deel*, **1** (5), 1–10.

Schneider, M. (1972): Kaltluftstau an Strassendämmen?——Nicht immer! *Met. Rdsch.*, **25** (6), 187–188.

Schneider-Carius, K. (1953): Die Grundschicht der Troposphäre. Geest & Portig, Leipzig, 1–168.

Schnelle, F. (1956): Ein Hilfsmittel zur Feststellung der Höhe von Forstlagen in Mittelgebirgstälern.

Met. Rdsch., **9**, 180–182.

Schweinfurth, U., Domrös, M. (1974): Local wind phenomena in the central highlands of Ceylon. *Bonner Met. Abh.*, (17), 387–401.

Schwerdtfeger, W. (1970): The climate of the Antarctic. In: Climates of the polar regions. *World Survey of Climatology*, **14**, 253–355.

Scorer, R. S. (1949): Theory of waves in the lee of mountains. *Q.J.R.M.S.*, **75**, 41–56.

Scorer, R. S. (1953, 1954, 1955): Theory of airflow over mountains. II. III. IV. *Q.J.R.M.S.*, **79**, 70–83; **80**, 417–428; **81**, 340–350.

Scorer, R. S. (1958): "Natural aerodynamics." Pergamon Press, London, 1–312.

Sekiguti, T. (1940): On the local names of wind in Japan (1) (2). *Geog. Rev. Japan*, **16** (6), 374–395; (7), 453–476.

Sekiguti, T. (1965):Geographical distribution of typhoon rains in Japan——Formation of orographic rain-bands. *Geog. Rev. Japan*, **38** (8), 501–518.

Sekiya,H. (1958):Kyokuchi no kion (Local air temperature). *Kishô Kenkyû Nôto*, **9** (2), 49–68.

Sendai Meteorological Observatory (1950): "Kiyokawadashi" wind damage report, 1–59.

Shimizu, S. (1964): On the diurnal variation of the winds accompanied by thermal high or low in the central region of Japan. *Tenki*, **11** (4), 138–141.

Shinohara, T. (1949): On the local front that appears in the neighbourhood of Utsunomiya in winter and on the climate of Tochigi Prefecture. *Jour. Met. Soc. Japan*, **27**, 119–122.

Shinohara, T. (1970): On the heavy rainfall of "Uetsu". *Tenki*, **17** (1), 23–28.

Shitara, H. (1958): On the winter weather divide in the Chûgoku region. *Geog. Rev. Japan*, **31** (11), 655–665.

Shitara, H. (1963): Meso-climatic divide seen from the discontinuity of the weather. *Sci. Rep. Tôhoku Univ. 7th Ser.* (12), 21–34.

Shitara, H. (1966): A climatological analysis of the weather distribution in Tôhoku District in winter. *Sci. Rep. Tôhoku Univ. 7th Ser.* (15), 35–54.

Shitara,H. (1967): The distribution of frequency of rainy or cloudy weather in the southern part of Tôhoku District under the "Yamase" cold wind. *Ann. Tôhoku Geog. Ass.*, **19** (3), 140.

Shitara, H. (1969):Distribution of cloudy or rainy weather in the northern Tôhoku under the easterly wind in Baiu season. *Ann. Tôhoku Geog. Ass.*, **21** (4), 220.

Shitara, H. et al. (1973): On the micro-scale distribution of the nocturnal cooling in a small basin. *Sci. Rep. Tôhoku Univ. 7th Ser.*, **23** (2), 163–185.

Sien, C. L. (1968): An analysis of rainfall patterns in Selangor. *Jour. Trop. Geog.*, **27**, 1–18.

Sien, C. L. (1970): Diurnal and seasonal variations of cloud patterns in Singapore. *Jour. Singapore Nat. Acad. Sci.*, **1** (3), 85–90.

Singer, I. A., Smith, M. E. (1953): Relation of gustiness to other meteorological parameters. *Jour. Met.*, **10**, 121–126.

Singer, I. A., Smith, M. E. (1966): Atmospheric dispersion at Brookhaven National Laboratory. *Air Wat. Pollutlon Int. J.*, **10**, 125–135.

Smith, M. E. (1951): The forecasting of micrometeorological variables. *Met. Monogr.*, **1** (4), 50–55.

Smithson, P. A. (1970): Influence of topography and exposure on airstream rainfall in Scotland. *Weather*, **25** (8), 379–386.

Solantie, R. (1968): The influence of the baltic sea and the gulf of Bothnia on the weather and climate of northern Europe, especially Finland, in autumn and in winter. *Finnish Met. Inst. Cont.* (*Helsinki*), (70), 1–28.

Soma, S. (1969): Dissolution of separation in the turbulent boundary layer and its applications to natural winds. *Pap. Met. Geoph.*, **20** (2), 111–174.

Starr, J. R., Browning, K. A. (1972): Observations of lee waves by high-power radar. *Q.J.R.M.S.*, **98**, 73–85.

Streten, N. A. (1963): Some observations of Antarctic katabatic winds. *Aust. Met. Mag.*, (42), 1–23.

Stringer, E. T. (1972 a): "Foundations of climatology". W. H. Freeman and Co., San Francisco, 1–586.

Stringer, E. T. (1972 b): "Techniques of climatology". W. H. Freeman and Co., San Francisco, 1–589.

Strong, A. E. (1972): The influence of a Great Lake anticyclone on the atmospheric circulation. *Jour. Appl. Met.*, **11** (4), 598–612.

Sugawara, Y. (1946): Über die durch topographische Einflüsse entstandenen Diskontinuitätslinien. *Geoph. Mag.*, **14**, 19–25.

Suzuki, H. (1956): An approach toward meso air mass climatology. *Geog. Rev. Japan*, **29**, 347–357.

Suzuki, S., Okanoue, M. (1943): Über den orographischen Einfluss auf die Luftdruckverteilung. *Proc. Physico-Mathematic Soc. Japan 3rd Ser.*, **25** (5), 368–370.

Suzuki, S. (1952): On the orographic influence on atmospheric pressures and current. *Geoph. Mag.*, **23** (4), 349–358.

Suzuki, S. (1954): On the orographic effect on atmospheric pressures and currents (VI). *Geoph. Mag.*, **25** (1–2), 151–158.

Suzuki, S., Yabuki, K. (1956): The air-flow crossing over the mountain range. *Geoph. Mag.*, **27**, 273–291.

Syono, S., Miyakoda, K., Manabe, S., et al. (1959): Broad scale and small scale analysis of a situation of heavy precipitation over Japan in the last period of bai-u season 1957. *Japanese Jour. Geoph.*, **2** (2), 59–103.

Szász, G. (1964): Bestimmung der nächtlichen Mikroadvektion durch Ausstrahlungsmessungen in der bodennahen Luftschicht. *Ang. Met.*, **5** (1/2), 7–12.

Szepesi, D. (1964): Influence of the temperature gradient in the lower 300m air layer on the dispersion of pollutants of industrial origin. *Idöjárás*, **68** (1), 10–17.

Tabukuro, K. (1919): Causes of rain and snowfall in association with monsoon on the coast of Sea of Japan. *Jour. Met. Soc. Japan*, *I*, **38**, 189–194.

Takahashi, K. (1940): On the local rain in Kwantô District due to northeasterly wind. *Jour. Met. Soc. Japan*, **18** (5), 158–160.

Takeda, K., Sakanoue, T., Motoda, Y. (1960): On the study of precipitation on mountains. *Rep. Met. Inst., Kyushu Univ.*, (4), 129–138.

Tamiya, H. (1972): Chronology of pressure patterns with bora on the Adriatic Coast. *Climat. Notes, Hosei Univ.*, (10), 52–63.

Tang, M.-ts. (1963): Diurnal variation of weather in Nan Shan Region. *Acta. Geog. Sinica*, **29** (3), 197–206.

Tatehira, R. (1961): Radar and meso-scale analysis of rainband in typhoon. *Jour. Met. Res.*, **13** (4), 264–279.

Tatehira, R. (1962): Radar and meso-scale analysis of rainband in typhoon Georgia in 1959. *Jour. Met. Res.*, **14** (9), 621–630.

Teichert, F., Greifenhagen, B. (1952): Demonstrationsversuche mit strömender Kaltluft. *Zeitsch. Met.*, **4**, 122–126.

Thambyaphillay, G. (1958): The Kachchan——a foehn wind in Ceylon. *Weather*, **13**, 107–114.

Thomas, T. M. (1963): Some observations of the chinook arch. *Weather*, **18** (6), 166–170.

Thompson, A. H., Dickson, C. R. (1958): Ground layer temperature inversions in an interior valley and canyon. Final. Rep. Quartermaster Research and Engineering Command, United States Army. DA 19-129-QM-399, Dep. Met., Univ. Utah., 1–136, +58p.

Thompson, A. H. (1967): Surface temperature inversions in a canyon. *Jour. Appl. Met.*, **6** (2), 287–296.

Thompson, R. D. (1973): Some aspects of the synoptic mesoclimatology of the Armidale District, New South Wales, Australia. *Jour. Appl. Met.*, **12** (4), 578–588.

Tohsha, M. (1953): Temperature inversion in the lower atmosphere. *Jour. Met. Res.*, **5** (8), 649–654.

Tsuchiya, K., Fujita, T. (1967): A satellite meteorological study of evaporation and cloud formation over the western Pacific under the influence of the winter monsoon. *Jour. Met. Soc. Japan*, **45**, 232–250.

Tyson, P. D. (1968): A note on the nomenclature of the topographically induced local winds of Natal. *S. African Geog. Jour.*, **50**, 133.

Tyson, P. D. (1969 a): Towards a regional model of local topographically-induced wind systems in Natal. *S. African Jour. Sci.*, **65** (7), 201–215.

Tyson, P. D. (1969 b): Statistical model of the mountain wind over Pietermaritzburg. *S. African Jour. Sci.*, **65** (9), 267–272.

Umeda, S. (1958): On the characteristic features of the temperature distribution in small areas at frosty night. *Mem. Ind. Met. Tokyo*, **22** (1), 15–21.

Urfer-Henneberger, Ch. (1969): Zur "warmen Hangzone" im Dischmatal bei Davos. *La Mét.*, **5** (10–11), 99–105.

Urfer-Henneberger, Ch. (1972): Mesoklimatische Temperaturverteilung im Dischmatal. *Verh.*

Schweiz. Naturforsch. Gesellschaft, 205–209.

Utagawa, K. (1964): Southerly wind at Matsumoto and inland high around Takayama in relation to the so-called Hokuriku front. *Jour. Met. Res.*, **16** (9), 498–504.

Utsunomiya Local Weather Observatory (1953): Studies on the counter measure of wind prevention in the Nasu region. Nasu Sangyo Shinkô-Kyôgi-Kai, Nasu, 1–43.

Vergeiner, I., Lilly, D. K. (1970): The dynamic structure of lee-wave flow as obtained from balloon and airplane observations. *Mon. Wea. Rev.*, **98**, 220–232.

Vergeiner, I. (1971 a): An operational linear lee wave model for arbitrary basic flow and two-dimensional topography. *Q.J.R.M.S.*, **97**, 30–60.

Vergeiner, I. (1971 b): Comments on: the chinook at Calgary (Canada) by W. A. R. Brinkmann. *Arch. Met. Geoph. Biokl.* (B), **19**, 339–341.

Vialar, J. (1948): Les vents régionaux et locaux. *Mem. Met. Nationale*, (31), 1–52.

Viezee, W., Collis, R. T. H. (1973): An investigation of mountain waves with lidar observations. *Jour. Appl. Met.*, **12** (1), 140–148.

Voigts, H. (1951): Experimentelle Untersuchungen über den Kaltluftfluss in Bodennähe bei verschiedenen Neigungen und verschiedenen Hindernissen. *Met. Rdsch.*, **4**, 185–188.

Vorontsov, P. A. (1958): Aerologicheskie issledovaniia pogranichnogo sloia atmosferi nad melkosopochnim repiefom zelinnik. *Trudy GGO*, (73), 61–86.

Vorontsov, P. A., Schelkovnikov, M. S. (1956): Opyt aerologicheskogo issledovanniia nizhnego sloia atmosfery v doline Azau. *Trudy GGO,* (63), 138–167.

Wagner, A. (1938): Theorie und Beobachtung der periodischen Gebirgswinde. *Beitr. Geoph.*, **52**, 408–449.

Wagner, R. (1963): Der Tagesgang der Lufttemperatur einer Doline im Bükk-Gebirge. *Acta Climat.*, **2–3**, 49–79.

Wagner, R. (1970): Kalte Luftseen in den Dolinen. *Acta Climat.*, **9** (1/4), 23–32.

Walter, E. (1938): Der Schweizerföhn. *Neujahrsbe. Narturf. Gesell. Zürich*, (140), 1–40.

Watts, I. E. M. (1955): The rainfall of Singapore Island. *Malayan Jour. Tropical Geog.* (7), 1–68.

Watts, I. E. M. (1959): Horizontal distribution of rainfall over a small area in the tropics. *Proc. 9th Pacific Sci. Congr. 1957,* **13** (Meteorology), 99–104.

Wexler, H. (1947): Structure of hurricanes as determined by radar. *Ann. New York Acad. Sci.*, **48** (8), 821–844.

Wilkins, E. M. (1955): A discontinuity surface produced by topographic winds over the Upper Snake River Plain, Idaho. *Bull. Amer. Met. Soc.*, **36**, 397–408.

Winter, F. (1956): Schornsteinrauch veranschaulicht das abendliche Einsetzen des Hangabwindes. *Met. Rdsch.*, **9**, 224.

Yabuki, K., Suzuki, S. (1967): A study on the airflow over mountain. *Bull. Univ. of Osaka Pref. Ser.* B, **19**, 51–193.

Yaji, M. (1969): Synoptic characteristics of the orographic rainband in the east of the Kii Peninsula, Central Japan. *Tokyo Geog. Papers*, (13), 9–23.

Yamashita, H. (1970): The synoptic-climatological characteristics of heavy rainfalls in the region of Pacific Ocean of Tôhoku District. *Ann. Tôhoku Geog. Ass.*, **22** (4), 177–184.

Yamashita, H. (1972): The synoptic-climatological characteristics of the most frequent heavy-rainfall area along the Pacific in Tôhoku District. *Ann. Tôhoku Geog. Ass.*, **24** (1), 1–9.

Yamashita, I. (1954): Surface winds in Tokyo. *Jour. Met. Res.*, **6**, 17–20.

Yanagisawa, Z. (1961): An analysis of stationary rainbands as observed by radar. *Pap. Met. Geoph.*, **12** (3–4), 294–309.

Yoshida, Y., Moriwaki, K. (1972): Some characteristics of the climate of Wright Valley, Victoria Land, Antarctica. In: Chirigaku no Shomondai, Hiroshima, 218–233.

Yoshimura, M. (1972): Chronology of the Bora-day at Ajdovščina and Trieste. *Climat. Notes, Hosei Univ.*, (10), 64–78.

Yoshimura, M., Nakamura, K., Yoshino, M. M. (1974): Local climatological observation in the Senj region, Croatia, and the Ajdovščina region, Slovenia, from November 1972 to January 1973. *Geog. Rev. Japan*, **47** (3), 143–154.

Yoshino, M. M. (1953): The change of local temperature distribution caused by the passing of cold front. *Geog. Rev. Japan*, **27**, 164–168.

Yoshino, M. M. (1955 a): Synoptic climatological study of the precipitation in the Kanto Plain and its surrounding mountainous region (1). *Geog. Rev. Japan,* **28** (8), 371–385.

Yoshino, M. M. (1955 b): Wind direction distribution in the city area of Tokyo. *Tenki,* **2**, 203–207.

Yoshino, M. M. (1958): A strong local wind along River Hohki, Tochigi Prefecture. *Geog. Rev. Japan,* **31** (10), 613–624.

Yoshino, M. M. (1961 a): "Shôkikô (Microclimate)." Chijin-shokan, Tokyo, 1–274.

Yoshino, M. M. (1961 b): Local climate in the Mt. Kirishima region. *Tokyo Geog. Papers,* **5**, 79–109.

Yoshino, M. M. (1968 a): Surface inversion at night in Japan. *Jour. Agricul. Met.,* **23** (4), 186–188.

Yoshino, M. M. (1968 b): Problems in local and microclimatology in relation to agriculture in Japan. In: Agroclimatological Methods, Proc. Reading Symp., UNESCO, 269–280.

Yoshino, M. M. (1969): Synoptic and local climatological study on Bora in Yugoslavia. *Geog. Rev. Japan,* **42** (12), 747–761.

Yoshino, M. M. (1970): Oroshi, ein starker Lokalwind in der Kanto-Ebene, Japan. *Colloquium Geographicum,* **12**, 43–57.

Yoshino, M. M. (1971): Die Bora in Jugoslawien; Eine synoptisch-klimatologische Betrachtung. *Ann. Met.,* (5), 117–121.

Yoshino, M. M. (1972): An annotated bibliography on bora. *Climat. Notes, Hosei Univ.,* (10), 1–22.

Young, F. D. (1921): Nocturnal temperature inversions in Oregon and California. *Mon. Wea. Rev.,* **49**, 138–148.

Chapter 6

Aulitzky, H. (1963 a): Bioklima und Hochlagenaufforstung in der subalpinen Stufe der Inneralpen. *Schweizerischen Zeitschrift für Forstwesen.* (1/2), 1–25.

Aulitzky, H. (1963 b): Grundlagen und Anwendung des vorläufigen Wind-Schnee-Ökogramms. *Mitt. d. forstl. Bundes-Vers. Mariabrunn,* (60), 763–834.

Aulitzky, H. (1965): Waldbau auf bioklimatischer Grundlage in der subalpinen Stufe der Innenalpen. *Centralblatt für das Gesamte Forstwesen, Wien,* **82** (4), 217–245.

Balchin, W. G. V., Pye, N. (1950): Observations on local temperature variations and plant responses. *Jour. Ecol.,* **38** (2), 345–353.

Barsch, D. (1963): Wind, Baumform und Landschaft. Eine Untersuchung des Windeinflusses auf Baumform und Kulturlandschaft am Beispiel des Mistralgebietes im französischen Rhônetal. *Freiburger Geog. Hefte,* (1), 21–130.

Black, R. F., Barksdale, W. L.(1949): Oriented lakes of Northern Alaska. *Jour. Geol.,* **57**, 105–118.

Borsy, Z. (1972): A szélerósió vizsgálata a magyarországi futóhomok területeken (Investigations of erosion by wind in the windblown sand areas of Hungary). *Földrajzi Közlemények,* **20** (2/3), 156–160.

Braun-Blanquet, J. (1961): "Die inneralpine Trockenvegetation. Von der Provence bis zur Steiermark." Gustav. Fischer Verl., Stuttgart, 1–273.

Büdel, J. (1944): Die morphologischen Wirkungen des Eiszeitklimas im gletscherfreien Gebiet. *Geol. Rdsch.,* **34**, 482–519.

Cabot, E. C. (1947): The Northern Alaskan coastal plain interpreted from aerial photographs. *Geog. Rev.,* **37**, 639–648.

Cantlon, J. E. (1953): Vegetation and microclimates on north and south slopes of Cushetunk Mountain, New Jersey. *Ecol. Monogr.,* **23**, 241–270.

Carson, C. E., Hussey, K. M. (1962): The oriented lakes of Arctic Alaska. *Jour. Geol.,* **70**, 417–439.

Chiba, T. (1949): Asymmetrical hill slopes in Dauria district. *Geog. Rev. Japan,* **22**, 106–112.

Chiba, T. (1950): On the hill-side slopes of the boundary districts of steppe and forest. *Geog. Rev. Japan,* **23** (2–5), 50–56.

Chung, D.-o. (1962): On the present bamboo groves of Cholla-nam-do and their proper treatment. *Jour. Korean Forest Soc.,* (2), 19–28.

Daubenmire, R. F. (1950): Plants and environment. Wiley, New York, 285–287.

Domes, N. (1950): Untersuchungen über die Windverbreitung der Früchte und Samen mitteleuropäischer Waldbäume. *Forstw. Cbl.,* **69** (10), 606–624.

Früh, J. (1902): Die Abbildung der vorherrschenden Winde durch die Pflanzenwelt. *Jahrb. Geog. Gesell., Zürich.* 1901–1902, 57–153.

Gilbert, G. K. (1904): Systematic asymmetry of crest lines in the high Sierra of California. *Jour. Geol.,*

12, 579–588.

Hartmann, F. K., Eimern, J. v., Jahn, G. (1959): Untersuchungen reliefbedingter kleinklimatischer Fragen im Gelände querschnitten der hochmontanen und montanen Stufe des Mittel und Süd-westharzes. *Ber. Deutsch. Wett.*, (50), 1–39.

Hartmann, F.-K., Schnelle, F. (1970): "Klimagrundlagen natürlicher Waldstufen und ihrer Wald-gesellschaften in deutscher Mittelgebirgen." Gustav Fischer Verl., Stuttgart, 1–176.

Hastings, J. D. (1971): Sand streets. *Met. Mag.*, **100**, 155–160.

Higuchi, K., Iozawa, T. (1971): Atlas of perennial snow patches in Central Japan. Nagoya Univ., Nagoya.

Holtmeier, F.-K. (1968): Entgegnung zu: "Über Schneeschliff in den Alpen" von Hans Turner. *Wett. Leben*, **20**, 201–205.

Holtmeier, F.-K. (1971): Der Einfluss der orographischen Situation auf die Windverhältnisse im Spiegel der Vegetation. *Erdkunde*, **25** (3), 178–195.

Ichikawa, M. (1960): Landslides and flood damage in the upper drainage basin of the Kano River, Izu Peninsula, Japan. *Geog. Rev. Japan*, **33**, 112–121.

Ishikawa, M., Suzuki, T. (1965): Frost action and soil stability on slope (II). *Ann. Rep. Hokkaido Branch, Forest Exp. St. 1964*, 218–237.

Iwatsuka, S. (1954): A physical geographical study on the landslides along the national railroads in Japan. *Tokyo Daigaku Chirigaku Kenkyu* (3), 97–114.

Karrasch, H. (1970): Das phänomen der klimabedingten Reliefasymmetrie in Mitteleuropa. *Göttinger Geog. Abh.*, (56), 1–299.

Kawada, M. (1933): "Shinrin-seitaigaku-kogi (Lectures on forest ecology)." Yôkendo, Tokyo, 99–106.

Kawamura, T. (1960): The distribution of heavy rainfall associated with the Kanogawa typhoon on the Izu Peninsula. *Geog. Rev. Japan*, **33** (3), 105–112.

Keay, R. W. J. (1957): Wind-dispersed species in a Nigerian forest. *Jour. Ecol.*, **45**, 471–478.

Kennedy, B. A., Melton, M. A. (1972): Valley asymmetry and slope forms of a permafrost area in the Northwest Territories, Canada. In: Polar geomorphology. ed. by R. J. Price and D. E. Sugden, Inst. British Geographers, Sp. Publ., London (4), 107–121.

Kim, D.-j. (1967): Die dreidimensionale Verteilung der Strukturböden auf Island in ihrer klimatischen Abhängigkeit. Inaug.-Diss. Erl. Doktorgr. Math.-Naturw. Fak. Univ. Bonn, 1–277.

King, R. B. (1971): Vegetation destruction in the sub-alpine and alpine zones of the Cairngorm Mountains. *Scott. Geog. Mag.*, **87** (2), 103–115.

Kušan, F. (1970): Velebitski botanički vrt (Stručni vodič). *Senjski Zbornik*, **4**, 71–98.

Lawrence, D. B. (1939): Some features of the vegetation of the Columbia River Gorge with special reference to asymmetry in forest trees. *Ecol. Monogr.*, **9**, 217–257.

Livingstone, D. A. (1954): On the orientation of lake basins. *Amer. Jour. Sci.*, **252** (9), 547–554.

Mackey, J. R. (1963): The Mackenzie Delta Area, N. W. T. *Dept. of Mines and Tech. Surveys, Canada, Geog. Branch Memoir*, (8), 1–202.

Maruyama, K. (1971): Effect of altitude on dry matter production of primeval Japanese beech forest communities in Naeba Mountains. *Mem. Fac. Agricul., Niigata Univ.* (9), 87–171.

Melton, M. A. (1960): Intravalley variation in slope angles related to microclimate and erosional environment. *Bull. Geol. Soc. Amer.*, **71**, 133–144.

Mino-Ishikawa, Y. (1972): Some aspects of the patterned ground. *Chiikikenkyu, Risshô Univ.*, **13** (2), 12–26.

Murakami, K. (1956): Landslips in volcano Akagi, Gumma Prefecture. *Geog. Rev. Japan*, **29**, 209–217.

Murayama, K. (1971): Effect of altitude on dry matter production of primeval Japanese beech forest communities in Naeba Mountains. *Mem. Fac. Agricul., Niigata Univ.*, (9), 87–171.

Nakano, T. (1956): "Nippon no heiya (Plains of Japan)." Kokonshoin, Tokyo, 249–256.

Naruse, T. (1971): Coastal sand dunes of Tanegashima Island. *Geog. Rev. Japan*, **44** (10), 697–706.

Numata, M., Mitsudera, M., Ogawa, K. (1957): Habitat conditions of the bamboo forest. *Chiba Univ., Fac. of Lit. and Sci., Kiyô*, **2** (2), 162–171.

Numata, M. (1958): "Seitaigaku no tachiba (Standpoint of ecology)." Kokonshoin, Tokyo, 168–170.

Obrebska-Starklowa, B. (1969): Stosunki mikroklimatyczne na pograniczu pieter leśnych i pól uprawnych w gorcach. (Microclimatic conditions on the border of forest belts and arable land in the Gorce Mts.). *Prace Geog.* (23), 1–141.

Ogasawara, K. (1964): Snow survey of Mt. Tateyama and Mt. Tsurugi of the Japanese North Alps. In: Nature of Kita Arupusu. ed. by Toyama-Daigaku Gakujutsu Chôsadan, Toyama, 123–152.

Olecki, Z., Widacki, W. (1970): Zwiazek mrozowych ruchów gruntu z warunkami termicznymi w Gaiku-Brzezowej (Pogórze Wielickie). (The course of frost movements in the soil at Gaik-Brzerowa, Wieliczka Foothills.) Studia Geomorph. Carpatho-Balcanica, 4, 107–120.

Oliver, J. (1960): Wind and vegetation in the Dale Peninsula. Field Studies, 1 (2), 1–12.

Ollier, C. D., Thomasson, A. J. (1957): Asymmetrical valleys of the Chiltern Hills. Geog. Jour., 123, 71–80.

Oosting, H. J., Hess, D. W. (1956): Microclimate and relic stand of Tsuga Canadensis in the lower piedmont of North Carolina. Ecology, 37, 28–39.

Oota, T. (1956): Consideration on the distribution of the subalpine deciduous broad-leaved forest zone. Jour. Japanese Forestry Soc., 38 (12), 482–487.

Owada, M., Yoshino, M. M. (1971): Prevailing winds in the Ishikari Plain, Hokkaido. Geog. Rev. Japan, 44 (9), 638–652.

Owada, M. (1973): Prevailing winds in the Konsen-Genya, southeastern Hokkaido. Geog. Rev. Japan, 46 (8), 505–515.

Perring, F. H. (1960): Climatic gradients of chalk grassland. Jour. Ecol., 48, 415–442.

Perring, F. H. (1962): The Irish Problem. Proc. Bounemouth Natural Sci. Soc., 52, 1–13.

Plesnik, P. (1971): Horná hranica lesa. +Mapová priloha vo Vysokých a v Belanských Tatrách (Die obere Waldgrenze in der Hohen und Belaner Tatra. +Karten). Vydavateľstvo Slov. Akad. Vied, Bratislava, 1–238. +Karten.

Plesnik, P. (1973): Some problems of the timberline in the Rocky Mountains compared with Central Europe. Arc. Alp. Res., 5 (3, pt. 2), A 77–84.

Poore, M. E. D., McVean, D. N. (1957): A new approach to Scottish mountain vegetation. Jour. Ecol., 45, 401–439.

Primault, B. (1971): Essai de comparaision des champs d'application de la méthode d'écologie appliquée et de l'analyse climatologique. Arbeitsberichte der Schweizerischen Meteorologischen Zentralanstalt, Zürich, (23), 1–7.

Putnam, P. C. (1948): "Power from the wind." Van Nostrand, New York, 1–224.

Rasche, H. H. (1958): Temperature differences in Harvard forest and their significance. Headq. Q. Res. & Eng. Command Tech. Rep. EP-80, 1–153.

Rice, E. L. (1960): The microclimate of a relict stand of sugar maple in Devils Canyon in Canadian County, Oklahoma. Ecology, 41, 445–453.

Rudberg, S. (1968): Wind erosion–preparation of maps showing the direction of eroding winds. Biuletyn Peryglacjalny, 17, 181–193.

Runge, F. (1955): Windgeformte Bäume und Sträucher und die von ihnen angezeigte Windrichtung. Met. Rdsch., 8, 177–179.

Runge, F. (1957): Windgeformte Bäume an der Italienischen Riviera. Met. Rdsch., 10, 47–48.

Russell, R. J. (1931): Geomorphological evidence of a climatic boundary. Science, 74, 484–485.

Sakanoue, T. (1972): Studies of the calamities caused by heavy rainfall (II). Jour. Agricul. Met., Tokyo, 27 (3), 85–92.

Sakharov, M. I. (1952): Influence of forest type on the temperature of spruce trunk. Akad. Nauk SSSR. Doklady Ser. Mat. Fiz., 85 (6), 1373–1376.

Sato, H., Kira, T. (1960): Plant production on slopes. I. Plant yield as related to total radiation received by different slopes. Physiology and Ecology, 9 (2), 70–78.

Sato, M. (1960): Range of the Japanese lichens (VI). Bull. Fac. Arts. and Sci., Ibaraki Univ., Nat. Sci., Japan, (11), 53–62.

Schmithüsen, J. (1968): "Allgemeine Vegetationsgeographie." Walter de Gruyter & Co., Berlin, 1–463.

Schweinfurth, U. (1956): Über klimatische Trockentäler im Himalaya. Erdkunde, 10, 297–302.

Scultetus, H. R. (1969): "Klimatologie." Westermann, Braunschweig, 1–163.

Shanks, R. E., Norris, F. H. (1950): Microclimatic variation in a small valley in Eastern Tennessee. Ecology, 31, 532–539.

Smith, H. T. U. (1949): Physical effects of pleistocene climatic changes in nonglaciated areas. Bull. Geol. Soc. Amer., 60, 1485–1516.

Soons, J. M., Rainer, J. N. (1968): Micro-climate and erosion processes in the Southern Alps, New

Zealand. *Geog. Ann.*, **50**A, 1–15.

Spurr, S. H. (1956): Forest associations in the Harvard Forest. *Ecol. Monogr.*, **26**, 245–262.

Spurr, S. H. (1957): Local climate in the Harvard forest. *Ecology*, **38**, 37–46.

Stamp, L. D. (1950): "The land of Britain. Its use and misuse". 261–263.

Strahler, A. N. (1950): Equilibrium theory of erosional slopes approached by frequency distribution analysis. *Amer. Jour. Sci.*, **248**, 670–696, 800–814.

Sugiyama, T., Saeki, M. (1965): Influence of the vegetation upon the occurrence of avalanches. Rep. Cooperative Res. f. Disaster Prevention, (National Res. Center Dis. Prev., Tokyo), (3), 29–41.

Svensson, H. (1972): Vindaktivitet på Laholmsslätten (Wind activity on the Laholm Plain). *Svensk Geografisk Årsbok*, **48**, 65–85.

Takeshita, K. (1971): Estimations of mountain disasters occurrence and their location analysis on the Kita-Kyushu. *Dept. of Forestry, Fukuoka Pref. Forest Exp. St., Fukuoka, Bull. Mountain Conservation*, (1), 1–85.

Tatewaki, M. (1953): Forms of trees in the district influenced by the sea fog along the Pacific side of eastern Hokkaido. *Bull. Soc. Plant Ecology, Sendai*, **2** (4), 162–169.

Temple, P. H. (1965): Some aspects of cirque distribution in the west-central Lake District, Northern England. *Geog. Ann.*, **42**A, 185–193.

Thorarinsson, S. (1951): Notes on patterned ground in Iceland with particular reference to the Icelandic "Flas". *Geog. Ann.*, **33**, 144–156.

Troll, C. (1944): Strukturböden, Solifluktion und Frostklimate der Erde. *Geol. Rdsch.*, **34** (7/8), 545–694.

Troll, C. (1952): Die Lokalwinde der Tropengebirge und ihr Einfluß auf Niederschlag und Vegetation. *Bonner Geog. Abh.*, (9), 124–182.

Troll, C. (1963): Landscape ecology and land development with special reference to the tropics. *Jour. Trop. Geog.*, **17**, 1–11.

Troll, C. (1969): Inhalt, Probleme und Methoden geomorphologischer Forschung. *Beih. Geol. Jb.*, *Geowiss. Tagung, Berlin*, 1967, **80**, 225–257.

Troll, C. (1973): The upper timberlines in different climatic zones. *Arc. Alp. Res.*, **5** (3, pt. 2), A 3–18.

Turner, H. (1968): Über "Schneeschliff" in den Alpen. *Wett. Leben*, **20**, 192–200.

Turner, H. (1971): Mikroklimatographie und ihre Anwendung in der Ökologie der subalpinen Stufe. *Ann. Met.*, (5), 275–281.

Walter, H. (1951): "Einführung in die Phytologie." III. Grundlagen der Pflanzenverbreitung. Stuttgart, 493–498.

Washburn, A. L. (1956): Classification of patterned ground and review of suggested origins. *Bull. Geol. Soc. Amer.*, **67**, 823–865.

Washburn, A. L. (1973): "Periglacial processes and environments." Edward Arnold, London, 1–320.

Weaver, J. E., Clements, F. E. (1938): "Plant ecology." McGraw-Hill, New York, 1–601.

Weischet, W. (1951): Die Baumneigung als Hilfsmittel zur geographischen Bestimmung der klimatischen Windverhältnisse. *Erdkunde*, **5**, 221–227.

Weischet, W. (1953): Zur systematischen Beobachtung von Baumkronendeformationen mit Klimatologischer Zielsetzung. *Met. Rdsch.*, **6**, 185–187.

Weischet, W. (1955): Die Geländeklimate der niederrheinischen Bucht und ihrer Rahmenlandschaften. *Münchner Geog. Hefte*, (8), 1–169.

Weischet, W. (1963): Grundvoraussetzungen, Bestimmungsmerkmale und klimatologische Aussagemöglichkeit von Baumkronendeformationen. *Freiburger Geog. Hefte*, (1), 5–19.

Whitehead, F. H. (1954): A study of the relation between growth form and exposure on Monte Maiella, Italy. *Jour. Ecol.*, **42**, 180–186.

Whitehead, F. H. (1957): Wind as a factor in plant growth. In: Control of the plant environment. Proc. Univ. Nottingham Fourth Easter School in Agricul. Sci., 1957, 84–94.

Whittaker, R. H. (1960): Vegetation of the Siskiyou mountains, Oregon and California. *Ecol. Monogr.*, **30**, 279–338.

Whittaker, R. H., Niering, W. A. (1965): Vegetation of the Santa Catalina Mountains, Arizona: a gradient analysis of the south slope. *Ecology*, **46** (4), 429–452.

Wilson, J. W. (1959): Notes on wind and its effects in arctic-alpine vegetation. *Jour. Ecol.*, **47**, 415–427.

Wolfe, J. N., Wareham, R. T., Scofield, H. T. (1949): Macroclimates and microclimates of Neotoma, a small valley in Central Ohio. *Bull. Ohio Biol. Survey*, **8** (1), 1–267.

Wolfe, J. N. (1951): The possible role of microclimate. *Ohio Jour. Science*, **51**, 134–138.

Yoshino, M. M., Yoshino, M. T. (1963): Lokalklima und Vegetation im Kirishima-Gebirge im südlichen Kyûshû, Japan. *Erdkunde*, **17** (3/4), 148–165.

Yoshino, M. M. (1964): Some local characteristics of the winds as revealed by wind-shaped trees in the Rhône Valley in Switzerland. *Erdkunde*, **18** (1), 28–39.

Yoshino, M. M. (1967): Wind-shaped trees as indicators of micro and local climatic wind situation. *Biometeorology*, **2**, 997–1005.

Yoshino, M. M. et al. (1972): On the local wind distribution and the wind break density in the Shari-Abashiri region, Hokkaido. *Jour. Agricul. Met., Tokyo*, **27** (4), 145–152.

Yoshino, M. M. (1973 a): Studies on wind-shaped trees: their classification, distribution and significance as a climatic indicator. *Climat. Notes, Hosei Univ.*, (12), 1–52.

Yoshino, M. M. (1973 b): Wind-shaped trees in the subalpine zone in Japan. *Arc. Alp. Res.*, **5** (3, pt. 2), A 115–126.

Yoshino, M. M. et al. (1973 c): Local climatological observation made in the Ajdovščina region, Slovenia, in November, 1970. *Climat. Notes, Hosei Univ.*, (14), 21–40.

Yoshino, M. M. et al. (1974): A study on the Bora region on the Adriatic Coast of Yugoslavia by means of wind-shaped trees. *Geog. Rev. Japan*, **47** (3), 155–164.

Zoller, H. (1954): Die Typen der *Bromus erectus*-Wiesen der Schweizer Juras; ihre Abhängigkeit von den Standortsbedingungen und wirtschaftlichen Einflüssen und ihre Beziehungen zur ursprünglichen Vegetation. *Beitr. z. geobotan. Landesaufnahme d. Schweiz, Bern* (33), 1–309.

ABBREVIATION OF NAMES OF PERIODICALS

Abh. Met. Dienst.	Abhandlungen des Meteorologischen und Hydrologischen Dienstes der Deutschen Demokratischen Republik (Berelin)
Acta Climat.	Acta Climatologica (Szeged, Hungary)
Acta Met. Sinica	Acta Meteorologica Sinica (Peking)
Advan. Geoph.	Advances in Geophysics
AF Geoph. Res. Pap.	Air Force Geophysics Research Papers, U.S.A.
Agricul. Met.	Agricultural Meteorology (Amsterdam)
Akad. Nauk S.S.S.R., Doklady. Ser. Mat. Fiz.	Akademiia Nauk S.S.S.R., Doklady. Seria Matematika, Fizika
Akad. Nauk S.S.S.R., Izv. Fiz. Atm. i Okeana	Akademiia Nauk S.S.S.R., Izvestiia. Fizika Atmosfery i Okeana
Akad. Nauk S.S.S.R., Izv. Ser. Geog.	Akademiia Nauk S.S.S.R., Izvestiia. Seria Geografiia
Amer. Jour. Sci.	American Journal of Science
Ang. Met.	Angewandte Meteorologie (Berlin)
Ann. Ass. Amer. Geog.	Annals of the Association of American Geographers
Ann. Disaster Prev. Res. Inst.	Annals of the Disaster Prevention Research Institute, Kyoto University
Ann. Geof.	Annali di Geofisica (Rome)
Ann. Géophy.	Annales de Géophysique (Paris)
Ann. Met.	Annalen der Meteorologie (Offenbach a.M.)
Ann. New York Acad. Sci.	Annals of the New York Academy of Science
Ann. Tôhoku Geog. Ass.	Annals of Tôhoku Geographical Association (Sendai, Japan)
Antarctic Meteorology	Antarctic Meteorology (New York)
Antarctic Rec.	Antarctic Record (Tokyo)
Arc. Alp. Res.	Arctic and Alpine Research (Boulder, Colorado)
Arch. Met. Geoph. Biokl. (A) and (B)	Archiv für Meteorologie, Geophysik und Bioklimatologie, Ser. A and B (Wien)
Arctic	Arctic (Montreal)
Arkiv Geofy.	Arkiv för Geofysik (Stockholm)
Atmosph. Env.	Atmospheric Environment (Oxford, England)
Atmosphere	Atmosphere (Toronto)
Aust. Geog.	Australian Geographer
Aust. Geog. Stud.	Australian Geographical Studies
Aust. Jour. Physics	Australian Journal of Physics (Melbourne)
Aust. Met. Mag.	Australian Meteorological Magazine (Melbourne)
Beitr. Geoph.	Beiträge zur Geophysik (Leipzig)
Beitr. z. geobotan. Landesaufnahme d. Schweiz.	Beiträge zur geobotanischen Landesaufnahme der Schweiz.
Beitr. zur Phys. d. fr. Atm.	Beiträge zur Physik der freien Atmosphäre
Ber. Deutsch. Wett.	Berichte des Deutschen Wetterdienstes (Offenbach a.M.)
Ber. Deutsch. Wett. US-Zone	Berichte des Deutschen Wetterdienstes in der US-Zone
Biokl. B	Bioklimatische Beiblätter der Meteorologischen Zeitschrift.
Bound. Layer Met.	Boundary Layer Meteorology (Dordrecht)
Bull. Amer. Met. Soc.	Bulletin of the American Meteorological Society
Bull. Geol. Soc. Amer.	Bulletin of the Geological Society of America
Canad. Geog.	The Canadian Geographer
Cen. Proc. Roy. Met. Soc.	Centenary Proceedings of the Royal Meteorological Society (London, 1950)
Comp. Met.	Compendium of Meteorology (Boston)

Contr. Atm. Phy. (Beitr. Phy. Atm.)	Contributions to Atmospheric Physics (Oxford, England) (formerly, Beiträge zur Physik der Atmosphäre)
Ecol. Monogr.	Ecological Monograph (Durham, N.C., U.S.A.)
Ecology	Ecology (Durham, N.C., U.S.A.)
Env. Res.	Environmental Research (New York)
Env. Sci. Tech.	Environmental Science and Technology (Washington)
Erde	Erde (Berlin)
Erdkunde	Erdkunde (Bonn)
Forstw. Cbl.	Forstwissenschaftliches Centralblatt
Geof. Met.	Geofisica e Meteorologia, Genoa
Geof. Pura Appl.	Geofisica Pura e Applecata
Geog. Ann.	Geografiska Annaler
Geog. Jour.	Geographical Journal (Royal Geographical Society, London)
Geog. Obshch. S.S.S.R., Izv.	Geograficheskoe Obshchestov S.S.S.R., Izvestiia
Geog. Rep.	Geographical Reports (Tokyo Metropolitan University)
Geog. Rev.	Geographical Review (American Geographical Society)
Geog. Rev. Japan	Geographical Review of Japan
Geog. Stud.	Geographical Studies
Geol. Rdsch.	Geologische Rundschau
Geoph. Jour.	Geophysical Journal, London
Geoph. Mag.	Geophysical Magazine, Tokyo
Geoph. Mem.	Geophysical Memoirs, London
Geophysica	Geophysica, Helsinki
Gerl. Beitr. Geoph.	Gerlands Beiträge zur Geophysik (Leipzig, Frankfult a.M.)
Idöjárás	Idöjárás, Budapest
Ind. Jour. Met. Geoph.	Indian Journal of Meteorology and Geophysics (Delhi)
Int. Ass. Sci. Hydr.	International Association of Scientific Hydrology, Bulletin
Int. Jour. Biomet.	International Journal Biometeorology (Amsterdam)
Japanese Jour. Geol. Geog.	Japanese Journal of Geology and Geography (Tokyo)
Japan. Prog. Climat.	Japanese Progress in Climatology
Jour. Agricul. Met.	Journal of Agricultural Meteorology (Tokyo)
Jour. Air Poll. Control Ass.	Journal of Air Pollution Control Association
Jour. Appl. Met.	Journal of Applied Meteorology (Boston)
Jour. Atm. Sci.	Journal of the Atmospheric Sciences (Boston)
Jour. Atm. Terrest. Phy.	Journal of Atmospheric and Terrestrial Physics, (Oxford, England)
Jour. Ecol.	Journal of Ecology
Jour. Geoph. Res.	Journal of Geophysical Research (Washington)
Jour. Glac.	Journal of Glaciology (Cambridge, England)
Jour. Hyd.	Journal of Hydrology (Amsterdam)
Jour. Met.	Journal of Meteorology (American Meteorological Society)
Jour. Met. Res.	Journal of Meteorological Research (Tokyo)
Jour. Met. Soc. Japan	Journal of Meteorological Society of Japan (Tokyo)
Jour. Trop. Geog.	Journal of Tropical Geography (Singapore)
La Mét.	La Météorologie (Paris)
Lund Stud. Geog. (A)	Lund Studies in Geography, Series A (Lund, Sweden)
Malayan Jour. Trop. Geog.	Malayan Journal of Tropical Geography
Mem. Ind. Met.	Memoirs of Industrial Meteorology (Tokyo)
Met. Ann.	Meteorologiske Annaler (Oslo)
Met. Geoastroph. Abstr. Bibli.	Meteorological and Geoastrophysical Abstracts and Bibliography (American Meteorological Society)
Met. Gidr.	Meteorologiia i Gidrologiia (Moscow)
Met. Mag.	Meteorological Magazine (London)
Met. Monogr.	Meteorological Monographs (Boston)
Met. Off. Geoph. Mem.	Meteorological Office, Geophysical Memoirs (Meteorological Office of the United Kingdom)

Met. Off. Prof. Notes	Meteorological Office, Professional Notes
Met. Rdsch.	Meteorologische Rundschau (Berlin)
Met. Zeitsch.	Meteorologische Zeitschrift
Met. Zpravy	Meteorogické Zpravy (Prague)
Mitt. d. Forstl. Burdes-Vers. Mariabrunn	Mitteilungen der Forsttichen Bundes-Versuchsanstalt Mariabrunn (Wien)
Mitt. D. W.	Mitteilungen des Deutschen Wetterdienstes (Offenbach)
Mitt. Schweiz. Anst. Forstl. Vers'wes	Mitteilungen der Schweizerlichen Anstalt für das Forstliche Versuchswesen (Birmensdorf)
Mon. Wea. Rev.	Monthly Weather Review (Washington)
Nature	Nature: Physical Science (London)
New Zealand Geog.	New Zealand Geographer
Oceanog. Mag.	Oceanographical Magazine (Tokyo)
Pap. Met. Geoph.	Papers in Meteorology and Geophysics (Tokyo)
Pet. Geog. Mitt.	Petermanns Geographische Mitteilungen
Phil. Trans. Roy Soc., Ser. A.	Philosophical Transactions of the Royal Society of London, Ser. A.
Prace Geog.	Prace Geograficzne
Proc. Int. Geog. Union	Proceeding of the International Geographical Union
Przegl. Geof.	Przeglad Geofizyezny (Warsaw)
Przegl. Geog.	Przeglad Geograficzny
Publ. Climat.	Publications in Climatology (Laboratory of Climatology, C.W. Thornthwaite Associates. New Jersey)
Pure Appl. Geoph.	Pure and Applied Geophysics (Basel)
Q.J.R.M.S.	Quarterly Journal of the Royal Meteorological Society
Rev. Géog. Alp.	Revue de Géographie Alpine (Grenoble)
Riv. Met. Aer.	Rivista di Meteorologia Aeronautica (Rome)
S. African Geog. Jour.	South African Geographical Journal
Science	Science (Washington)
Sci. Rep. Tôhoku Univ. 7th Ser. Geog.	Science Report, Tôhoku University, 7th Ser., Geography (Sendai, Japan)
Scott. Geog. Mag.	Scottish Geographical Magazine
Smith. Misc. Coll.	Smithsonian Miscellaneous Collections (Smithsonian Institution, Washington, D.C.)
Soviet Geography	Soviet Geography
Studia Geoph. Geod.	Studia Geophysica et Geodaetica (Prague)
Tellus	Tellus (Stockholm)
Tokyo Jour. Climat.	Tokyo Journal of Climatology
Trans. Amer. Geoph. Union	Transactions of the American Geophysical Union
Trans. Papers Inst. Brit. Geog.	Transactions of the Institute of British Geographers
Univ. Calif. Publ. Geog.	University of California, Publications in Geography.
Weather	Weather (London)
Weatherwise	Weatherwise (Boston)
Wett. Leben	Wetter und Leben (Wien)
Zeitsch. Geomorph.	Zeitschrift für Geomorphologie.
Zeitsch. Met.	Zeitschrift für Meteorologie (Berlin)

AUTHOR INDEX

A

Abe, M., 207, 403
Abels, H. F., 27
Agazy, L., 21
Ageta, Y., 273, 274
Aichele, H., 35, 145, 246, 247, 304, 408
Aigner, S., 257
Aizenshtat, B. A., 50, 51
Akiyama, T., 388
Alaka, M. A., 400, 401
Albrecht, F., 298
Aldrich, J. H., 173
Alissow (Alisov), B. P., 38, 67, 135, 136, 143, 144, 185, 186, 191, 263, 367
Andersson, T., 116
Angell, J. K., 108
Ångström, A., 42, 45, 77
Arai, T., 225
Arakawa, H., 83, 109, 180, 220
Arakawa, S., 390, 391, 402
Asai, T., 36, 157, 158, 170, 331
Asakawa, T., 179, 299
Asakuno, K., 111
Ashburn, E. V., 417
Ashwell, I. Y., 34, 360, 361
Aslyng, H. C., 48, 49
Atkinson, B. W., 102, 103
Atlas, D., 173
Aubert de la Rüe, E., 263, 355, 358
Aulitzky, H., 34, 198–201, 236, 248, 252, 254, 255, 267, 433, 434, 444
Austin, J. M., 417–419
Azuma, S., 68, 78

B

Bach, W., 84, 88, 106
Baier, W., 34
Balchin, W. G. V., 34, 250, 285, 304, 464
Barksdale, W. L., 480
Barnes, F. A., 102
Barry, R. G., 9, 13, 211
Barsch, D., 449, 450
Baumgartner, A., 35, 46, 67, 120, 123, 124, 140, 191, 208, 239, 240, 242, 248, 260, 407, 432–434
Baum, W. A., 5, 38
Baynton, H. W., 130, 422, 423

van Bemmelen, W., 159
Beran, D. W., 358, 361
Bérenger, M., 35, 398, 399
Berényi, D., 6, 35, 54
Bergen, J. D., 412
Bergeron, T., 180, 211
Berg, H., 408, 413
Berlyant, T. G., 38, 45
Berman, S., 65
Bernot, F., 301
Berson, F. A., 176
Best, A. C., 60, 208
Biggs, W. G., 169
Billwiller, R., 37, 277
Bjelanović, M. M., 6
Blackadar, A. K., 70, 71, 421
Black, R. F., 480
Blanford, H. F., 28, 33
Bleibaum, I., 191, 327, 408
Blom, J., 164
Bluskowa, D., 249
Böer, W., 7, 35, 250
Bogdan, O., 7, 37
Bogolepow, M. A., 23
Böhm, H., 244
Bornstein, R. D., 86
Boros, J., 54, 261
Borsy, Z., 479
Bosart, L. F., 353
Bouët, M., 277
Boyko, H., 7
Bradley, E. F., 63
Braun-Blanquet, J., 462
Brinkman, W. A. R., 358, 359, 361
Brockmann-Jerosch, H., 37
Brodie, 24
Brooks, C. F., 38
Brooks, C.E.P., 34, 111
Brown, K. W., 147
Brunt, D., 13, 34, 45, 52, 417
Büdel, J., 472
Budyko, M. I., 26, 38, 41–43, 45, 46, 48
Buettner, J. K., 77, 280
Burckhardt, H., 35, 260
Burgos, J. J., 128
Busch, N. E., 65
Byers, H. R., 7
Byzova, N. L., 75

C

Caborn, J. M., 142, 143

Cabot, E. C., 480
Çachera, P., 398
Čadež, M., 380, 391–393, 413
Cagliolo, A., 251
Cantlon, J. E., 37, 120, 251, 440, 441
Carpenter, A., 24
Carson, C. E., 480
Chambers, F., 28
Chanakya, 17
Chandler, T. J., 34, 80, 93, 102, 111
Chang, Jen-hu, 50, 53
Changnon, S. A., 100, 103, 104
Chernigovskii, N. T., 77
Chia, L. S., 44
Chiba, T., 470
Chopra, K. P., 312
Christ, H., 22
Chromow, S. P., 5, 38
Chudnovskii, A. F., 61
Chung, D. -o., 464
Church, J. E., 38
Clarke, J. F., 86, 88
Clarke, R. H., 374
Clements, F. E., 444
Cohen, A., 399
Collis, R. T. H., 399
Conrad, V., 34, 77, 193, 218
Coolinge, V. K., 293
Corby, G. A., 401, 403
Cotte, L., 23
Court, A., 5
Cowles, H. C., 37
Cox, H. J., 38
Cramer, H. E., 69
Crowe, P. R., 400
Currie, B. W., 360
Czelnai, R., 293

D

Daigo, Y., 36, 80
Dammann, W., 35, 268
Daubenmire, R. F., 59, 448
Daubert, K., 423
Davidson, B., 60, 408
Davis, N. E., 105
Davis, W. M., 28
Davitaya, F. F., 38
DeWard, R., 28
Deacon, E. L., 61, 63, 71
Defant, A., 34, 161, 276, 277, 284, 408, 412

531

DeMarraid, G. A., 86
DeMarrais, A., 269
Derham, W., 18
Devaux, J., 77
Devyatova, V. A., 55–57, 66, 71
Djunin, A. K., 137
Diamond, M., 193
Dimitrov, D., 289
Dingle, A. N., 13
Dirmhirn, I., 34, 44, 123, 124, 258
Dobosi, Z., 35
Dodd, A. V., 422
Dokuchayev., V. V., 38
Domes, N., 462
Domrös, M., 116, 258, 325, 358
Döös, B. R., 399
Dorno, C., 77
Doumani, G. A., 273
Dove, H. W., 32
Drewes, W. V., 193
Drozdov, O. A., 38, 67
Druyan, L. M., 108
Duckworth, F. S., 85, 92, 97
Dunbar, G. S., 430
Dzerdzervski, B. L., 381

E

East, C., 87
Ebara, S., 211
Ebermayer, E. W. F., 33
Eckel, O., 34, 77
Edagawa, H., 317
Edwards, R. S., 182
Egli, E., 150
Ehrlich, A., 379, 380
van Eimern, J., 7, 35, 123, 142, 248–250, 264, 266
Ekhart, E., 34, 207, 425
Elliott, R. D., 77
Elomaa, E., 250
Elsasser, W., 417
Emond, H., 80, 93
Endrödi, G., 13
Eoaz, J., 21
Eriksen, W., 7, 80, 96, 98–100, 332
Ertel, H., 69
Eskuche, U., 142
Estoque, M. A., 162, 168
Evelyn, J., 16, 18
Evseev, P. K., 381, 382

F

Federov, E. E., 38
Fedorov, S. F., 133
Feldmann, G., 421, 423, 425
Fels, E., 24
Fernow, B. E., 33
v. Ficker, H., 358
Fiedler, F., 9, 64

Findlater, J., 312
Fisher, E. L., 162, 176
Fleagle, R. G., 282, 411
Flemming, G., 7, 35, 119, 137, 140, 145, 234, 268, 270, 272, 309, 332
Fletcher, R. D., 287
Flint, R. F., 224
Fliri, F., 34, 187, 188, 207, 213, 214, 357
Flohn, H., 7, 180, 188, 209, 263, 282, 418
Flower, W. D., 60
Flowers, E., 418
Fobbes, J., 21
Förchtgott, J., 34, 232, 396, 397, 400
Forel, A., 28
Frankenberger, E., 35, 50
Franken, E., 35, 248, 249, 264
Frenkiel, F. N., 114
Frenkiel, J., 268
Friedel, H., 219
Frimescu, M., 114
Frizzola, J. A., 176
Früh, J., 445
Fuggle, R. F., 96
Fuh, B.-p., 268
Fujimoto, J., 299
Fujimura, I., 185, 192, 203, 204, 211, 234
Fujita, T., 198
Fujiwara, S., 19
Fukaishi, K., 177, 343, 344
Fukuda, K., 340
Fukui, E., 5, 36, 83, 88, 97, 118, 330
Fukuoka, Y., 54, 55, 113, 196
Fukuoka, S., 115
Fukutomi, T., 140
Funatsu, Y., 181, 182
Futi, H., 163

G

Galahkov, N. N., 157
Garnett, A., 84, 88, 244
Garnier, B. J., 34, 240–242
Gavrilova, M. K., 51, 77, 262
Gavrilović, D., 298
Geiger, R., 4–7, 41, 43, 46, 55, 77, 90, 121, 122, 127, 194, 248, 258, 263, 411, 429, 433
Geisler, J. E., 162, 174
Gelmgoltz, N. F., 248
Georgii, H. W., 110, 234
Gerbier, N., 35, 398, 399
Gesner, C., 21
Gifford, F. A., Jr., 114, 268, 422, 436
Gilbert, G. K., 470
Gilbert, O. L., 116

Glasspool, J., 34
Gleeson, T. A., 282
Glenn, C. L., 358, 360
Gocho, Y., 214, 215, 317
Godske, C. L., 37, 288
Goldreich, Y., 246
Gol'tsberg, I. A., 6, 39
Götz, G., 36
Gradmann, R., 22
Grah, R. F., 134
Graves, M. E., 169
Gregory, S., 155
Grillo, J. N., 104
Grunow, J., 35, 51, 139, 208
Gugiuman, I., 245
Gunji, T., 434
Gutermann, T., 356, 357

H

Haase, G., 10
Hall, W. F., 7
Hamberg, H. E., 33
Hamm, J. M., 80, 98, 114
Hammond, W. H., 23
Hand, J. F., 77
Hanna, S. R., 114
Hannell, F. G., 97
v. Hann, J., 23, 24, 28, 29, 32, 34, 193
Hare, F. K., 34, 44, 151
Hartmann, F. K., 120, 122, 123, 131, 251, 439, 440
Hasegawa, T., 53
Hastenrath, S., 209
Hastings, J. D., 479
Hatakeyama, H., 36
Hawke, E. L., 34, 258
Hazen, H. A., 28
Heckert, L., 129, 141
Heck, W. J., 293
Heer, O., 22
Heigel, K., 76, 301, 303, 304
Hellmann, G., 23, 26, 35, 60
Hendl, M., 190
Henning, D., 208, 209, 326
Henning, I., 208, 326
Henry, A. J., 38, 210, 430, 434
Hentschel, G., 425
Henz, J. F., 232
Hess, D. W., 445
Hess, M., 37, 200, 201
Hesselberg, T., 69
Hewson, E. W., 38, 168
Heywood, G. S. P., 34, 372, 410
Higuchi, K., 198, 320, 475
Hinds, W. T., 269
Hirata, T., 210
Hirt, M. S., 176
Hitsuma, M., 347, 348
Hoćevar, A., 258
Hoinkes, H., 226

Holland, J. Z., 268, 269, 436
Holmboe, J., 397, 399
Holloway, L., 7
Holmes, R. M., 13, 151, 361
Holmgren, B., 64, 272, 373
Holtmeier, F.-K., 299, 448, 452
Homén, T., 26, 34
Hopkins, A. D., 155
Hoppes, E., 33
Höschele, K., 114
Hoshiai, M., 225
Houghton, D. D., 401, 402
Howard, L., 23, 80
Howell, W. E., 373
Hubert, L. F., 312
Huff, F. A., 103, 293
Hufty, A., 117
von Humboldt, A., 31
Huschke, R. E., 5
Huss, E., 165, 166
Hussey, K. M., 480
Huttenlocher, F., 244
Huzimura, I., 184

I

Iaroslavtsev, I. N., 77
Ichikawa, M., 472, 473
Iizuka, H., 142, 143
Imahori, K., 139
Imai, K., 63
Ino, H., 344
Inoue, E., 59, 63, 403
Inoue, K., 173
Iozawa, T., 475
Ishida, T., 231
Ishihara, K., 326, 327
Ishihara, Y., 134, 148, 298
Ishii, M., 146, 243
Ishikawa, M., 138, 468
Ishikawa, N., 428
Ishikawa, Z., 21
Ishimaru, Y., 204
Itoo, K., 112
Iudin, M. I., 7
Ives, R. L., 38, 207, 361, 410
Iwatsuka, S., 472

J

Jacobs. W. C., 307
Jahn, G., 123
James, D., 73
Johnson, A. Jr., 175
Jones, P. M., 77, 109
Julian, L. T., 360
Junghaus, I. H., 54

K

Kabasawa, M., 352, 353
Kaempfert, W., 35
Kai, K., 84, 97
Kaiser, H., 35, 264

Kalb, M., 80
Kalitin, N. H., 77
Kalma, J. D., 141
Kamiko, T., 347
Kaminskiy, A. A., 38
Kaps, E., 35, 157, 250, 267
Karapiperis, P. P., 159
Karrasch, H, 471
Kasahara, A., 401, 402
Kataoka, K., 138
Kato, T., 36, 210
Katsumi, S., 148
Katsuya, M., 36, 339, 344
Kauf, H., 286
Kauper, E. K., 173
Kawamura, T., 59, 80, 84, 90,
    91, 93, 116, 307–309, 312,
    313, 349, 472, 473
Kayane, I., 6, 82, 83, 95, 171
Keay, R. W. J., 462
Kendrew, W. G., 360, 380
Keil, K., 4
Keller, Hs. M., 148, 151
Kennedy, B. A., 471
Kern, H., 248
Kerner, A., 29
Kerner, F., 29, 30
Kerner von Marilaun, 22
Kerr, D., 380
Khemani, L. T., 101
Kimball, H. H., 77
Kim, D.-j., 475, 476
Kikuchi, R., 410
Kimura, H., 415, 416
Kimura, K., 78
King, R. B., 479
Kinoshita, S., 208
Kira, T., 463
Kitazawa, Y., 123
Kittredge, J., 37
Klinov, F. Ya., 65
Kluge, M., 48
Knipping, E., 32
Knoch, K., 6, 35, 193
Knudsen, J., 37, 165, 282
Kobayashi, T., 292
Koch, H. G., 35, 120, 127, 128,
    231, 357, 431, 432, 434
Kodama, R., 211
Kohlbach, W., 160
Koide, F., 36
Koloskov, P. I., 38
Konda, K., 128
Kondo, J., 48, 118, 119
Kondratyev, K. Ya., 240
Konček, M., 34, 218, 238
Konstantinov, A. P., 143, 144
Kopec, R. J., 100, 155
Köppen, W., 34
Koschmieder, H., 397, 399
Kosiba, A., 37, 218

Kratzer, A., 24, 35, 80, 85, 88
Kraus, G., 35
Krawczyk, B., 48
Kremser, 24, 25
Kreutz, W., 35, 52, 53, 142, 247,
    264, 265, 327
Kristensen, K. J., 48
Kronfuss, H., 200
Kubo, 157
Küettner, J., 320, 397, 408
Kurashige, K., 92
Kušan, F., 441, 442, 444
Kusano, K., 314

L

Laitinen, E., 48
Lamb, H. H., 34
Landsberg, H. E., 4, 38, 80, 110,
    111
Larsson, L., 397, 399
Lauer, W., 368
Lauscher, F., 34, 80, 90, 189,
    194–196, 358
Lavoie, R. L., 180
Lawrence, D. B., 37, 448, 453
Lawrence, E. N., 34, 97, 101,
    153, 247, 408, 411
Laykhtman, D. L., 39, 61, 67
Lee, Ch. W., 349, 350
Lee, R., 239, 240
Lehmann, P., 35, 414
Leidreiter, W., 425
Leighly, J., 26, 38
Leopold, L. B., 175, 176, 326
Lessmann, H., 35
Lettau, H. H., 60, 171
Levi, M., 377, 378
Lied, N. T., 372, 373
Lilly, D. K., 399, 400, 402
Lindqvist, S., 81, 104
Lingova, S., 45
Linsley, R. K., 292
Livingstone, D. A., 480
Löbner, A., 110
Loewe, F., 275
Lomas, J., 36, 182, 246
Long, I. F., 63, 403
Longley, R. W., 34, 359
López, M. E., 373
v. Lorenz-Liburnau, 33
Louw, W. J., 81
Lowry, W. P., 132, 332
Lüdi, W., 37, 120
Ludwig, F. L., 96
Lumb, F. E., 180
Lunelund, H.,77
Lynboslavsky, G. A., 38
Lyons, W. A., 167
Lyra, G., 400

M

MacHattie, L. B., 120, 283, 428, 434
Mackey, J. R., 480
Mäde, A., 5, 35, 147, 250
Magata, M., 162, 402
Magono, C., 320, 351
Makino, Y., 347, 348
Malsch, W., 408
Manley, G., 34, 258, 379, 397, 399
Mano, H., 180, 409, 410, 414, 415, 427
Marsch, J. S., 359
Marsh, G. P., 32
Marsh, K. J., 114
Martin, H. C., 29, 141
Maruyama, I., 147
Maruyama, K., 443
Maruyasu, T., 222–224
Masatsuka, A., 291, 292, 327, 328, 421, 424
Mashkova, G. B., 63, 417
Mason, B. J., 9
Máthé, I., 35
Mather, K. B., 372, 374
Matoda, M., 23
Matsumoto, S., 317, 340
Matsuno, M., 36
Mattsson, J. O., 54
Maurain, C., 34
Maury, M. F., 27
Mayama, T., 138, 139
McCaul, C. C., 32
McCormick, R. A., 114
McDowell, S., 430
McIlroy, I. C., 55
McVean, D. N., 466
Meiklejohn, J., 379
Melton, M. A., 470, 471
Mendonca, B. G., 284
Meyer, J. A., 81
Miara, K., 48
Michalczewski, J., 248
Mihai, E., 37
Mihăilescu, V., 7, 37
Miller, D. H., 30, 38, 46, 48–50, 77, 78, 88, 89, 120, 121, 126, 150, 242
Miller, G. S., 374
Milosavljević, K., 190, 203
Milosavljević, M., 203, 380, 381
Mino-Ishikawa, Y., 477
Misawa, K., 36, 275
Mitchell, J. M., 97, 379
Mitscherlich, G., 121
Miyazawa, S., 341
Mizukoshi, M., 59, 81, 94, 117
Möller, F., 417
Monin, A. S., 9, 63
Montefinale, A. C., 302

Monteith, J. L., 48, 49, 124
Morawetz, S., 226, 250
Morgen, A., 35
Mörikofer, W., 6, 37
Morita, Y., 177
Moriwaki, K., 331
Moroz, W. J., 168
Müller, W., 297
Munn, R. E., 34, 86, 87, 94, 109, 422
Murakami, K., 474
Murakami, M., 19
Musaelyan, Sh. A., 400
Müttrich, A., 33
Myrup, L. O., 88

N

Nagao, T., 80
Nägeli, W., 37, 121, 142, 266, 267
Nagel, F. W., 208
Nakahara, M., 36, 304
Nakajima, C., 214, 270, 317, 328
Nakamura, K., 340
Nakamura, S., 51
Nakano, T., 477
Nakayama, A., 317, 347, 353
Nakaya, U., 36, 220
Namekawa, T., 36
Namie, T., 36
Napoleon III, 22
Naruse, T., 477, 478
Neamu, G., 37
Neef, E., 10
Neiburger, M., 178
Nemoto, S., 404
Neuber, E., 153
Nicholls, J. M., 400
Niedźwiedź, T., 248, 434
Nishimura, D., 351
Nishina, N., 36
Nishizawa, T., 48, 53, 80, 90, 94, 332
Nitze, F. W., 408
Nitzschke, A., 154
Nkemdirim, L. C., 49
Nogami, M., 224, 225
Nomoto, S., 337
Nordlund, G., 162, 163
Norris, F. H., 438
Nose, A., 257
Nosek, M., 34, 332
Numata, M., 437, 438
Nutting, 77
Nyberg, A., 76

O

Obukhov, A. M., 63
Obrebska-Starklowa (Obrebska-Starkel), B., 247, 248, 252, 265, 266, 433

Ôdaira, T., 113
Ogasawara, K., 466
Ogden, T. L., 104
Oguntoyinbo, J. S., 45
Ogura, Y., 62
Ohmura, A., 241, 242
Ohta, Y., 173, 174
Okabayashi, T., 321, 351
Okanoue, M., 152, 239
Oke, T. R., 87, 98
Okita, T., 106
Okołowicz, W., 5, 37
Olecki, Z., 117, 199
Oliver, H. R., 54, 120
Oliver, J., 34, 53, 54, 153, 250, 298
Ollier, C. D., 472
Olsson, A., 77
Onodera, S., 140
Ono, S., 209
Onuma, T., 221, 295
Oobayashi, T., 402
Ooi, S., 334, 346
Oosawa, S., 132
Oosting, H. J., 445
Oota, T., 466
Ooura, H., 139
Orlicz, M., 374
Ortolani, M., 209
Orvig, S., 34
Ôta, M., 421
Otani, T., 394
Owada, M., 196, 197, 199, 450, 455–458
Ozaki, K., 317, 318, 319
Ozawa, Y., 36, 172, 249, 250

P

Pack, D. H., 108
Padmanabhamurty, B., 87
Paeschke, W., 60
Paffen, K.- H., 10, 195
Panofsky, H. A., 9, 65, 164
Paradiž, B., 361
Parry, M., 34, 92, 102, 114
Pascal, B., 28
Paszýnski, J., 37, 47, 48, 117
Patton, C. P., 177
Peace, R. L., 180
Pearce, R. P., 162
Peattie, R., 38, 190
Pédrabolde, P., 35
Peppler, W., 166
Perlewitz, P., 23
Pernter, J. M., 30
Perring, F. H., 445, 465
Perry, A. H., 9
Peterson, J. T., 80, 88
Peterson, P., 286
Petkovšek, Z., 301
Piéry, M., 34

Plaetschke, J., 414
Pleiß, H., 234
Plesnik, P., 34, 452
Pockels, F., 208, 209
Poggi, A., 219
Polli, S., 364
Poltavskii, V. V., 65
Pooler, F. Jr., 106
Poore, M. E. D., 466
Preston-Whyte, R. A., 94, 160, 324
Priestley, C. H. B., 59
Primault, B., 459
Prošek, P., 5
Prügel, H., 320
Prutzer, E., 296, 297
Putnam, P. C., 449
Putnins, P., 372
Pye, N., 34, 250, 285, 304, 464

Q
Queney, P., 400
Quitt, E., 34, 98, 298

R
Raethjen, P., 7
Rainer, J. N., 475
Rao, P. K., 282, 408
Rasche, H. H., 37, 438
Rassow, L., 110
Ratzel, R., 31
Record, F. A., 69
Reid, D. C., 176
Reifsnyder, W. E., 121
Reiher, M., 411
Reiter, R., 267, 436
Renou, E., 23
Reuter, H., 34
Reynolds, R. D., 399
Rice, E. L., 445
Rider, M. E., 63
Riehl, H., 361
Rink, J., 421, 425
Ritchie, J. C., 44, 151
Rodewald, M., 320
Rosenkranz, F., 304
Ross, P., 37, 120
Rossby, C. G., 37
Rossi, V., 34, 159, 160, 163, 324
Rouse, W. R., 240
Roux, L., 218
Rubenson, R., 26
Rubinstein, E. S., 39, 67
Rubner, K., 138
Rudberg, S., 479
Runge, F., 448, 449
Russel, F. A., 25
Russell, R. J., 470
Russler, B. H., 291
Rutkovskii, V. I., 147

S
Saeki, M., 466, 468
Sahashi, K., 410
Saito, M., 178
Saito, R., 336
Sakagami, J., 75, 76
Sakanoue, T., 328, 474
Sakharov, M. I., 465
Sammadar, J. N., 17
Samokhvalov, N. F., 381
Sanderson, M., 34, 103
Sapozhnikova, S. A., 5, 12, 39, 61, 143
Sarker, R. P., 214
Sarson, P. B., 55
Sasaki, N., 170, 260
Sasaki, Y., 90
Sasakura, K., 5, 23, 36, 153
Sato, G., 18
Sato, H., 463
Sauberer, F., 34, 44, 77, 226, 258, 288
de Saussure, H. B., 21, 29–32
Savinov, S. I., 42
Scaëtta, M. H., 4
Schamp, H., 358, 374
Schelkovnikov, M. S., 279, 408
Schindler, G., 301
Schirmer, H., 35, 287, 288
Schmauß, A., 408
Schmidt, F. H., 394, 395
Schmidt, G., 110
Schmidt, W., 4, 5, 34, 39, 258
Schmithüsen, J., 10, 438
Schmitz, H. P., 63
Schneider, K., 101
Schneider, M., 35, 260, 410
Schneider-Carius, K., 416
Schnelle, F., 7, 35, 120, 245, 246, 257, 260, 304, 410
Schöne, V., 35, 59, 164
Schroeder, M. J., 176
Schubach, K., 247, 257
Schubert, J., 33, 35
Schüepp, M., 37, 234, 284
Schultz, L. G., 28
Schuppert, K., 132
Schweinfurth, U., 279, 358, 462
Schwerdtfeger, W., 372
Scofield, H. T., 37
Scorer, R. S., 34, 396, 400, 401
Scultetus, H. R., 249, 264, 450
Seemann, J., 35
Sekiguti, T., 5, 13, 36, 81, 82, 85, 100, 116, 117, 131, 252, 275, 304, 305, 323, 376, 377
Sekine, T., 109
Sekiya, H., 431, 434
Selleck, G. W., 132
Sellers, W. D., 45, 48
Selyaninov, G. T., 38, 39

Seppänen, M., 137
Shanks, R. E., 130, 438
Sharon, D., 81
Shaw, R. W., 176
Shieh, L. J., 114
Shimizu, S., 334
Shinohara, T., 319, 348
Shitara, H., 37, 152, 155, 156, 167, 172, 246, 259, 260, 329, 330, 410
Sien, C. L., 323
Simonis, R., 303
Simpson, J. E., 173, 176
Singer, I. A., 435
Skoczek, J., 48
Slatyer, R. O., 55
Smirnov, V. A., 39
Smith, H. T. U., 471, 258
Smith, L. P., 34
Smith, M. E., 435
Solantie, R., 338
Soma, S., 403–405
Sonntag, K., 101
Soons, J. M., 475
Soó, R., 35
Spinnangr. F., 37
Spinnangr, G., 37
Spreen, W. C., 291
Spurr, S. H., 438
Stamp, L. D., 34, 448
Stanev, S., 219
Stanhill, G., 124, 141
Starr, J. R., 399
Steinhauser, F., 34, 80, 219, 234, 243, 277
Sterten, A. K., 282, 283
Stevenson, T., 26
Stewart, I. M., 86
Stewart, J. B., 120, 124, 125
Stoenescu, S. M., 191
Stone, K. H., 10
Stout, G. E., 293
Strachey, H., 27, 31
Strahler, A. N., 471
Stranz, D., 165, 166
Streten, N. A., 372
Stringer, E. T., 14, 34, 320, 399, 400
Strong, A. E., 335, 336
Sugawara, Y., 191, 345
Sugaya, J., 36, 220, 221, 285, 286, 295
Sugiyama, T., 466–468
Sundborg, Å., 37, 80
Sutton, O. G., 34, 59, 65, 68, 75
Suzuki, B., 21
Suzuki, H., 320
Suzuki, S., 5, 36, 369, 370, 389, 403
Suzuki, T., 228–230
Svensson, H., 479

Sverdrup, H. U., 297
Sykes, R. B. Jr., 180
Syôno, S., 319
Szász, G., 407
Szepesi, D., 288, 289, 423

T

Tabata, T., 140
Tabukuro, K., 338
Tadeda, S., 36
Takahashi, Kihei, 296
Takahashi, Kôichiro, 9, 161, 346
Takahashi, M., 80, 81
Takasaki, M., 224, 226
Takasu, K., 36, 68, 69, 78–80
Takeda, K., 36, 289, 290, 328
Takeshita, K., 474
Takeuchi, Y., 232
Tamate, M., 210
Tamiya, H., 97, 364
Tanaka, M., 181
Tanaka, Y., 33
Tang, M.-ts., 324
Tani, N., 63, 64, 142
Tatehira, R., 323
Tateishi, Y., 179, 294, 295
Tatewaki, M., 448
Taylor, J. A., 52
Teichert, F., 407
Temple, P. H., 474
Teodoreanu, E., 7, 37
Tepper, M., 7
Terjung, W. H., 48, 117
Thambyaphillay, G., 358
Thams, C., 77
Thiele, A., 116
Thom, A. S., 64, 120, 124, 125
Thomas, M. K., 34
Thomas, T. M., 360
Thompson, A. H., 420, 428
Thompson, R. D., 332
Thorarinsson, S., 477
Thornthwaite, C. W., 6, 13, 38, 46, 60, 151
Thyer, N. (H.), 280, 282
Timofejev, M. P., 67
Toba, Y., 181
Tohsha, M., 419, 420
Tollner, H., 34
Torricelli, E., 28
Troll, C., 6, 10, 23, 192, 195, 231, 279, 437, 452, 462, 469, 477
Tsuboi, Y., 249, 250
Tsuchiya, I., 88
Tsujimura, T., 36
Tsuneoka, Y., 317
Tsutsumi, K., 109
Turnage, W. V., 53
Turner, D. B., 114

Turner, H., 34, 149, 186, 187, 220, 452, 463
Tyndal, J., 21, 29
Tyson, P. D., 271, 324

U—V

Uchijima, Z., 59, 63, 64, 69, 80
Uehara, M., 251
Ueno, T., 36, 304
Uhlig, S., 35
Újvárossy, M., 35
Ukita, N., 10
Umeda, S., 247, 431
Undt, W., 34, 80
Urfer-Henneberger, Ch., 281, 282, 432, 434
Utagawa, K., 335
Utsumi, T., 176
Vekilska, B., 289
Verber, J. L., 154, 155
Vergeiner, I., 361, 399, 402
Vialar, J., 375, 376
Vidie, 28
Viezee, W., 399
Violle, J., 30
Visher, S. S., 38, 154
Vitruvius, M., 15
Voeikov, A. I., 26, 27, 33, 38, 76, 77
Voigts, H., 411
Volkovitskaya, Z. I., 63
Vorontsov, P. A., 39, 58, 65, 68, 73, 74, 167, 168, 279, 408, 412, 415, 426, 427
Vowinckel, E., 34
Voznesenskiy, A. V., 38
Vysotskiy, G. N., 38

W

Wada, N., 88
Wagner, A., 34, 35, 210, 276, 414
Wagner, R., 12, 35, 54, 258, 413, 429
Walter, E., 356
Walter, F., 27
Walter, H., 446, 452
Wareham, R. T., 37
Warner, J., 73
Wartena, L., 164
Washburn, A. L., 477, 480
Watanabe, K., 7
Watanabe, S., 135, 300
Watts, I. E. M., 324, 328
Weaver, J. E., 444
Webb, E. K., 60, 71, 72
Weger, N., 35
Weickmann, H., 103
Weischet, W., 9, 35, 209, 210, 450, 452
Wendler, G., 252, 261–263

Whitehead, F. H., 462
Whittaker, R. H., 441
van Wijk, W. R., 55
de Wilde, J., 55
Wilkins, E. M., 112, 335, 354
Williams, P. Jr., 241, 242, 293
Wilson, C. C., 134
Wilson, J. W., 452, 453
Winiger, M., 207
Winter, F., 412
Wittwer, W. C., 23
Woelfle, M., 140, 142
Wolfe, J. N., 37, 120, 257, 438
Wollny, E., 26

Y

Yabuki, K., 369, 370, 389, 403
Yaji, M., 323
Yamamoto, G., 70, 71, 118
Yamanaka, T., 36
Yamashita, H., 312
Yamashita, I., 166, 312, 436
Yamashita, S., 49, 117
Yamazaki, H., 21
Yamazaki, T., 320
Yanagisawa, Z., 321, 322
Yazawa, T., 36
Yokoi, C., 211
Yokoi, S., 109
Yokoyama, O., 73
Yoshida, S., 211
Yoshida, Yoshio, 274, 275, 331
Yoshida, Yoshinobu, 295
Yoshimura, M., 364, 399
Yoshino, M. M., 5, 36, 63, 64, 69, 83, 84, 93, 97, 100, 108, 115, 123, 125, 127, 130, 135, 137, 142, 144–146, 148, 159–161, 164, 166, 179, 185, 186, 190–193, 202, 204, 205, 210, 212, 213, 216–219, 221, 227, 233, 235, 237, 238, 251–253, 268, 270–272, 278, 283, 302, 304, 310, 311, 315, 316, 332, 333, 337, 339, 345–347, 352, 362, 363, 365–367, 369, 372, 383, 384, 386, 387, 389, 394, 399, 404–406, 409, 424, 428, 430, 431, 436, 443, 446, 447, 449–455, 459–461, 464, 465, 469, 478
Yoshino, M. T., 464
Yoshioka, H., 328
Young, F. D., 38, 430, 434
Yuyama, Y., 206, 207

Z

Zimmermann, H., 117, 118
Zoller, H., 37, 120, 443
Zólyomi, B., 35

SUBJECT INDEX

A

ablation form, 231
—— of snow, 225
—— period, 226
absolute humidity, 98, 130
active surface, 38, 42, 192
actual snow line, 225
advection fog, 178
aeolian sand, 477, 479
aerial photograph, 224, 226, 228
aerodynamic resistance, 64, 124, 125
aerosol, 118
afforestation, 21
agricultural meteorology, 36
ai, 376
air-avalanche, 408
air flow over mountains, 396
air pollution, 16, 25, 78, 110, 114, 116
air pressure, 183
airstream, 267, 328
air temperature, 27, 28, 33, 53, 55, 56, 80, 126, 128, 151, 170, 189, 245
Akagi-oroshi, 371
albedo, 42, 43, 47, 77, 188
Albtalwind, 374
alpine zone, 466
anabatic wind, 372
anaji, 376
anaze, 376
andro, 374
annual range, 193, 195
anti-land breeze, 158
anti-sea breeze, 158
apple, 452
arable land, 16, 43
arashi, 285, 299
areal fluctuation of wind direction, 311
area rainfall, 291, 292
Arthastra, 17
asagiri, 299
ash, 121, 452
assimilation, 453
association, 437
asymmetry index, 471
—— of valley, 469
atmospheric pressure, 28
—— turbulence, 65
Austausch coefficient, 67
avalanche, 31, 226, 227, 231, 467

B

back radiation, 42
bai-u, 330
Balaton-wind, 375
bamboo, 390, 437, 438, 464
band structure, 319, 322
banner cloud, 402
barchanoid, 273
Bayerischer Wind, 374
beech, 121, 123, 135, 146, 443, 452
Belat, 374
birch, 121, 135, 452
Bishop wave, 395
blizzard, 395
boarèn, 375
bohorok, 394
bora, 361, 363, 391–393, 395, 405
boraccia, 361
Bôsô front, 345, 347, 368
botany, 37
Br, 111
breva, 284, 374
brezza di mare, 375
brisas, 375
brises solaires, 375
broad-leaved tree, 122, 123, 136, 147, 448
Brüscha, 374
brushing, 449
budding, 303
bulk physiological resistance, 125
bulk Richardson number, 71
bura, 361
Büsserschnee, 231

C

calcium, 116
canopy, 128
cap cloud, 205
cap-form cloud, 207
Catalina eddy, 117
catchment area, 292, 298
cedar, 127, 135
chanduy, 374
charachaicha, 367
cherry, 305, 452
chinook, 32, 358–361, 392
—— arch, 360, 395
chlorine, 116
Chôkai-oroshi, 385

cirques, 474
citrus tree, 148
city, 15, 23
—— climate, 25, 34, 35
—— fog, 24, 25, 104, 105, 110
—— planning, 16
—— temperature, 23, 80, 88
—— wind, 24
climatic landscape, 36
—— snow-line, 224
climatological resistance, 125
climatology, 3
cloud amount, 30, 42, 204
—— band, 320
—— bridge, 207
—— chain, 320
—— particle, 204
—— rolls, 320
—— series, 320
—— streaks, 321
—— street, 320
cold air drainage, 407, 410, 411, 416
—— invasion, 392
—— lake, 299, 392, 407, 413–416, 428, 429, 444
—— pond, 413
—— run-off, 407, 409, 410, 413
—— sea, 413
—— stream, 410, 412
colloid meteorology, 9
condensation level, 189, 204, 207
conflict zone, 207
coniferous tree, 120
constant, 71
contaminant, 116
contessa del vento, 207
convectional type, 210
convergence, 317, 319, 323, 325, 342, 372
—— line, 342
—— zone, 173, 207
Coromell, 374
CO_2, 110
—— concentration, 78, 79
counterflow, 272
country breeze, 106
—— wind, 24
creeping pine, 466, 469
crest cloud, 207, 356, 395
cross-over effect, 85

537

cross-shelterbelt, 143, 145
cross-valley wind, 236
cuckoo wind, 377
cumulus, 204, 324
cushion-shaped tree, 448
cutting, 148
cypress, 448

D

daily maximum rainfall, 474
—— rainfall, 319
—— range, 195, 196
—— variation, 56
damp haze, 204
dashi, 19, 20
dashinarai, 19
deciduous broad-leaved tree, 132, 135, 136
—— trees, 121, 127
decreasing rate, 250
deflation, 479
deformation grade, 450, 453
degree of vegetation difference, 440
den of wind, 285
density of snow, 137
—— of the observation points, 292
deposit gauge, 111
depth of snow, 137
—— of the inversion layer 418
desert vegetation, 48
diffusion, 75
—— parameter, 75, 76
Dimbula blowing, 358
discontinuity line, 152
dissipation of inversion, 419
disturbance, 317
diurnal change, 184, 192, 234, 251, 255
—— variation, 56, 252, 254, 276
divergence, 336
doline, 442–444
domai, 388
down-slope wind, 372, 412, 414–416
drift, 274
drizzle, 101, 102, 110
dry adiabat, 393
dry adiabatic lapse rate, 189, 416
dry valley, 275, 330, 331
Dunsthaube, 85
duration of inversion, 424
dust, 110
dyke, 249

E

ecoclimate, 7

ecology, 37
eco-system, 437
ecotope, 10
edaphic range, 437
eddy, 264, 272, 283, 284
—— conductivity, 69, 90
—— viscosity, 69
effective radiation, 45
—— Reynolds number, 403
—— ridge altitude, 276
—— roughness length, 64, 65
80% flowering, 305
Ekman layer, 70
Elbtalwind, 374
elephanta, 291, 395
embatis, 375
emissivity, 76, 96
emvatis, 375
energy balance, 117
—— budget, 124
—— output, 89
environmental scale, 438
epidemics, 16
Erler Wind, 374
erosion, 472
etesia, 159, 377, 395
Etschwind, 374
evaporation, 48, 130, 131, 133, 296
evapotranspiration, 297, 298
exchange coefficient, 66, 67
experiment, 39
exposure, 291
extratropical cyclone, 316, 336
eye-brow-shaped tree, 466

F

fallwind, 147, 345, 368, 394, 395, 402
film of cold air, 415
fir, 16, 121, 123, 135, 452
first flowering, 303, 304
fish hook, 376
fjord, 380
flag-form cloud, 207
flagging, 449
—— tree form, 265
flag-shaped tree, 448, 459
Fliese, 10
flowering, 302, 304
fluctuation, 310
—— angle, 271
—— of wind direction, 436
Flurwind, 24, 106
flux Richardson number, 71
fog, 104, 176, 186, 204, 299
—— belt, 299
—— collector, 208
—— content, 139
—— days, 203, 205
—— drop, 204

—— hollow, 301
—— layer, 299
—— particles, 139, 208, 238
—— precipitation, 135, 138' 208
—— prevention forest, 139, 140
fog water content, 140
—— quantity, 208
föhn, 32, 267, 355, 391, 392, 395
—— effect, 32, 325, 356, 379, 386, 393
—— gasse, 357
—— guards, 357
—— wächter, 357
—— wall, 356, 395
—— wave, 397
folklore, 379
footprint, 273
forest, 21, 32, 33, 43, 119
—— canopy, 129
—— fire, 37
—— floor, 127
—— line, 452
—— meteorology, 119
forestry, 37
forest structure, 437
formation, 437
fossil evidence, 445
freeze-free date, 154
—— season, 155
freeze-thaw effect, 475
—— phenomena, 477
freezing, 51, 470
frequency of precipitation, 313
friction velocity, 64
front action, 476
frontal inversion, 418
frost, 105, 153, 256, 258, 260, 474, 477
—— action, 475
—— damage, 18, 260
—— hollow, 256
—— path, 256
—— pocket, 256–258
—— probability, 257
frost-free period, 258
frostless period, 157, 257, 258
—— zone, 152
Froude number, 404
fudoki, 17
full blooming, 302
—— leaves, 303

G

gallego, 374
garbé, 375
gas, 176
Gassan-oroshi, 385
Geländeklima, 6, 10
ghibli, 479

Gilbert's law of divide, 470
glacial, 471
glacier, 225, 242
glaze storm, 453
global irradiation, 187
—— radiation, 188, 240, 242
god of wind, 388
Görlitzer Wind, 374
grade of wind-shaped tree, 451
grassland, 43, 465
gravity wind, 372, 407
ground inversion, 299, 417, 421, 422
ground temperature, 26, 33, 51, 53–55, 130, 196, 198, 261, 262
ground water table, 437
growing season, 257
gust, 109, 361
—— factor, 72
—— probability, 206
gustiness, 270, 357, 435, 436
—— factor, 72

H

Haboro-cyclone, 337
hae, 376, 377
hai, 376
half leaves, 303
halny, 374
hanaboro, 239
hanging cloud, 206, 207
harvest, 304
Haruna-oroshi, 371
haze, 89
—— cap, 85
heat balance, 26, 41, 46–49, 51, 124, 126, 150, 188, 189, 263
—— budget, 48, 118, 171, 226
—— capacity, 47
—— conductivity, 47
—— flux, 41, 63
—— island, 23, 82, 84, 85, 88
—— transfer, 126
heath, 466
heavy rainfall, 317–319
—— rain zone, 315
—— snowfall, 340, 341
Heckenlandschaft, 145
Heidelberger Talwind, 374
helm, 395
—— bar, 379, 395, 397
—— wind, 379, 397
hemlock, 444, 445
Hidaka-shimokaze, 390
high-level inversion, 418
high mountain, 200
hikata, 376
hikata-kaze, 391
Hirahakko, 389
Hiroto-kaze, 385
hoarfrost, 237, 238

Hokuriku front, 338, 341, 342, 344, 345, 347
Höllentäler, 374
Höllenwind, 374
horologium, 15
horse chestnut, 452
hotokoro-kaze, 385
house density, 81
human activity, 97
humidity, 33, 98, 130, 202, 463
—— island, 100
hydraulic jump, 372, 395, 401, 402, 405, 406
hydrologic rainfall, 287, 290, 327, 328

I

ice, 43
—— cap, 372
ice-tongue, 225
icing, 51, 237, 238
imbat, 375
imbatto, 375
Inakaze, 166
inasa, 19
industrial complex, 104
industrialization, 101
inferno, 374
insolation, 25, 30, 116, 117, 121, 239, 185
—— coefficient, 261
instability line, 323, 324
instantaneous wind velocity, 232
intensity of gust, 72
—— of inversion, 422
—— of turbulence, 73, 270, 271
interception, 133, 134
intrastand, 437
inverna, 374
inversion, 191, 246, 248, 252, 256, 301, 373, 425
—— layer, 57, 194, 219, 247, 248, 256, 335, 370, 407, 416, 417, 421, 425–428, 433
—— of vegetation, 444
—— rate, 418, 425, 428
—— top, 427
irregular pattern, 273
Ishikari front, 349, 351
isohyetal map, 291

J

Japanese cedar, 127, 135, 137, 138, 147, 390, 469
—— cypress, 134
—— persimmon, 260
joran, 166
jump, 402
junkwind, 375

K

kachchan, 325, 358
kapa, 395, 405, 406
Karif, 374
Kármán constant, 71
Kashima weather, 346
katabatic effect, 379
—— wind, 273, 355, 372, 373, 407, 410, 412
kazenomiya, 385
kazenomiya-no-kaze, 385
khamsin, 378
—— depression, 378
kibana, 239
kisame, 138
kita-kaze, 385
kitewind, 375
Kiyokawa-dashi, 384, 385
Knicklandschaft, 145
kochi, 19, 20, 377
koschava, 380, 381
kossava, 380
kudari, 376
kuroshio, 314
kynuria, 374

L

lapse rate, 250, 464
—— of ground temperature, 196
larch, 450, 451, 463
latent heat, 41
lateral gustiness, 270
lake anticyclone, 336
—— breeze, 28, 165–167, 180
—— front, 176
—— index, 169
lake-effect storm, 180
lake fog, 179
—— high, 336
lakeshore, 18, 27, 151
laminar streaming, 397
land breeze, 27, 28, 152, 153, 158, 159, 161, 166, 175, 374, 408
landscape, 437
Landschaft, 10
landslide, 472–474
landslip, 472, 474
lapse rate, 130
—— state, 416
larch-pine, 299
larch (trees), 138, 146
Laykhtman graph, 62
lee eddies, 397
—— wave, 35, 361, 389, 395–400, 405, 406
lenticular cloud, 399
lichen, 116, 459, 465, 479
lifetime, 319
light, 122, 123, 462

lightning, 232
linden, 452
line of discontinuity, 175
Liptowski, 374
local airstream, 307
—— ancityclone, 333
—— climate, 3
—— climatic division, 439
—— climatology, 13, 36
—— cyclone, 336, 338
—— discontinuous line, 334
—— front, 175, 207, 338, 339, 351, 353, 354
—— high, 339
—— Richardson number, 71
—— wind, 355, 358, 374, 383, 390, 395
Loewe's phenomenon, 372
logarithmic law, 60–62, 272
Lokalsynoptik, 7
London smog, 111
low-level (air) trajectory, 108, 436
—— inversion, 417
——jet (stream), 317, 421, 427

M
macroclimate, 3
Mahākaĺa, 16
mai-mai-kaze, 387
maji, 376, 377
Malojaschlange, 299
Maloja wind paradox, 284
Mānasāra Śilpaśāstra, 15
maple, 445, 452
marinada, 375
maskat, 374
mass balance, 225
matsubori-kaze, 385, 389, 390
matinière, 374
maximum temperature zones, 210
maze, 376
McMurdo oasis, 274, 330
meadow, 44
medlar tree, 452
meltemi, 377
mesoclimate, 3
mesometeorology, 7, 9
meteorological rainfall, 287, 289, 290, 327, 328
microadvection, 407
microclimate, 3
microclimatology, 13
micrometeorology, 9, 34
micro-topography, 469
minami, 19, 20, 376
miniature front, 354, 355
mist, 204
mistral, 32, 376, 395, 448
Mitternachtswind, 374

Mn, 111
Moazagotl, 397
model experiment, 403
momentum transfer, 124
morget, 166, 374
morning glory, 374
morphogenous layer, 272
moss, 366, 465
mountain, 20, 28, 183, 263
—— and valley breeze, 236, 276, 281, 283
—— breeze, 20, 31, 354, 374
—— climatology, 36
—— climbing, 20
—— snow, 338
—— wave, 392, 399
—— wind, 407, 408, 410
moving laboratory, 39
mulberry, 260

N
Nachtwind, 374
nagi, 160
narai, 19, 20, 376
Nasu-oroshi, 371
needle tree, 121, 127, 136
—— ice erosion, 479
negative cloud street, 320
NE-monsoon, 325
net radiation, 41, 151
neutral state, 416
new town, 97
nieve de los penitentes, 231
nightime temperature, 245
nimbostratus, 204
nishi, 19
Nivale inversion, 418
nivation cirque, 474
—— hollow, 474
nocturnal circulation, 408
—— cooling, 90
northers, 367, 368
norder, 367
north föhn, 356
NW-monsoon, 330, 344, 345

O
oak, 121, 129, 135, 141, 452
oasis-effect, 151
Oberwind, 284, 374
occurrence frequency, 423
ohkawara, 347
okiage, 178
okinishi, 344
opening of leaves, 302
ora, 284, 374
orchard, 148
organismic level, 438
orientation, 291
oriented lake, 480
orographic snow line, 224, 225

oroshi, 21, 345, 346, 368
oxidant, 115
ozone, 117, 118

P
paesano, 284, 374
pampero, 396
pampeiro, 396
patterned ground, 477
pear, 452
perennial snow patch, 225
periglacial, 471, 478, 479
permafrost, 198
permeability, 143
persimmon tree, 275
Perth doctor, 375
phenological event, 304
—— gradient, 303, 304
—— stage, 303
phenology, 302
photosynthesis, 462
pine, 121, 123, 126, 134, 135, 146, 148, 299, 448, 449, 452
pit, 273
plain snow, 338
plant ecology, 437
polar dry desert, 275
pollutant, 114
polygon, 477
pontias, 374
poplar, 452
population, 97
—— density, 97
Portuguese trade, 395
possible sunshine, 244
—— sunshine duration, 187
potassium, 116
power law, 60, 270
precipitation, 33, 100, 133, 179, 208, 285, 312
—— distribution, 211, 315
—— maximum, 209, 327
—— probability, 206
—— variability, 213
—— zone, 209
prefrontal squall line, 342
pressure jump, 344
primary vortex, 404

Q
quarajel, 374
quarnero, 361

R
radar echo, 317
—— rainband, 322
radiation, 462
—— balance, 41, 46–48, 51, 124, 188, 189, 226, 261
—— budget, 242
—— flux, 41

—— fog, 301
—— index, 240
—— inversion, 417, 422
rain, 30
rainband, 323, 353
rainfall, 285
—— intensity, 289–291, 216
—— stripe, 320
—— zone, 208–210, 288
rain island, 101
—— shadow, 31, 288
Râjapatha, 15
Rausu-kaze, 391
rebat, 166, 375
reffoli, 361
refoli, 361
regional snow line, 225
—— synoptik, 7
relative humidity, 130
relict, 444
relief, 38
relative humidity, 27, 98, 202, 203
remaining snow, 226, 228–230, 243
representativity, 293
research method, 39
Reynolds number, 403, 405
Rhönwind, 374
Richardson number, 65, 71, 268, 269, 436
ripple mark, 273
rise, 291
riverbank, 151
river fog, 179
rotor, 400, 401
—— streaming, 397
roughness, 47
—— height, 164
—— length, 164
—— parameter, 60–62, 90, 109, 272, 373, 462
runoff, 147

S
Sagami-wan cyclone, 337
saganishi, 19
salt particles, 181
salty wind, 181, 182
sand dune, 477, 478
—— ripple, 477, 479
San'in front, 344
sarma, 367
sastrugi, 273, 274, 374
satoyuki, 338, 340, 343—345
Schlernwind, 374
Schneefresser, 357
scirocco, 395
Scorer number, 403
Scorer's parameter, 215, 401
sea breeze, 27, 28, 108, 113, 158,

159, 161, 170, 171, 175, 334, 354, 378
—— circulation, 176, 326
—— cloud, 175
—— forerunner, 174
—— front, 117, 173, 174, 175, 326, 354
sea fog, 176—178
seashore, 18, 27, 151
secondary orographic precipitation, 352
secular change, 83, 104
separation, 404
serai, 345
sericultural industry, 36
severe local rain storm, 336
south föhn, 357
shallow water equation, 401
sharquieh, 378
Sharidake-oroshi, 147
shelterbelt, 136, 142–144, 147
shirasu, 474
shower, 286
showery rainfall, 328
shûchû-gô-u, 336
silver thaw, 238, 453
silvimeteorology, 119
sky radiation, 42
slope gradient, 470
small hill, 249
smog front, 173
—— layer, 89
smoke, 110, 111, 112, 115
—— control area, 115
snow, 26, 31, 43, 218, 465, 474
—— accumulation, 136, 137, 147, 218, 220, 223, 224, 273, 293, 294, 296, 469
—— band, 323
—— cover, 27, 38, 76, 77, 135, 219, 295, 350, 465
—— depth, 179, 295
—— drift, 136, 137
—— eater, 357, 358
snowfall, 218, 342, 349
—— band, 320
—— intensity, 345
snowflake, 293
snow-free season, 199
snow line, 31, 224, 225
—— melting, 51, 126
snow of penitent, 231
snow transport, 137
soil climate, 52, 437
—— formation, 437
solar irradiation, 463
—— radiation, 41, 117, 118, 120, 150, 185, 240, 241, 463
—— spectrum, 43
solaures, 374
somma, 389

sopero, 284, 374
SO_2, 110, 112—114, 116, 117
sover, 284, 374
south föhn, 356
specific run-off, 147
spillover, 287
spiral band, 322
spot climate, 4
spruce, 33, 121, 135, 299, 438, 444, 452, 465
spruce-mountain pine, 442
squall line, 395
squamish, 380
stable state, 416
standard atmosphere, 183, 184
standing eddy streaming, 397
—— jump, 373
steadiness of wind, 234
stem inclination, 468
stone ring, 476
Stösse, 361
stratus, 204
streamline, 222, 273, 307, 310, 311, 314, 319, 350
stripe, 478
structural soil, 475
subalpine zone, 452, 466
submontane zone, 199
subshelterbelt, 145
subsidence inversion, 418, 419
submicron particle, 111
sukhovey (suhovei), 381, 382, 395,
sulphur, 116
sumatras, 323, 396
summit zone, 247
sunshine, 25, 186, 187, 242
surface inversion, 417—419
—— temperature, 53
—— winds, 308, 309
Suruga-wan cyclone, 337, 338
Suttsu-dashikaze, 390
synoptic condition, 39
—— meteorology, 9
SW-monsoon, 325

T
tachiyazawa-dashi, 385
Takayama high, 334, 337
tamakaze, 376
tamarack, 438
Tauernwind, 374
Teine-oroshi, 390
temperature, 463
—— lapse rate, 189, 190, 194
teresa, 395
terral, 375
thermal belt, 191, 247, 429–434
—— diffusion, 67
—— eddy, 312
—— stratification, 269

—— thunderstorm, 232
thermic, 73, 74
thermion, 73
thermocline, 373
thermodynamic cold anticyclone, 335
thermoisopleth-diagram, 195
throwing, 449
thunder, 345
—— cloud, 102
thunderstorm, 102, 103, 168,
  180, 231, 232, 286, 291, 292,
  293, 324
timber line, 444, 452
tivano, 284, 374
Tokachi-kaze, 390, 391
topoclimate, 6, 35
topoclimatology, 13
topographic effect, 325
topography, 183
torbole, 284
total radiation, 42
trade wind, 175
tramontana, 395
transpiration, 130, 131, 133
tree crown, 150
—— carpet, 449
tresa, 395, 405
Troll-effect, 279
tropaia, 375
tsukiyama, 385
Tsukuba-oroshi, 371
tundra, 43, 151
turbidity factor, 117, 189
turbulence, 89, 418
turbulent flow, 436
turf scarp, 479

U

ultraviolet ray, 118
umzansi, 324
unstable state, 416
Unterwind, 284, 374
up-and down-slope wind, 236
updraft, 322

—— area, 323
up-slope wind, 279
up-valley wind, 279, 284
uranishi, 344, 345
urban heat plume, 86
urban temperature, 80
Utsunomiya front, 347, 348, 371
Uva blowing, 358

V

valley asymmetry, 469
—— breeze, 31, 279, 282, 284,
  374
valley wind, 334, 355, 378
—— effect, 279
Vamana, 16
vapour pressure, 98
vaudaire, 375
vegetation, 33, 38, 44, 116, 257,
  467
—— inversion, 442
ventilation, 267
vertical velocity, 322
Viehtaner wind, 374
village planning, 15
—— snow, 338
Vintschger, 374
viracao, 375
virazon, 375
visibility, 302
volcanic ash, 198
vortex street, 312
—— system, 312

W

Wakasa-wan front, 344
Walliser Talwind, 374
warm slope zone, 247
Wasatch wind, 374
watakushi-kaze, 18
water balance, 49, 296
water budget, 48
water content of snow, 220, 221,
  295
waterfall, 157

—— phenomenon, 369
water resource, 296
—— surface, 44
—— vapour, 202
—— vapour pressure, 202,
  203
wavelength, 397, 400, 405
wave streaming, 397
weather divide, 328–330
wet adiabatic lapse rate, 189,
  190, 416
willow, 452
wind, 31, 33, 106, 140, 158, 232,
  262, 269
wind break, 142, 147, 265, 389,
  390
—— clipping, 449
—— direction, 235, 307
—— distribution, 263
—— erosion, 478
—— rose, 478
wind-shaped tree, 145, 237, 366,
  367, 445–447, 450, 457
wind-snow ecogram, 444
wind speed, 463
wind velocity, 26, 56, 235, 266
wisperwind, 374
world record of daily rainfall,
  317

Y

yamaji (yamaji-kaze), 387–389
yamase, 172, 173, 178, 330–332
yamashita-kaze, 385
yamayuki, 338, 344, 345
yokozekaze, 385
yukiokoshi, 345
yû-nagi, 160

Z

zero-plane displacement, 62,
  109, 141
zonda, 395

GEOGRAPHIC INDEX

A

Abashiri, 144–146
Aberporth, 49
Abukuma River, 260
Adriatic Coast, 301, 356, 361, 363, 364, 366, 405, 406, 444, 448, 461
—— Sea, 361, 380
Africa, 195, 196, 374, 378, 395
Aichi, 18
Aizu, 18
Ajdovščina, 364, 453, 454
Akaho, 275, 304, 305
Akita, 424
—— Prefecture, 22
Alaska, 480
—— Range, 379
Alberta, 358–360
Albuquerque, 96
Alma-Ata, 185, 186, 191
Alps, 28, 29, 34, 35, 37, 77, 190, 195, 196, 207, 213, 218, 219, 226, 232, 236, 297, 356, 358, 374, 391, 413, 444, 475
Altdorf, 356, 357
Amagi Mountains, 472
Amundsen–Scott Base, 418
Amur River, 470
Andes, 193, 373, 462
Angala River, 157
Angola, 198
Antarctica (Antarctic continent), 273, 330, 372, 373, 395, 418
Aoki, 250
Apalachian Mountains (Appalachia), 150, 445
Arabia, 374
Arakawa Canal, 105
Aral Sea, 381
Arava, 378
Arctic Ocean, 77
Argentina, 128, 231, 395
Argun River, 470
Arizona, 470
Armidale district, 332
Aruik-Baluiksky, 408
Asahikawa, 106, 245
Aschersleben, 250
Ashdod, 81
Asia, 378
—— Minor, 375
Assam, 33, 209

B

Astrakhan, 381
Athens, 15, 159
Atsukeshi, 448
Austria, 34, 186, 194, 195, 214, 219, 244, 284, 375
Axel Heiberg Island, 479
Azau, 279, 280, 408
Azusa River, 267

Baar Basin, 247
Bad Kissingen, 117, 118
Bad Ragaz, 356, 357
Bad Tölz, 190
Baffin Island, 241, 242
Bagnères, 29
Bakar, 365
Bakony, 289
Baku, 367
Balaton Lake, 375
Balkan, 248, 289, 413
Baltic Sea, 153, 338
Balzers, 356
Bandai, 409
Banff, 359
Barbados, 44, 242
Batavia, 159, 160
Bath, 250, 285
Bavaria, 33, 187, 214, 374
Bavaria-Drau, 210
Baye de montreux, 287
Beit Shean Basin, 246
Belgrade, 381
Bengal, 28
Bergell, 284
Bergen, 165
Berlin, 23–25, 80, 88
Beskiden, 248
Bhutan, 462
Big Delta, 379, 380
Bishop, 403
Black Forest, 218, 374
Black Mountains, 399
Black Rock Forest, 120
Black Sea, 361, 367
—— Coast, 374
Blyde Valley, 271
Boden See (see Lake Constance)
Bohemia, 232
Bolivia, 210, 462
Bombay, 101
Bonn, 80, 92, 93

C

Bordeaux, 22
Börzsöny, 288, 289
Bôsô Peninsula, 322, 347
Bôsô region, 345
Bosson Glacier, 30
Boston, 353
Boulder, 360, 402
Brasil, 135
Bremen, 100
British Columbia, 380
—— Isles (Islands), 28, 48, 399
Brno, 98, 332
Brookhaven, 435, 436
Brussels, 29
Bucheben, 277
Buchs, 356
Budapest, 100, 423
Buenos Aires, 251
Bükk (Mountain), 35, 54, 258, 289, 428, 429
Bulgaria, 219, 289
Buramaputra, 353
Burrard inlet, 380

Cairngorm Mountains, 479
Calgary, 359, 360
California, 173, 176, 177, 269, 419, 434, 471
Cambridge, 465
—— Bay, 151
Canada, 34, 44, 210
Canadian Rockies, 395
Carbon River, 282
Carpathia (Carpathian Mountains or Ranges), 191, 199, 200, 245, 413
Carpathian Basin, 289
Cape Denison, 372
—— Guardafui, 312
Cape of Erimo, 176
Caspian Plain, 381
—— Sea, 381
Catalonia, 375
Caucasus (Mountains), 279, 408
Central America, 209, 367
—— Asia, 188
Ceylon, 209, 258, 325, 358
Chamonix, 29
Chapel Hill, 100
Cherrapunji, 209, 353
Chiba, 322

Chiba Prefecture, 347
Chicago, 37, 100, 104, 167
Chijiwa Bay, 317
Chile, 210, 231, 373
Chogo Lungma Gl., 188
Cholla-nam-do, 464
Chongjin, 176
Chôshi, 346
Chubu district, 353
Chûbetsu River, 220, 221
Chûgoku Range, 376
Chûgûshi, 217
Chukhung, 188
Cincinnati, 86
Col de Géant, 21, 29
Cologne, 100
Colombia, 130, 210, 373
—— Front Range, 207, 211
Colombian Andes, 209
Colorado, 361, 412
—— Front Range, 207, 211
Columbia River, 448, 453
Columbus, 108
Como Lake, 374
Cona, 175
Coney Island, 27
Copenhagen, 49
Cordillera de los Andes, 231
Cornwallis Island, 452
Cres Island, 461
Crimea Peninsula, 367
Croatia, 365
Crossfell, 379, 399
—— Range, 395
Cushetunk Mountain, 120, 251, 440, 441
Cypress Hills, 151
Czarna Woda, 434
Czechoslovakia, 34, 238, 397

D

Dale Fort, 153, 154
Dale Peninsula, 153
Dallas, 96
Dalmatia, 375
Damascus, 15
Danzig Bay, 160
Danube Basin, 22
—— River, 244, 248
Daunmoräne, 220
Dauria, 470
Davos, 187, 235, 284
Dawson Creek, 151
Death Valley, 48
Deisch, 277
Denton, 96
Detroit, 103, 423
Devils Canyon, 445
Devon Island, 272
Dinar Alps, 356, 361, 365, 395, 399, 406

Dionisia district, 251
Dischma Valley, 187, 266, 267, 281, 282, 432, 434, 463
Divcibare, 203
Doi, 387
Dorset, 465
Douglas Point, 422
Drau, 297
Durban, 94, 160
Dutch Coast, 211
Dyjskosvratecky úval, 298

E

East Africa, 462
East Europe, 54
East Frisian Island, 449
East Riding, 465
Ebetsu, 455
Echigo Mountains, 466
Ecuador, 188, 210
Eden River, 379
Edo, 16
—— River, 105
Eisfluh, 278
El Adem, 479
Elbe River, 157
Elibrus, 408
Emi Koussi, 207
Engadine, 284
England, 34, 111, 448
Erz Mountains, 218
Esdrealon, 378
Ethiopia, 195, 196, 462
Etsch Valley, 374
Europe, 106, 193, 232, 248, 258, 378, 471
European Russia, 147

F

Fairbanks, 261, 262
Falkenstein, 242
Faulhorn, 29
Fedchenko Glacier, 188, 201
Fichtelberg, 234
Finkenbach, 245, 257
Finland, 26, 34, 48, 77, 153, 162, 324
Fjardarheidi, 476
Fläscherberg, 356
France, 28, 34, 214
French Alps, 398, 399
Frankop, 248
Ft Worth, 96
Fukagawa, 105
Fukui, 466
Fukuoka, 83, 119, 152, 153, 354, 474
Fukushima, 247
—— Basin, 431, 434
Funatsu, 191, 409

G

Gaik Brzezowa, 199, 475
Garda Lake, 374, 375
Garmisch-Partenkirchen, 267
Gatersleben, 250
Geneva, 29
Germany, 35, 195, 214, 264, 304
Gifu Prefecture, 68
Giessen, 52
Glands Mulets, 30
Glarus, 22
Gleissberge, 434
Gonohe, 172
Gorce Mountains, 265, 266, 459
Gotemba, 234
Gr. Falkenstein (see Mt. Gr. Falkenstein)
Gran Canaria, 312
Graz, 250
Great Basin, 410
Great Britain, 445
Great Lakes, 27, 104, 154, 155, 167, 180
Great Smoky Mountains, 130, 192
Greece, 375, 395
Greenland, 78, 193, 372
Green Ridge, 120, 251
Grenoble, 374
Grison, 355
Gstettner, 257
Gulf of Siam, 375
Guayaquil, 374
Gulf of Aqaba, 378
Gulf of Carpentaria, 374
Gumma, 369, 370
Gurgl Valley (see Obergurgl)

H

Hachijôjima, 322
Hachinohe, 172
Haggen, 200
Hakkôda, 295
Hakata Bay, 153
Hakatajima Island, 415
Halle, 110
Hamasaka, 68
Hamburg, 49, 320
Hamilton, 97
Harburger, 250
Harima, 17
Harvard Forest, 438, 445
Hary, 320
Harz (Mountains), 218, 251, 439, 440
Hausen, 264
Havelländisches Luch, 59
Hawaii Islands, 175, 176, 208, 284, 326
Hayashida, 191
Heidelberg, 247

Helsinki, 100
Herefordshire, 247
Heta, 347
Hibariga-oka, 97
Hiji River, 299
Hikone, 218
Himalayas, 195, 196, 279, 284, 462
Hiratsuka, 347
Hiroshima, 119, 152
Hirschberg, 397
Hiruno, 68
Hohenpeissenberg, 208, 270, 303, 304
Hohen Venn, 408, 413
Hohki River, 372
Hokkaido, 97, 123, 139, 148, 172, 176, 218, 226, 307, 308, 312, 313, 337, 351, 390, 391, 402, 428, 450
Hokuriku district, 222, 338, 340–343
Höllen Valley, 374
Holstein, 320
Hong Kong, 286, 324
Honjo, 105
Honshu, 226, 344, 346, 347
Howe Sound, 380
Hungary, 35, 288, 477

I

Ibaraki Prefecture, 292, 347
Ice Cap Station, 64
Iceland, 475–477
Idaho, 354
Ilan, 352, 353
—— Plain, 353
Illinois, 293
Ilmala, 159, 160, 163
Ina Valley, 166
Inawashiro, 409, 410
India, 15, 17
Inland Sea (region) (of Japan), 19, 160, 176, 414
Innsbruck, 357, 358
Inn Valley, 374
Iraq, 377
Ireland, 445
Isahaya, 318, 319
Isawa River, 286
Ise Bay, 347
Ishikari Bay, 349
—— Plain, 320, 321, 349–351, 450, 455
Ishizuchi Mountain Range, 387
Israel, 36, 377
Italy, 207, 214, 232, 375
Iwai River, 285, 286
Iwamizawa, 349
Iwo Island, 270
Iyo-mishima, 387

Izumo (Plain), 21, 146
Izu Peninsula, 19, 152, 153, 347, 472, 473

J

Jägerberg, 434
Jamne, 265, 267, 459
Jämtland Mountains, 397, 399
Japan, 16, 17, 36, 48, 51, 53, 55, 80, 97, 101, 106, 135, 136, 158, 159, 186, 193, 194, 196, 197, 202, 203, 205, 210, 233, 256, 275, 285, 293, 295, 296, 304, 309, 317, 320, 321, 323, 331, 333, 335, 338, 339, 368, 376, 383, 448, 452, 466, 474, 475, 477
Japanese Alps, 225
Japan Sea (see Sea of Japan),
—— coast, 21, 33, 159, 340, 377
Jaszcze, 248, 252, 265, 267, 459
Java, 210
Jávorkút, 54
Jena, 286
Johannesburg, 96, 246
Jordan (Rift) Valley, 246, 377, 399
Jukskei River, 271
Jutland, 320

K

Kaczawa Hills, 48
Kagawa Prefecture, 251
Kagoshima, 424
Kakioka Basin, 431
Kakuda (Basin), 246, 259, 260, 410
Kaleto, 249
Kalteneber, 250
Kaltenordheim, 250
Kamabuchi, 148
Kamenskoje Highland, 191
Kananaskis (Valley), 283, 359, 428, 434
Kanogawa River, 473
Kanto district, 171, 267, 322, 327, 345, 346, 434
—— Plain, 157, 181, 182, 232, 316, 317, 337, 346, 347, 353, 368, 369, 371, 474
Karakoram, 188
Karakul Lake, 188
Karlobag, 365, 367
Karlsruhe, 80, 114
Karnali Valley, 462
Kärnten, 374
Karo, 345
Kashima-nada, 171
Kashima dune, 477
Kasprowy Wierch, 374

Kawaguchi, 270, 419–421
Kawanakajima, 299
Kawanoe, 387
Kawasaki, 116
Kazakhstan, 381
Keihin industrial region, 113, 114, 116
Keiyö industrial region, 113
Khartoum, 53, 54, 298
Khasi Hills, 353
Kichijôji, 83, 106
Kiel, 80, 96, 98–100
Kii Channel, 328
Kii Mountains, 323
Kimberley, 27
Kinki district, 317
Kirgiz, 248
Kirigamine, 446, 478
Kirishima, 213
Kiyosato, 146
Klagenfurt, 425
Klützer Winkel, 164
Kobe, 119, 159, 389
Kôfu, 434
—— Basin, 431
Kokchetavska, 408, 415, 426, 427
Köln, 80
Kongo River, 375
Konsen-genya, 451
Korea, 464
Korean Peninsula, 176
Koshagyl, 188
Koshimizu, 146
Középbérc, 258
Kråkenes Fyr, 165
Krakow, 47
Kuala Langat, 323
Kuala Lumpur, 323
Kujû, 328
Kumagaya, 90, 91–94
Kumamoto, 217, 218
Kum-Bel-Pass, 188
Kure, 152
Kuriyama, 349
Kurobe River, 222–224, 389
Kushiro, 177
Kyoto, 199
Kyushu, 97, 474

L

Laholm Plain, 479
Lake Baykal, 367
Lake Biwa, 328, 351, 389
Lake Como, 284
Lake Constance, 27, 165, 245, 246, 408
Lake District, 474
Lake Erie, 154, 155, 169, 170, 180
Lake Garda, 284

Lake Huron, 422
Lake Hill, 240
Lake Inawashiro, 155, 156, 167
Lake Ladoga, 167, 168
Lake Léman, 28, 166, 277, 374, 375
Lake Maggiore, 374
Lake Michigan, 28, 103, 167, 168, 335
Lake Ontario, 180
Lake Suwa, 69, 166
Lake Towada, 181
Lammi, 250
Lanai, 175
Lanchow, 324
Landquart, 356
Langenseifen, 264, 265
La Paz River, 279
La Plata, 396
La Porte, 103
Laramie Range, 470
Leicester, 96
Leipzig, 110
Lena River, 51
Lethbridge, 359
Libia, 17
Liège, 117
Lindenberg, 421, 424
Linz, 80
Little Carpathian Ridge, 400
Little Hampden Valley, 472
Liverpool Bay, 480
Livinenthal, 355
Ljubljana, 301
Lohit Valley, 462
Lomnitzer Spitze, 238
London, 16, 23–25, 80, 88, 92, 93, 96, 97, 101–105, 110–112, 115
Los Angeles, 88, 117
Louisville, 86, 106
Lübeck, 164
Lund, 81
Lunz, 288
Lyon, 34

M

Mackenzie, 359
—— River, 471
Maderia Island, 312
Madras, 27
Maebashi, 54, 55, 157
Makisono, 191
Malabar Coast, 395
Malakka Strait, 396
Malaya, 33
Malay Peninsula, 323
Maloja, 284
—— Pass, 299
Malmö, 54
Marcus Is., 424

Margottet, 30
Martigny, 277, 278
Massachusetts, 28
Mátra, 289
Matsumoto, 116
Maui, 175
Mauna Kea, 175
Mauna Loa, 284
Mawson, 372
Mawsynram, 209, 353
McMurdo, 331
—— Oasis, 274
Mediterranean Sea, 377, 395
Mexican Highland, 368
Middle East, 399
Milfield, 379
Mirny, 372
Mishima, 234
Missouri, 358
Mitagawa, 243
Mito, 217, 322
Mittelgebirge, 120, 440
Miyagi Prefecture, 304, 314
Modau Valley, 257
Modric Dolac, 441–443
Mogami River, 384
Monschau, 146, 413, 414
Montana, 232
Mont Blanc, 21, 30
Monte Rosa, 30
Montgomery Ridge, 120, 251
Monti della Luna, 205
Montreal, 86–88, 96, 423
Moravia, 232
Mörel, 277
Morioka, 218
Moscow, 23, 100, 135
Mouna Kea, 326
Mouna Loa, 326
Mouskorbé, 207
Mpanga, 124
Mt. Akagi, 21, 69, 474
Mt. Asama, 75
Mt. Aso, 190, 216–218, 233, 389, 421, 424
Mt. Azuma, 237, 447, 460, 461
Mt. Balinovac, 444
Mt. Bandai, 414, 415
Mt. Bucegi, 191
Mt. Daisetsu, 220, 221
Mt. Erz, 138, 139
Mt. Fuji, 20, 157, 184, 185, 190, 192, 193, 196, 198, 199, 203, 204, 206, 207, 211, 219, 229, 231, 233, 234, 237, 347, 402–405
Mt. Gassan, 477
(Mt.) Gr. Falkenstein, 120, 191, 208, 248, 432–434
Mt. Hachibuse, 328
Mt. Hakone, 190

Mt. Haruna, 210, 211
Mt. Hakusan, 21
Mt. Hohenpeissenberg, 301
Mt. Ibuki, 190, 211, 216–218, 220, 233, 237, 238
Mt. Iwaki, 230
Mt. Iwate, 190, 211, 218, 228, 230, 237
Mt. Kilimanjaro, 209
Mt. Kirishima, 123, 191, 212, 315, 316, 464, 465
Mt. Kobushi, 204
Mt. Kohala, 326
Mt. Komagatake, 21
Mt. Kurohime, 469
Mt. Kusatsu-Shirane, 477
Mt. Majella, 462
Mt. Maljen, 190, 203
Mt. Mitsumine, 210, 211
Mt. Naeba, 21
Mt. Nagi, 385
Mt. Nantai, 210, 211
Mt. Neko, 460, 461
Mt. Nisekoannupuri, 208
Mt. Odaigahara (Oodaigahara), 138, 216, 217
Mt. Ohnaminoike, 235
Mt. Ontake, 20
Mt. Rainier, 280, 282
Mt. Rishiri, 230
Mts. Gr. Arber, 248
Mt. Shari, 146, 147
Mt. Shiragami, 210, 211
Mt. Sonnblick, 208, 226, 235
Mt. Takabotchi, 291, 328
Mt. Tateyama, 20
Mt. Tibesti, 299
Mt. Toyouke, 387
Mt. Tsukuba, 190, 191, 210, 211, 217, 232, 233, 304, 430, 431
Mt. Unzen, 216–218, 233
Mt. Velika Kosa, 443
Mt. Veliki Zavižan, 441, 444
Mt. Waialeale, 209
Mt. Waita, 328
Mt. Washigamine, 478
Mt. Washington, 233
Mt. Yake, 461
Mt. Yokomine, 291
Mt. Yôtei, 228–230
Mt. Zugspitze, 190
Munich (München), 23, 100, 101, 116, 123, 421, 423, 425
Muskingum Basin, 291
Muttenz, 120
Myôgi, 127, 130
Myôjin-ike, 179

N

Naeba Mountains, 443

Nagahama, 299, 300
Nagano Prefecture, 305
Nagasaki, 217, 218
Nagoya, 16, 80, 119, 347
Nakao Pass, 461
Naka River, 105, 371
Nakayama, 261
Nan Shan, 324
Naoetsu, 344, 347
Nasu, 348, 408
—— alluvial fan, 371
Natal, 324, 462
Natanya, 182
Nayoro Basin, 307
Near East, 395
Neckar Valley, 374
Neotoma, 120, 257, 438
Nepal-Himalaya, 188
New England, 353
New Haven, 121
New Mexico, 232, 358
New Zealand, 475
New York, 86, 114
Niger, 207
Nigeria, 45
Niigata, 137, 342, 358, 462, 466
—— Prefecture, 343, 344
Nikkô, 127, 130
Nishiura, 270, 271, 409
North America, 77, 358
North Carolina, 430, 434, 445
—— Europe, 54
—— Sea, 449
Northway, 379
Norway, 37, 282, 283, 288, 466
Novaia Ladoga, 167
Novaya Zemlya, 367
Novi Sad, 381
Novi Vinodolski, 365
Novorossiysk, 367
Nowy Targ, 248
Nozu River, 298
Nuwara Eliya, 258
Nyons, 374

O

Oak Ridge, 436
Oasis Bardai, 299
Oberengadin, 299, 374
Obergurgl, 120, 198–201, 252,
    254, 255, 433, 434
Obergurgl-Poschach, 199, 296,
    297
Obihiro, 218
Ochiishi, 139, 140
Ôchô, 414, 427
Odenwald, 410
Ogaki, 80, 81, 85, 100
Ohio, 37
Ohmenheim, 145
Ohzu Basin, 299, 300, 389

Oirase River, 170
Okayama, 17, 385, 386
Okhotsuk, 367
Okinawa, 270
Oklahoma, 445
Onehira, 177
Oregon, 175, 332, 453
Os, 120
Osaka, 80, 83, 97, 113, 118, 119
—— Bay, 328, 351
—— Plain, 351
Ôsumi Peninsula, 123
Ota River, 152
Ottawa, 423
Ötz Valley (Ötztal), 186, 219,
    220, 236, 248, 267, 296, 358,
    374
Owada, 257
Owase, 217

P

Pacific (Ocean), 369, 376, 423
Pahta-Arale, 68
Palmyra, 15
Palo Alto, 85, 92, 96
Pannonia Plain, 301, 380
Paramo de Cotopaxi, 188
Paris, 16, 23, 24, 30, 34
—— Basin, 35
Paznaun Valley, 244
P. Bagna, 205
Peak of Doppyo, 232
Peary Land, 275
Pedro, 258
Pennines, 379
Persia, 377
Persian Gulf, 374
Peru, 193
Peru-Bolivian Andes, 209
Pfalz, 304
Pic du Midi, 29
Pietermaritzburg, 271
Pilatus, 21
Pilis, 289
Pinczów, 47
Pizalum, 356
Pleven, 45
Point Barrow, 480
Poland, 37, 47, 48, 374, 397
Pomona, 430
Portugal, 395
Poschach (see Obergurgl-Pos-
    chach)
Poti, 367
Potsdam, 129, 141
Po Valley, 302
Predeal, 191
Pretoria, 81
Princess Elizabeth Land, 372,
    373
Provence, 375

Providence, 353
Pru del Vent, 448
Prussia, 33
Pulkobo, 38
Pyrenees, 77

Q

Quickborn, 247

R

Raba Valley, 248, 434
Ras Asir, 312
Rauris Valley, 277
Reading, 92, 102, 114, 115
Reckingen, 277, 279
Rhein, 452
Rheinland, 304
Rhein-Main, 268
Rhein Valley, 356, 357, 374
Rhön, 191, 327
Rhône Valley (River), 142, 277,
    278, 374, 395, 448, 449
Rickmansworth, 258
Riesengebirge, 397, 399
Risø, 164
River Danube, 44
River Edo, 347
River Fuji, 347
River Shô, 285
River Tone, 347
River Yodo, 351
Riviera, 448
Rocky Mountains (Rockies),
    232, 356, 358, 359, 361, 395,
    399, 410, 452
Ross Sea, 330
Rötgen, 413, 414
Rothamsted, 258
Rouen, 465
Rt Čardak, 406
Ruhr, 116
Rumania, 37, 114, 380
Russia, 27, 38, 188

S

Sabi River, 371
Sado Island, 369
Sagami Bay, 337, 345, 347
Saghalien, 36
Sahara, 207
Saigawa, 179
Saigo, 317
Sai River, 299
Salt Lake, 420
Sambongi Plain, 172
S-America, 375
San Francisco, 85, 96
—— Bay, 177
Sanhsing, 353
San'in district, 342
San Jose, 96, 430

Sanriku Coast, 178
Santa Catalina Mountain, 441
Sapporo, 83, 119, 424
Saratov, 381
Sarma River, 367
Sary-Tasch, 188
S-Asia, 379
Sawtooth, 269
Saxony, 234, 309, 374
Scandinavia, 211, 324, 364
Schleswig-Holstein, 145
Schneekoppe, 397
Schwäbische Alps, 22
Scotland, 448, 466
Sea of Japan, 338, 339, 341–347,
    351, 369, 376, 383
Sea of Okhotsk, 147, 176, 332
Selangor, 323
Sendai, 119, 315
Senj, 365, 366, 405–407
Sesia, 210
Seto Inland Sea (see Inland Sea)
Shari, 144–146
Shari-Abashiri region, 450
    —— River, 146
Sheffield, 84
Shibecha, 177
Shiga Island, 153
Shimada, 317, 353
Shimonoseki, 119
Shinano-Ōmachi, 80
Shiozawa, 466, 467
Shiraito Waterfall, 157
Shirakawa River, 389
Shiretoko Peninsula, 391
Shizuoka, 109, 353
    —— Prefecture, 20
Shōnai (Plain), 22, 384
Shunjo, 217
Siberia, 51
Sierra Nevada, 77, 78, 126, 150,
    395, 397, 399
Sierra Range, 403
Silesia, 47, 48, 117, 218
Singapore, 323, 324, 328
Sinn Valley, 374
Sirre, 277
Sirte, 207
Slovenia, 258, 301
Smarna Gora, 301
Snake River, 355
Somalia, 312
Somalian Coast, 374
Sonnblick (see Mt. Sonnblick)
Sotodomari, 164
South Bass Island, 155, 156
Soviet Union (see U.S.S.R.)
Spain, 374, 375
Sparta, 374
Spearfish, 360
Spitzbergen, 77

Split, 363, 364
St. Ann's Head, 153, 154
Staufenberg, 120, 122, 123, 131,
    439
Steiermark, 250
Stillberg(alp), 267, 463
St. Louis, 23, 96, 100
Stockholm, 100
Straits of Malacca, 323
Stuttgart, 80. 98
Sugadaira 64, 251–253, 271,
    272, 283, 294, 295, 409
Sula Fyr, 165
Sumatra, 394
Sumida River, 105
Suruga Bay, 337, 345
Sutlej Valley, 462
Suwa, 80
Sveldrovsk, 27
Swabian Alb, 471
Sweden, 33, 37, 77, 324, 479
Swiss Jura, 443
Switzerland, 22, 37, 214

T
Tadami River, 292
Tagokura, 292
Taiwan, 352, 353
Takada, 342, 347
Takayama, 334
Tanegashima Island, 477, 478
Tanjong Malim, 323
Tateno, 322, 369, 389, 424
Tatra (Tatras Mountains), 374,
    452
Tenerife, 312
Tennessee, 438
Tenryū Valley, 275
Tharandt, 270
Thetford (Forest), 120, 124
Thüringen, 232, 250, 357
Tibesti, 207
Tochigi Prefecture, 348
Tōbetsu, 349
Tochiomata, 227
Tohoku district, 172, 176, 180,
    218, 312, 314, 332
Tokachi Plain, 390
Tōkamachi, 135, 219, 296
Tokyo, 16, 23, 36, 80, 83, 84, 86,
    88, 94, 95, 97, 98, 100–109,
    111, 112, 114–119, 127, 130,
    171, 173, 174, 179, 192, 309,
    354, 436
Tokyo Bay, 160, 171, 172
Tomakomai, 178
Tone River, 157
Toronto, 87, 94, 117, 176
Tottori, 170, 345, 385
    —— Prefecture, 68
Toulon, 28

Touro, 177
Toussidé, 207
Towada Lake, 295
Toyama, 160, 161, 285
Transilenian Ala-Tau, 248
Transvaal Valley, 271
Transylvanian Alps, 380
Traun Lake, 374
Trieste, 361, 364
Tsrikovka, 408
    —— Village, 427
Tsuganomori, 295
Tsugaru (Strait), 21, 159, 341
Tsurui, 177
Tübingen, 423
Tucson, 53
Tuktoyaktuk Peninsula, 480
Turkestan (Mountains), 77, 188
Turkey, 198, 377
Tyne, 116, 293
Tyrol, 29, 187, 188, 236, 374,
    452
Tzara Bârsei Basin, 245

U
Uchihara, 131
Uetsu region, 319
Ukraine, 380
Ulakhan-Taryn, 51
Unggi, 176
Unhang River, 268
United States of America, (see
    U.S.A.)
Untertaunus, 264
Uppsala, 76, 80, 92, 93, 116
Uruguay, 375
U. S. A., 37, 101, 111, 193, 195,
    196, 210, 444
U.S.S.R., 38, 43, 73, 77, 136
Utah, 211, 293, 374
Utsunomiya, 217, 348

V
Vågsøy, 165
Valsetz Basin, 332
Vancourver, 380
Velebit Mountain, 444
Veliko Gradiste, 381
Venetia, 187
Venezuela, 209, 462
Vestfold Hills, 372, 373
Vermont, 408
Victoria Land, 275
Vienna, 24, 34, 39, 80, 88
Virful Omul, 191
Vogelsberg, 327
Volga River, 181, 381, 395
Volgograd, 381
Vrsac, 381

W

Wadi Bardagué, 298
Wajima, 119, 424
Wakasa Bay, 344
Wales, 182
Washington, 453
Watari Hills, 260
Weisseritz, 270
Wermsdorfer forest, 120, 127
Wernigerode, 250
West Gahts, 214
Whitesands Missile Range, 399
Wieliczka, 475

Wilmington, 292
Winnepeg, 96
Wismar, 164
Wojcieszów, 47, 48
Wright Valley, 274, 275, 331
Wurmberg, 439

Y

Yakutia, 51
Yama, 422
Yamagata, 209, 358
Yangtze, 268
Yesre'el Valley, 378

Yokkaichi, 117
Yokohama, 119
Yonezawa, 81, 82
Yosasa River, 372
Yûfutsu Plain, 178
Yugoslavia, 361

Z

Zavižan, 442
Zayul, 462
Zingst, 154
Zrnovrtica, 406
Zugspitze, 185, 186, 190